Birkhäuser

Operator Theory: Advances and Applications

Volume 229

Founded in 1979 by Israel Gohberg

Alexandre Almeida
Luís Castro
Frank-Olme Speck
Editors

Advances in Harmonic Analysis and Operator Theory

The Stefan Samko Anniversary Volume

Birkhäuser

Editors
Alexandre Almeida
Luís Castro
Departamento de Matemática
Universidade de Aveiro
Aveiro, Portugal

Frank-Olme Speck
Instituto Superior Tecnico
Departamento Matematica
Universidade Tecnica Lisboa
Lisboa, Portugal

ISBN 978-3-0348-0793-7 ISBN 978-3-0348-0516-2 (eBook)
DOI 10.1007/978-3-0348-0516-2
Springer Basel Heidelberg New York Dordrecht London

Mathematics Subject Classification (2010): 31-XX , 43-XX , 47-XX

Printed on acid-free paper

Springer Basel is part of Springer Science+Business Media (www.birkhauser-science.com)

Contents

Preface

Harmonic Analysis and Applications is one of the most rapidly developing areas of mathematics at the beginning of the twenty-first century. It has plenty connections to Operator Theory, Functional Analysis, modern Potential Theory, Partial Differential Equations, Boundary Value Problems, Integral and Pseudodifferential Equations, Complex Analysis and, last not least the Theory of Function Spaces. This intersects with the areas of Fractional Calculus and the most recent Variable Exponent Analysis – in $L^{p(\cdot)}$ spaces.

Stefan G. Samko is an exponent of these areas and their developments. His impressive lifework becomes evident in the survey article of Vakhtang Kokilashvili which leads the sequence of articles of this book. The paper of Sergei Rogosin treats some of Stefan Samko's partial interests in a rather compact form. Other contributions of this volume reflect the wide-range and cross connections between the areas mentioned before. Thus, these peer-reviewed research and survey papers will be of interest of a wide-range of readership in pure and applied mathematics.

The seventieth birthday of Stefan Samko on March 28, 2011, has been commemorated in a series of international conferences in Russia and Portugal, here particularly during the IDOTA 2011 on Integral and Differential Operators and Their Applications, held at Aveiro, June 30–July 2, organized by the *Center for Research and Development in Mathematics and Applications*, University of Aveiro. The contributions of this volume are related to lectures of this meeting and a summer school and workshop, STOP 2011 on Selected Topics of Operator Theory, held at Lisbon, June 24–30, which was organized by the *Center for Functional Analysis and Applications*, the research center in Portugal that hosted Stefan Samko within the last fifteen years. These last two scientific events beneficiated from the financial support of Portuguese Science and Technology Foundation ("FCT – Fundação para a Ciência e a Tecnologia") – which we also would like to acknowledge in here.

The editors like to express their gratitude to Stefan Samko for a fruitful and pleasant cooperation within that period of time. We feel happy to honour our hard-working colleague and humorous friend by the edition of this volume, and wish him good health and ever greater success in his work.

Aveiro and Lisbon, in March 2012

Alexandre Almeida, Luís Castro and Frank-Olme Speck

Stefan G. Samko

Operator Theory:
Advances and Applications, Vol. 229, 1–48

Stefan G. Samko – Mathematician, Teacher and Man

V. Kokilashvili

1. Introduction

It is a great honour for me, to contribute this article on the occasion of a remarkable anniversary. The goal of this paper is to present, in a concise manner, the scientific achievements of a leading authority in mathematical analysis, Professor Stefan Samko.

I will mention the following trends of his research.

2. Scientific origin from BVP and SIE, 1965–1974

Stefan Grigorievich Samko started his first research in 1964 in the scientific school of Prof. F.D. Gakhov in the area of Boundary Value Problems (BVP) and Singular Integral Equations (SIE), who was known by the solution of the Riemann boundary value problem. As a graduate student of Rostov State University he defended his diploma (equivalent to Master Degree) under the supervisorship of Prof. Yu. Cherski, it was related to the so-called exceptional cases of an abstract singular integral equation $(A_1 + A_2S)\varphi = f$ in a Banach space, where S satisfies the property $S^2 = I$. Such an equation studied by Cherskii when the operators $A_1 \pm A_2$ were invertible, was now investigated under the assumption that they are not invertible, this study being published in [123], 1965.

One of the main topics of studies in S. Samko's PhD theses was the investigation of solvability of the so-called generalized Abel integral equation

$$u(x)\int_a^x \frac{\varphi(t)\,dt}{(x-t)^\mu} + v(x)\int_x^b \frac{\varphi(t)\,dt}{(t-x)^\mu} = f(x), \quad a < x < b, \tag{2.1}$$

used in particular in mixed boundary value problems. It involves the well-known expressions of the left- and right-hand sided forms of the fractional integration. He found various explicit relations [125], 1967, connecting such forms with each other via singular integral operators. One of them is given in the sequel in (3.1). Relations of such a type allowed to easily reduce equations of form (2.1) and others to weighted singular integral equations. However, in this way there arose a problem of a precise characterization of the space of functions representable as fractional integrals of function in this or other weighted space used in the theory of singular integral equations. Several of his publications were devoted to the solution of this problem together with the study of the solvability of the equations (2.1) and more general integral equations of the first kind in the corresponding spaces of functions, see [124], [127], 1967; [128], 1968; [129], [131], 1969; [133], [134], 1970; [135], [136], [137], 1971; [140], 1975. Some of these results were later summarized in his books [197], 1987, and [198], 1993. The paper [126], 1967, on the reduction of certain integral equations of the first kind arising in elasticity and hydrodynamics to equations of the second kind, is also close to this topic. A detailed study of integral equations of the first kind with a logarithmic type kernel was undertaken in [143], 1976; [154], 1978, where such equations were reduced to singular integral equations jointly with a certain condition of orthogonality to delta-type functions supported at the end points of the interval, which allowed to give a complete picture of solvability of such equations of the first kind.

Another point of the interest, related to his PhD studies was a search of forms of singular integral equations, different from the characteristic singular equations, which admit a solution in closed form. Results of this kind may be found in [130], [132], 1969; [139], 1974.

A brief overview on integral equations of the first kind related to the Riemann boundary problem (or equivalently to singular integral equations), together with an overview of some multidimensional integral equations of the first kind, studied afterwards, was later presented in [162], 1981.

Two papers [15], 1973, and [16], 1975, of that period, inspired by applications, were related to the generalized argument principle for analytic functions vanishing at the boundary.

In this period there appeared a short note [141], 1975, where S. Samko observed that the Babenko-Stein theorem on the boundedness of singular integral operators in L^p with power weight is an immediate consequence of the non-weighted case and the known boundedness of integral operators with kernels homogeneous of order $-n$ of Hardy-Littlewood type.

Two later papers [217], 1984, and [218], 1985, were also related to his interests in one-dimensional singular integrals. In [217] there was studied the classical Hilbert operator

$$Sf(x) = \frac{1}{\pi}\int_a^b \frac{f(t)\,dt}{t-x}$$

in the weighted generalized Hölder space $H_0^\omega([a,b];\varrho)$ with a power weight $\varrho(x) = (x-a)^\alpha(b-x)^\beta$ and obtained a weighted Zygmund type estimate for the continuity modulus of Hf :

$$\omega(\varrho Sf,h) \le c\int_0^h \frac{\omega(\varrho f,t)}{t^\gamma}\,dt + ch\int_h^{b-a} \frac{(\varrho f,t)}{t(t+h)}\,dt$$

in the case $1 \le \alpha < 2, 1 \le \beta < 2$ (the easier case $0 < \alpha < 1, 0 < \beta < 1$ was known earlier), which was applied to obtain the boundedness conditions of the operator S in $H_0^\omega([a,b];\varrho)$. In [218] there was proved a Zygmund type estimate of a conjugate function (the singular operator along a circumference) without weight, but in a more fine setting of moduli of continuity of fractional order of type, see (3.2) in the next section. Similar estimates for hypersingular integrals (HSI) were given in [219], 1986.

3. Research in Fractional Calculus (FC), 1967–1996

FC is a topic which in fact was always of permanent interest of S. Samko. In this section we touch the period of 1967–1996. His studies in this topic after 1996, when he was already in Portugal, are outlined in Section 6.1.

3.1. One-dimensional Fractional Calculus

3.1.1. Relations between left- and right-hand sided fractional integration. The first studies in the area of boundary value problems, singular integral equations and integral equations of the first kind, led S. Samko to a general interest to the Fractional Calculus.

In the sequel S. Samko's interest in this topic covered wide areas in FC, starting from some properties of classical fractional one-dimensional Riemann-Liouville derivatives to multidimensional fractional integration and differentiation, fractional powers of operators, fractional Sobolev type spaces and others. Some of his one-dimensional results in FC were related to the study of equations of the type (2.1). In particular, when studying the equation (2.1), he arrived at various types of operator relations between the operator of the left-hand sided Riemann-Liouville fractional integration and that of the right-hand sided one. One of them has the form

$$\begin{aligned}\int_x^b \frac{\varphi(t)\,dt}{(x-t)^{1-\alpha}} &= \int_a^x \frac{\cos(\alpha\pi)\varphi(t)\,dt}{(x-t)^{1-\alpha}} \\ &\quad + \frac{1}{\pi}\int_a^x \frac{\sin(\alpha\pi)\,dt}{(t-a)^\alpha(x-t)^{1-\alpha}}\int_a^b \frac{(s-a)^\alpha\varphi(s)\,ds}{s-t},\end{aligned} \tag{3.1}$$

being one of the first results in this area. We refer to [125], 1967; [127], 1968; [129], 1969; [133], 1970, for this and other relations of such a type.

3.1.2. Estimates of moduli of continuity. Let

$$\omega_\gamma(f,h) := \sup_{|t|<h} \|\Delta_t^\gamma f\|_X, \quad \gamma > 0, \tag{3.2}$$

be the modulus of continuity of fractional order, where $\Delta_t^\gamma f = (I - \tau_t)^\gamma f$, $\tau_h f(x) = f(x-h)$, and $X = L_p, 1 \leq p < \infty$ or $X = C$. It was used in the study of singular integrals in [218], 1985, and HSI in [219], 1986.

We mention the weighted estimate

$$\omega\left(\varrho I_{a+}^\alpha f, h\right) \leq ch^{\alpha+\nu-1} \int_0^h \frac{\omega(\varrho f, t)}{t^\nu}\, dt + ch \int_h^{b-a} \frac{\omega(\varrho f, t)}{t^{2-\alpha}}\, dt$$

for the continuity modulus of the Riemann-Liouville fractional integral $I_{a+}^\alpha f(x)$, where $0 < \alpha < 1, \varrho(x) = (x-a)^\mu$, $0 \leq \mu < 2 - \alpha$, $\nu = \max(1, \mu)$ and $\omega(f,t)$ is the usual continuity modulus (the choice $\gamma = 1$ and $X = C$ in (3.2)), which was obtained in [200], 1987 (the proof in the non-weighted case $\mu = 0$ may be found in the book [198], p. 249; in this case the above estimate includes only the first term). This estimate, together with a similar estimate for fractional derivatives, allowed in [200] to obtain the statement that the Riemann-Liouville-fractional integration operator I_{a+}^α maps the weighted generalized Hölder space $H_0^\omega([a,b],\varrho)$ with such a weight, exactly onto the space $H_0^{\omega_\alpha}([a,b],\varrho)$ with the same weight and better dominant $\omega_\alpha(h) := h^\alpha \omega(h)$ of the continuity moduli, under the assumption that ω belongs to a certain Bari-Stechkin type class.

An extension of such weighted results in the spaces $H_0^\omega([a,b],\varrho)$ to the operators of the form of $\int_0^x k(x-t)f(t)\,dt$ with more general kernels may be found in [201], 1993.

3.1.3. In collaboration with Bertram Ross. In 1992–1993 S. Samko was a Fulbright Professor in the USA, at the University of New Haven, where he obtained several essential results in collaboration with Prof. Bertram Ross, known by his studies in FC and famous as the organizer of the first Conference on FC held in USA in 1974.

In their first paper [202], 1993, they introduced fractional integrals $I_{a+}^{\alpha(x)}$ and derivatives $D_{a+}^{\alpha(x)}$ of Riemann-Liouville type with variable order $\alpha(x)$ (see also [169]), posed some open questions and studied the compositions $I_{a+}^{\alpha(x)} I_{a+}^{\beta(x)}$ with special calculations for the particular case where $\alpha(x)$ is a step function. A Fourier transforms approach via the multiplier $(-i\xi)^{-\alpha(\xi)}$ was also suggested there and conditions on $\alpha(\xi)$ given, which guarantee the existence of a locally integrable kernel of the corresponding convolution operator. This paper in fact was a start of his general interest to what is now called Variable Exponent Analysis, see Section 6.

The papers [116], [170], 1995, are also in the same spirit. In the former, there were studied the mapping properties of the operator $I_{a+}^{\alpha(x)}$ in the Hölder spaces

$$H^{\lambda(\cdot)}([a,b]) := \{f : |f(x+h) - f(x)| \leq ch^{\alpha(x)} \leq Ch^{\alpha(x)}\}$$

also of variable order; already here the log-condition on $\lambda(x)$ appeared under which it was shown that $I_{a+}^{\alpha(\cdot)}$ maps $H_0^{\lambda(\cdot)}$ into $H_0^{\lambda(\cdot)+\alpha(\cdot)}$, $\sup_x[\lambda(x)+\alpha(x)]<1$. In the latter there were given conditions on $\alpha(x)$ and $\beta(x)$, under which the difference $I_{a+}^{\alpha(\cdot)}I_{a+}^{\beta(\cdot)} - I_{a+}^{\alpha(\cdot)+\beta(\cdot)}$ with $a=-\infty$ is a compact operator in $L^p(\mathbb{R})$.

Another study in FC worth of mentioning, was made in the paper [117], 1994, made at the same period, was related to a question of existence of nowhere differentiable functions which have fractional derivatives. For any $\nu_0>0$ which may be integer in particular, there was constructed a Weierstrass type function $W_{\nu_0}(x)$ which *nowhere has derivative of the order* ν_0, *but has fractional derivatives of every order* $0<\nu<\nu_0$ (*which even satisfy the Hölder condition of order* $\nu_0-\nu$).

3.1.4. Other. As is known, the Liouville forms of fractional integro-differentiation (the left-hand side and right-hand side ones) for functions defined on the whole real line, do not admit functions growing either at $-\infty$ or $+\infty$. In [215], 1992, for Chen's non-convolution type modification I_c^α of the operators of fractional integro-differentiation, related to a given point $c\in\mathbb{R}$ and applicable to functions with an arbitrary growth at $\pm\infty$, there was constructed the corresponding Marchaud form of fractional differentiation $\mathbb{D}_c^\alpha$ and shown that it generates a real left inverse to Chen fractional integrals. This result was developed in [216], 2001, where one can also find a version of non-convolution type identity approximation with L^p-convergence globally or locally, under natural assumptions.

In [19], 1997, S. Samko studied the possibility of influence of the weight on the asymptotics of singular values of the Riemann-Liouville fractional integration operator I_0^α in L^2-spaces and shown that the presence of the weight may cause an asymptotic behaviour different from what might be expected from the smoothness of the kernel of the operator, in the case of both finite or infinite interval. A similar question was also touched for the multidimensional case of the Riesz potential operator.

We also mention his result on the coincidence of the domains of Liouville and Grunwald-Letnikov fractional differentiation operators in [161], 1885; [164], 1990, see also its formulation below for the multidimensional case.

3.2. Multidimensional FC

References to the studies in multidimensional FC are dispersed in the next sections, mainly given in Section 5. Here we single out only the following multidimensional statement with directional multidimensional derivatives, but in one-dimensional direction, obtained in [163], 1990. By X and Y here is denoted one of the spaces $X=L^p(\mathbb{R}^n), 1<p<\infty$ or $X=C?(\dot{\mathbb{R}}^n), Y=L^r(\mathbb{R}^n), 1<r<\infty$, or $Y=C(\dot{\mathbb{R}}^n)$ with $\dot{\mathbb{R}}^n$ denoting the compactification of $\mathbb{R}^n$ by the unique infinite point.

Let $f\in Y$. *The Grünwald-Letnikov derivative*

$$\lim_{\substack{|h|\to 0\\(X)}} \frac{(\Delta_h^\alpha f)(x)}{|h|^\alpha}$$

and the directional Liouville fractional derivative

$$\lim_{\substack{\varepsilon\to 0\\(X)}} \frac{1}{\varkappa(\alpha,\ell)} \int_{\varepsilon}^{\infty} \frac{(\Delta^{\alpha}_{th'} f)(x)}{t^{1+\alpha}}\, dt, \quad \ell > \alpha,$$

in the direction $h' = \frac{h}{|h|}$ *exist simultaneously and coincide with each other.*

Another multidimensional result in FC we mention here is the realization of the fractional powers $(-|x|^2\Delta)^{\frac{\alpha}{2}}$ of the operator $-|x|^2\Delta$ invariant with respect to dilations which was suggested in [1], 1996, and given with proofs in [2], 2000.

Stefan Samko's research in FC made him known as one of the top experts worldwide known in FC, especially after the publication of the book [197], 1987 (Russian edition) and [198], 1993 (English edition), jointly with A. Kilbas and O. Marichev. He was Chairman of the International Programme Committee of the 1st IFAC Workshop on Fractional Differentiation and its Applications (FDA-04), Bordeaux, France, July 19–21, 2004, and Honorable Chairman of the International Programme Committee of the 2nd IFAC Workshop on Fractional Differentiation and its Applications, Porto, Portugal, 19–21 July, 2006.

4. Equations with involutive operators, 1970–1977

This research was made in collaboration with Nikolai Karapetiants, S. Samko's colleague and close friend. It started in 1969, being the most intensive in 1970–1977 and afterwards continued with intervals till the untimely death of Nikolai in 2005. In this section we overview their results obtained in 1970–1977. In Subsection 6.2 the reader may find the results of the later period 1997–2002.

They started with some problems in the theory of convolution type integral equations, a topic popular in 70s. In particular, they weakened conditions on a function $a(t), t \in \mathbb{R}$ ensuring the compactness of the convolution operator $ak * f$ or $k * (af)$ in the space $L^p(\mathbb{R}), 1 \leq p \leq \infty$ (as known, the condition $a(x) \to 0$ as $x \to \infty$ is sufficient for that). They showed [27], 1970, that this tendency to zero may hold in a very weak sense, expressed in terms of measure of the set where $a(x) \neq 0$. It is interesting to note that this topic was also studied more or less at the same period by Frank Speck, with whom Stefan Samko met 25 years later in Portugal. This result and its discrete analogue immediately originated applications to various types of convolution type equations, continual and discrete, which they presented in [29], 1970; [30], [31], 1971; [37], 1973. Their papers [41], [42], 1975, adjoining this topic, concerns the study of Fredholmness of a class of convolution type equations with discontinuous symbol.

The next cycle of their research was related to singular integral equations with Carleman shift and later with a general approach to equations with the so-called involutive operator Q such that $Q^2 = I$. First, in [28], 1970, and [38], 1973, they studied the functional equation $\psi(x-\alpha) - b(x)\psi(x) = g(x)$ with the irrational (non-Carleman) shift and also with rational shift in the case of the degeneracy of the symbol of the equation in [28] and a certain boundary value problem [32], 1972, also with a non-Carleman shift.

Then they turned to SIE with Carleman type shifts, i.e.,

$$(A_1 + A_2 Q)\varphi = f \tag{4.1}$$

in the space $L^p(\Gamma)$. Here $A_i = a_i I + b_i S, i = 1, 2$, are SIOs and $Q\varphi = \varphi[\alpha(t)]$ is the Carleman shift operator, $Q^2 = I$. In their first paper [33], 1972, on such SIOs on a closed curve, they showed that the known theory of the Fredholmness of such equations is a simple consequence of the relation between the given operator $A_1 + A_2 Q$ and the so-called associate operator $A_1 - A_2 Q$. This relation is realized via the operator $U = S$, when the shift $\alpha(t)$ changes the orientation on the curve and $Uf(t) = [\alpha(t) - t]f(t)$ when the shift preserves it. They were the first to obtain the criterion of Fredholmness of SIO with Carleman shift on an open curve [36], 1972, and in the case of discontinuous coefficients $a_i(t), b_i(t)$ [39], 1973, where the results already heavily depend on the behaviour of the coefficients and the shift $\alpha(t)$ at the end-points of the curve and the points of discontinuity of the coefficients. These results were developed in [43]–[44], 1975.

In this study they realized that their approach is applicable in a more general setting of equations of form (4.1) in an abstract Banach space X with an arbitrary involutive operator Q and the linear operators A_1, A_2 in X subject to some simple axioms. Studies of such equations in an abstract form were known only in the case which reflected only the nature of singular integral equations without shifts; in an abstract form this means that the operators A_i commute with the involutive operator Q up to a compact term. They developed a general scheme to investigate the Fredholmness of such equations, not assuming the above condition. The first version of this scheme appeared in [35], 1972, where it was applied to SIE on the whole line with fractional-linear Carleman shift

$$\alpha(x) = \frac{\delta x + \beta}{\gamma x - \delta}, \tag{4.2}$$

the case of infinite curves having some peculiarities in the theory of equations with shift, the finite point $x_0 = \delta$ being shifted to the infinite point. To cover such equations, in the space of p-integrable functions, they replaced the space $L^p(\mathbb{R})$ by the weighted space $L^p(\mathbb{R}, |x - \delta|^{\frac{p}{2}-1})$ with a special weight. Later, in [52], 2000, they returned to this topic and dealt with the case of a general power weight $|x - \delta|^\gamma, -1 < \gamma < p - 1$, which however required another technique and more complicated terms in which the Fredholmness criterion is given, see also the presentation of these results in the book [53], 2001.

Then in [40], 1974, they applied this scheme to obtain the Fredholmness criterion and formula for the index for convolution equations of Wiener-Hopf type with the reflection $(Q\varphi)(t) = \varphi(\nu - t)$.

The next step of their steps was to develop such a general scheme to equations with a generalized involutive operator Q, i.e., the operator which satisfies the condition $Q^n = I$. Such a scheme was first presented in [45], 1976. In this case the initial operator

$$K = A_1 + QA_2 \cdots Q^{n-1}A_n \tag{4.3}$$

has $n-1$ associate operators $K^{(j)} = \sum\limits_{s=1}^{n} e^{2\pi i \frac{j(s-1)}{n}} Q^{s-1}A_s$ and the key moment in that paper was finding the generalization of the Gohberg-Litvinchuk matrix identity from dimension 2 to dimension n. This identity, together with some simple axioms, led them to a general statement on reducing the Fredholmness property of the operator K to that of its matrix extension without the involutive operator, which included also the formula for the index. A version of this n-dimensional matrix extension identity is also contained in [46], 1977, where it was used to obtain the criterion of Fredholmness and formula for the index in $\ell^p, 1 \le p \le \infty$, of discrete Wiener-Hopf equations

$$\varphi_n - \sum_{i=1}^{N} \gamma_n^i \sum_{k=0}^{\infty} a_{n-k}^i \varphi_k = f_n, \quad n = 0, 1, 2, 3, \dots ,$$

where $\{a_n\}_{n=0}^{\infty} \in \ell^1$ and γ_n^i may stabilize at infinity to different values at infinity depending on i.

A "non-matrix" version of this general scheme for the two-terms equation $(A + QB)\varphi = f$ with an involutive operator Q of order n was developed in [47], 1977, together with applications to various concrete integral equations such as convolution type equations with reflection, singular and convolution type integral equations with complex conjugate unknown function.

This cycle of studies was presented in the book [48], 1988 (Russian version) and in its enlarged English edition [53], 2001.

5. Function spaces of fractional smoothness, influence of Steklov Mathematical Institute

As already mentioned in Section 2, in S. Samko's first studies related to one-dimensional integral equations of the first kind, Stefan Samko was interested in the description of the space of functions representable as fractional integrals of functions from a certain well-known space. Nowadays, this is well known that the range of such an operator over, for instance, a weighted Hölder space is in the scale of the same spaces, while this is not the case for Lebesgue type spaces. In the one-dimensional case he gave such a characterization of the range $I^\alpha(L^p)$ in [138], 1973, in terms of the convergence of the Marchaud fractional derivatives. An attempt to cover the multidimensional case naturally led him to the theory of fractional Sobolev type spaces and to contacts with P.I. Lizorkin, after which there started his contacts with Steklov Mathematical Institute in Moscow and his further interests and scientific activity was much influenced by the Seminar of Academician Serguey M. Nikolskii. He defended his Doctor Theses (2nd degree in the Soviet Union) in Steklov Mathematical Institute and his contacts with this institute continued in fact almost till he moved first to USA and then to Portugal.

5.1. Hypersingular integrals and spaces of the type of Riesz potentials

For the classical operator of harmonic analysis

$$I^\alpha \varphi(x) = \frac{1}{\gamma_n(\alpha)} \int\limits_{\mathbb{R}^n} \frac{\varphi(y)\,dy}{|x-y|^{n-\alpha}}, \qquad x \in \mathbb{R}^n, \tag{5.1}$$

known as the Riesz potential or Riesz fractional integral, it was well known that the inverse operator $(I^\alpha)^{-1}$ may be formally constructed via Fourier transforms. However, a direct realization of such a construction, which could be applied in this or other concrete function spaces, required a development of the apparatus of HSIs. It was undertaken by S.Samko in [144], [145], 1976, and [149], 1977. In particular, in [145] it was shown that the Marchaud-Lizorkin approach to realize the regularization of HSI via finite differences:

$$\mathbb{D}^\alpha f(x) := \int\limits_{\mathbb{R}^n} \frac{\Delta_h^\ell f(x)}{|h|^\alpha}\, dh := \lim_{\varepsilon \to 0} \mathbb{D}_\varepsilon^\alpha f(x), \quad \mathbb{D}_\varepsilon^\alpha f(x) := \int_{|h|>\varepsilon} \frac{\Delta_h^\ell f(x)}{|h|^\alpha}\, dh$$

is effective for application to potentials of Riesz type. One of the results of [145], 1976, states that such a HSI interpreted in the proper way, generates the operator left inverse to I^α in the space $L^p(\mathbb{R}^n), 1 \le p < \frac{n}{\alpha}$. Moreover, the range of the operator I^α over $L^p(\mathbb{R}^n), 1 < p < \frac{n}{\alpha}$ may be exactly characterized as consisting of those functions $f \in L^q(\mathbb{R}^n)$ with the Sobolev exponent $q = \frac{np}{n-\alpha p}$ for which there converge truncated HSIs $\mathbb{D}_\varepsilon^\alpha f(x)$. An extension of such a description to the case of weighted L^p-spaces with Muckenhoupt-Wheeden weights was given in [96], 1985.

In [148], 1977, a similar description of the range was shown in terms of the uniform boundedness of $\|\mathbb{D}_{\varepsilon_k}^\alpha f\|_p$ along any sequence $\varepsilon_k \to 0$. In [204], 1980, there may be found also a description of $I^\alpha(L^p)$ in similar terms via Riesz potentials of order $\{\alpha\}$ of higher derivatives of order $[\alpha]$. Results of such type were earlier known in the easier case of Bessel potential operators when one may take $q = p$. A similar characterization of the range of I^α over Orlicz spaces was given in [196], 1975.

In [146], 1977, there was suggested a certain characterization of the space of pointwise multipliers in $I^\alpha(L^p)$ in terms of the uniform L^p-boundedness of a certain family of integral operators. From this description there were derived sufficient conditions, in terms of the L^p-modulus of continuity of a function to be such a pointwise multiplier.

The study of the space of Riesz potentials led him to the introduction of new spaces

$$L^\alpha_{p,r}(\mathbb{R}^n) := \{f \in L^r(\mathbb{R}^n),\ \mathbb{D}^\alpha f \in L^p(\mathbb{R}^n)\}, \quad 1 \le r < \infty,\ 1 \le p < \infty,$$

of the type of fractional Sobolev spaces in which the functions f themselves and their fractional derivatives $\mathbb{D}^\alpha f$ belong to Lebesgue spaces with in general different exponents p and r, the case $r = p$ corresponding to the Bessel potential space $B^\alpha(L^p)$ and the case $r = \frac{np}{n-\alpha p}, 1 < p < \frac{n}{\alpha}$, to the Riesz potential space $I^\alpha(L^p)$. It was shown in [145], 1976, and [149], 1977, that

$$L^{\alpha}_{p,r}(\mathbb{R}^n) = L^r \cap I^{\alpha}(L^p)$$

with the properly interpreted Riesz potential operator I^{α} in the overcritical case $p \geq \frac{n}{\alpha}$. One of the difficult problems to treat in such spaces was a simultaneous approximation of functions f in L^r-norms and their derivatives $\mathbb{D}^{\alpha} f$ in L^p-norms with different p and r. It was solved in [145], 1976, in the under-critical case $p < \frac{n}{\alpha}$ and in [94], 1981, in the over-critical case.

In the paper [163], 1990, where it was shown that for $f \in L^r(\mathbb{R}^n)$, the convergence in $L^p(\mathbb{R}^n)$ of the truncated HSIs $\mathbb{D}^{\alpha}_{\varepsilon} f(x)$ is equivalent to the existence of $(-\Delta)^{\alpha} f$ defined in terms of the Poisson semigroup P_{ε}:

$$(-\Delta)^{\alpha} f = \lim_{\substack{\varepsilon \to 0 \\ (L^p)}} \frac{1}{\varepsilon^{\alpha}} (I - P_{\varepsilon})^{\alpha} f \tag{5.2}$$

with p and r independent of each other, and moreover,

$$f \in L^{\alpha}_{p,r}(\mathbb{R}^n) \quad \Longleftrightarrow \quad f \in L^r(\mathbb{R}^n) \text{ and } \|(I - P_t) f\|_p \leq C t^{\alpha} \tag{5.3}$$

for all $\alpha > 0, 1 < r < \infty$ and $1 < p < \infty$.

5.2. Potential type operators with homogeneous kernels

Influenced by his interests in the theory of integral equations, he turned to applications of the method of HSI for solution of some multidimensional integral equations of the first kind. In [142], 1976, [147], 1977, and [157], 1980, he studied potential type operators

$$(K^{\alpha}_{\theta} \varphi)(x) = \int_{\mathbf{R}^n} \frac{\theta\left(\frac{x-y}{|x-y|}\right)}{|x-y|^{n-\alpha}} \varphi(y)\, dy, \quad 0 < \alpha < n, \tag{5.4}$$

with a homogeneous kernel, the function θ called the characteristics of the operator (5.4). For the symbol of this operator, i.e., the Fourier transform $\widehat{\theta_{\alpha}}(\xi)$ of the kernel $\theta_{\alpha}(x) := \frac{\theta(x')}{|x|^{n-\alpha}}$, $x' = \frac{x}{|x|}$, there was proved the formula

$$\widehat{\theta}_{\alpha}(\xi) = \Gamma(\alpha) p.f. \int\limits_{\mathbb{S}^{n-1}} \frac{\theta(\sigma)}{(-i\sigma\xi)^{\alpha}}\, d\sigma,$$

where the integral is treated in the sense of the Hadamard finite part when $\alpha \geq 1$ and it is assumed that $\theta \in C^{\lambda}$, $\lambda > \max(0, \alpha - 1)$ and $\alpha \neq 1, 2, 3, \dots$ (this formula has a modification in the case α is an integer). In [157] there was also studied the inverse problem of constructing the function θ by the given homogeneous symbol of the potential operator.

In [157] the following theorem on the smoothness properties of the symbol was also proved:

$$\theta \in C^{\lambda}(\mathbb{S}^{n-1}) \quad \Longrightarrow \quad \widehat{\theta_{\alpha}} \in C^{\lambda - \alpha + 1}(\mathbb{S}^{n-1})$$

where the Hölder space $C^{\lambda-\alpha+1}(\mathbb{S}^{n-1})$ should be introduced with the logarithmic factor in the case of integer α. (Note that this corresponds to the limiting case $p=\infty$, if compared with a setting of smoothness in $L_p^\lambda(\mathbb{S}^{n-1})$-terms.)

The following result was also proved:

$$\theta \in C^\lambda(\mathbb{S}^{n-1}),\ \lambda \geq \alpha+n-1 \quad \Longrightarrow \quad K_\theta^\alpha(L^p) \subset I^\alpha(L^p),\ 1<p<\frac{n}{\alpha}$$

and $K_\theta^\alpha(L^p) = I^\alpha(L^p)$ in the elliptic case.

To invert potential type operators of form K_θ^α, he introduced the generalized HSI of the form

$$\mathbb{D}_\Omega^\alpha f)(x) = \int\limits_{\mathbf{R}^n} \frac{(\Delta_y^\ell f)(x)}{|y|^{n+\alpha}} \Omega\left(\frac{y}{|y|}\right) dy$$

and showed that in application to such an inversion the choice of the type of the finite difference $(\Delta_y^\ell f)(x)$ plays an important role, especially in the case where α is an integer. It was proved in the elliptic case, that for sufficiently smooth $\theta(x')$, under the appropriate choice of the type of the difference, there exists an Ω, such that $\mathbb{D}_\Omega^\alpha$ is the left inverse to K_θ^α in the space $L^p(\mathbb{R}^n)$ and the function Ω may be explicitly constructed:

$$\Omega(x') = \text{const}\, p.f. \int\limits_{\mathbb{S}^{n-1}} \frac{d\sigma}{\widehat{\theta_\alpha}(ix'\sigma)^{n+\alpha}}.$$

In a special case where $\theta(x') = P_2(x', x')$ is a restriction of a quadratic form onto $\mathbb{S}^{n-1}$, in [167], 1993, the above integral was explicitly calculated supposing that there are known the eigenvalues of the matrix P and the matrix W transforming it to the diagonal form.

As is well known, various potential kernels serve as fundamental solutions to differential operators in partial derivatives. As a general statement of such a kind, in [153], 1978, there was proved that *every homogeneous differential operator with constant coefficients may be represented in the form of a HSI with a certain function Ω, which is also a homogeneous polynomial of the same order and it is constructed explicitly.*

As a by-product of his studies of potential operators with homogeneous kernel, in [156], 1978, he gave a formula for the Fourier transform of functions of the form $Y_m(x)|x|^{-\beta}$, where Y_m is a harmonic polynomial, for some "exceptional" values of γ, and in [152], 1978, he gave the solution of the following problem: given a homogeneous polynomial

$$P_m(x) = \sum_{|j|=m} a_j x^j, \quad x^j = x_1^{j_1} \cdots x_n^{j_n},$$

what are necessary and sufficient conditions (imposed on the coefficients a_j) for $P_m(x)$ to be harmonic?

The problem of convergence of HSI with a homogeneous function $\Omega(x')$ was studied in [95], 1982, where it was shown that if for a function f there converges its HSI with $\Omega \equiv \text{const}$, then it converges with any bounded Ω; the inverse is also true under a certain ellipticity condition on Ω. Similar questions in the case of non-homogeneous functions Ω stabilizing at the origin and infinity, were earlier studied in [149], 1977.

An application of the method of HSI to the inversion of potential type operators with a non-homogeneous radial characteristic $\theta(|x-y|)$ may be found in [206], 1985.

We also refer to an overview of applications of this method to potential-type integral transforms in the paper [166], 1993.

In contrast to the case of Bessel type potentials, Riesz fractional integrals $I^\alpha f$ when interpreted within the frameworks of distributions, need a test functions space invariant with respect to the operator I^α. Such a space of Schwartz test functions which are orthogonal to all the polynomials, is known as the Lizorkin test function space. In relation with studies of potential type operators whose symbol may be zero or infinity on a set different from the origin, S. Samko in [151], 1977, introduced a Lizorkin type space of Schwartz test functions whose Fourier transforms vanish on a given set $V \subset \mathbb{R}^n$:

$$\Phi_V := \{\varphi \in \mathcal{S}(\mathbb{R}^n) : \widehat{\varphi}(\xi) = 0 \text{ for } \xi \in V\} \tag{5.5}$$

and gave their description in the case of conic sets V corresponding to convolution operators with homogeneous kernel. For a large variety of sets V with $|V| = 0$, in [158], 1982, he proved that such a test function space is dense in $L^p(\mathbb{R}^n), 1 < p < \infty$, which proved to be essential in many papers on inverting potential type operators, where it was necessary to approximate by nice functions from a space invariant with respect to the operator. See also [168], 1995, for similar denseness in the spaces $L^{\bar{p}}(\mathbb{R}^n)$ with mixed norm.

5.3. Spherical HSI and potentials

After the investigation of symbols of convolution operators with homogeneous kernel, and also influenced by the studies of P.I. Lizorkin and S.M. Nikolskii on approximation of functions on the sphere $\mathbb{S}^{n-1}$, S. Samko investigated potential operators on $\mathbb{S}^{n-1}$; in [150], 1977, he introduced some new forms of spherical fractional type integrals and HSI. One of them proved to have an important role in his paper [102], 2010, more than 30 years later, where such a construction of a spherical potential operator arose in application to an inverse problem of aerodynamics. In the paper [159], 1983, he presented a survey on spherical singular and potential operators, including the related problem of restoring the kernel of the spatial potential by its symbol. In [100] there was given a characterization of the fractional Sobolev type space $L^\alpha_p(\mathbb{S}^{n-1})$ of functions $f \in L_p(\mathbb{S}^{n-1})$ such that the fractional power of the Beltrami-Laplace operator is also in $L_p(\mathbb{S}^{n-1})$, via convergence of the spherical HSIs

$$\int\limits_{\mathbb{S}^{n-1}} \frac{f(x)-f(\sigma)}{|x-\sigma|^{n-1-\alpha}} d\sigma, \qquad x \in \mathbb{S}^{n-1}, \;\; 0<\alpha<1.$$

A shorter and more direct proof of the same description was later given in [187], 2003.

Related to the studies of spaces of fractional smoothness of functions on the sphere, in the paper [211], 1987, there was called attention to the fact that different definitions of the space $C^1(\mathbb{S}^{n-1})$ are not equivalent, since their comparison involves a singular type operator, not bounded in the space $C(\mathbb{S}^{n-1})$. This study was developed in [212], 2000, where a detailed consideration of function spaces of the type $C^\lambda(\mathbb{S}^{n-1})$ and $H^\lambda(\mathbb{S}^{n-1}), \lambda > 0$, was presented with a special emphasis on the non-equivalence of their definitions in the case of integer λ. The spaces $C^\lambda(\mathbb{S}^{n-1})$ are defined in terms of fractional differentiation on the base of the space $C(S^{n-1})$, the space $H^\lambda(\mathbb{S}^{n-1})$ denotes a Hölder type space. The averaged type Hölder spaces $H^\lambda(S^{n-1})$ were also dealt with. It was shown that these spaces coincide, or on the contrary are different, depending on the fact what kind of definition of the differentiation of integer order is used: in terms invariant with respect to rotations or in terms connected with differentiation with respect to Cartesian coordinates.

Many of these results were later presented in his book [160], 1984 (Russian edition) and later in its enlarged English version [183], 2002.

6. Portugal period; after 1995

In January of 1995 he moved to Portugal, where he was first an invited researcher for half a year at the University of Algarve and then took a position there. He learned Portuguese language and taught various courses in this language, which took a time from his research in the first years of living there. In 1996–2000 he mostly continued research in the topics developed before his departure to Portugal, but after 2000 mainly concentrated on variable exponent analysis (VEA), an occasional interest to which was seen already in his papers [202], 1993, [116], [170], 1995, on fractional operators of variable order and Hölder spaces with variable exponent.

6.1. FC continued; constant exponents

In 1998 S. Samko returned to his old interest in multidimensional potential type operators and HSI, the Riesz fractional differentiation, in particular. In 1999 there were published two surveying papers [97] and [98] with an overview of the application of the method of approximative inverses to potential type operators and realization of fractional powers of various types of differential operators in partial derivatives.

6.1.1. Approximative inverses for the fractional type operators. In the method of approximative inverses, the HSI is replaced by the limit of convolutions with "nice" kernels. The standard way to use this approach in application to potential type operators was to introduce a factor $e^{-|\xi|}$ or $e^{-|\xi|^2}$ in the symbol of the potential type operator. However, this always led to complicated expressions in Fourier pre-images. In [174], 1998, [177], 1999, he suggested and realized this approach in an opposite direction, by introducing the parameter ε directly in the kernel of the kernel of inverting operation, not caring about the simplicity of its Fourier transform. He found a family of sequences $q_\alpha(x)$ of kernels, admissible for the Riesz potential operator (5.1), for approximative inverse operators

$$\frac{1}{\varepsilon^{n+\alpha}} q_\alpha\left(\frac{x}{\varepsilon}\right) * f \tag{6.1}$$

the limit of which coincide with the corresponding HSI in the frameworks of $L^p(\mathbb{R}^n)$-densities. In particular, this led to the following very simple formula of the inversion for $I^\alpha \varphi = f$ in the case $0 < \Re\alpha < 2$:

$$\varphi(x) = \frac{1}{\gamma_n(-\alpha)} \lim_{\varepsilon\to 0} \int_{R^n} \left[\frac{1}{(|y|^2+\varepsilon^2)^{\frac{n+\alpha}{2}}} - \frac{n+1}{\alpha} \frac{\varepsilon}{(|y|^2+\varepsilon^2)^{\frac{n+\alpha}{2}+1}} \right] f(x-y)dy. \tag{6.2}$$

This approach via approximation (6.1) was extended in [99], 2002, to the case of $L^p(\mathbb{R}^n, \varrho)$-densities with Muckenhoupt-Wheeden $A_{p,q}$-weights and in [61], 2003, to the two-weight setting.

An application of a similar approach to the one-dimensional Liouville fractional differentiation D^α_+ may be found in [118], 1999, and to fractional powers $(-A)^\alpha$ in an abstract Banach space in [178], 2000, see also an expository article [181], 2001, on approximative inverses in problems of such a kind.

6.1.2. Local nature of Riesz potential operators. The goal of S. Samko's paper [176], 1999, was to reveal the nature of Riesz potentials (5.1) with L^p-densities in the case of large α. The potential $I^\alpha\varphi$ of a function $\varphi \in L_p(\mathbb{R}^n)$ (with complex α in general), in the case $\Re\alpha \geq \frac{n}{p}$ is interpreted as a distribution over the Lizorkin test function space Φ (the space (5.5) with $V = \{0\}$). Although $f = I^\alpha\varphi$ of $\varphi \in L^p(\mathbb{R}^n)$ is in general a distribution, the finite differences $\Delta^\ell_h(I^\alpha\varphi)$ are always regular functions belonging to $L^p(\mathbb{R}^n)$) for $\ell > \Re\alpha$ and

$$I^\alpha(L^p) \subset L^p_{loc}(R^n),\ 1 \leq p < \infty,\ \Re\alpha > 0.$$

6.1.3. Miscellaneous. S. Samko's paper [180], 2000, of that period, presents an overview of the application of HSI to construction of fractional powers of operators, based on his talk at the 1st International Conference on Semigroups of Operators, Newport Beach, California, in 1998.

His paper [187], 2003, deals with the periodization of the Riesz fractional integral of two variables and its connection with the Weyl type differentiation

$$I^\alpha f(x,y) = \sum_{|m|+|n|\neq 0} \frac{f_{mn}}{(m^2+n^2)^{\frac{\alpha}{2}}}\, e^{i(mx+ny)}.$$

of doubly periodic functions. It reveals different effects, depending on whether one uses the repeated periodization or the double one. It was shown that doubly-periodic Weyl-Riesz kernel of order $0 < \alpha < 2$ in general coincides with the periodization of the Riesz kernel, the repeated periodization being possible for all $0 < \alpha < 2$, while the double one is applicable only for $0 < \alpha < 1$. The well-known reason is that the Weyl periodic fractional integral of a 2π-periodic function $f(x)$, generally speaking, coincides with the properly interpreted Liouville fractional integral of f. This research was inspired by the fact that in the "real" multidimensional fractional integration, i.e., not repeated (mixed) one, it is not possible to separate variables.

In [108], 2005, there was given a generalization of the Marchaud formula for fractional differentiation to the multidimensional case of functions on a domain in $\mathbb{R}^n$:

$$\mathbb{D}_\Omega^\alpha f(x) = c(\alpha)\left[a_\Omega(x)f(x) + \int\limits_\Omega \frac{f(x)-f(y)}{|x-y|^{n+\alpha}}dy\right], \quad x \in \Omega, \quad 0<\alpha<1,$$

where $a_\Omega(x) = \int_{R^n\setminus\Omega} \frac{dy}{|x-y|^{n+\alpha}} \sim [\mathrm{dist}(x,\partial\Omega)]^{-\alpha}$ and $a_\Omega(x)$ was calculated explicitly in terms of the Gauss hypergeometric function in the case where Ω is a ball. It was also shown there that the operator $\mathbb{D}_\Omega^\alpha$ acts boundedly from the range of the Riesz potential operator $I_\Omega^\alpha(L_p(\Omega))$ to $L_p(\Omega), 1 < p < \frac{1}{\alpha}$. The considerations related to the function $a_\Omega(x)$ generated some technical trick used later in [113], 2009, in the study of the Hardy inequality in domains in the variable exponent setting.

In [188], 2005, he obtained the best constant C for the Hardy-Rellich-type inequality for the fractional powers of the minus Laplace operator:

$$\int_{\mathbb{R}^n} |f(x)|^p|x|^\mu \, dx \le C^p \int_{\mathbb{R}^n} |(-\Delta)^{\frac{\alpha}{2}} f(x)|^p|x|^\gamma \, dx$$

where $\alpha > 0$, $\alpha - n$ is not an integer, $1 \le p < \infty, \alpha p - n < \gamma < n(p-1)$ and $\mu = \gamma - \alpha p$; the best constant being

$$C = 2^{-\alpha}\frac{\Gamma\left(\frac{n(p-1)-\gamma}{2p}\right)\Gamma\left(\frac{n+\gamma-\alpha p}{2p}\right)}{\Gamma\left(\frac{n+\gamma}{2p}\right)\Gamma\left(\frac{n(p-1)+\alpha p-\gamma}{2p}\right)}.$$

In [90], 2010, there were studied mapping properties of the mixed Riemann-Liouville fractional integrals of order (α, β)

$$\left(I_{0+,0+}^{\alpha,\,\beta}\varphi\right)(x,y) = \frac{1}{\Gamma(\alpha)\Gamma(\beta)}\int\limits_0^x\int\limits_0^y \frac{\varphi(t,\,\tau)dtd\tau}{(x-t)^{1-\alpha}(y-\tau)^{1-\beta}}, \quad x>0,\ y>0, \quad (6.3)$$

in the Hölder spaces defined by mixed differences with the main goal to reveal the mapping properties in dependence on the usage of usual or mixed differences in the definition of Hölder spaces, both in non-weighted and weighted spaces (with power weights).

In [11], 2011, there were studied potential operators

$$I_\Omega^\alpha f(x) := \int\limits_\Omega \frac{f(y)\,dy}{|x-y|^{n-\alpha}}$$

over a domain $\Omega \subset \mathbb{R}^n$ in usual Hölder spaces $H^\lambda(\Omega)$ and generalized Hölder spaces $H^\omega(\Omega)$. As is known, the difficulty in establishing mapping properties of this operator in such spaces is caused by the fact that the so-called cancellation property does not hold for domains Ω (except for the case $\Omega = \mathbb{R}^n$), i.e., the property that the potential of a constant function is constant. The potential of a constant behaves like $[\mathrm{dist}(x, \partial\Omega)]^\alpha$ near the boundary, and a result of the type $I_\Omega^\alpha : H^\lambda(\Omega) \to H^{\lambda+\alpha}(\Omega), \quad \lambda + \alpha < 1$ is not possible in domains. In [11] it was proved that the result of the type

$$I_\Omega^\alpha : H_0^\lambda(\Omega) \to H^{\lambda+\alpha}(\Omega),$$

where $H_0^\lambda(\Omega)$ is the subspace in $H^\lambda(\Omega)$ of functions vanishing at the boundary, (and more generally, for spaces of the type $H_0^\omega(\Omega)$) holds at the least for the so-called uniform domains (known also as satisfying the banana condition).

6.2. Equations with involutive operators, continued

In 1997 there was renewed the collaboration with Nikolai Karapetiants who visited Portugal on a sabbatical in 1997–1998 and some periods later. The book [53], 2001, on general methods to study equations containing generalized involutive operators and their applications, was written at that period. It was an updated and essentially enlarged version of the previous Russian edition [48], 1988. Before it was published, a series of their papers appeared. In [49], 1999, they studied the boundedness in L^p of the operator

$$K\varphi(x) := \int_{|y|<a} k(x,y)\varphi(y)dy$$

(including necessary conditions for positive kernels), and the Fredhomness properties of multidimensional integral equations

$$\lambda\varphi(x) = K\varphi(x) + f(x), \quad |x| < a$$

where $0 < a \leq \infty$ and $k(x,y)$ is homogeneous of degree $-n$, i.e., $k(\lambda x, \lambda y) = \lambda^{-n} f(x,y)$ or has some other types of homogeneity.

In [50], 1999, they presented a modified and generalized their general approach of investigating the Fredholmness properties of equations with involutive operators, discussed in Section 4, this development of the method included also equations

$$\sum_{j=1}^{m}\sum_{k=1}^{n} P^{j-1}Q^{k-1}A_{jk}\varphi = f$$

with two independent generalized involutive operators P and Q, $Q^n = P^m = I$ (compare with (4.3)).

In the paper [51], 2000, there was studied the Fredholmness in the space $L^p(\mathbb{R}^1, |x|^\gamma)$, $1 < p < \infty$, $-1 < \gamma < p-1$, of the equation

$$\lambda\varphi(x) - P_+ a P_- Q\varphi = f(x), \quad x \in R^1,$$

where $P_\pm = \frac{1}{2}(I \pm S)$, $S\varphi = \frac{1}{\pi i}\int_{-\infty}^{\infty} \frac{\varphi(t)}{t-x}dt$, $a = a(x)$ is a piece-wise continuous function and $Q\varphi = \varphi(-x)$. Such equations arise in applications in diffraction theory. Since $Q^2 = I$, it was possible to use their approach to equations with involutive operators which allowed to obtain Fredholmness conditions and a formula for the calculation of the index in effective terms. It was shown that the essential spectrum of this equation is described in simple terms of the generalized lemniscates defined by

$$r^2 = \frac{4}{\sin^2 2\alpha\pi}\cos(\varphi + \pi\alpha)\cos(\varphi - \pi\alpha), \quad \alpha = \frac{1+\gamma}{p},$$

unilateral in case of a jump of $a(x)$ at the origin or infinity and bilateral in case of jumps at other points.

In [52], 2000, they obtained a criterion of Fredholmness and a formula for the index of singular integral equations with shift (4.2) in the space

$$L^p(\mathbb{R}, |x-\delta|^\gamma), \quad 1 < p + \infty, \quad -1 < \gamma < p-1,$$

where it was revealed that the unboundednes of $\mathbb{R}$ in the presence of the shift moving one of the points to infinity implies an essential complication of the theory in the case $\gamma \neq \frac{p}{2} - 1$. To overcome the arising difficulties, the Fredholmness of systems of singular integral equations perturbed by operators with homogeneous kernels, was preliminarily developed.

An application of their general approach to equations with involutive operators to integral equations on the real line with homogeneous kernels and shift $\frac{1}{x}$ was given in [55], 2001.

Their paper [54], 2001, seems to be the first paper where multidimensional integral equations with shift (transformation of variables) were studied. Applying their method, in this paper they gave criteria of invertibility or Fredholmness of some multidimensional integral operators with linear transformation

$$\alpha(x) = Ax + \beta$$

in the Euclidean space, which satisfies the generalized Carleman condition of order n (this is equivalent to say that $A^n = I$ and β is he root of the equation $(I + A + \cdots + A^{N-1})\beta = 0$). Such transformations have a rotational structure and the application of the general approach in the multidimensional case even in the case of linear transformations proved to be heavily depending on the specific properties of the matrix A. The considered equations involved convolution type operators, singular Calderón-type operators and operators with a homogeneous kernel.

6.3. Variable Exponent Analysis: 1993–2003

S. Samko's first papers [202], 1993, [116], [170], 1995, in this field have already been mentioned above.

As is well known, the interest in the Lebesgue spaces $L^{p(\cdot)}(\Omega)$ with variable exponent does not come only from their mathematical curiosity, but also from their importance in applications. The difficulties arising in the study of these spaces and operators in them are caused by the fact that these spaces are not invariant with respect to translation. During the last two decades, especially the last one, these spaces have been intensively studied.

In 1998 there appeared the paper [173] written already in the pre-Portugal period, where some properties of fractional integration operators of variable orders were summarized, the equivalence of different norms in the Lebesgue spaces with variable exponent $p(x)$ was proved and it was observed that the Minkowsky inequality holds in these spaces.

Two next papers [172], [171], 1998, on convolution type operators in variable exponent Lebesgue spaces $L^{p(\cdot)}(\Omega)$, one a continuation of another, were written already in the Portugal period. In the first one, already in 1998, for a bounded set Ω and log-continuous exponent $p(x)$ there were obtained the relations

$$\|\chi_{B(x,r)}(\cdot)| \cdot -x|^{\beta(x)}\|_{p(\cdot)} \sim r^{\beta(x)+\frac{n}{p(x)}}, \quad \beta(x)p(x)+n>0, \tag{6.4}$$

$$\|\chi_{\Omega\setminus B(x,r)}(\cdot)| \cdot -x|^{\beta(x)}\|_{p(\cdot)} \sim r^{\beta(x)+\frac{n}{p(x)}}, \quad \beta(x)p(x)+n>0, \tag{6.5}$$

which were of importance in various applications, especially the particular case

$$\|\chi_{B(x,r)}\|_{p(\cdot)} \sim r^{\frac{n}{p(x)}}.$$

Estimate (6.5) allowed to prove in [171], via the Hedberg trick, the Sobolev $p(x) \to q(x)$-theorem, $\frac{1}{q(x)} = \frac{1}{p(x)} - \frac{\alpha(x)}{n}$, for the potential operators (also of variable order) under the assumption that the maximal operator

$$Mf(x) = \sup_{r>0} \frac{1}{|B(x,r)|} \int\limits_{B(x,r)} |f(y)|\, dy \tag{6.6}$$

is bounded in $L^{p(\cdot)}(\Omega)$.

In [171], 1998, it was also shown that the Young theorem for convolutions is no more valid for variable exponents, caused by the non-invariance of the spaces with respect to translations. For a class of kernels there were given some necessary conditions on $p(x)$ for this theorem to hold. Roughly speaking, a convolution operator may be bounded in $L^{p(\cdot)}$ without additional restrictions on the exponent $p(x)$, if its kernel has a singularity at the origin only.

The next important step was made in [175], 1999 (see also [179], 2000) where he had shown that it is possible to use mollifiers

$$\frac{1}{\varepsilon^n} \int \mathcal{K}\left(\frac{x-y}{\varepsilon}\right) f(y)\, dy$$

in variable exponent Lebesgue spaces with log-continuous exponent $p(x)$, and proved that $C_0^\infty(\mathbb{R}^n)$ is dense in the Sobolev space $W^{m,p(\cdot)}(\mathbb{R}^n)$.

In the case of bounded domains in $\mathbb{R}^n$ and $x_0 \in \overline{\Omega}$, in [185], 2003, there was proved the Hardy type inequality

$$\left\| |x-x_0|^{\beta-\alpha} \int\limits_{\Omega} \frac{f(y)\,dy}{|y-x_0|^{\beta}|x-y|^{n-\alpha}} \right\|_{L^{p(\cdot)}(\Omega)} \leq C\,\|f\|_{L^{p(\cdot)}(\Omega)}, \quad 0<\alpha<n, \;\; x_0\in\overline{\Omega},$$

where $-\frac{n}{p(x_0)} < \beta < \frac{n}{q(x_0)}$ and in [186], 2003, there was obtained the generalization

$$\left\| I^{\alpha(\cdot)} f \right\|_{L^{q(\cdot)}(\Omega,|x-x_0|^{\mu})} \leq C\,\|f\|_{L^{p(\cdot)}(\Omega,|x-x_0|^{\gamma})} \tag{6.7}$$

of the Hardy-Littlewood-Stein-Weiss inequality, where $\frac{1}{q(x)} \equiv \frac{1}{p(x)} - \frac{\alpha(x)}{n}$,

$$\alpha(x_0)p(x_0) - n < \gamma < n[p(x_0)-1] \quad \text{and} \quad \mu = \frac{q(x_0)}{p(x_0)}\,\gamma.$$

The proof of (6.7) was based on the estimate

$$\left\| |x-y|^{\nu(x)} \chi_{\Omega\setminus B(x,r)}(y) |y|^{\beta(x)} \right\|_{p(\cdot)} \leq C r^{\nu(x)+\frac{n}{p(x)}} (r+|x|)^{\beta(x)}, \tag{6.8}$$

obtained in the same paper, which generalizes the estimate (6.5) to the case of power weights.

6.4. Variable Exponent Analysis in collaboration with V. Kokilashvili, 2001–present

In 2002 there started the collaboration with the author of this article. At the first stage there were obtained the conditions of the boundedness of the maximal operator M, potential type operators and operators with fixed singularity (of Hardy and Hankel type) in the spaces $L^{p(\cdot)}(\varrho,\Omega)$ over a bounded open set in $\mathbb{R}^n$ with a power weight $\rho(x) = |x-x_0|^{\gamma}$, $x_0 \in \overline{\Omega}$, and a log-continuous exponent $p(x)$. These results announced in [68], 2002, were published with proofs in [74], 2004. In particular, for bounded sets Ω there was proved that the maximal operator M is bounded in the space $L^{p(\cdot)}$ with the power weight $|x-x_0|^{\beta}, x_0 \in \Omega$, if and only

$$-\frac{n}{p(x_0)} < \beta < \frac{n}{p'(x_0)}. \tag{6.9}$$

In [69], 2002, [72], 2003, there were introduced variable exponent spaces $\Lambda^{p(\cdot)}$ it terms of the rearrangements:

$$\Lambda^{p(\cdot)} := \{f \in L^1_{\text{loc}}(\Omega) : f^{**} \in L^{p(\cdot)}(\mathbb{R}^1_+)\}$$

where the boundedness of various operators was investigated.

Later, in [12], 2008, this approach was extended to more general Lorentz type spaces $L^{p(\cdot),q(\cdot)}(\Omega,w)$, $\Omega \subseteq \mathbb{R}^n$, with weight. In this generalization it was possible to avoid the local log-condition, assuming only the log-decay conditions at the origin and infinity, which was based on the results for the one-dimensional Hardy inequalities in variable exponent Lebesgue spaces proved in [10], 2007, see Subsection 6.5.5. Apart from the study of some basic properties of the space itself, in this paper there were obtained conditions of the boundedness of various operators including ergodic maximal and singular operators.

The above result of the paper [74] on the weighted boundedness of the maximal operator was immediately used in [73], 2003, to obtain a similar boundedness of Calderón-Zygmund type singular operators and the corresponding supremal operator. As a corollary very important for application to singular integral equations and boundary value problems, there was shown that a condition of type (6.9) is necessary and sufficient for the Cauchy singular integral operator

$$S_\Gamma f(t) = \int_\Gamma \frac{f(\tau)d\tau}{\tau - t}, \quad t \in \Gamma,$$

to be bounded in the variable exponent Lebesgue space with a power weight, where it was supposed that Γ is a Lyapunov curve or a curve of bounded rotation. Such an application to singular integral equations with piecewise continuous coefficients was made in [71], 2003, where the authors proved a criterion of Fredholmness and an index formula in the Lebesgue spaces $L^{p(\cdot)}(\Gamma)$ on a finite closed Lyapunov curve Γ or a curve of bounded rotation. To this end, they first gave a certain abstract version of the Gohberg-Krupnik scheme to study the Fredholmness of SIO with piece-wise continuous coefficients, applicable to arbitrary Banach function spaces. This abstract version was then applied to the case of variable exponent Lebesgue spaces. The obtained criterion showed that Fredholmness in such spaces and the index depend only on the values of the function $p(t)$ at the discontinuity points of the coefficients of the operator.

The Sobolev type estimate

$$\left\| (1+|x|)^{-\gamma(x)} I^{\alpha(\cdot)} f \right\|_{L^{q(\cdot)}(\mathbb{R}^n)} \le c \, \|f\|_{L^{p(\cdot)}(\mathbb{R}^n)} \tag{6.10}$$

for the Riesz potential $I^{\alpha(\cdot)}$ of variable order, was obtained in [70], 2003, under the local log and decay conditions on $p(x)$ and $1 < p(\infty) \le p(x) \le P < \infty$, where $\frac{1}{q(x)} = \frac{1}{p(x)} - \frac{\alpha(x)}{n}$ and $\gamma(x) = A_\infty \alpha(x) \left[1 - \frac{\alpha(x)}{n}\right]$, A_∞ being the constant from the decay condition.

In [62], 2005, there was studied the Riemann boundary value problem

$$\Phi^+(t) = G(t)\Phi^-(t) + g(t)$$

for analytic functions in the class of analytic functions represented by the Cauchy type integral with density in the spaces $L^{p(\cdot)}(\Gamma)$ with variable exponent, with piecewise coefficient $G(t)$, admitting also more general oscillating coefficients. There were obtained solvability conditions and formulas for the solution in the explicit form, reflecting their dependence on local values of the variable exponent. The related boundary singular integral equations in $L^{p(\cdot)}(\Gamma)$ were also treated. That solution of the boundary value problem allowed to obtain the weight results for Cauchy singular integral operators in $L^{p(\cdot)}(\Gamma)$-spaces, among them an extension of the Helson-Szegö theorem.

These applications naturally led to the necessity to extend the result on the boundedness of the Cauchy singular operator to more general curves. In case of a constant exponent p the most general curves admitted for this, are Carleson curves. In [76], 2005, there were announced results on the weighted boundedness of the maximal, singular and potential operators on such curves, and in [75], 2005, there was given an overview on facts related to the proof on Carleson curves together with the sketch of the main ideas of the proof. In [63], 2006, there was given the complete proof of the weighted boundedness of the singular operator S_Γ together with the proof of the necessity for Γ to be a Carleson curve. The main result of [63] runs as follows.

Let Γ be a simple Carleson curve, $w(t) = |t-t_0|^\mu$, $t_0 \in \Gamma$, $1 < p_- \leq p_+ < \infty$ and $p(x)$ be log-continuous. Then the operator S_Γ is bounded in $L^{p(\cdot)}(\Gamma, w)$, if and only if $-\frac{1}{p(t_0)} < \mu < \frac{1}{p'(t_0)}$. If S_Γ is bounded, then Γ is a Carleson curve.

The proof of a similar weighted boundedness of Sobolev type for potential operators on Carleson curves in the case of power weights appeared in [80], 2008.

In [65], 2006, in the case of bounded domains Ω in $\mathbb{R}^n$ there were obtained conditions for the weighted boundedness of the maximal operator, in $L^{p(\cdot)}(\Omega, w)$, in the case of weights w in the Bari-Stechkin class (they oscillate between different power functions, in general). These conditions were formulated in terms of bounds for the so-called Matuszewska-Orlicz lower and upper indices $m(w)$ and $M(w)$ of the weight. In a similar settings of oscillating weight, in [67], 2007, there was studied the boundedness of singular Calderón-Zygmund type operators in bounded domains in $\mathbb{R}^n$ and singular Cauchy integral operator S_Γ along a Carleson curve Γ.

In the paper [66], 2007, also because of applications, there were given conditions of the boundedness of the maximal operator M on Carleson curves with similar weights from the Bari-Stechkin-class. In this paper there was also shown that, both in the case of Carleson curves and bounded domains in $\mathbb{R}^n$, it is possible to cover the case of weight functions satisfying a certain similarity with the general Muckenhoupt-looking condition

$$\begin{aligned}\sup_{x\in\Omega, r>0} &\left(\frac{1}{|B(x,r)|} \int\limits_{B(x,r)} |\rho(y)|^{p(y)} dy\right) \\ &\times \left(\frac{1}{|B(x,r)|} \int\limits_{B(x,r)} \frac{dy}{|\rho(y)|^{\frac{p(y)}{p_- -1}}}\right)^{p_- -1} < \infty,\end{aligned} \tag{6.11}$$

(which coincides with the usual Muckenhoupt class A_p in the case of constant p, but is narrower than what is necessary in the case of variable p).

The next natural step was the extension of the developed technique of weighted estimation of operators to the case of more general settings when the underlying space is an arbitrary quasimetric measure space (X, d, μ) equipped with a quasidistance $d(x, y)$ and a Borel measure μ. For the maximal operator M such an extension in case of doubling measure was done in the papers [78], 2007; [81], 2008, where in particular there was shown that the condition of type (6.11) may be also admitted as well as the case of the oscillating radial type weights.

In [77], 2007 and [82], 2008 there was proved a certain $p(\cdot) \to q(\cdot)$-version of Rubio de Francia's extrapolation theorem for weighted spaces $L^{p(\cdot)}_{\varrho}$ on metric measure spaces. By means of this extrapolation theorem and known theorems on the boundedness with Muckenhoupt weights in the case of constant p, there were derived results on the weighted $p(\cdot) \to q(\cdot)$- or $p(\cdot) \to p(\cdot)$-boundedness – in the case of variable exponent $p(x)$ – of various classical operators of harmonic analysis: potential type operators, Fourier multipliers, multipliers of trigonometric Fourier series, singular integral operators on Carleson curves, majorants of partial sums of trigonometric Fourier series etc.

See also another version of presentation of the above results in [85], 2009.

The paper [83], 2008, was devoted to the problem of the boundedness of the so-called Vekua generalized singular integral operator

$$\widetilde{S}_\Gamma f(t) = \frac{1}{2\pi i} \int_\Gamma \Omega_1(t,\tau) f(\tau)\, d\tau - \Omega_2(t,\tau)\overline{f}(\tau)\, d\overline{\tau}$$

which arises in the theory of Vekua's generalized analytical functions defined by the equation $\partial_{\bar{z}}\Phi(z) + A(z)\Phi(z) + B(z)\overline{\Phi}(z) = 0$. The boundedness of this operator along an arbitrary Carleson curve Γ was established in weighted Lebesgue spaces $L^{p(\cdot)}(\Gamma, \varrho)$ with variable exponent $p(\cdot)$ and oscillating radial weights; the obtained result even in the case of constant p was an improvement of what was known.

6.5. Variable Exponent Analysis, continued: 2004–present

A number of the results in VEA in this period was obtained in the research joint with S. Samko's PhD students and collaborators in Portugal, A. Almeida and H. Rafeiro. At that period he had collaboration with many colleagues, L. Diening, V. Guliev, J. Hasanov, M. Hajibayev, A. Karapetyants, L.-E. Persson, V. Rabinovich, E. Shargorodsky, S. Umarkhadzhiev and B. Vakulov among them.

6.5.1. More on weighted estimates of potential operators. In [213], [214], 2005, there was proved the weighted $L^{p(\cdot)}(\mathbb{R}^n, \rho_{\gamma_0,\gamma_\infty}) - L^{q(\cdot)}(\mathbb{R}^n, \rho_{\mu_0,\mu_\infty})$-Sobolev theorem with $\frac{1}{q(x)} = \frac{1}{p(x)} - \frac{\alpha}{n}$ for the Riesz potential operator I^α where

$$\rho_{\gamma_0,\gamma_\infty}(x) = |x|^{\gamma_0}(1+|x|)^{\gamma_\infty-\gamma_0}, \mu_0 = \frac{q(0)}{p(0)}\gamma_0,\ \mu_\infty = \frac{q(\infty)}{p(\infty)}\gamma_\infty,$$
$$\alpha p(0) - n < \gamma_0 < n[p(0)-1], \qquad \alpha p(\infty) - n < \gamma_\infty < n[p(\infty)-1],$$

under some additional condition relating the values of the exponents at the points 0 and ∞, which was removed in [203], 2007. Similar versions of weighted Sobolev theorems for spherical potential operators were also obtained there via stereographical projection.

In the papers [119], 2007, and [122], 2010, an estimate

$$\|I^\alpha f\|_{L^{q(\cdot)}\left(\Omega, w^{\frac{q}{p}}\right)} \leq C\,\|f\|_{L^{p(\cdot)}(\Omega,w)},$$

of such a kind was obtained for $\Omega \subseteq \mathbb{R}^n$, in the case of more general weights w in the Bari-Stechkin class. This result was based on a generalization in [119], 2007, of the estimate (6.8), to weights more general from the Bari-Stechkin class. It runs as follows

$$\left\| |x-y|^{\nu(x)} \chi_{\Omega \setminus B(x,r)}(y)\varphi(|y|) \right\|_{p(\cdot)} \leq C r^{\nu(x)+\frac{n}{p(x)}} \varphi(r+|x|) \tag{6.12}$$

where $0 \in \overline{\Omega}$, under the assumptions that $\sup\limits_{x\in\Omega}[\nu(x)p(x)+n] < 0$.

In [23], 2011, variable exponent estimates for potential operators were obtained in the case of more general kernels

$$\frac{a([\varrho(x,y)])}{[\varrho(x,y)]^N}$$

on bounded quasimetric measure space (X,μ,d) with doubling measure μ satisfying the upper growth condition $\mu B(x,r) \leq K r^N$, $N \in (0,\infty)$. Under some assumptions on $a(r)$ in terms of almost monotonicity, there was proved that such potential operators are bounded from the variable exponent Lebesgue space $L^{p(\cdot)}(X,\mu)$ into a certain Musielak-Orlicz space $L^\Phi(X,\mu)$ with the N-function $\Phi(x,r)$ defined by the exponent $p(x)$ and the function $a(r)$. A reformulation of the obtained result in terms of the Matuszewska-Orlicz indices of the function $a(r)$ was also given. In [22], 2010, these results were extended to the weighted case with radial type power weight $w = [\varrho(x,x_0)]^\nu, x_0 \in X$, for such generalized potential operators acting from the weighted variable exponent Lebesgue space $L^{p(\cdot)}(X,w,\mu)$ into a certain weighted Musielak-Orlicz space $L^\Phi(X, w^{\frac{1}{p(x_0)}}, \mu)$.

6.5.2. Studies related to HSI and the range $I^{\alpha(\cdot)}(L^{p(\cdot)})$ in case of variable exponents. In [4], 2006, there were introduced and studied Riesz and Bessel potential spaces of constant order α with densities in the variable exponent Lebesgue spaces $L^{p(\cdot)}(\mathbb{R}^n)$. It was shown that the spaces of these potentials can be characterized in terms of convergence of hypersingular integrals, under the assumption that the maximal operator M is bounded in the space $L^{p(\cdot)}(\mathbb{R}^n)$. This generalized the corresponding description of the range $I^\alpha(L^p)$ mentioned in Subsection 5.1, to the variable exponent case, representing the range as the space of $f \in L^{q(\cdot)}(\mathbb{R}^n)$ with the Sobolev exponent $q(x) = \frac{np(x)}{n-\alpha p(x)}$ for which there converge truncated HSIs $\mathbb{D}^\alpha_\varepsilon f(x)$.

As a consequence of this characterization, there was given a relation between such spaces of Riesz or Bessel potentials and the variable Sobolev spaces:

$$\begin{aligned}\mathcal{B}^\alpha[L^{p(\cdot)}(\mathbb{R}^n)] &= L^{p(\cdot)}(\mathbb{R}^n)\bigcap I^\alpha[L^{p(\cdot)}(\mathbb{R}^n)]\\ &= \{f\in L^{p(\cdot)}(\mathbb{R}^n) : \mathbb{D}^\alpha f\in L^{p(\cdot)}(\mathbb{R}^n)\}\end{aligned} \tag{6.13}$$

and it was shown that in the case of integer α the Bessel potential space $\mathcal{B}^\alpha[L^{p(\cdot)}(\mathbb{R}^n)]$ with the variable exponent $p(x)$ coincides with the Sobolev spaces.

In [110], 2008, it was shown that the method of approximative inverses discussed in Subsection 6.1.1, is also applicable to Riesz potential operators $I^\alpha\varphi$ for densities φ in the variable exponent spaces $L^{p(\cdot)}(\mathbb{R}^n)$ under natural assumptions on $p(x)$.

The paper [109], 2007, appeared as a kind of return of interest to one-dimensional integral equations

$$\int_a^b \frac{c(x,t)}{|x-t|^{1-\alpha}}\varphi(t)\,dt = f(x),$$

of the first kind, where $c(x,t)$ has a jump at the diagonal $x=t$, for which (2.1) serves as a model equation. The state of art in VEA allowed to deal with the well-posedness of such equations with solutions φ in the weighted variable exponent Lebesgue space $L^{p(\cdot)}([a,b],w)$ and right-hand sides f in the weighted variable exponent Sobolev type space $L^{\alpha,p(\cdot)}([a,b],w)$ of fractional smoothness. There was obtained a criterion of Fredholmness and formula for the Fredholm index from $L^{p(\cdot)}([a,b],w)$ to $L^{\alpha,p(\cdot)}([a,b],w)$ and given formulas in closed form for solutions $\varphi\in L^{p(\cdot)}([a,b],w)$ of the equation (2.1) in dependence on the values of the variable exponent $p(x)$ at the endpoints $x=a$ and $x=b$.

In relation with the above applications to integral equations of the first kind, the characterization of the variable exponent Riesz or Bessel potential spaces (see (6.13)) in [111], 2008, a similar problem of characterization was solved in the one-dimensional case $n=1$, but in a more difficult situation of the interval where the influence of the end points come into the game. The corresponding range of fractional integrals over the space $L^{p(\cdot)}(\Omega,\varrho)$ with variable exponent $p(\cdot)$ and a power type weight ϱ was described. in terms of convergence of the corresponding Marchaud derivatives and that this range may be also obtained as the restriction on Ω of Bessel potentials with densities in $L^{p(\cdot)}(\Omega,\varrho)$.

As a by-product result in the proofs in [111] there was also proved a Hardy type inequality (see (6.18) in the sequel) adjusted for the finite interval (a,b) with the weight $(x-a)^{\mu(x)}(b-x)^{\nu}(x)$ where the influence of the both end points on the conditions on the exponents had to be separated in terms of the decay conditions at the end-points.

A similar equation

$$\int_{\mathbb{R}^n} \frac{c(x,y)}{|x-y|^{n-\alpha}}\varphi(y)dy \;=\; f(x)\,, \qquad x\in\mathbb{R}^n, \qquad 0<\alpha<1,$$

of the 1st kind over the whole space $\mathbb{R}^n$ was later investigated in [207], 2010, and [208], 2011, where it was shown that the method of reducing such equations with L^p-solutions to equations of the second kind with an operator compact in $L^p(\mathbb{R}^n)$ remains valid for the variable exponent Lebesgue spaces $L^{p(\cdot)}(\mathbb{R}^n)$ under the usual local and decay log-conditions on $p(x)$.

In [5], 2007, for functions in the Sobolev space $W^{1,p(\cdot)}(\Omega)$ with variable exponent $p(x)$, on a bounded domain in $\mathbb{R}^n$ with Lipschitz boundary, there was proved the pointwise estimate

$$|f(x)-f(y)| \le \frac{c}{\min[p(x),p(y)]-n}\,\||\nabla f|\|_{p(\cdot),\Omega}\,|x-y|^{1-\frac{n}{\min[p(x),p(y)]}} \tag{6.14}$$

for all $x,y\in\Omega$ such that $p(x)>n, p(y)>n$, and the embedding $W^{1,p(\cdot)}(\Omega)\hookrightarrow C^{0,1-\frac{n}{p(\cdot)}}(\Omega)$ proved. The pointwise estimate allowed to obtain a result on the boundedness of the hypersingular operator

$$\mathcal{D}^{\alpha(\cdot)}f(x) = \int_\Omega \frac{f(x)-f(y)}{|x-y|^{n+\alpha(x)}}\,dy \tag{6.15}$$

over a bounded domain from $W^{1,p(\cdot)}(\Omega)$ into $L^{q(\cdot)}(\Omega)$ under a certain natural choice of the variable exponents $p(x)$ and $q(x)$.

When the underlying space is a quasimetric measure space (X,d,μ) with doubling condition, the pointwise estimate

$$|f(x)-f(y)| \le C(\mu,\alpha,\beta)\,\Big[d(x,y)^{\alpha(x)}\mathcal{M}^\sharp_{\alpha(\cdot)}f(x) + d(x,y)^{\beta(y)}\mathcal{M}^\sharp_{\beta(\cdot)}f(y)\Big]$$

of a nature similar to (6.14), with the fractional sharp maximal function

$$\mathcal{M}^\sharp_{\alpha(\cdot)}f(x) = \sup_{r>0}\frac{r^{-\alpha(x)}}{\mu B(x,r)}\int_{B(x,r)} |f(y)-f_{B(x,r)}|\,d\mu(y)$$

of variable order and $0<\alpha_-\le\alpha_+<\infty, 0<\beta_-\le\beta_+<\infty$, was obtained in [6], 2009. With the aid of this estimate there was proved the embedding

$$M^{1,p(\cdot)}(X)\hookrightarrow H^{1-\frac{N}{p(\cdot)}}(X)$$

of the Hajłasz-Sobolev space $M^{1,p(\cdot)}(X)$ with $p_->N$ into the Hölder space of variable order, where the lower dimension N of X comes from the estimate $\mu B(x,r)\ge cr^N$.

The paper [7], 2009, performed also in the context of quasimetric function spaces (X,d,μ) dealt with potential and hypersingular integrals in the variable exponent setting. For potential operators of forms

$$\mathfrak{I}^\alpha f(x) = \int_\Omega \frac{[d(x,y)]^\alpha}{\mu B(x,d(x,y))} f(y)\,d\mu(y), \quad I_n^{\alpha(\cdot)}f(x) = \int_\Omega \frac{f(y)\,d\mu(y)}{[d(x,y)]^{n-\alpha(x)}}, \tag{6.16}$$

where $\Omega \subset X$ is an open bounded set and n comes from the upper estimate for $\mu B(x,r)$, Sobolev type estimates were proved. HSI were studied there in two forms, one as in (6.15) and another

$$\mathfrak{D}^{\alpha(\cdot)} f(x) = \int_\Omega \frac{f(x)-f(y)}{\mu B(x,d(x,y))[d(x,y)]^{\alpha(x)}}\, d\mu(y),$$

for which were found conditions of their boundednes from the Hajłasz-Sobolev space $M^{1,p(\cdot)}(X)$ into a Lebesgue $L^{q(\cdot)}(\Omega)$, this being an extension of the results in [5] for the Euclidean case.

In [121], 2010, there were studied Hölder spaces $H^{\lambda(\cdot)}(X)$ on a quasimetric measure space (X,d,μ), whose Hölder exponent $\lambda(x)$ is variable. There were obtained Zygmund type estimates of the local continuity modulus of potential operators $I_n^{\alpha(\cdot)}$ of form (6.16) and HSI $\mathcal{D}^{\alpha(\cdot)} f$ of form (6.15), of variable complex-valued orders $\alpha(x)$, via such modulus of their densities. These estimates allowed to treat not only the case of the spaces $H^{\lambda(\cdot)}(X)$, but also the generalized Hölder spaces $H^{w(\cdot,\cdot)}(X)$ of functions whose continuity modulus is dominated by a given function $w(x,h), x \in X, h > 0$.

By means of those estimates there were established theorems on mapping properties of potential operators of variable order $\alpha(x)$, from such a variable exponent Hölder space with the exponent $\lambda(x)$ to another one with a "better" exponent $\lambda(x)+\alpha(x)$, and similar mapping properties of hypersingular integrals of variable order $\alpha(x)$ from such a space into the space with the "worse" exponent $\lambda(x)-\alpha(x)$ in the case $\alpha(x) < \lambda(x)$. In the case of potentials, it was also admitted that $\Re\alpha(x)$ may vanish at a set of measure zero and in this case the action of the potential operator was considered to weighted generalized Hölder spaces with the weight $\alpha(x)$.

These results were also discussed in [120], 2009, where the mapping properties for HSI were given with $\Re\alpha(x)$ vanishing on a set of measure zero.

In [114], 2010, the results of type (5.2) and (5.3) were generalized for variable exponents in the case of the Bessel potential space $\mathcal{B}^\alpha[L^{p(\cdot)}(\mathbb{R}^n)]$, see (6.13), corresponding to the case $p = r$. Namely, for $f \in L^{p(\cdot)}(\mathbb{R}^n)$ with $p(\cdot)$ satisfying the log- and decay conditions, the limits $\lim_{\varepsilon\to 0+} \frac{1}{\varepsilon^\alpha}(I-P_\varepsilon)^\alpha f$ and $\lim_{\varepsilon\to 0+} \mathbb{D}_\varepsilon^\alpha f$ exist in $L^{p(\cdot)}(\mathbb{R}^n)$ simultaneously and coincide with each other and

$$\mathcal{B}^\alpha[L^{p(\cdot)}(\mathbb{R}^n)] = \left\{f \in L^{p(\cdot)}(\mathbb{R}^n), \|(I-P_\varepsilon)^\alpha f\|_{p(\cdot)} \le C\varepsilon^\alpha\right\}.$$

6.5.3. Morrey and Campanato spaces. In [3], 2008, there were introduced variable exponent Morrey spaces

$$L^{p(\cdot),\lambda(\cdot)}(\Omega) = \left\{ f : \sup_{x\in\Omega,\, r>0} \frac{1}{r^{\lambda(x)}} \int\limits_{B(x,r)} |f(y)|^{p(y)} dy < \infty \right\}, \quad \Omega \subseteq \mathbb{R}^n,$$

for which there was shown the equivalence of various types of norms and embedding theorems proved. The main results obtained in this paper provided conditions on the boundedness, in $L^{p(\cdot),\lambda(\cdot)}(\Omega)$ over bounded Ω, of the Hardy-Littlewood maximal operator M and potential type operators

$$I^{\alpha(\cdot)}f(x) = \int\limits_{\Omega} \frac{f(y)\,dy}{|x-y|^{n-\alpha(x)}} \tag{6.17}$$

of variable order $\alpha(x)$. In the case of constant α, there was also proved the boundedness theorem in the limiting case $p(x) = \frac{n-\lambda(x)}{\alpha(x)}$, when the potential operator I^α acts from $L^{p(\cdot),\lambda(\cdot)}$ into BMO.

The results of the paper [3] found their development in the papers [21] and [20], 2010. In [21] there were considered more general Morrey spaces $\mathcal{M}^{p(\cdot),\omega}(\Omega)$ defined by the norm

$$\|f\|_{\mathcal{M}^{p(\cdot),\omega}} = \sup_{x\in\Omega, r>0} \frac{r^{-\frac{n}{p(x)}}}{\omega(x,r)} \|f\|_{L^{p(\cdot)}(B(x,r))}$$

and obtained conditions of the boundedness of the Hardy-Littlewood maximal operator M, the potential operator (6.17), the fractional maximal operator $M^{\alpha(x)}f(x) = \sup_{r>0}|B(x,r)|^{\frac{\alpha(x)}{n}-1}\int_{B(x,r)}|f(y)|dy$, also of variable order $\alpha(x)$ and Calderón-Zygmund type singular operator. For the potential operator there the generalizations of both the Spanne and Adams type results known for constant exponents and a generalization of the Sobolev-Adams exponent were obtained.

In [20] there was introduced another type of spaces, namely the Morrey-Adams type spaces defined by the norm

$$\|f\|_{\mathcal{M}^{p(\cdot),\theta(\cdot),\omega(\cdot)}(\Omega)} = \sup_{x\in\Omega}\left\| \frac{\omega(x,r)}{r^{\frac{n}{p(x)}}}\|f\|_{L^{p(\cdot)}(\widetilde{B}(x,r))}\right\|_{L^{\theta(\cdot)}(0,\ell)}, \qquad \ell = \operatorname{diam}\Omega,$$

the case $\theta(r)\equiv\infty$ corresponding to the studies in [21], where the same classical operators of harmonic analysis were studied.

Results obtained in [21] and [20] proved to be new even in the case of constant p, because they did not impose monotonicity type conditions on $\omega(x,r)$.

6.5.4. PDO in variable exponent setting. Although the first cited paper below is not related to variable exponents, it is mentioned here because it was also in the areas SIO and PDO.

In [103], 2006, there were studied the Fredholm spectra and Fredholm properties of SIO on composed Carleson curves with discontinuous coefficients in weighted Hölder spaces $H^\lambda(\gamma,\varrho)$, with the admission of coefficients of SIO and also curves Γ and weights ϱ, which may slowly oscillate at the nodes of the curve. To this end there was used the technique of PDO, which allowed to reveal the influence of the oscillations on the appearance of massive Fredholm spectra, and obtain a criterion of Fredholmness and index formula for SIO in $H^\lambda(\gamma,\varrho)$.

In [104], 2007, and [105], 2008, the results of [73], 2003, on the weighted boundedness of Calderón-Zygmund singular operators in $L^{p(\cdot)}(\mathbb{R}^n,w)$ with power weight w were generalized to more general operators of the form

$$\mathbb{A}f(x) = \int_{\mathbb{R}^n} k(x, x-y) f(y) dy$$

where it was shown that this boundedness is governed by the estimate

$$\|\mathbb{A}\|_{L^{p(\cdot)}(\mathbb{R}^n,w)} \le c(n,p,w)\,[\lambda_1(\mathbb{A}) + \lambda_2(\mathbb{A}) + \nu(\mathbb{A})]$$

where $\lambda_1(\mathbb{A}) := \sup\limits_{|\alpha|=1} \sup\limits_{x,z\in\mathbb{R}^n\times\mathbb{R}^n} |z|^{n+1} |\partial_x^\alpha k(x,z)|$, $\lambda_2(\mathbb{A}) := \sup\limits_{|\beta|=1} \sup\limits_{x,z\in\mathbb{R}^n\times\mathbb{R}^n} |z|^{n+1} \times \left|\partial_z^\beta k(x,z)\right|$ and $\nu(\mathbb{A})$ is the constant from the weak type estimate $|\{x \in \mathbb{R}^n : |\mathbb{A}f(x)| > t\}| \le \frac{\nu(\mathbb{A})}{t} \|f\|_1$.

In this paper there was also made an important observation that the famous Krasnoselski theorem on unilateral interpolation of compactness remains valid in variable exponent Lebesgue spaces.

Both these two facts allowed in [105] to obtain a necessary and sufficient condition for pseudodifferential operators of the Hörmander class $OPS^m_{1,0}$ with symbols slowly oscillating at infinity, to be Fredholm in weighted Sobolev spaces $H^{s,p(\cdot)}_w(\mathbb{R}^n)$ with constant smoothness s, variable $p(\cdot)$-exponent, and exponential weights w.

These studies were continued in [106], 2011, where there was obtained the boundedness of singular integral operators in the variable exponent Lebesgue spaces $L^{p(\cdot)}(\Gamma, w)$ on a class of *composed Carleson curves* Γ with the weights w having a finite set of oscillating singularities. The proof was based on the boundedness of Mellin PDO in the spaces $L^{p(\cdot)}(\mathbb{R}_+, d\mu)$, $d\mu = \frac{dr}{r}$.

This allowed in [106] to obtain a criterion of local invertibility of singular integral operators with piecewise slowly oscillating coefficients acting on the space $L^{p(\cdot)}(\Gamma, w)$ via the corresponding criteria of local invertibility at the point 0 of Mellin PDO and local invertibility of SIO on $\mathbb{R}$.

On the base of the Simonenko local theory for studying Fredholmness of operators, adjusted for the variable exponent setting, there was obtained a criterion of Fredholmness of singular integral operators in the variable exponent Lebesgue spaces $L^{p(\cdot)}(\Gamma, w)$ where Γ belongs to a class of composed Carleson curves slowly oscillating at the nodes, and the weight w has a finite set of slowly oscillating singularities. Apart from this criterion, another essential advance in [106], was that the explicit formula for the Fredholmness index was given in a general setting. In the case of variable exponent spaces the formula for the index was known before that in the case of non-oscillating coefficients and non-oscillating curves, as given in [71].

6.5.5. Miscellaneous in variable exponent analysis. In [10], 2007, there was proved a statement important for various applications, showing that bounded kernels $k(x)$ with the estimate

$$|k(x)| \le \frac{C}{(1+|x|)^\nu} \quad \text{with} \quad \nu > n\left(1 - \frac{1}{p(\infty)} + \frac{1}{q(\infty)}\right)$$

generate a convolution operator $k * f$ bounded from $L^{p(\cdot)}(\mathbb{R}^n)$ to $L^{q(\cdot)}(\mathbb{R}^n)$ under the only assumption that $1 \le p_- \le p_+ < \infty, 1 \le q_- \le q_+ < \infty$, $p(x), q(x)$ satisfy the decay condition and $q(\infty) \ge p(\infty)$ (i.e., the local log-condition is not required). This allowed to obtain the Hardy inequality

$$\left\| x^{\alpha(x)+\mu(x)-1} \int_0^x \frac{f(y)\,dy}{y^{\alpha(y)}} \right\|_{L^{q(\cdot)}(\mathbb{R}^1_+)} \le C \, \|f\|_{L^{p(\cdot)}(\mathbb{R}^1_+)} \tag{6.18}$$

(and a similar inequality for the dual Hardy operator) in the variable exponent setting under the natural assumptions

$$0 \le \mu(0) < \frac{1}{p(0)}, \quad 0 \le \mu(\infty) < \frac{1}{p(\infty)}, \qquad \alpha(0) < \frac{1}{p'(0)}, \quad \alpha(\infty) < \frac{1}{p'(\infty)},$$
$$\frac{1}{q(0)} = \frac{1}{p(0)} - \mu(0), \qquad \frac{1}{q(\infty)} = \frac{1}{p(\infty)} - \mu(\infty),$$

not requiring local log-condition, but supposing that only the decay condition holds for $\alpha(x), \mu(x)$ and $p(x)$ at the points $x = 0$ and $x = \infty$.

Within the frameworks of L^p-spaces with constant p the following dominated compactness theorem was known. Let $\mathbf{K}$ be a linear integral operator (LIO) with the kernel $K(x, y)$; if the LIO with the kernel $|K(x, y)|$ is bounded in L^p and the LIO with the kernel $L(x, y) \ge |K(x, y)|$ is compact in L^p, then $\mathbf{K}$ is also compact. In the paper [112], 2008, it was shown that this is valid for every Banach function space with absolutely continuous norm, in particular for weighted variable exponent Lebesgue spaces $L^{p(\cdot)}(\Omega, \mu, \varrho)$. An application to potential type operators in such spaces was given there.

Compactness of operators in variable exponent Lebesgue spaces was also discussed in [191], 2010, where in particular it was shown that integral operators

$$Kf(x) = \int_\Omega \mathcal{K}(x, y) f(y)\, dy,$$

over an open set $\Omega \subset \mathbb{R}^n$ with bounded measure, $|\Omega| < \infty$, whose kernel $K(x, y)$ is dominated by a difference kernel: $|K(x, y)| \le \mathcal{A}(|x - y|)$, is compact in the space $L^{p(\cdot)}(\Omega, \varrho)$ whenever the maximal operator is bounded in this space, and ϱ is an arbitrary weight such that $L^{p(\cdot)}(\Omega, \varrho)$ is a Banach function space. In the non-weighted case $\varrho = 1$ there was also given a modification of this statement for convolution operators over $\mathbb{R}^n$ with vanishing coefficients, which were already discussed in the beginning of Section 4 in the case of constant p.

In [113], 2009, there was proved that in variable exponent spaces $L^{p(\cdot)}(\Omega)$, where $p(\cdot)$ satisfies the log-condition and Ω is a bounded domain in $\mathbf{R}^n$ with the property that $\mathbf{R}^n \backslash \overline{\Omega}$ has the cone property, the validity of the Hardy type inequality

$$\left\| \frac{1}{\delta(x)^\alpha} \int_\Omega \frac{\varphi(y)}{|x-y|^{n-\alpha}}\, dy \right\|_{p(\cdot)} \leqq C\|\varphi\|_{p(\cdot)}, \quad 0<\alpha<\min\left(1, \frac{n}{p_+}\right)$$

where $\delta(x) = \operatorname{dist}(x, \partial\Omega)$, is equivalent to a certain property of the domain Ω expressed in terms of α and χ_Ω (close to the property that χ_Ω is a pointwise multiplier in the space $I^\alpha(L^{p(\cdot)})$ of Riesz potentials). This led to a result which is even new in the case of constant exponents p that such a Hardy inequality holds for the domains satisfying the Strichartz condition.

In the paper [56], 2010, which seems to be the first paper in VEA in complex analysis, there was introduced and described the space $\mathrm{BMO}^{p(\cdot)}(\mathbb{D})$, $1 \leq p(z) < \infty$ of functions of bounded mean oscillation on the unit disc in the complex plane in the hyperbolic Bergman metric with respect to the Lebesgue measure. In the definition of this space, the Berezin transform played a role of an average, being a kind of substitution of the Poisson transform in complex analysis.

In [193], 2011, there were obtained weighted estimates of type (6.12) in case of truncation to the ball $B(x,r)$ itself, not to its exterior $\Omega\backslash B(x,r)$. Such estimates are of importance in application to the study of operators in Morrey type spaces with variable exponents. A non-weighted case of such an estimate was given in (6.4). The estimate obtained in [193] covered the case of radial type weights from the Bari-Stechkin class (such weights may oscillate between power functions with different exponents). In the case of power weights this estimate runs as follows:

$$\left\| |x-y|^{\nu(x)} \chi_{B(x,r)}(y) |y|^{\beta(x)} \right\|_{p(\cdot)} \leq c \frac{r^{\frac{n}{p(x)}+\nu(x)} |x|^{\min\left(\frac{n}{p(x)}+\nu(x)+\beta(x),0\right)}}{(r+|x|)^{\min\left(\frac{n}{p(x)}+\nu(x),-\beta(x)\right)}} \tag{6.19}$$

under the assumption that $\inf_{x\in\Omega}[\nu(x)p(x)+\beta(x)p(0)+n] \neq 0$. (In the case of more general weights in such an estimation one should distinguish the cases where the norm remains finite when x approaches the singular point $x_0 = 0$ of the weight or becomes infinite).

In [115], 2011, there were introduced and studied variable exponent Campanato spaces $\mathcal{L}^{p(\cdot),\lambda(\cdot)}X$ on spaces X of homogeneous type, defined by the modular form

$$\sup_{x\in X, r>0} \frac{1}{(\mu B(x,r))^{\lambda(x)}} \int_{B(x,r)} \left| f(y) - f_{B(x,r)} \right|^{p(y)} d\mu(y).$$

There was proved the embedding

$$\mathcal{L}^{q(\cdot),\nu(\cdot)}X \hookrightarrow \mathcal{L}^{p(\cdot),\lambda(\cdot)}X \quad \text{whenever} \quad \frac{1-\lambda(x)}{p(x)} \geq \frac{1-\nu(x)}{q(x)},$$

and also proved the equivalence of these spaces, up to norms, to variable exponent Morrey spaces $L^{p(\cdot),\lambda(\cdot)}X$, when $\lambda_+ < 1$ and for X with constant dimension, i.e.,

$B(x,r) \sim r^m$, there was also proved the equivalence

$$\mathcal{L}^{p(\cdot),\lambda(\cdot)}X \cong H^{\alpha(\cdot)}(X), \quad \alpha(x) = m\frac{\lambda(x)-1}{p(x)}$$

to the variable exponent Hölder space, when $\lambda_- > 1$, this coincidence with the Hölder space in the case of X with non-constant dimension replaced by two-sided embeddings.

Finally we note that surveys of selected topics in VEA may be found in [189], 2005, and [84], 2008, [190], 2009.

7. Miscellaneous

In [57], 1977, there were studied mapping properties of one-dimensional integral operators $K\varphi \equiv \int_0^a K(x,t)\varphi(t)\,dt$ with kernels $K(x,t)$ homogeneous of degree -1, in the generalized Hölder space

$$H^\omega(0,a) = \{\varphi(x)\colon |\varphi(x+h)-\varphi(x)| \le A\omega(h)\}, \quad 0 < a < \infty.$$

There were obtained the conditions

$$\int_0^\infty |K(1,t)|\,dt < \infty, \quad \int_0^a |K(h,t)|\omega(t)\,dt \le c\omega(h),$$

$$\int_0^x |K(x,1)|t^{-1}\,dt \in H^\omega(0,a),$$

ensuring the boundedness of the operator K in $H^\omega(0,a)$.

In comparison with the theory of fractional type operators acting in $L^p(\mathbb{R}^n)$, there naturally arise Fourier multipliers, or more specifically, multipliers with absolutely integrable Fourier transforms, when one wishes to cover the cases $p=1$ and $p=\infty$, the class of such multipliers being also known as the Wiener algebra

$$W_0(\mathbb{R}^n) = \{f : \ f = \widehat{\varphi}, \ \varphi \in L^1(\mathbb{R}^n)\}.$$

In [89], 1988, in relation with the study of potential operators with homogeneous kernel, there were given certain tests for functions to belong to this algebra. The interest in this topic led to writing a survey paper [199], 1994, on such tests. We refer also to Subsection 4.1 of Chapter 1 of the book [183] in this relation.

In [210], 1996, there were introduced the spaces $L^{\mathrm{rad}}_{p_2}(L^{\mathrm{ang}}_{p_1})$ defined by a mixed norm with different exponents p_2 and p_1 in polar and angular variables, for which there was obtained the corresponding anisotropic interpolation inequality, and proved the Riesz-Thorin theorem in such spaces. The latter was also applied to prove the Hausdorff-Young type theorem for the Fourier transforms in $L^{\mathrm{rad}}_{p_2}(L^{\mathrm{ang}}_{p_1})$.

When living in USA in 1992–1993, he had scientific discussions with Professor Kenneth Miller, known in FC by his book on fractional differential equations, joint with Bertram Ross. As a result, later there appeared two papers on completely monotonic papers, joint with K. Miller, one was two-page short [92], 1997, where it was shown that the complete monotonicity of the generalized Mittag-Leffler function $E_{\alpha,\beta}(-x)$ with $0 < \alpha \leq 1$ is an immediate consequence of the known case $\beta = 1$ of the usual Mittag-Leffler function. Another one [93], 2001, was an expository article on completely monotonic functions.

In connection with the development of the method of approximative inverses in papers [174], 1998, discussed in Subsection 6.1, in [182], 2001, he extended the well-known presentation $K_{\nu+1}(z) - K_{\nu-1}(z) = 2\nu\frac{K_\nu(z)}{z}$ for the Bessel-McDonald function $K_\nu(z)$, and also for other Bessel functions, to the case of the difference $K_{\nu+m}(z) - K_{\nu-m}(z)$.

In the paper [26], 2001, there was studied the Volterra nonlinear integral equation

$$\varphi^m(x) = a(x)\int_0^x k(x-t)b(t)\varphi(t)dt + f(x),\ 0 < x < d \leq \infty,$$

with $m > 1$ and nonnegative functions $a(x)$, $k(u)$, $b(t)$ and $f(x)$. In the general case some upper bounds were given for the averages $\frac{1}{x}\int_0^x \varphi(t)dt$ of the solution. In the case when $a(x)$, $k(u)$, $b(t)$ and $f(x)$ have lower power estimates near the origin, lower power type bounds directly for solutions $\varphi(x)$ were also given and conditions for the uniqueness of the solution in some weighted space of continuous functions have been proved.

Three papers [194], [8], 2003; [9], 2008, were devoted to the solvability of convolution type integral equations of the first kind

$$K\varphi(x) :\equiv \int_a^x k(x-t)\varphi(t)\,dt = f(x), \quad -\infty \leq a < x < b \leq \infty,$$

with a locally integrable kernel $k(x) \in L_1^{loc}(\mathbb{R}^1_+)$ known as Sonine equation. The latter means that there exists a locally integrable kernel $\ell(x)$ such that

$$\int_0^x k(x-t)\ell(t)\,dt \equiv 1$$

(locally integrable divisors of the unit, with respect to the operation of convolution). The formal solution of such an equation is well known, but it does not work, for example, in the case of solutions in the spaces $X = L^p(\mathbb{R}^1)$ and is not defined on the whole range $K(X)$. In [194] in the case $a = -\infty, b = \infty$, there were discovered various properties of Sonine kernels which allowed to construct the real inverse operator, within the framework of the spaces $X = L^p(\mathbb{R}^1)$, in a form similar to the Marchaud form for the inversion of fractional integration, and the description of the range $K(X)$ was given. The paper [8] dealt with a more difficult case $-\infty < a < b \leq \infty$, when the influence of the end points on the solvability and convergence of the inverse operator comes into a play. In [9], 2008, for such equations with $-\infty < a < b < \infty$ in the case of almost decreasing Sonine kernels $k(t)$ there are proved weighted estimates of continuity moduli $\omega(\mathbb{K}\varphi, h)$ and $\omega(\mathbb{K}^{-1}f, h)$ which allowed to show that the weighted generalized Hölder spaces

$H_0^\omega(\rho)$ and $H_0^{\omega_1}(\rho)$ are suitable well-posedness classes for these integral equations of the first kind under the choice $\omega_1(h) = hk(h)\omega(h)$.

In [101], 2010, there was proved the following Stein type pointwise inequality

$$\left| \ell(x) \int_{|y|<\mu(x)} k(|x-y|) f(y)\, dy \right| \leq C_0 M f(x),$$

for truncated convolutions, where $Mf(x)$ is the maximal function, the radial kernel $k(t)$ has some quasi-monotonicity property and the coefficient $\ell(x)$ and the truncation function $\mu(x)$ satisfy the condition $\sup_x |\ell(x)| \int_0^{\mu(x)} k(t) t^{n-1} dt < \infty$. A similar dual inequality was also proved and both applied to obtain various Hardy-type norm inequalities in an arbitrary Banach function spaces X, for instance:

$$\left\| v(x) \int\limits_{|y|<\mu(x)} k(|x-y|) \frac{f(y)}{w(y)}\, dy \right\|_X \leq A \|f\|_X ,$$

with application in particular to the case where X is the variable exponent Lebesgue spaces.

In [87], 2010; [88], 2011, there was studied the boundedness of the singular operator

$$S_\Gamma^\gamma f(t) = |t-t_0|^\gamma \int\limits_\Gamma \frac{f(\tau)\, d\tau}{|\tau - t_0|^\gamma (\tau - t)}, \quad t \in \Gamma,$$

with a power weight, $t_0 \in \Gamma$, in grand Lebesgue spaces defined by the norm

$$\|f\|_{L^{p),\theta}} = \sup_{0<\varepsilon<p-1} \left(\varepsilon^\theta \int_\Gamma |f(t)|^{p-\varepsilon} d\mu(t) \right)^{1/(p-\varepsilon)}, \quad 1 < p < \infty,\ \theta > 0,$$

(considered on a Carleson curve Γ because of application) and the corresponding vanishing grand Lebesgue spaces.

It was shown that the condition $-\frac{1}{p} < \gamma < \frac{1}{p'}$ remains necessary and sufficient for the boundedness of the operator S_Γ^γ in both the vanishing and nonvanishing grand spaces. In this relation note that *weighted boundedness of an operator in grand Lebesgue space is not the same as the boundedness in weighted grand Lebesgue space.*

In [209], 2011, there were introduced grand Lebesgue spaces on open sets Ω of infinite measure in $\mathbb{R}^n$, via controlling the integrability of $|f(x)|^{p-\varepsilon}$ at infinity by means of a weight (depending also on ε); in general, such spaces are different for different ways to introduce dependence on the weight on ε. One of the ways was to introduce such space $L^{p),\theta}(\Omega, \langle x\rangle^{-\lambda})$ with $\langle x\rangle = \sqrt{1+|x|^2}$ via the norm

$$\sup_{0<\varepsilon<p-1} \varepsilon^{\frac{\theta}{p-\varepsilon}} \|f\|_{L^{p-\varepsilon}(\Omega, \langle x\rangle^{-\lambda(\varepsilon)})} \quad \text{with} \quad \lambda(\varepsilon) = \lambda + \frac{n-\lambda}{p}\varepsilon + \nu(\varepsilon)$$

and the choice of ν such that $\nu(\varepsilon) > 0$ for $\varepsilon > 0$ and $\nu(0) = 0$.

There was also given a version of weighted grand Lebesgue spaces, different from the usual ones, for bounded sets, in which together with the passage from p to $p-\varepsilon$ a weight also depending on ε was introduced. In both the versions, by means of the Stein-Weiss interpolation theorem with change of measure, it was shown that linear operators bounded in a Lebesgue space with Muckenhoupt weights are also bounded in the corresponding grand Lebesgue space with a Muckenhoupt weight.

In two papers [17], 2010, and [192], 2011, in relation to some applied problems in the elasticity theory he dealt with some model hypersingular integral equations arising in problems with cracks in elastic media, two-dimensional and three-dimensional, respectively. In particular, in [192] there were constructively found axisymmetric solutions of the equation

$$\int\limits_S \frac{\partial^2}{\partial n_{\mathbf{x}} \partial n_{\mathbf{y}}} \left(\frac{1}{|\mathbf{y}-\mathbf{x}|} \right) u(\mathbf{y}) dS_{\mathbf{y}} = f(\mathbf{x}), \quad \mathbf{x} \in S,$$

(where $\mathbf{n_x}, \mathbf{n_y}$ stand for the normal vectors at the points $\mathbf{x}, \mathbf{y}$ of S) in the case of plane circular cracks.

In the paper [102], 2010, also related to application, this time in aerodynamics, there was given an application of the spherical harmonic analysis to the solution of a certain problem which arises in aerodynamics of a rarefied medium of non-interacting point masses moving at unit velocity in all directions. Given the density of the velocity distribution, one easily calculates the pressure created by the medium. In [102] there was solved the inverse problem: given the pressure distribution, determine the density. This led to the problem of solving an integral equation over semisphere, which was solved by making use of some spherical convolution operators introduced and studied in [150], see Subsection 5.3, which allowed to construct the inverting operator in the explicit form:

$$\mathcal{L}f(x') = \frac{1}{16\pi^3}\delta^2(\delta+6) \int\limits_{S^2} \frac{\pi - \arccos(x' \cdot \sigma)}{\sqrt{1-(x' \cdot \sigma)^2}} \left(\int\limits_{\sigma s < 0} f(s) ds \right) d\sigma \qquad (7.1)$$

where δ is the Beltrami-Laplace operator.

Finally, some words should be said about Professor Stefan Samko's art of communication with colleagues and students, and his creative attitude to teaching. His natural friendly way of communication and feeling for humor in any situation attract people, both students and colleagues. His successful work as a teacher and scientific supervisor is well known. He had twenty-one defended PhD students, eighteen of them in his "previous life" in Russia, as he himself likes to say and three in the later period in Portugal. His research group is known in the international scientific community and enjoys the esteem of his colleagues. He taught various courses in universities of Russia, USA and Portugal, organizing also online lectures. In different countries and for different audiences, he has a talent to handle the material easily and to present the most difficult parts as something interesting and naturally appearing. The assistance professors and PhD students working with him know very well his generosity of consulting and giving advices. His plenary lectures in international conferences are always very popular and appreciated.

I feel that this brief description and assessment of Stefan Samko's scientific and pedagogical activities and results would be by no means be complete if I did not illuminate his personality from other sides as well. He has been always known as a very honest person, very human, very social and of high moral qualities, in work and in everyday's life. He was a mountain climber when young, he likes to travel, enjoys good books, good wine and good company. His colleagues and friends know his enthusiasm for mathematics and a strong will and ability that he manifests in his research. Stefan Samko's exceptional capacity in hard work and bright talent enabled him to became an outstanding mathematician. With seventy years of life he continues to carry out interesting research and is obviously full of further plans.

References

[1] A.V. Abramyan, V.A. Nogin, and S.G. Samko, *Fractional powers of the operator* $-|x|^2\Delta$ *in* L_p*-spaces.* Differ. Uravn. **32**:2 (1996), 275–276.

[2] A.V. Abramyan, V.A. Nogin, and S.G. Samko, *Fractional powers of the operator* $-|x|^2\Delta$ *in* L_p*-spaces.* Proc. A. Razmadze Math. Inst. **124** (2000), 1–22.

[3] A. Almeida, J. Hasanov, and S. Samko, *Maximal and potential operators in variable exponent Morrey spaces.* Georgian Math. J. **15**:2 (2008), 195–208.

[4] A. Almeida and S. Samko, *Characterization of Riesz and Bessel potentials on variable Lebesgue spaces.* J. Function Spaces and Applic. **4**:2 (2006), 113–144.

[5] A. Almeida and S. Samko, *Pointwise inequalities in variable Sobolev spaces and applications.* Z. Anal. Anwend. **26**:2 (2007), 179–193.

[6] A. Almeida and S. Samko, *Embeddings of variable Hajlasz-Sobolev spaces into Hölder spaces of variable order.* J. Math. Anal. and Appl. **353** (2009), 489–496.

[7] A. Almeida and S. Samko, *Fractional and hypersingular operators in variable exponent spaces on metric measure spaces.* Meditter. J. Math. **6** (2009), 215–232.

[8] R. Cardoso and S. Samko, *Sonine integral equations of the first kind of in* $L^p(0,b)$. Fract. Calc. and Appl. Anal. **6**:3 (2003), 235–258.

[9] R. Cardoso and S. Samko, *Weighted generalized Hölder spaces as well-posedness classes for Sonine integral equations.* J. of Integral Equat. Appl. **20**:4 (2008), 437–480.

[10] L. Diening and S. Samko, *Hardy inequality in variable exponent Lebesgue spaces.* Frac. Calc. Appl. Anal. **10**:1 (2007), 1–18.

[11] L. Diening and S. Samko, *On potentials in generalized Hölder spaces over uniform domains* R^n. Revista Matem. Compl., 2011. DOI: 10.1007/s13163-010-0043-6.

[12] L. Ephremidze, V. Kokilashvili, and S. Samko, *Fractional, maximal and singular operators in variable exponent Lorentz spaces.* Fract. Calc. Appl. Anal. **11**:4 (2008), 407–420.

[13] Ya. Erusalimskii, A. Kilbas, V. Kokilashvili, V. Rabinovich, and S. Samko, *In memoriam: Nikolai Karapetyants.* Fract. Calc. Appl. Anal. **7**:4 (2004), 475–478.

[14] F.D. Gakhov and Samko S.G., *General singular equation on a disclosed contour.* page 41. The international Congress of Mathematicians. Moscow. Abstracts of short communications, Functional Analysis, 5, Nauka, 1966.

[15] F.D. Gakhov, E.I. Zverovich, and Samko S.G., *Increment of the argument, logarithmic residue and a generalized principle of the argument.* Dokl. Akad. Nauk SSSR, **213** (1973), 1233–1236.

[16] F.D. Gakhov, E.I. Zverovich, and Samko S.G., *A generalized argument principle.* Vesci Akad. Navuk BSSR Ser. Fiz.-Mat. Navuk **1975**:1 (1975), 5–16, 139.

[17] A. Gil and S. Samko, *On a hypersingular equation of a problem, for a crack in elastic media.* Fract. Calc. Appl. Anal. **13**:1 (2010), 1–8.

[18] H.-J. Glaeske, V. A. Il′in, A. A. Kilbas, E. I. Moiseev, M. Saigo, and S.G. Samko, *Obituary: Anatolii Platonovich Prudnikov.* Integral Transform. Spec. Funct. **8**:1-2 (1999), 1–2.

[19] R. Gorenflo and S. Samko, *On the dependence of asymptotics of s-numbers of fractional integration operators on weight functions.* Integral Transform. Spec. Funct. **5**:3-4 (1997), 191–212.

[20] V. Guliev, J. Hasanov, and S. Samko, *Boundedness of the maximal, potential and singular integral operators in the generalized variable exponent Morrey spaces* $M^{p(\cdot),\theta(\cdot),\omega(\cdot)}(\Omega)$. J. Math. Sci. **170**:4 (2010), 1–21.

[21] V. Guliev, J. Hasanov, and S. Samko, *Boundedness of the maximal, potential and singular operators in the generalized variable exponent Morrey spaces.* Math. Scand. **107** (2010), 285–304.

[22] M. Hajiboyev and S. Samko, *Weighted estimates of generalized potentials in variable exponent Lebesgue spaces on homogeneous spaces.* In Operator Theory: Advances and Applications, vol. 210, eds. R. Duduchava, I. Gohberg, S. Grudsky, V. Rabinovich, pages 107–122. Birkhäuser, 2010.

[23] M. Hajiboyev and S. Samko, *Generalized potentials in variable exponent Lebesgue spaces on homogeneous spaces.* Math. Nachr. **284**:1 (2011), 53–66.

[24] D. Israfilov, V. Kokilashvili, and S. Samko, *Approximation in weighted Lebesgue and Smirnov spaces with variable exponents.* Proc. A. Razmadze Math. Inst. **143** (2007), 25–35.

[25] V.A. Kakichev, S.G. Samko, and A.I. Kheifits, *Nikolai Vasilievich Govorov. In memory of a scholar, friend and teacher.* Izv. Severo-Kavkaz. Nauchn. Tsentra Vyssh. Shkoly Estestv. Nauk. **1988**:4 (1988), 65–70, 143.

[26] N.K. Karapetiants, A.A. Kilbas, M. Saigo, and S.G. Samko, *Upper and lower bounds for solutions of nonlinear Volterra convolution integral equations with power nonlinearity.* J. Integral Equations Appl. **12**:4 (2000), 421–448.

[27] N.K. Karapetiants and S.G. Samko, *A certain class of convolution type integral equations, and its application.* (Russian) Dokl. Akad. Nauk SSSR **193**:5 (1970), 981–984. Transl. in Soviet Math. Dokl., **11** (1970), 1050–1054.

[28] N.K. Karapetiants and S.G. Samko, *The functional equation* $\psi(x-\alpha)-b(x)\psi(x) = g(x)$. (Russian) Izv. Akad. Nauk Armyan. SSR, Matematika **5**:5 (1970), 441–448.

[29] N.K. Karapetiants and S.G. Samko, *The index of certain classes of integral operators.* (Russian) Dokl. Akad. Nauk SSSR, 194(3):504–507, 1970. Transl. in *Soviet Math. Dokl.* **11**:5 (1970), 1229–1233.

[30] N.K. Karapetiants and S.G. Samko, *A certain class of convolution type integral equations and its application.* (Russian) Izv. Akad. Nauk SSSR, ser. Mat. **35**:3 (1971), 714–726. Transl. in Math. USSR, Izvestija, **5**:3 (1971), 731–744.

[31] N.K. Karapetiants and S.G. Samko, *The discrete Wiener-Hopf operators with oscillating coefficients.* (Russian) Dokl. Akad. Nauk SSSR **200**:1 (1971), 17–20. Transl. in Soviet Math. Dokl., **12**:5 (1971), 1303–1307.

[32] N.K. Karapetiants and S.G. Samko, *A certain boundary value problem with a shift in the theory of analytic functions.* Izv. Vysch. Uchebn. Zaved. Matematika **1972**:11(126) (1972), 18–22.

[33] N.K. Karapetiants and S.G. Samko, *A certain new approach to the investigation of singular integral equations with a shift.* (Russian) Dokl. Akad. Nauk SSSR **202**:2 (1972), 273–276. Transl. in *Soviet Math. Dokl.* **13**:1 (1972), 79–83.

[34] N.K. Karapetiants and S.G. Samko, *On a class of involutive equations.* (Russian) In Collection of research papers "Matem. Analis i ego prilozh.", pages 29–35. Rostov State University, Rostov-na-Donu, 1972.

[35] N.K. Karapetiants and S.G. Samko, *Singular integral operators on the axis with a fractional- linear shift of Carleman type and the Noether property for operators containing an involution.* (Russian) Izv. Akad. Nauk Armyan. SSR, Matematika **7**:1 (1972), 68–77.

[36] N.K. Karapetiants and S.G. Samko, *Singular integral operators with shift on an open contour.* (Russian) Dokl. Akad. Nauk SSSR **204**:3 (1972), 536–539. Transl. in *Soviet Math. Dokl.* **13**:3 (1972), 691–696.

[37] N.K. Karapetiants and S.G. Samko, *The index of certain classes of integral operators.* (Russian) Izv. Akad. Nauk Arm. SSR, Matematika **8**:1 (1973), 26–40.

[38] N.K. Karapetiants and S.G. Samko, *On $\alpha(t)$-factorization of functions and functional equations with degenerate symbol.* (Russian) Izv. Severo-Kavk. Nauchn. Tsentra Vysshei Shkoly, Ser. Estestv. Nauk **1973**:4 (1973), 41–43.

[39] N.K. Karapetiants and S.G. Samko, *Singular integral equations with Carleman shift in the case of discontinuous coefficients, and a study of whether a certain class of linear operators with involution is Noetherian.* (Russian) Dokl. Akad. Nauk SSSR **211**:2 (1973), 281–284. Transl. in Soviet Math. Dokl. **1**:4 (1973), 1006–1011.

[40] N.K. Karapetiants and S.G. Samko, *Convolution operators and singular integral operators with reflection.* (Russian) Trudy Tbil. Matem. Inst. Akad. Nauk Gruz. SSR **44** (1974), 100–112.

[41] N.K. Karapetiants and S.G. Samko, *Singular convolution operators with a discontinuous symbol.* (Russian) Dokl. Akad. Nauk SSSR **221**:6 (1975), 1260–1263. Transl. in Soviet Math. Dokl. **16**:2 (1975), 510–514.

[42] N.K. Karapetiants and S.G. Samko, *Singular convolution operators with a discontinuous symbol.* (Russian) Sibirsk. Mat. Zh **16**:1 (1975), 44–61. Transl. in Siberian Math. J. **16**:1 (1975), 35-48.

[43] N.K. Karapetiants and S.G. Samko, *Singular integral operators with Carleman shift in the case of piece-wise continuous coefficients,* I. (Russian) Izv. Vysch. Uchebn. Zaved., Matematika. **1975**:2 (1975), 43–54.

[44] N.K. Karapetiants and S.G. Samko, *Singular integral operators with Carleman shift in the case of piece-wise continuous coefficients,* II. (Russian) Izv. Vysch. Uchebn. Zaved., Matematika. **1975**:3 (1975), 34–42.

[45] N.K. Karapetiants and S.G. Samko, *Investigation of Noether nature of linear equations with generalized involutive operators and its applications.* (Russian) In Collection of research papers "Metody Teorii Funktsii i Funkts. Anal.", pp. 14–21. Checheno-Ingushskii University, Groznyi, 1976.

[46] N.K. Karapetiants and S.G. Samko, *Discrete convolution operators with coefficients that become almost stable.* (Russian) Mat. Zametki **22**:3 (1977), 339–344.

[47] N.K. Karapetiants and S.G. Samko, *A study of the Noethericity of operators with an involution of order n, and its applications.* (Russian) Izv. Vysch. Uchebn. Zaved., Matematika. **1977**:11 (1977), 15–26.

[48] N.K. Karapetiants and S.G. Samko, *Equations with involutive operators and their applications.* (Russian) Rostov-na-Donu, Izdat. Rostov Univ., 1988. 192 pages.

[49] N.K. Karapetiants and S.G. Samko, *Multidimensional integral operators with homogeneous kernels.* Fract. Calculus & Applied Analysis **2**:1 (1999), 67–96.

[50] N.K. Karapetiants and S.G. Samko, *On a certain approach to the investigation of equations with involutive operators.* Intern. J. Math. Stat. Sci. **8**:2 (1999), 66–93.

[51] N.K. Karapetiants and S.G. Samko, *On Fredholm properties of a class of Hankel operators.* Math. Nachr. **217** (2000), 75–103.

[52] N.K. Karapetiants and S.G. Samko, *Singular integral equations on the real line with a fractional-linear Carleman shift.* Proc. Razmadze Math. Inst. **124** (2000), 73–106.

[53] N.K. Karapetiants and S.G. Samko, *Equations with Involutive Operators.* Birkhäuser, Boston, 2001.

[54] N.K. Karapetiants and S.G. Samko, *On multi-dimensional integral equations of convolution type with shift.* Integral Equations Operator Theory **39**:3 (2001), 305–328.

[55] N.K. Karapetiants and S.G. Samko, *Singular integral equations on the real line with homogeneous kernels and the inversion shift.* J. Nat. Geom. **19**:1-2 (2001), 1–26.

[56] A. Karapetyants and S. Samko, *Spaces* $\mathrm{BMO}^{p(\cdot)}(\mathbb{D})$ *of a variable exponent* $p(z)$. Georgian Math. J. **17**:3 (2010), 529–542.

[57] A. Kilbas and S. Samko, *Integral operators with homogeneous kernels in the Hölder H-space.* (Russian) Izv. Akad. Nauk Biel. SSR, Ser. Fis.-Mat. Nauk **21**:1 (1977), 5–8.

[58] A. Kilbas and S. Samko, *Anatolii Platonovich Prudnikov: a note in memoriam.* Fract. Calc. Appl. Anal. **2**:1 (1999), 107–108.

[59] A. Kilbas and S. Samko, *Professor Eric Russell Love* [1912–2001]. *A note in memoriam.* Fract. Calc. Appl. Anal. **4**:4 (2001), 552–554.

[60] V. Kiryakova, A.A. Kilbas, S.G. Samko, and O.I. Marichev, 70*th anniversary of Professor Rudolf Gorenflo.* Fract. Calc. Appl. Anal. **3**:4 (2000), 339–342.

[61] V. Kokilashvili, A. Meskhi, and Samko S., *On the inversion and characterization of the Riesz potentials in the weighted Lebesgue spaces.* Memoirs on Differential Equations and Mathematical Physics, **29** (2003), 31–45.

[62] V. Kokilashvili, V. Paatashvili, and Samko S., *Boundary value problems for analytic functions in the class of Cauchy-type integrals with density in* $L^{p(\cdot)}(\Gamma)$. Bound. Value Probl. **2005**:1 (2005), 43–71.

[63] V. Kokilashvili, V. Paatashvili, and Samko S., *Boundedness in Lebesgue spaces with variable exponent of the Cauchy singular operators on Carleson curves.* In Ya. Erusalimsky, I. Gohberg, S. Grudsky, V. Rabinovich, and N. Vasilevski, editors, "Operator Theory: Advances and Applications", dedicated to the 70th birthday of Prof. I.B. Simonenko, vol. 170, pp. 167–186. Birkhäuser Verlag, Basel, 2006.

[64] V. Kokilashvili, V. Paatashvili, and Samko S., *The Riemann problem in the class of functions representable by a Cauchy-type integral with density in* $L^{p(\cdot)}(\Gamma)$. Dokl. Akad. Nauk **421**:2 (2008), 164–167.

[65] V. Kokilashvili, N. Samko, and S. Samko, *The maximal operator in variable spaces* $L^{p(\cdot)}(\Omega, \rho)$ *with oscillating weights.* Georgian Math. J. **13**:1 (2006), 109–125.

[66] V. Kokilashvili, N. Samko, and S. Samko, *The Maximal Operator in Weighted Variable Spaces* $L^{p(\cdot)}$. J. Function spaces and Appl. **5**:3 (2007), 299–317.

[67] V. Kokilashvili, N. Samko, and S. Samko, *Singular operators in variable spaces* $L^{p(\cdot)}(\Omega, \rho)$ *with oscillating weights.* Math. Nachr. **280**:9-10 (2007), 1145–1156.

[68] V. Kokilashvili and S. Samko, *Maximal and Fractional Operators in Weighted* $L^{p(x)}$*-Spaces.* Proc. A. Razmadze Math. Inst. **129** (2002), 145–149.

[69] V. Kokilashvili and S. Samko, *Singular Integrals and Potentials in Some Banach Spaces with Variable Exponent.* Proc. A. Razmadze Math. Inst. **129** (2002), 150–155.

[70] V. Kokilashvili and S. Samko, *On Sobolev Theorem for the Riesz type Potentials in Lebesgue Spaces with Variable Exponent.* Zeit. Anal. Anwend. **22**:4 (2003), 899–910.

[71] V. Kokilashvili and S. Samko, *Singular Integral Equations in the Lebesgue Spaces with Variable Exponent.* Proc. A. Razmadze Math. Inst. **131** (2003), 61–78.

[72] V. Kokilashvili and S. Samko, *Singular Integrals and Potentials in Some Banach Spaces with Variable Exponent.* J. Function Spaces and Appl. **1**:1 (2003), 45–59.

[73] V. Kokilashvili and S. Samko, *Singular Integrals in Weighted Lebesgue Spaces with Variable Exponent.* Georgian Math. J. **10**:1 (2003), 145–156.

[74] V. Kokilashvili and S. Samko, *Maximal and fractional operators in weighted* $L^{p(x)}$ *spaces.* Revista Matematica Iberoamericana **20**:2 (2004), 493–515.

[75] V. Kokilashvili and S. Samko, *Boundedness in Lebesgue spaces with variable exponent of maximal, singular and potential operators.* Izvestija VUZov. Severo-Kavkazskii region. Estestvennie nauki, Special issue "Pseudodifferential equations and some problems of mathematical physics", dedicated to the 70th birthday of Prof. I.B. Simonenko, pp. 152–158, 2005.

[76] V. Kokilashvili and S. Samko, *Weighted Boundedness in Lebesgue spaces with Variable Exponent of Classical Operators on Carleson Curves.* Proc. A. Razmadze Math. Inst. **138** (2005), 106–110.

[77] V. Kokilashvili and S. Samko, *A general approach to weighted boundedness of operators of harmonic analysis in variable exponent Lebesgue spaces.* Proc. A. Razmadze Math. Inst. **145** (2007), 109–116.

[78] V. Kokilashvili and S. Samko, *The maximal operator in weighted variable spaces on metric measure spaces.* Proc. A. Razmadze Math. Inst. **144** (2007), 137–144.

[79] V. Kokilashvili and S. Samko, *On a generalization of Calderón-Zygmund's theorem in weighted Lebesgue spaces with variable exponent.* Bull. Georgian Natl. Acad. Sci. **175**:1 (2007), 34–39.

[80] V. Kokilashvili and S. Samko, *Boundedness of maximal operators and potential operators on Carleson curves in Lebesgue spaces with variable exponent.* Acta Mathematica Sinica **24**:11 (2008), 1775–1800.

[81] V. Kokilashvili and S. Samko, *The maximal operator in weighted variable exponent spaces on metric spaces.* Georgian Math. J. **15**:4 (2008), 683–712.

[82] V. Kokilashvili and S. Samko, *Singular operators and Fourier multipliers in weighted Lebesgue spaces with variable exponents.* (Russian) Vestn. St-Peterbg. Univ., Ser. I, Mat. Mekh. Astron. **2008**:2 (2008), 56–68. Transl. in Vestnik St. Petersburg Univ. Math. **41**:2 (2008), 134–144.

[83] V. Kokilashvili and S. Samko, *Vekua's generalized singular integral on Carleson curves in weighted variable Lebesgue spaces.* In Operator Algebras, Operator Theory and Applications, vol. 181 of Oper. Theory Adv. Appl., pp. 283–293. Birkhäuser Verlag, Basel, 2008.

[84] V. Kokilashvili and S. Samko, *Weighted boundedness of the maximal, singular and potential operators in variable exponent spaces.* In A.A. Kilbas and S.V. Rogosin, editors, Analytic Methods of Analysis and Differential Equations, pp. 139–164. Cambridge Scientific Publishers, 2008.

[85] V. Kokilashvili and S. Samko, *Operators of harmonic analysis in weighted spaces with non-standard growth.* J. Math. Anal. Appl. **352** (2009), 15–34. doi:10.1016/j.jmaa.2008.06.056.

[86] V. Kokilashvili and S. Samko, *A refined inverse inequality of approximation in weighted variance exponent Lebesgue spaces.* Proc. A. Razmadze Math. Inst. **151** (2009), 134–138.

[87] V. Kokilashvili and S. Samko, *Boundedness of weighted singular integral operators on a Carleson curves in Grand Lebesgue spaces.* In ICNAAM 2010: Intern. Conf. Numer. Anal. Appl. Math., vol. 1281, pp. 490–493. AIP Confer. Proc., 2010.

[88] V. Kokilashvili and S. Samko, *Boundedness of weighted singular integral operators in Grand Lebesgue spaces.* Georgian Math. J. **18**:2 (2011), 259–269.

[89] G.S. Kostetskaya and S.G. Samko, *A condition for the absolute integrability of Fourier integrals.* (Russian) Izv. Vyssh. Uchebn. Zaved. Mat. **1988**:3 (1988), 72–75. Transl. in Soviet Math. (Iz. VUZ) **32**:3 (1988), 103–108.

[90] T. Mamatov and S. Samko, *Mixed fractional integration operators in mixed weighted Hölder spaces.* Frac. Calc. Appl. Anal. **13**:3 (2010), 245–259.

[91] K. Miller, A. Prudnikov, B. Sachdeva, and S. Samko, *Obituary: Bertram Ross* [1917–1993]. Integral Transform. Spec. Funct. **2**:1 (1994), 81–82.

[92] K.S. Miller and S.G. Samko, *A note on the complete monotonicity of the generalized Mittag-Leffler function.* Real Anal. Exchange **23** (1997), 753–755.

[93] K.S. Miller and S.G. Samko, *Completely monotonic functions.* Integral Transform. Spec. Funct. **12**:4 (2001), 389–402.

[94] V.A. Nogin and S.G. Samko, *On the simultaneous approximation of functions and their Riesz derivatives* (Russian) Dokl. Akad. Nauk SSSR **261**:3 (1981), 548–550. Transl. in *Soviet Math. Dokl.* **24**:3 (1981), 552–554.

[95] V.A. Nogin and S.G. Samko, *Convergence in $L_p(R^n)$ of hypersingular integrals with a homogeneous characteristic.* (Russian) In Collection of papers "Differ. i integr. uravn. i ikh prilozh.", pp. 119–131. Elista, Kalmyskii universitet, 1982.

[96] V.A. Nogin and S.G. Samko, *Inversion and description of Riesz potentials with densities in weighted L_p-spaces.* (Russian) Izv. Vyssh. Uchebn. Zaved. Mat. **1985**:1 (1985), 70–72. Transl. in Sov. Math. **29**:1 (1985), 95–99.

[97] V.A. Nogin and S.G. Samko, *Method of approximating inverse operators and its applications to inversion of potential type integral operators.* Integr. Transform. and Special Funct. **8**:1–2 (1999), 205–228.

[98] V.A. Nogin and S.G. Samko, *Some applications of potentials and approximative inverse operators in multi-dimensional fractional calculus.* Fract. Calc. Appl. Anal. **2**:2 (1999), 205–228.

[99] V.A. Nogin and S.G. Samko, *Inversion and characterization of Riesz potentials in weighted spaces via approximative inverse operators.* Proc. A. Razmadze Math. Inst. **129** (2002), 99–106.

[100] P.M. Pavlov and S.G. Samko, *A description of the space $L_p^\alpha(S_{n-1})$ in terms of spherical hypersingular integrals.* (Russian) Dokl. AN SSSR, Ser. Mat. **276**:3 (1984), 546–550.

[101] L.-E. Persson and S.G. Samko, *Some new Stein and Hardy type inequalities.* J. Math. Sci. **169**:1 (2010), 113–129.

[102] A. Plakhov and S. Samko, *An inverse problem of Newtonian aerodynamics.* Math.Eng. Sci. Aerospace. **1**:4 (2010), 351–369.

[103] V.S. Rabinovich, N.G. Samko, and S.G. Samko, *Local Fredholm spectrums and Fredholm properties of singular integral operators on Carleson curves acting on weighted Hölder spaces.* Integr. Equat. Oper. Theory **56**:2 (2006), 257–283.

[104] V.S. Rabinovich and S.G. Samko, *Essential spectra of pseudodifferential operators in Sobolev spaces with variable smoothness and variable Lebesgue indices.* Doklady Akademii Nauk **417**:2 (2007), 167–170. Transl. in Doklady Mathematics **76**:3 (2007), 835–838.

[105] V.S. Rabinovich and S.G. Samko, *Boundedness and Fredholmness of pseudodifferential operators in variable exponent spaces.* Integr. Eq. Oper. Theory **60**:4 (2008), 507–537.

[106] V.S. Rabinovich and S.G. Samko, *PDO approach to singular integral operators in weighted variable exponent Lebesgue spaces on Carleson curves.* Integr. Eq. Oper. Theory, 2011, to appear.

[107] V.S. Rabinovich and S.G. Samko, *Singular integral operators on a class of composed Carleson curves acting on weighted variable exponents $L^{p(\cdot)}$-spaces.* Funct. Anal., 2011, to appear.

[108] H. Rafeiro and S. Samko, *On multidimensional analogue of Marchaud formula for fractional Riesz-type derivatives in domains in* R^n. Fract. Calc. and Appl. Anal. **8**:4 (2005), 393–401.

[109] H. Rafeiro and S. Samko, *On a class of fractional type integral equations in variable exponent spaces.* Fract. Calc. Appl. Anal. **10**:4 (2007), 399–421.

[110] H. Rafeiro and S. Samko, *Approximative method for the inversion of the Riesz potential operator in variable Lebesgue spaces.* Fract. Calc. Appl. Anal. **11**:3 (2008), 269–280.

[111] H. Rafeiro and S. Samko, *Characterization of the range of one-dimensional fractional integration in the space with variable exponent.* In Operator algebras, operator theory and applications, volume 181 of Oper. Theory Adv. Appl., pp. 393–416. Birkhäuser Verlag, Basel, 2008.

[112] H. Rafeiro and S. Samko, *Dominated compactness theorem in Banach function spaces and its applications.* Compl. Anal. Oper. Theory **2**:4 (2008), 669–681.

[113] H. Rafeiro and S. Samko, *Hardy type inequality in variable Lebesgue spaces.* Annal. Acad. Scient. Fenn. Mathem. **34** (2009), 279–289. Corrigendum in *Annal. Acad. Scient. Fenn. Mathem.*, vol. 35, 2010.

[114] H. Rafeiro and S. Samko, *Characterization of the variable exponent Bessel potential spaces via the Poisson semigroup.* J. Math. Anal. Appl. **365**:2 (2010), 483–497.

[115] H. Rafeiro and S. Samko, *Variable exponent Campanato spaces.* J. Math. Sci. **172**:1 (2011), 143–164.

[116] B. Ross and S.G. Samko, *Fractional integration operator of variable order in the Hölder spaces* $H^{\lambda(x)}$. Internat. J. Math. Math. Sci. **18**:4 (1995), 777–788.

[117] B. Ross, S.G. Samko, and E.R. Love, *Functions that have no first-order derivative might have fractional derivatives of all orders less than one.* Real Anal. Exchange **20**:1 (1994/95), 140–157.

[118] N. Samko and S. Samko, *On approximative definition of fractional differentiation.* Fract. Calc. Appl. Anal. **2**:3 (1999), 329–342.

[119] N. Samko, S. Samko, and B. Vakulov, *Weighted Sobolev theorem in Lebesgue spaces with variable exponent.* J. Math. Anal. and Appl. **335** (2007), 560–583.

[120] N. Samko, S. Samko, and B. Vakulov, *Potential type operators and hypersingular integrals in variable order Hölder spaces on homogeneous spaces.* (Russian) Izv. Severo-Kavk. Nauchn. Tsentra Vysch. Schkoly (special issue), 2009, 40–45.

[121] N. Samko, S. Samko, and B. Vakulov, *Fractional integrals and hypersingular integrals in variable order Hölder spaces on homogeneous spaces.* J. Funct. Spaces and Appl. **8**:3 (2010), 215–244.

[122] N. Samko, S. Samko, and B. Vakulov, *Weighted Sobolev theorem in Lebesgue spaces with variable exponent: corrigendum* [MR2340340]. Armen. J. Math. **3**:2 (2010), 92–97.

[123] S.G. Samko, *General singular equation in the exceptional case.* (Russian) Differ. Uravnen. **1**:8 (1965), 1108–1116.

[124] S.G. Samko, *A general singular equation on an open contour and a generalized Abel equation.* (Russian) Dokl. Akad. Nauk SSSR **177**:1 (1967), 44–47.

[125] S.G. Samko, *A generalized Abel equation and an equation with Cauchy kernel.* (Russian) Dokl. Akad. Nauk SSSR **176** (1967), 1019–1022.

[126] S.G. Samko, *On the reduction of certain integral equations of the first kind in the theory of elasticity and hydrodynamics to equations of the second kind.* (Russian) J. Appl. Math. Mech. **31** (1967), 368–371.

[127] S.G. Samko, *The generalized Abel equation and fractional integration operators.* (Russian) Diffentsial'nye Uravneniya **4**:2 (1968), 298–314.

[128] S.G. Samko, Noether's theory for the generalized Abel integral equation (in Russian). Differencial'nye Uravneniya **4** (1968), 315–326.

[129] S.G. Samko, *Abel's generalized equation, Fourier transform, and convolution type equations.* (Russian) Dokl. Akad. Nauk SSSR **187** (1969), 743–746.

[130] S.G. Samko, *The general singular operator and an integral operator with an automorphic kernel.* (Russian) Izv. Vysch. Uchebn. Zaved., Matematika. **1969**:1 (1969), 67–77.

[131] S.G. Samko, *The integral modulus of continuity of potentials with densities that are summable on the real line with weight.* In Mathematical Analysis and its Applications (Russian), pp. 175–184. Izdat. Rostov. Univ., Rostov-na-Donu, 1969.

[132] S.G. Samko, *The solvability in closed form of singular integral equations.* (Russian) Dokl. Akad. Nauk SSSR **189** (1969), 483–485.

[133] S.G. Samko, *Abel's generalized integral equation on the line.* (Russian) Izv. Vysš. Učebn. Zaved. Matematika **1970**:8(99) (1970), 83–93.

[134] S.G. Samko, *Integral equations of the first kind with a kernel of potential type.* (Russian) In Proc. Conf. on Boundary value problems (Kazan, 1969), pp. 216–220. Izdat. Kazan. Univ., Kazan, 1970.

[135] S.G. Samko, *A certain class of potential type operators on the line.* (Russian) Izv. Vyssh. Uchebn. Zaved. Mat. **1971**:5 (1971), 92–100.

[136] S.G. Samko, *Integral equations of the first kind with kernel of potential type.* (Russian) Izv. Vyssh. Uchebn. Zaved. Mat. **1971**:4 (1971), 78–86.

[137] S.G. Samko, *Operators of potential type.* (Russian) Dokl. Akad. Nauk SSSR **196**:2 (1971), 299–301. Transl. in Soviet Math. Dokl. **12**:1 (1971), 125–128.

[138] S.G. Samko, *The space $I^\alpha(L_p)$ of fractional integrals, and operators of potential type.* (Russian) Izv. Akad. Nauk Armyan. SSR **7**:5 (1973), 359–383.

[139] S.G. Samko, Singular integral and integrodifferential equations with analytic kernels. Izv. Severo-Kavkaz. Naucn. Centra Vysch. Shkoly Ser. Estestv. Nauk. **1974**:4 (1974), 86–94, 138–139.

[140] S.G. Samko, *Integral equations of convolution type of the first kind with a power kernel.* (Russian) Izv. Vysš. Učebn. Zaved. Matematika **1975**:4(155) (1975), 60–67.

[141] S.G. Samko, *Proof of the Babenko-Stein theorem.* (Russian) Izv. Vysš. Učebn. Zaved. Matematika **1975**:5 (1975), 47–51.

[142] S.G. Samko, *Generalized Riesz potentials.* (Russian) Seminar Inst. Prikl. Mat. Tbil. Univ. **11** (1976), 35–44.

[143] S.G. Samko, *Integral equations of the first kind with a logarithmic kernel.*. In Mapping methods (Russian), pp. 41–69. Checheno-Ingush. Gos. Univ., Groznyi, 1976.

[144] S.G. Samko, *The spaces $L^{\alpha}_{p,r}(R^n)$ and hypersingular integrals.* (Russian) Vesci Akad. Navuk BSSR Ser. Fiz.-Mat. Navuk **1976**:2 (1976), 34–41.

[145] S.G. Samko, *Spaces of Riesz potentials.* (Russian) Izv. Akad. Nauk SSSR, Ser. Mat. **40**:5 (1976), 1143–1172. Transl. in Math. USSR Izvestija **10**:5 (1976), 5, 1089–1117.

[146] S.G. Samko, *The classes $C^{\lambda}(R^n)$, and multipliers in the space $I^{\alpha}(L^p)$ of Riesz potentials.* (Russian) Izv. Severo-Kavkaz. Naučn. Centra Vysš. Školy Ser. Estestv. Nauk. **1977**:3 (1977), 13–17.

[147] S.G. Samko, *Generalized Riesz potentials and hypersingular integrals, their symbols and inversion.* (Russian) Dokl. Akad. Nauk SSSR **232**:3 (1977), 528–531. Transl. in Soviet Math. Dokl. **18**:1 (1977), 97–101.

[148] S.G. Samko, *On the description of the image $I^{\alpha}(L^p)$ of Riesz potentials.* (Russian) Izv. Akad. Nauk Armyan. SSR, Matematika **12**:5 (1977), 329–334.

[149] S.G. Samko, *The spaces $L^{\alpha}_{p,r}(R^n)$ and hypersingular integrals.* (Russian) Studia Math. (PRL) **61**:3 (1977), 193–230.

[150] S.G. Samko, *Spherical potentials, spherical Riesz differentiation and their applications.* (Russian) Izv. Vysš. Učebn. Zaved. Matematika **1977**:2(177) (1977), 135–139. Transl. in Soviet Math. (Izvestiya VUZ. Matematika) **21**:2 (1977), 106–110.

[151] S.G. Samko, *Test functions that vanish on a given set, and division by a function.* (Russian) Mat. Zametki **21**:5 (1977), 677–689.

[152] S.G. Samko, A *criterion of harmonicity of polynomials.* (Russian). Differ. Uravnenia **14**:5 (1978), 938–939.

[153] S.G. Samko, *Hypersingular integrals with homogeneous characteristic.* (Russian) Trudy Inst. Prikl. Mat. Tbil. Univ. **5-6** (1978), 235–249.

[154] S.G. Samko, *Integral equation of the first kind with a logarithmic kernel,* II. (Russian) In Math. Analysis and its Appl., no. 1, 103–120. Rostov. Univ., Rostov-na-Donu, 1978.

[155] S.G. Samko, *Methods of inversion of potential type operators and equations with involutive operators and their applications.* (Russian) Doctor of Science Thesis. Moscow, Steklov Math. Inst., 300 pp., 1978.

[156] S.G. Samko, *The Fourier transform of the functions $Y_m(x/x)/x^{n+\alpha}$.* (Russian) Izv. Vyssh. Uchebn. Zaved. Mat. **1978**:7(194) (1978), 73–78.

[157] S.G. Samko, *Generalized Riesz potentials and hypersingular integrals with homogeneous characteristics; their symbols and inversion.* (Russian) Trudy Mat. Inst. Steklov **156** (1980), 157–222. Transl. in Proc. Steklov Inst. Mat. (1983), Issue 2, 173–243.

[158] S.G. Samko, *Denseness in $L^p(R^n)$ of spaces Φ_V of Lizorkin type.* (Russian) Mat. Zametki **31**:6 (1982), 855–865.

[159] S.G. Samko, *Singular integrals over a sphere and the construction of the characteristic by a symbol.* (Russian) Izv. Vyssh. Uchebn. Zaved. Mat. **1983**:4 (1983), 28–42. Transl. in *Soviet Math. (Iz. VUZ)* **27**:4 (1983), 35–52.

[160] S.G. Samko, *Hypersingular integrals and their applications.* (Russian). Rostov-on-Don, Izdad. Rostov Univ, 1984. 208 pages.

[161] S.G. Samko, *The coincidence of Grunwald-Letnikov differentiation with other forms of fractional differentiation. the periodic and the nonperiodic case.* (Russian) In

Reports of the extended sessions of a seminar of I.N. Vekua Inst. Appl. Math. **1**, 183–186. Tbilisi, Tbilis. Gos. Univ., 1985.

[162] S.G. Samko, *One- and multidimensional integral equations of the first kind with a weak singularity in the kernel* (Russian) In Commemorative seminar on boundary value problems, Minsk (1981), pp. 103–115. University of Belarus, Izdad. "Universitetskoe", 1985.

[163] S.G. Samko, *Hypersingular integrals and differences of fractional order.* (Russian) Trudy Mat. Inst. Steklov **192** (1990), 164–182. Transl. in Proc. Steklov Inst. Math. **1992**:3 (1992), 175–194.

[164] S.G. Samko, *The multidimensional fractional integrodifferentiation and the Grunwald-Letnikov approach to fractional calculus.* In Intern. Conf. "Fract. Calculus and its Applic." Tokyo, 1989, pp. 221–225. College of Engineering Nihon Univ., 1990.

[165] S.G. Samko, *The denseness of Lizorkin type spaces* Φ_V *in the spaces* $L^{\bar{p}}(R^n)$ *with mixed norm.* (Russian) Dokl. Akad. Nauk SSSR **319** (1991), 567–569.

[166] S.G. Samko, *Inversion theorems for potential-type integrals transforms in* R^n *and on* S^{n-1}. Integr. Transf. and Special Funct. **1**:2 (1993), 145–163.

[167] S.G. Samko, *On the inversion of the generalized Riesz potential with a quadratic characteristic.* (Russian) In Collection of research papers "Differential and integral equations and complex analysis", pp. 77–90. Kalmytsky State University, Elista, 1993.

[168] S.G. Samko, *Denseness of the spaces* Φ_V *of Lizorkin type in the mixed* $L^{\bar{p}}(R^n)$*-spaces.* Studia Math. **113**:3 (1995), 199–210.

[169] S.G. Samko, *Differentiation and integration of variable (fractional) order.* Dokl. Akad. Nauk **342**:4 (1995), 458–460.

[170] S.G. Samko, *Fractional integration and differentiation of variable order.* Anal. Math. **21**:3 (1995), 213–236.

[171] S.G. Samko, *Convolution and potential type operators in* $L^{p(x)}$. Integr. Transf. and Special Funct. **7**:3-4 (1998), 261–284.

[172] S.G. Samko, *Convolution type operators in* $L^{p(x)}$. Integr. Transf. and Special Funct. **7**:1-2 (1998), 123–144.

[173] S.G. Samko, *Differentiation and integration of variable order and the spaces* $L^{p(x)}$. In Operator Theory for Complex and Hypercomplex Analysis (Mexico City, 1994), volume 212 of Contemp. Math., pp. 203–219. Amer. Math. Soc., Providence, RI, 1998.

[174] S.G. Samko, *A new approach to the inversion of the Riesz potential operator.* Fract. Calc. Appl. Anal. **1**:3 (1998), 225–245.

[175] S.G. Samko, *Denseness of* $C_0^\infty(R^n)$ *in the generalized Sobolev spaces* $W^{m,p(x)}(R^n)$. (Russian) Dokl. Ross. Acad. Nauk **369**:4 (1999), 451–454. Transl. in Dokl. Math. **60** (1999), 382–385.

[176] S.G. Samko, *On local summability of Riesz potentials in the case* $\operatorname{Re}\alpha > 0$. Anal. Math. **25**:3 (1999), 205–210.

[177] S.G. Samko, *Some remarks to the author's paper "A new approach to the inversion of the Riesz potential operator".* Fract. Calc. and Appl. Anal. **2**:1 (1999).

[178] S.G. Samko, *Approximative approach to fractional powers of operators*. In Proceedings of the Second ISAAC Congress, Vol. 2 (Fukuoka, 1999), volume 8 of Int. Soc. Anal. Appl. Comput., pp. 1163–1170. Kluwer Acad. Publ., 2000.

[179] S.G. Samko, *Denseness of $C_0^\infty(R^N)$ in the generalized Sobolev spaces W^M, $P(X)(R^N)$*. In Intern. Soc. for Analysis, Applic. and Comput., vol. 5, "Direct and Inverse Problems of Math. Physics", ed. by R.Gilbert, J. Kajiwara and Yongzhi S. Xu, pp. 333–342. Kluwer Acad. Publ., 2000.

[180] S.G. Samko, *Fractional powers of operators via hypersingular integrals*. In Semigroups of operators: theory and applications (Newport Beach, CA, 1998), volume 42 of Progr. Nonlinear Differential Equations Appl., pp. 259–272. Birkhäuser, Basel, 2000.

[181] S.G. Samko, *Approximative approach to construction of fractional powers of operators*. (Russian) Izvestia Vysch. Uchebn. Zaved., Severo-Kavkazskii Region, Estestv. Nauki **2001**:4 (2001),41–44.

[182] S.G. Samko, *On some index relations for Bessel functions*. Integral Transform. Spec. Funct. **11**:4 (2001), 397–402.

[183] S.G. Samko, *Hypersingular Integrals and their Applications*. London-New York: Taylor & Francis, Series "Analytical Methods and Special Functions", vol. 5, 2002.

[184] S.G. Samko, *Fractional Weyl-Riesz integrodifferentiation of periodic functions of two variables via the periodization of the Riesz kernel*. Appl. Anal. **82**:3 (2003), 269–299.

[185] S.G. Samko, *Hardy inequality in the generalized Lebesgue spaces*. Frac. Calc. and Appl. Anal **6**:4 (2003), 355–362.

[186] S.G. Samko, *Hardy-Littlewood-Stein-Weiss inequality in the Lebesgue spaces with variable exponent*. Frac. Calc. and Appl. Anal. **6**:4 (2003), 421–440.

[187] S.G. Samko, *On Inversion of FractionalSpherical Potentials by Spherical Hypersingular Operators*. Operator Theory, Advances and Appl. **142** (2003), 357–368.

[188] S.G. Samko, *Best constant in the weighted Hardy inequality: the spatial and spherical version*. Fract. Calc. Appl. Anal. **8**:1 (2005), 39–52.

[189] S.G. Samko, *On a progress in the theory of Lebesgue spaces with variable exponent: maximal and singular operators*. Integr. Transf. and Spec. Funct. **16**:5-6 (2005), 461–482.

[190] S.G. Samko, *Some classical operators of variable order in variable exponent analysis*. In Operator Theory: Advances and Applications, vol. 193, pp. 281–301. Birkhäuser, 2009.

[191] S.G. Samko, *On compactness of operators in variable exponent Lebesgue spaces*. In Operator Theory: Advances and Applications, Vol. 202, pp. 497–507. Birkhäuser, 2010.

[192] S.G. Samko, *On a 3d-hypersingular equation of a problem for a crack*. Fract. Calc. Appl. Anal. **14**:1 (2011), 19–30.

[193] S.G. Samko, *Weighted estimates of truncated potential kernels in the variable exponent setting*. Compl. Variabl. Ellipt. Equat., 2011.
DOI: 10.1080/17476931003728396.

[194] S.G. Samko and R.P. Cardoso, *Integral equations of the first kind of Sonine type*. Int. J. Math. Math. Sci. **2003**:57 (2003), 3609–3632.

[195] S.G. Samko and R.P. Cardoso, *Sonine integral equations of the first kind in $L_p(0,b)$*. Fract. Calc. Appl. Anal. **6**:3 (2003), 235–258.

[196] S.G. Samko and A.F. Chuvenkov, *On Riesz potentials in Orlicz space.* (Russian) Collection of papers "Math. Anal. and its Appl.", vol. 7, Izdat. Rostov Univ, Rostov–on–Don, 1975.

[197] S.G. Samko, A.A. Kilbas, and O.I. Marichev, *Fractional Integrals and Derivatives and Some of Their Applications.* Nauka i Tekhnika, Minsk, 1987.

[198] S.G. Samko, A.A. Kilbas, and O.I. Marichev, *Fractional Integrals and Derivatives. Theory and Applications.* Gordon & Breach. Sci. Publ., London-New York, 1993.

[199] S.G. Samko and G.S. Kostetskaya, *On absolute summability of Fourier integrals.* Vestnik Ross. Univ. Druzhby Narodov (Bull. of Russian University of People's Friendship), Moscow **1**:1 (1994), 138–168.

[200] S.G. Samko and Kh.M. Murdaev, *Weighted Zygmund estimates for fractional differentiation and integration, and their applications.* Trudy Matem. Inst. Steklov **180** (1987), 197–198. Transl. in Proc. Steklov Inst. Math. AMS, 1989, issue 3, 233–235.

[201] S.G. Samko and Z. Mussalaeva, *Fractional type operators in weighted generalized Hölder spaces.* Proc. Georgian Acad. Sci., Mathem. **1**:5 (1993), 601–626.

[202] S.G. Samko and B. Ross, Integration and differentiation to a variable fractional order. Integral Transform. Spec. Funct. **1**:4 (1993), 277–300.

[203] S.G. Samko, E. Shargorodsky, and B. Vakulov, *Weighted Sobolev theorem with variable exponent for spatial and spherical potential operators,* II. J. Math, Anal. Appl. **325**:1 (2007), 745–751.

[204] S.G. Samko and S.M. Umarkhadzhiev, *Description of a space of Riesz potentials in terms of higher derivatives.* (Russian) Izv. Vyssh. Uchebn. Zaved. Mat. **1980**:11 (1980), 79–82. Transl. in Sov. Math. **24**:11 (1980), 95–98.

[205] S.G. Samko and S.M. Umarkhadzhiev, *On a functional equation with an infinite group of shifts.* (Russian) In Collection of papers "Mat. anal. i ego prilozh.", pp. 11–14. Groznyi, Checheno-Ingushskii Univ., 1984.

[206] S.G. Samko and S.M. Umarkhadzhiev, *Applications of hypersingular integrals to multidimensional integral equations of the first kind.* (Russian) Trudy Mat. Inst. Steklov **172** (1985), 299–312. Transl. in *Proc. Steklov Inst. Math.*, 1987, Issue 3, 325–339.

[207] S.G. Samko and S.M. Umarkhadzhiev, *Regularization of multidimensional integral equations of the* 1*st kind in variable exponent spaces.* In Intern. Conf. Numer. Anal. Appl. Math. 2010, AIP Conf. Proc., Vol. 1281, pp. 506–508, 2010.

[208] S.G. Samko and S.M. Umarkhadzhiev, *On a regularization of a multidimensional integral equation of the first kind in Lebesgue spaces with variable exponent.* Mat. Zametki, 2011, to appear.

[209] S.G. Samko and S.M. Umarkhadzhiev, *On Iwaniec-Sbordone spaces on sets which may have infinite measure.* Azerb. J. Math. **1**:1 (2011), 67–84.

[210] S.G. Samko and E.E. Urnysheva, *The spaces $L^{\mathrm{rad}}_{p_2}(L^{\mathrm{ang}}_{p_1})$ with a radial-spherical mixed norm.* Izv. Vyssh. Uchebn. Zaved. Mat. **1996**:4 (1996), 50–58.

[211] S.G. Samko and B.G. Vakulov, *Equivalent normings in the spaces of functions of fractional smoothness on a sphere of the type $C^\lambda(S^{n-1}), H^\lambda(S^{n-1})$.* (Russian) Izv. Vysch. Uchebn. Zaved., Matematika **1987**:4 (1987), 68–71.

[212] S.G. Samko and B.G. Vakulov, *On equivalent norms in fractional order function spaces of continuous functions on the unit sphere.* Fract. Calc. Appl. Anal. **3**:4 (2000), 401–433.

[213] S.G. Samko and B.G. Vakulov, *A weighted Sobolev theorem for spatial and spherical potentials in Lebesgue spaces with variable exponents.* Dokl. Akad. Nauk **403**:1 (2005), 7–10.

[214] S.G. Samko and B.G. Vakulov, *Weighted Sobolev theorem with variable exponent.* J. Math, Anal. Appl. **310** (2005), 229–246.

[215] S.G. Samko and M.U. Yakhshiboev, *A modification of Riemann-Liouville fractional integro-differentiation applied to functions on R^1 with arbitrary behaviour at infinity.* Izv. Vysch. Uchebn. Zaved., Mat. **1992**:4 (1992), 96–99.

[216] S.G. Samko and M.U. Yakhshiboev, *On a class of identity approximation operators of a non-convolution type.* Fract. Calc. Appl. Anal. **4**:4 (2001), 523–530.

[217] S.G. Samko and A.Ya. Yakubov, *Zygmund's estimate for a singular integral with a rapidly decreasing power weight.* Izv. Vyssh. Uchebn. Zaved. Mat. **1984**:12 (1984), 42–51.

[218] S.G. Samko and A.Ya. Yakubov, *Zygmund estimate for moduli of continuity of fractional order of a conjugate function.* Izv. Vyssh. Uchebn. Zaved. Mat. **1985**:12 (1985), 49–53, 77.

[219] S.G. Samko and A.Ya. Yakubov, *Zygmund estimate for hypersingular integrals in the case of moduli of continuity of fractional order.* Izv. Severo-Kavk. Nauchn. Tsentra Vysshei Shkoly, Ser. Estestv. Nauk **1986**:3 (1986), 42–47.

V. Kokilashvili
A. Razmadze Mathematical Institute
of I. Javakhishvili Tbilisi State University

and

International Black Sea University
Georgia
e-mail: `kokil@rmi.ge`

Operator Theory:
Advances and Applications, Vol. 229, 49–63

The Role of S.G. Samko in the Establishing and Development of the Theory of Fractional Differential Equations and Related Integral Operators

Sergei V. Rogosin

To Stefan Samko on the occasion of his 70th birthday

Abstract. The aim of this work is to describe main aspects of the modern theory of fractional differential equations, to present elements of classification of fractional differential equations, to formulate basic components of investigations related to fractional differential equations, to pose some open problems in the study of fractional differential equations.
A survey of results by S.G. Samko on different problems of modern mathematical analysis is given. Main results of S.G. Samko having an essential influence on the establishing and development of the theory of fractional differential equations are singled out.

Mathematics Subject Classification (2010). Primary 26A33, 34A08, 35R11, 45P05, 47G40; Secondary 26-03, 45-03, 47-03.

Keywords. Fractional calculus, fractional differential equations, fractional integral operators, potential type operators.

1. Main aspects of the modern theory of fractional differential equations

1.1. Elements of the classification

In this section we present a brief introduction to the theory of fractional differential equations. Since it is not the main goal of the article, we restrict our attention only to the elements of classification of those types of problems which belong to the discussion in the framework of this theory. For wider expositions, which include many historical facts and extended bibliography, we refer to the main book by

S.G. Samko (published jointly with his co-authors A.A. Kilbas and O.I. Marichev) [43], to recent monographs [6], [23], and to survey paper [24].

It should be mentioned that in earlier time (see, e.g., [43]) mainly fractional differential equations with Riemann-Liouville fractional derivative (see definition (*R-L*) below) were investigated. Whereas later, motivated by needs of applications, the use of the Caputo fractional derivative (see definition (*Caputo*) below) became popular. Equations with Caputo derivatives for the first time in the book form were presented in [44]. Both types of derivatives were discussed in [23], see also [6].

Ordinary fractional differential equations
This collection includes all equations of the form

$$F(x, y(x), \mathcal{D}_{a_1}^{\alpha_1}\omega_1(x)y(x), \mathcal{D}_{a_2}^{\alpha_2}\omega_2(x)y(x), \ldots, \mathcal{D}_{a_n}^{\alpha_n}\omega_n(x)y(x)) = g(x), \tag{1.1}$$

where $\mathcal{D}_{a_j}^{\alpha_j}$ is a right/left-sided derivative of one of the known types (see Sec. 2 below and/or [43]), $\alpha_1, \ldots, \alpha_n$ are arbitrary positive real numbers, and $\omega_1, \ldots, \omega_n$ are certain weight functions. Such equations appeared due to direct generalizations of different applied models (description of a number of these applications can be found in the Proceedings of IFAC Workshops on Fractional Differentiation and its Applications (2004, Bordeaux, France; 2006, Porto, Portugal; 2008, Ankara, Turkey; 2010, Badajoz, Spain; 2012, Nanjing, China)). The main achievements of the theory of ordinary fractional differential equations are connected with basic results of the fractional calculus (see, e.g., [37], [43]).

We have to point out the following kinds of equations, which are particular cases of equation (1.1):

- linear homogeneous fractional ordinary differential equations,
- linear inhomogeneous fractional ordinary differential equations,
- linear homogeneous fractional ordinary differential equations with variable coefficients,
- linear inhomogeneous fractional ordinary differential equations with variable coefficients,
- nonlinear fractional ordinary differential equations,
- sequential differential equations of fractional order,
- ordinary fractional differential equations on the whole real line,
- multidimensional ordinary fractional differential equations.

The multidimensional theory is less developed because of several reasons. First of all, the corresponding elements of the theory of multidimensional fractional integro-differentiation are not completely investigated. Second, researchers are still in the process of discovering of the physical, mechanical, chemical, biological phenomenon, which can be adequately described by certain multidimensional models involving fractional derivatives. Therefore, one can single out only few types of multidimensional ordinary fractional differential equations developed up to the same level as the corresponding one-dimensional equations. For discussion of these problems we refer to the book [25] and references therein.

Fractional partial differential equations

Several models involving fractional partial differential equations are known. Anyway, a complete classification of this class of equations does not exist. The most cited source on this subject is the book [23]. We can also mention the recent books [33], [69] and extended lists of references presented in [23] and in [33]. Here we present several examples of known fractional partial differential equations. It should be mentioned that fractional derivatives with respect to the time variable are taken in these equations either in the form of Riemann-Liouville (see definition (*R-L*) below in Sec. 2) or in the form of Caputo (see definition (*Caputo*) below in Sec. 2). As for fractional derivatives with respect to spatial variables, they are taken in some appropriate sense including Riesz-Feller type fractional derivatives (inverses to fractional potential type operators) (see Sec. 2 below and/or [23], [43]).

- Gerasimov's equation:

$$\rho \frac{\partial^2 u}{\partial t^2} = k \left(D^{\alpha}_{-,t} \left[\frac{\partial^2 u}{\partial x^2} \right] \right) (x, t).$$

- "Hyperbolic" fractional differential equation:

$$\left(D^{\alpha}_{0+,x} D^{\beta}_{0+,y} u \right) (x, y) = f \left[x, y, u, D^{\alpha}_{0+,x} u, D^{\alpha-1}_{0+,x} D^{\beta}_{0+,y} u, D^{\beta}_{0+,y} u, D^{\alpha-1}_{0+,x} u \right].$$

- Differential equation of semi-integer order:

$$\sum_{k=0}^{n} a_k \left(D^{k/2}_{0+,x} D^{(n-k)/2}_{0+,y} u \right) (x, y) = f(x, y).$$

- Fractional diffusion-wave equation:

$$\left(D^{\alpha}_{0+,t} u \right) (x, t) = \left(D^{\beta}_{x} u \right) (x, t) \quad (1 \leq \alpha \leq 2,\ 0 \leq \beta \leq 2).$$

- Multidimensional diffusion-wave equation:

$$\left(D^{\alpha}_{0+,t} u \right) (\mathbf{x}, t) = \lambda^2 \left(\Delta_{\mathbf{x}} u \right) (\mathbf{x}, t),\ \mathbf{x} \in \mathbb{R}^n \quad (0 < \alpha < 2).$$

1.2. Methods of investigation

In this subsection we mention problems which are usually considered for fractional differential equations. We also single out the types of solutions as well as indicate methods for the solution.

Treating problems:

- Cauchy problem: values of standard derivatives at some points are given.
- Cauchy-type problem: values of fractional derivatives and/or integrals at some points are given.
- Dirichlet-type problem (for partial fractional differential equations): values of standard and/or fractional derivatives at end-points of an interval are given.
- Initial-boundary value problem with local and/or non-local initial conditions (i.e., Cauchy-type initial conditions).

Types of solutions:

- Solutions in classical functional spaces such as Schauder-type spaces or Lebesgue-type spaces, or corresponding weighted spaces.
- Generalised or weak solutions (i.e., distributions on certain spaces of test functions).

Methods of solution:

- Compositional methods based on certain known formulas for different kind of fractional integrals and derivatives.
- Methods of integral transforms, which allow to reduce fractional differential equations to functional equations.
- Integral equations method. Since, by definition, the fractional derivatives are compositions of differential operators and certain integral operators, then it is possible to find (for special types of equations) integral equations equivalent to considered fractional differential equations.
- Numerical methods. Numerical methods for fractional differential equations differ essentially as from the numerical methods for classical differential equations as from those performed for integral equations. Anyway one can find in the literature several numerical algorithms which are especially worked up for fractional equations (see, *e.g.*, [44]).

2. Basic components of investigations related to fractional differential equations

In this section we briefly outline certain investigations which form a base for further development of the theory of fractional differential equations.

2.1. Development of fractional calculus

Starting point for any investigation concerning fractional differential equations is to establish the corresponding results for fractional derivatives and integrals. Below we single out some collections of the properties which are usually in discussion.

- Formal properties (including semi-group properties), composition formulas for different types of fractional derivatives, calculation of fractional derivatives of elementary and special functions. Below one can find a list of some definitions of fractional derivatives. Surely this list is not complete and contains only most frequently cited notations.
  - Riemann-Liouville derivatives

$$\left(D^{\alpha}_{a+}u\right)(x) = \frac{1}{\Gamma(n-\alpha)}\left(\frac{d}{dx}\right)^{n}\int\limits_{a}^{x}\frac{u(t)dt}{(x-t)^{\alpha-n+1}} \quad (n = [\mathrm{Re}\,\alpha]+1, \quad \alpha > 0); \tag{R-L}$$

– Caputo (or Gerasimov-Caputo) derivatives

$$\left({}^C D^\alpha_{a+} u\right)(x) = \left(D^\alpha_{a+} \left[u(t) - \sum_{k=0}^{n-1} \frac{u^{(k)}(a)}{k!}(t-a)^k \right] \right)(x); \qquad \text{(Caputo)}$$

– Erdelyi-Kober derivatives

$$\left(D^\alpha_{a+;\sigma,\eta} u\right)(x) = x^{-\sigma\eta} \left(\frac{1}{\sigma x^{\sigma-1}} \frac{d}{dx} \right)^n \frac{\sigma x^{\sigma(n-\alpha)}}{\Gamma(\alpha)} \int_a^x \frac{t^{\sigma(\eta+\alpha+1)-1} u(t)dt}{(x^\sigma - t^\sigma)^{n-\alpha}}; \qquad \text{(E-K)}$$

– Hadamard derivatives

$$\left(\mathcal{D}^\alpha_{a+} u\right)(x) = \left(x \frac{d}{dx} \right)^n \frac{1}{\Gamma(n-\alpha)} \int_a^x \left(\log \frac{x}{t} \right)^{n-\alpha+1} \frac{u(t)dt}{t}; \qquad \text{(Hadamard)}$$

– Grünwald-Letnikov derivatives

$$u^{\alpha)}_+(x) = \lim_{h\to+0} \frac{\left(\Delta^\alpha_h u\right)(x)}{h^\alpha} = \lim_{h\to+0} \frac{\sum_{k=0}^{\infty} (-1)^k \binom{\alpha}{k} u(x-kh)}{h^\alpha}; \qquad \text{(G-L)}$$

– Riesz derivatives

$$\mathbf{D}^\alpha u(\mathbf{x}) = \mathcal{F}^{-1} |\boldsymbol{\omega}|^\alpha \mathcal{F} u(\mathbf{x}), \quad \mathbf{x} \in \mathbb{R}^n; \qquad \text{(Riesz)}$$

– other definitions of fractional derivatives via pseudo-differential operators method (see, e.g., [20]).

- Acting properties of operators of fractional integration (differentiation) in different functional spaces.
- Representation of functions by fractional integrals of functions which belong to different functional spaces (in particular, characterization of spaces $I^\alpha(L_p)$).
- Operator properties (relations to singular integral operators, inversion formulas, relations to operators of integral transforms).

2.2. Development of the theory of first-order integral equations

Here we mention several kinds of integral equations which are closely related to the problems of fractional differential equations. The study of their solvability as well as their solution in closed form are important components of the theory of fractional differential equations.

– Abel integral equations and their applications.
– Integral equations with weak singularities.
– Volterra integral equations.
– Convolution type integral equations.

One of the most import aims here is to prove an equivalence of differential equations and integral equations.

2.3. Development of methods of integral transforms

As it was already mentioned, the method of integral transforms is one of the basic methods at the investigation of fractional differential equations. The most used integral transforms in this branch of mathematics are the following transforms:

- Fourier integral transforms.
- Laplace integral transforms.
- Mellin integral transforms (see, *e.g.*, [34]).
- Integral transforms with special functions in the kernel.
- Integral **G**-transforms

$$(\mathbf{G}f)(x) = \int_0^\infty G^{m,n}_{p,q}\left[xt \,\middle|\, \begin{matrix}(a_i)_{1,p}\\(b_j)_{1,q}\end{matrix}\right] f(t)dt,$$

where

$$G^{m,n}_{p,q}\left[z \,\middle|\, \begin{matrix}(a_i)_{1,p}\\(b_j)_{1,q}\end{matrix}\right] = \frac{1}{2\pi i}\int_{\mathcal{L}} \frac{\prod\limits_{j=1}^{m}\Gamma(b_j+s)\prod\limits_{i=1}^{n}\Gamma(1-a_i-s)}{\prod\limits_{i=n+1}^{p}\Gamma(a_i+s)\prod\limits_{j=m+1}^{q}\Gamma(1-b_j-s)} z^{-s}ds.$$

- Integral **H**-transforms (see, *e.g.*, [21])

$$(\mathbf{H}f)(x) = \int_0^\infty H^{m,n}_{p,q}\left[xt \,\middle|\, \begin{matrix}(a_i,\alpha_i)_{1,p}\\(b_j,\beta_j)_{1,q}\end{matrix}\right] f(t)dt,$$

where

$$H^{m,n}_{p,q}\left[z \,\middle|\, \begin{matrix}(a_i,\alpha_i)_{1,p}\\(b_j,\beta_j)_{1,q}\end{matrix}\right] = \frac{1}{2\pi i}\int_{\mathcal{L}} \frac{\prod\limits_{j=1}^{m}\Gamma(b_j+\beta_j s)\prod\limits_{i=1}^{n}\Gamma(1-a_i-\alpha_i s)}{\prod\limits_{i=n+1}^{p}\Gamma(a_i+\alpha_i s)\prod\limits_{j=m+1}^{q}\Gamma(1-b_j-\beta_j s)} z^{-s}ds.$$

2.4. Development of the theory of special functions

Methods and results of special function theory are very important for the problems which we are discussing. First of all this is due to the fact that a number of fractional differential equations admits closed form solutions. These solutions are represented usually via special functions of specific types. Special functions mostly applied in the theory of fractional differential equations are the following (see, *e.g.*, [35]):

- generalizations of Mittag-Leffler special functions

$$E_{\alpha,\beta}(z) = \sum_{k=0}^{\infty}\frac{z^k}{\Gamma(\alpha k+\beta)};$$

– generalizations of Wright special functions

$$ {}_p\Psi_q \left[\begin{array}{c} (a_i,\alpha_i)_{1,p} \\ (b_j,\beta_j)_{1,q} \end{array} \middle| z \right] = \sum_{k=0}^{\infty} \frac{\prod_{i=1}^{p} \Gamma(a_i+\alpha_i k)}{\prod_{j=1}^{q} \Gamma(b_j+\beta_j k)} \frac{z^k}{k!}; $$

– general types of special functions:

– G-function: $G^{m,n}_{p,q}\left[z \middle| \begin{array}{c} (a_i)_{1,p} \\ (b_j)_{1,q} \end{array} \right]$,

– H-function (see, *e.g.*, [36]): $H^{m,n}_{p,q}\left[z \middle| \begin{array}{c} (a_i,\alpha_i)_{1,p} \\ (b_j,\beta_j)_{1,q} \end{array} \right]$.

2.5. Development of multidimensional fractional calculus

At last we have to mention several multidimensional operators whose properties are now forming the base for developing a theory of multidimensional fractional differential equations:

– polypotential type operator

$$ (\mathcal{K}^{\alpha}\varphi)(\mathbf{x}) = \frac{1}{2^n \prod_{k=1}^{n} \Gamma(\alpha_k)\cos\alpha_k\pi/2} \int_{\mathbb{R}^n} \frac{\varphi(\mathbf{t})dt}{\prod_{k=1}^{n} |x_k-t_k|^{\alpha_k}}; $$

– Riesz potential

$$ (\mathbf{I}^{\alpha}\varphi)(\mathbf{x}) = \frac{1}{\gamma_n(\alpha)} \int_{\mathbb{R}^n} \frac{\varphi(\mathbf{y})d\mathbf{y}}{|\mathbf{x}-\mathbf{y}|^{\alpha}}; $$

– hypersingular operator

$$ (\mathbf{D}^{\alpha}\varphi)(\mathbf{x}) = \frac{1}{d_{n,l}(\alpha)} \int_{\mathbb{R}^n} \frac{(\Delta^l_{\mathbf{y}}\varphi)(\mathbf{x})}{|\mathbf{y}|^{n+\alpha}} d\mathbf{y}, \quad (\Delta^l_{\mathbf{y}}\varphi)(\mathbf{x}) = \sum_{k=0}^{l} (-1)^k \binom{l}{k} \varphi(\mathbf{x}-k\mathbf{y}). $$

3. The role of Professor S.G. Samko in the creation and development of the theory of fractional differential equations

First of all we have to point out that it is problematic to mention all results by S.G. Samko having influence on the development of the discussed theory. Therefore, we restrict ourselves only on the main results which we will briefly outline. In particular, the list of S.G. Samko's books and articles, cited here, is surely not complete.

Main directions in which S.G. Samko obtained results having an essential influence on the establishing and development of the theory of fractional differential equations are the following:

– singular integral equations and boundary value problems;
– Abel integral equations and their applications;
– integral equations with weak singularities;

- integral equations of convolution type;
- fractional calculus;
- fractional powers of operators;
- one- and multi-dimensional theory of potential type operators;
- hypersingular operators;
- functional spaces with variable exponents.

One can compare these results with the above-described aims of the theory and applications of fractional differential equations. It makes the reader an impression on the great importance of the results by S.G. Samko.

The rest of this section is organized as follows. We mention certain directions as titles of subsections, specify some results in these directions and present the corresponding references.

3.1. Singular integral equations and boundary value problems

- Abstract theory (Noether theory) of complete (or general) singular integral equations and singular integral equations with a shift [15], [16], [17] (including equations with integrals along an open arc Γ [7], [14])

$$a(t)\varphi(t) + b(t)\int\limits_{\Gamma} \frac{\varphi(\tau)}{\tau - t} d\tau + \int\limits_{\Gamma} k(t,\tau)\varphi(\tau)d\tau = f(t).$$

- Classes of complete singular integral equations solvable in closed form [53], [57].
- Generalized argument principle [8], [9].
- Acting properties of singular integral operators (SIO) (including SIO on Carleson curves) in spaces with variable exponents [27], [29], [30], [31].
- Applications of the theory of singular integral equations to the investigation of boundary value problems in classes of analytic functions represented by the Cauchy type integrals [26], [28].

3.2. Abel integral equations and their generalizations

- Solution in closed form of generalized Abel type integral equations [49], [50], [55].
- Proof of the relation formulas of fractional integration operators with singular integral operators [51], in particular,

$$I_-^{\alpha}\varphi = \cos\,\alpha\pi I_+^{\alpha}\varphi + \sin\,\alpha\pi S I_+^{\alpha}\varphi.$$

- Noether theory of generalized Abel type integral equations [52], [54]:

$$u(x)\int\limits_a^x \frac{\varphi(t)dt}{(x-t)^{\mu}} + v(x)\int\limits_x^b \frac{\varphi(t)dt}{(t-x)^{\mu}} = f(x).$$

3.3. Integral equations with weak singularities

- Solution in closed form of integral equations with logarithmic kernel [60].
- Normal solvability, asymptotic method of solution to integral equations with logarithmic and homogeneous kernel [22].
- Asymptotic behaviour of singular values of integral operators with weak singularities [10].
- Inversion of integral equation

$$\int_{-\infty}^{x} k(x-t)\varphi(t)dt = f(x)$$

  with Sonine's kernel $k \in L_1^{loc}(\mathbb{R}_+^1)$ (i.e., such that there exists $l \in L_1^{loc}(\mathbb{R}_+^1)$ such that $\int_0^x k(x-t)l(t)dt \equiv 1$) [3], [4].
- Classes of correctness in the weighted Hölder spaces for Sonine's integral equation [5].

3.4. Convolution type integral equations

- Solvability of convolution type integral equations [12], [13]

$$a_0(t)\varphi(t) + \sum_{j=1}^{n} a_j(t) \int_{-\infty}^{\infty} b_j(\tau)h_j(t-\tau)\varphi(\tau)d\tau = f(t).$$

- Solution in a closed form of special kinds of convolution type integral equations [18].
- Noether theory of convolution type integral equations [19].
- Convolution type operators in spaces with variable exponent [64], [65].
- Nonlinear convolution type operators and equations [11].

3.5. Fractional integro-differentiation

One of the most essential achievements in this area were gathered in the books written by S.G. Samko jointly with his friends and colleagues A.A. Kilbas and O.I. Marichev [42], [43]. These books are now standard reference works on fractional calculus. They constitute the very precise and careful analysis of all essential questions of this subject. It is not a surprise that the English version [43] is called by many scientists as "THE BIBLE OF FRACTIONAL CALCULUS". Inspite of the great importance of [43] and its encyclopedic nature, investigations of fractional differential equations, conducted after this book and inspired by it, still deserve a separate survey article.

Below we briefly describe a number of precise results obtained by S.G. Samko in this area.

- Description of the image $\mathbf{I}^{\alpha}(L_p)$ of the Riesz potential [56], [58].
- Isomorphism of weighted Hölder spaces under the acting of the operator of fractional derivative [39].

- Marchaud-type formula for the operator of fractional derivative in domains $\Omega \subset \mathbb{R}^n$ [45]

$$D^{\alpha}_{\Omega}\varphi(\mathbf{x}) = c(\alpha)\left[\varphi(\mathbf{x}) \int\limits_{\mathbb{R}^n\setminus\Omega} \frac{d\mathbf{y}}{|\mathbf{x}-\mathbf{y}|^{n+\alpha}} + \int\limits_{\Omega} \frac{\varphi(\mathbf{x})-\varphi(\mathbf{y})}{|\mathbf{x}-\mathbf{y}|^{n+\alpha}} d\mathbf{y}\right].$$

- Description of functions that have no first-order derivative, but have fractional derivatives of all orders less than one [32].
- Approximative definition of a fractional differentiation [48].
- Characterization of the range of one-dimensional fractional integration in the space with variable exponent [47].
- Description of the acting properties of generalized Riemann-Liouville operator of fractional integration of variable order

$$I^{\alpha(x)}_{a+}\varphi(x) = \frac{1}{\Gamma(\alpha(x))} \int\limits_a^x \frac{\varphi(t)dt}{(x-t)^{1-\alpha(x)}}$$

  in generalized Hölder spaces [38], [39].
- Formula for a left inverse to Liouville operator of fractional integration (i.e., with $a = -\infty$) of variable order [63].
- Fractional differentiation and integration of variable order in spaces with variable exponent [41], [62].

3.6. Fractional powers of operators

- Application of the hypersingular integrals' method to construction of effective formulas for fractional powers of the classical operators of mathematical physics

$$I - \Delta, \qquad -\Delta_x + \frac{\partial}{\partial t}, \qquad I - \Delta_x + \frac{\partial}{\partial t},$$

  where I is an identity operator, and Δ is the Laplace operator [1], [66], [67].

3.7. The theory of (one- and multidimensional) potential type operators

- Inversion of the Riesz type fractional potential [40], [59]

$$(\mathcal{K}^{\alpha}\varphi)(x) = \frac{1}{\gamma_{n-1}(\alpha)} \int\limits_{|\mathbf{x}|=1} \frac{\varphi(\sigma)d\sigma}{|\mathbf{x}-\sigma|^{n-1-\alpha}}.$$

- Investigation of properties of operators generalizing fractional integration operators

$$\left(I^{\alpha(\cdot)}_{m}\varphi\right)(x) = \int\limits_{\Omega} \frac{\varphi(\mathbf{y})d\mu(\mathbf{y})}{d(\mathbf{x},\mathbf{y})^{m-\alpha(x)}}$$

  in the Lebesgue spaces with variable exponents [2].
- Application to multidimensional integral equations of the first kind [68], [46].
- Relation to Grünwald-Letnikov's approach to fractional calculus [61].

4. Conclusion

The theory of fractional differential equations is a highly developing branch of mathematics. These equations are arising either from theoretical considerations and generalizations or from many applications for which the fractional machinery is a natural tool for the description of the corresponding phenomenon.

In Section 2 we tried to describe briefly the main aims and ideas which are used at the formulation and investigation of fractional differential equations. Surely such a description is far from completeness since the subject is too wide.

In order to establish and develop the theory of fractional differential equations one needs to create background, to develop instruments for investigation and to introduce new innovative ideas. In all these directions Professor S.G. Samko obtained many important results. In Section 3 we have seen only few of these results, which lay in the core of the considered theory. A lot of fresh and innovative ideas can be found in the cited papers and books by S.G. Samko.

We hope that our article helps the reader to find new ideas and to show the ways for further development of a very interesting branch of science – the theory and applications of fractional differential equations.

Acknowledgment

The author would like to express his gratitude to Prof. Luis de Castro for the brilliant organization of the Conference in honour of Prof. Stefan Samko.

The author is grateful to the reviewers for valuable remarks concerning the text of the article. This work is partially supported by the Belarusian Fund for Fundamental Scientific Research under grant F10MC-024.

References

[1] A. Abramyan, V.A. Nogin, S. Samko, *Fractional powers of the operator* $-|x|^2\Delta$ *in* L^p*-spaces.* Differ. Equations **32**, No. 2 (1996), 281–283.

[2] A. Almeida, S. Samko, *Characterization of Riesz and Bessel potentials on variable Lebesgue spaces.* J. Function Spaces and Applic., **4**, No. 2 (2006), 113–144.

[3] R. Cardoso, S. Samko, *Integral equations of the first kind of Sonine type.* Intern. J. Math. and Math. Sci. **57** (2003), 3609–3632.

[4] R. Cardoso, S. Samko, *Sonine integral equations of the first kind of in* $L^p(0,b)$. Fract. Calc. and Applied Anal. **6**, No. 3 (2003), 235–258.

[5] R. Cardoso, S. Samko, *Weighted generalized Hölder spaces as well-posedness classes for Sonine integral equations.* J. Integr. Equat. Appl. **20**, No. 4 (2008), 437–480.

[6] K. Diethelm, *The Analysis of Fractional Differential Equations.* Springer Verlag: Lecture Notes in Mathematics **2004**, Heidelberg etc., 2010.

[7] F.D. Gakhov, S.G. Samko, *General singular equation on an open contour* (Russian). Abstracts of Intern. Congress of Math. (1966), Moscow, Section 5 (Functional.Analysis), p. 41.

[8] F.D. Gakhov, E.I. Zverovich, S.G. Samko, *Increment of the argument, logarithmic residue and a generalized principle of the argument.* Soviet Math. Dokl. **14**, No. 6 (1973), 1856–1860.

[9] F.D. Gakhov, E.I. Zverovich, S.G. Samko, *A generalized argument principle* (Russian). Izv. Akad. Nauk Biel. SSR, ser. fis.-mat. nauk, No. 1 (1975), 5–16.

[10] R. Gorenflo, S. Samko, *On the dependence of asymptotics of s-numbers of fractional integration operators on weight functions.* Integr. Transform. and Special Funct. **5** (1997), 191–212.

[11] N.K. Karapetiants, A.A. Kilbas, M. Saigo, S.G. Samko, *Upper and lower bounds for solutions of nonlinear Volterra convolution equations with power nonlinearity.* J. of Integr. Equat. and Appl. **12**, No. 4 (2001), 421–448.

[12] N.K. Karapetiants, S.G. Samko, *A certain class of convolution type integral equations and its application* (Russian). Soviet Math. Dokl. **11** (1970), 1050–1054.

[13] N.K. Karapetiants, S.G. Samko, *A certain class of convolution type integral equations and its application* (Russian). Math. USSR, Izvestija **5**, No. 3 (1971), 731–744.

[14] N.K. Karapetiants, S.G. Samko, *Singular integral operators with shift on an open contour.* Soviet Math. Dokl. **13**, No. 3 (1972), 691–696.

[15] N.K. Karapetiants, S.G. Samko, *Singular integral operators on the axis with a fractional-linear shift of Carlemanian type and the Noether property for operators containing an involution* (Russian). Izv. Akad. Nauk Armyan. SSR, **7**, No. 1 (1972), 68–77.

[16] N.K. Karapetiants, S.G. Samko, *Singular integral operators with Carleman shift in the case of piece-wise continuous coefficients, I* (Russian). Izv. Vuzov. Matematika, No. 2 (1975), 43–54.

[17] N.K. Karapetiants, S.G. Samko, *Singular integral operators with Carleman shift in the case of piece-wise continuous coefficients, II* (Russian). Izv. Vuzov. Matematika, No. 3 (1975), 34–42.

[18] N.K. Karapetiants, S.G. Samko, *Integral equations of convolution type of the first kind with a power kernel* (Russian). Izv. Vuzov. Matematika, No. 4 (1975), 60–67.

[19] N.K. Karapetiants, S.G. Samko, *Singular convolution operators with a discontinuous symbol* (Russian). Soviet Math. Dokl. **16**, No. 2 (1975), 510–514.

[20] S. Kempfle, I. Schaefer and H. Beyer, *Fractional calculus via functional calculus: theory and applications.* Nonlinear Dynamics, **29** (2002), 99–127.

[21] A.A. Kilbas, M. Saigo, *H-Transform. Theory and Applications.* Boca Raton-London-New York-Washington, D.C., Chapman and Hall/CRC, 2004.

[22] A.A. Kilbas, S.G. Samko, *Integral operators with homogeneous kernels in the Hölder H-space* (Russian). Dokl. Akad. Nauk Biel. SSR **21**, No. 1 (1977), 5–8.

[23] A.A. Kilbas, H.M. Srivastava, and J.J. Trujillo, *Theory and Applications of Fractional Differential Equations.* North-Holland Mathematics Studies. **204**. Elsevier, Amsterdam, etc., 2006.

[24] A.A. Kilbas, and J.J. Trujillo, *Differential equations of fractional order: methods, results and problems – II.* J. Appl. Anal., **81** (2002), 435–494.

[25] V. Kiryakova, *Generalized Fractional Calculus and Applications.* Harlow, Longman, 1994 [Pitman Research Notes in Mathematics, **301**].

[26] V. Kokilashvili, V. Paatashvili, S. Samko, *Boundary value problems for analytic functions in the class of Cauchy type integrals with density in* $L^{p(\cdot)}$. Bound. Value Probl. **2005**, No. 1-2 (2005), 43–71.

[27] V. Kokilashvili, V. Paatashvili, S. Samko, *Boundedness in Lebesgue spaces with variable exponent of the Cauchy singular operators on Carleson curves.* "Operator Theory: Advances and Applications", **170** (dedicated to the 70th birthday of Prof. I.B. Simonenko). Editors Ya. Erusalimsky, I. Gohberg, S. Grudsky, V. Rabinovich, N. Vasilevski, Birkhäuser, 2006, 167–186.

[28] V. Kokilashvili, V. Paatashvili, S. Samko, *Riemann Problem in the Class of Cauchy-Type Integrals with Density in* $L^p(.)(G)$. Doklady Mathematics **78**, No. 1 (2008), 510–513.

[29] V. Kokilashvili, N. Samko, S. Samko, *Singular operators in variable spaces* $L^{p(\cdot)}(\Gamma,\rho)$ *with oscillating weights.* Math. Nachrichten **280**, No. 9-10 (2007), 1145–1156.

[30] V. Kokilashvili, S. Samko, *Singular operators and Fourier multipliers in weighted Lebesgue spaces with variable exponents.* Vestnik St. Petersburg University. Mathematics **41**, No. 2 (2008), 134–144.

[31] V. Kokilashvili, S. Samko, *Boundedness of weighted singular integral operators in Grand Lebesgue spaces.* Georg. J. Math. **18**, Issue 2 (2011), 259–269.

[32] E.R. Love, B. Ross, S. Samko, *Functions that have no first-order derivative might have fractional derivatives of all orders less than one.* Real Analysis Exchange **20**(2) (1994/95), 140–157.

[33] F. Mainardi, *Fractional calculus and waves in linear viscoelasticity.* London, Imperial College Press, 2010.

[34] O.I. Marichev, *Handbook of Integral Transforms of Higher Transcendental Functions, Theory and Algorithmic Tables.* Chichester, Ellis Horwood, 1983.

[35] A.M. Mathai, H.J. Haubold, *Special Functions for Applied Scientists.* New York, Springer, 2008.

[36] A.M. Mathai, R.K. Saxena, H.J. Haubold, *The H-function. Theory and applications.* Dordrecht, Springer, 2010.

[37] K.S. Miller, B. Ross, *An Introduction to the Fractional Calculus and Fractional Differential Equations*, New York, John Wiley and Sons, 1993.

[38] H.M. Murdaev, S.G. Samko, *Action of fractional integrodifferentiation in weighted generalized Hölder spaces* $H_0^{\omega}(\rho)$ *with the weight* $\rho(x) = (x-a)^{\mu}(b-x)^{\nu}$ (Russian). Deponierted in VINITI May 11, 1986, No. 3350-B86 DEP.

[39] Z.U. Mussalaeva, S.G. Samko, *Fractional type operators in weighted generalized Hölder spaces.* Proc. of the Georgian Academy of Sci., Mathematics **1**, No. 5 (1993), 601–626.

[40] V.A. Nogin, S.G. Samko, *Inversion and description of Riesz potentials with densities in weighted* L^p*-spaces* (Russian). Izv. Vuzov. Matematika, No. 1 (1985), 70–72 (Transl. in Soviet Math. (Izv.VUZ)).

[41] B. Ross, S.G. Samko, *Fractional integration operator of variable order in the Hölder spaces* $H^{\lambda(x)}$. Intern. J. of Math. and Math. Sci. **18**, No. 4 (1995), 777–788.

[42] S.G. Samko, A.A. Kilbas, O.I. Marichev, *Fractional Integrals and Derivatives and Some of Their Applications* (Russian). Minsk, Nauka i Tekhnika. 1987.

[43] S.G. Samko, A.A. Kilbas, O.I. Marichev, *Fractional Integrals and Derivatives: Theory and Applications.* New York and London, Gordon and Breach Science Publishers, 1993.

[44] I. Podlubny, *Fractional Differential Equations.* New York, Academic Press, 1999.

[45] H. Rafeiro, S. Samko, *On multidimensional analogue of Marchaud formula for fractional Riesz-type derivatives in Domains in* $\mathbb{R}^n$. Fract. Calc. Appl. Anal. **8**, No. 4 (2005), 393–401.

[46] H. Rafeiro, S. Samko, *On a class of fractional type integral equations in variable exponent spaces.* Fract. Calc. and Appl. Anal. **10**, No. 4 (2007), 399–421.

[47] H. Rafeiro, S. Samko, *Characterization of the range of one-dimensional fractional integration in the space with variable exponent.* "Operator Theory: Advances and Applications", **181** (eds. M.A. Bastos, I. Gohberg, A.B. Lebre), 2008, 393-416.

[48] N. Samko, S.G. Samko, *On approximative definition of a fractional differentiation.* Fract. Calc. and Applied Anal. **2**, No. 3 (1999), 329–342.

[49] S.G. Samko, *A generalized Abel equation and an equation with Cauchy kernel* (Russian). Dokl. Akad. Nauk SSSR, **176**, No. 5 (1967), 1019–1022.

[50] S.G. Samko, *A general singular equation on an open contour and a generalized Abel equation* (Russian). Dokl. Akad. Nauk SSSR, **177**, No. 1 (1967), 44–47.

[51] S.G. Samko, *A generalized Abel equation and fractional integration operators* (Russian). Differentsial'nye Uravnenija, **4**, No. 2 (1968), 298–314.

[52] S.G. Samko, *Noether's theory for the generalized Abel integral equation* (Russian). Differentsial'nye Uravnenija **4**, No. 2 (1968), 315–326.

[53] S.G. Samko, *The solvability in closed form of singular integral equations.* Soviet Math. Dokl. **10**, No. 6 (1969), 1445–1448.

[54] S.G. Samko, *Abel's generalized equation, Fourier transform and convolution type equations.* Soviet Math. Dokl. **10**, No. 4 (1969), 942–946.

[55] S.G. Samko, *Abel's generalized integral equation on the line* (Russian). Izv. Vyssh. Uchebn. Zaved. Math. No. 8 (1970), 83–93.

[56] S.G. Samko, *The space* $I^\alpha(L^p)$ *of fractional integrals, and operators of potential type* (Russian). Izv. Akad. Nauk Armyan. SSR **7**, No. 5 (1973), 359–383.

[57] S.G. Samko, *Singular integral and integro-differential equations with analytic kernels* (Russian). Izv. Severo-Kavk. Tsentra Vysshei Shkoly, ser. estestv. nauk. No. 4 (1974), 86–94.

[58] S.G. Samko, *On the description of the image* $I^\alpha(L^p)$ *of Riesz potentials* (Russian). Izv. Akad. Nauk Armyan. SSR, Matematika **12**, No. 5 (1977), 329–334.

[59] S.G. Samko, *Generalized Riesz potentials and hypersingular integrals, their symbols and inversion.* Soviet Math. Dokl. **18**, No. 1 (1977), 97–101.

[60] S.G. Samko, *Integral equation of the first kind with a logarithmic kernel, II* (Russian). Collection of papers "Math. Analysis and its Appl.", Rostov-on-Don, Izdat. Rostov. Univ., 1978, 103–120.

[61] S.G. Samko, *The multidimensional fractional integrodifferentiation and the Grunwald-Letnikov approach to fractional calculus.* Proc. Intern. Conf. "Fract. Calculus and its Applic.", Tokyo, 1989, College of Engineering, Nihon Univ., 1990, 221–225.

[62] S.G. Samko, *Fractional differentiation and integration of variable order*. Dokl. Math. **51**, No.3 (1995), 401–403.

[63] S.G. Samko, *Fractional integration and differentiation of variable order*. Analysis Mathematica **21** (1995), 213–236.

[64] S.G. Samko, *Convolution type operators in* $L^{p(x)}$. Integr. Transform. and Special Funct. **7**, No. 1-2 (1998), 123–144.

[65] S.G. Samko, *Convolution and potential type operators in the space* $L^{p(x)}$. Integr. Transform. and Special Funct. **7**, No. 3-4 (1998), 261–284.

[66] S.G. Samko, *Fractional powers of operators via hypersingular operators*, In "Progress in Nonlinear Diff. Eq. and Their Applic.", Birkhäuser, **42**, 2000, 259–272.

[67] S.G. Samko, *Approximative approach to construction of fractional powers of operators* (Russian). Izvestia Vysch. Uchebn. Zaved., Severo-Kavkazskii Region, Estestv. Nauki, No. 4 (2001), 41–44.

[68] S.M. Umarkhadzhiev, S.G. Samko, *Applications of hypersingular integrals to multidimensional integral equations of the first kind*. Proc. Steklov Inst. Math. Issue 3 (1987), 325–339.

[69] V.V. Uchaikin, *Method of fractional derivatives* (Russian). Ulyanovsk, Artishok, 2008.

Sergei V. Rogosin
Nezavisimosti ave 4
Belarusian State University
BY-220030 Minsk
Belarus
e-mail: rogosinsv@gmail.com

Operator Theory:
Advances and Applications, Vol. 229, 65–76

Energy Flow Above the Threshold of Tunnel Effect

F. Ali Mehmeti, R. Haller-Dintelmann and V. Régnier

Abstract. We consider the Klein-Gordon equation on two half-axes connected at their origins. We add a potential that is constant but different on each branch. In a previous paper, we studied the L^∞-time decay via Hörmander's version of the stationary phase method. Here we apply these results to show that for initial conditions in an energy band above the threshold of the tunnel effect a fixed portion of the energy propagates between group lines. Further we consider the situation that the potential difference tends to infinity while the energy band of the initial condition is shifted upwards such that the particle stays above the threshold of the tunnel effect. We show that the total transmitted energy as well as the portion between the group lines tend to zero like $a_2^{-1/2}$ in the branch with the higher potential a_2 as a_2 tends to infinity. At the same time the cone formed by the group lines inclines to the t-axis while its aperture tends to zero.

Mathematics Subject Classification (2010). Primary 34B45; Secondary 47A70, 35B40.

Keywords. Networks, Klein-Gordon equation, stationary phase method, L^∞-time decay, energy flow.

1. Introduction

In this paper we study the energy flow of waves in two coupled one-dimensional semi-infinite media having different dispersion properties. Results in experimental physics [8, 9], theoretical physics [7] and functional analysis [4, 6] describe phenomena created in this situation by the dynamics of the tunnel effect: the delayed reflection and advanced transmission near nodes issuing two branches. Our purpose is to describe the influence of the height of a potential step on the energy flow of wave packets above the threshold of tunnel effect.

Parts of this work were done, while the second author visited the University of Valenciennes. He wishes to express his gratitude to F. Ali Mehmeti and the LAMAV for their hospitality.

We consider the following setting: let N_1, N_2 be disjoint copies of $(0, +\infty)$. Consider numbers a_k satisfying $0 \leq a_1 \leq a_2 < +\infty$. Find a vector (u_1, u_2) of functions $u_k : [0, +\infty) \times \overline{N_k} \to \mathbb{C}$ satisfying the Klein-Gordon equations

$$[\partial_t^2 - \partial_x^2 + a_k]u_k = 0 \text{ on } N_k, \ k = 1, 2,$$

on N_1, N_2 coupled at zero by usual Kirchhoff conditions and complemented with initial conditions for the functions u_k and their derivatives.

Reformulating this as an abstract Cauchy problem, one is confronted with the self-adjoint operator $A = (-\partial_x^2 + a_1, -\partial_x^2 + a_2)$ in $L^2(N_1) \times L^2(N_2)$, with a domain that incorporates the Kirchhoff transmission conditions at zero. For an exact definition of A, we refer to Section 2. The problem described above can be reformulated as

$$\begin{aligned} \ddot{u}(t) + Au(t) &= 0, \\ u(t) &\in D(A), \end{aligned} \tag{1}$$

for all $t > 0$ together with initial conditions. It is well known that the following expression is invariant with respect to time for solutions of (1):

$$E(u(t, \cdot)) = \frac{1}{2}\Big(\|\dot{u}(t, \cdot)\|_H^2 + (Au, u)_H\Big). \tag{2}$$

In Section 2 we recall the solution formula that was proved in [2] by an expansion in generalized eigenfunctions in the more general setting of a star shaped network with semi-infinite branches.

In Section 3 we recall our result on L^∞-time decay proved in [3]. There we obtained the exact L^∞-time decay rate $c \cdot t^{-1/2}$ in the group velocity cones together with an expression for the coefficient c for initial conditions in a compact energy band in (a_2, ∞). In the present work we refine an estimation from below in this context.

In Section 4 we use the preceding results to estimate the L^2-norm of the outgoing solution on N_2, which is a part of the total energy given in (2). We suppose that the initial condition belongs to an energy band above the threshold of the tunnel effect, so that the solution propagates in the branch with the higher potential N_2. We consider the L^2-norm of the solution at time t both on the whole branch N_2 and inside the cone delimited by the group lines corresponding to the bounds of the energy band. It turns out that the first norm has an upper asymptotic bound and the second one an upper and lower asymptotic bound which behave as $a_2^{-1/2}$, if the height of the potential step a_2 tends to infinity. This implies that the ratio between the energy on the whole branch and the energy between the group lines is time asymptotically confined in a finite interval above 1 which is independent of a_2.

These results might be interpreted in terms of quantum mechanics as follows: a relativistic, massive particle without spin in a one-dimensional world is submitted to a potential step at the origin. It is supposed to have enough kinetic energy to overcome the step with a fixed remaining energy. Classically, the particle should leave the potential step with a velocity which is independent of the height of the step. Our results show that in the quantum mechanical model, when the height of the potential step tends to infinity, the velocity of the outgoing component of the particle tends to zero while the particle is more and more localized. At the same time the total outgoing energy tends to zero while its ratio to the energy inside the cone delimited by the group lines remains time asymptotically in a constant finite interval above one. Similar estimates should be possible for the other parts of the total energy (2).

The last observation might lead to the idea of viewing the total energy as being subdivided in a part inside the group line cones and a part outside these cones. The ratio of these two energies is quite independent of the experimental configuration, but depends mainly on the chosen energy band.

The results of this paper are related to results in experimental physics, theoretical physics and functional analysis (spectral theory, asymptotic estimates, analysis on networks, cf. [3] and the references cited there). For example in [4] we obtained information on the splitting of the energy flow near zero. In [1] a (not optimal) estimation for the L^∞-time decay rate has been obtained but without any information on the localization of the energy. In [5] the relation of the eigenvalues of the Laplacian in an L^∞-setting on infinite, locally finite networks to the adjacency operator of the network is studied. In [11], the authors consider general networks with semi-infinite ends. They give a construction to compute some generalized eigenfunctions but no attempt is made to construct explicit inversion formulas.

2. A solution formula

The aim of this section is to recall the tools we used in [2] as well as the solution formula of the same paper for a special initial condition and to adapt this formula for the use of the stationary phase method in the next section.

Definition 2.1 (Functional analytic framework).

i) Let N_1, N_2 be identified with $(0, +\infty)$. Put $N := \overline{N_1} \times \overline{N_2}$, identifying the endpoints 0.

ii) Two transmission conditions are introduced:

$$(T_0)\text{: } (u_1, u_2) \in C(\overline{N_1}) \times C(\overline{N_2}) \text{ satisfies } u_1(0) = u_2(0).$$

This condition in particular implies that (u_1, u_2) may be viewed as a well-defined function on N.

$$(T_1)\text{: } (u_1, u_2) \in C^1(\overline{N_1}) \times C^1(\overline{N_2}) \text{ satisfies } \sum_{k=1}^{2} \partial_x u_k(0^+) = 0.$$

iii) Define the Hilbert space $H = L^2(N_1) \times L^2(N_2)$ with scalar product

$$(u, v)_H = \sum_{k=1}^{2} (u_k, v_k)_{L^2(N_k)}$$

and the operator $A : D(A) \longrightarrow H$ by

$$D(A) = \Big\{(u_1, u_2) \in H^2(N_1) \times H^2(N_2) : (u_1, u_2) \text{ satisfies } (T_0) \text{ and } (T_1)\Big\},$$

$$A(u_1, u_2) = (A_1 u_1, A_2 u_2) = \Big(-\partial_x^2 u_k + a_k u_k\Big)_{k=1,2}.$$

Note that, if $a_1 = a_2 = 0$, then A is the Laplacian in the sense of the existing literature, cf. [5, 11].

Definition 2.2 (Fourier-type transform V).

i) For $k = 1, 2$ and $\lambda \in \mathbb{C}$ let $\xi_k(\lambda) := \sqrt{\lambda - a_k}$ as well as

$$s_1(\lambda) := -\frac{\xi_2(\lambda)}{\xi_1(\lambda)}, \lambda \neq a_1 \quad \text{and} \quad s_2(\lambda) := -\frac{\xi_1(\lambda)}{\xi_2(\lambda)}, \lambda \neq a_2.$$

Here, and in all what follows, the complex square root is chosen in such a way that $\sqrt{r \cdot e^{i\phi}} = \sqrt{r} e^{i\phi/2}$ with $r > 0$ and $\phi \in [-\pi, \pi)$.

ii) For $\lambda \in \mathbb{C}$ and $j, k \in \{1, 2\}$, we define generalized eigenfunctions $F_\lambda^{\pm,j} : N \to \mathbb{C}$ of A by $F_\lambda^{\pm,j}(x) := F_{\lambda,k}^{\pm,j}(x)$ with

$$\begin{cases} F_{\lambda,k}^{\pm,j}(x) = \cos(\xi_j(\lambda)x) \pm i s_j(\lambda) \sin(\xi_j(\lambda)x), & \text{for } k = j, \\ F_{\lambda,k}^{\pm,j}(x) = \exp(\pm i \xi_k(\lambda)x), & \text{for } k \neq j. \end{cases}$$

for $x \in \overline{N_k}$.

iii) For $l = 1, 2$ and $\lambda \in \mathbb{R}$ let

$$q_l(\lambda) := \begin{cases} 0, & \text{if } \lambda < a_l, \\ \frac{\xi_l(\lambda)}{|\xi_1(\lambda) + \xi_2(\lambda)|^2}, & \text{if } a_l < \lambda. \end{cases}$$

iv) Considering q_1 and q_2 as weights for our L^2-spaces, we set $L_q^2 := L^2((a_1, +\infty), q_1) \times L^2((a_2, +\infty), q_2)$. The corresponding scalar product is

$$(F, G)_q := \sum_{k=1}^{2} \int_{(a_k, +\infty)} q_k(\lambda) F_k(\lambda) \overline{G_k(\lambda)}\, d\lambda$$

and its associated norm $|F|_q := (F, F)_q^{1/2}$.

v) For all $f \in L^1(N, \mathbb{C})$ we define $Vf : [a_1, +\infty) \times [a_2, +\infty) \to \mathbb{C}$ by

$$(Vf)_k(\lambda) := \int_N f(x) \overline{(F_\lambda^{-,k})}(x)\, dx, \ k = 1, 2.$$

In [2], we show that V diagonalizes A and we determine a metric setting in which it is an isometry. Let us recall these useful properties of V as well as the fact that the property $u \in D(A^j)$ can be characterized in terms of the decay rate of the components of Vu.

Theorem 2.3. *Endow $C_c^\infty(N_1) \times C_c^\infty(N_2)$ with the norm of $H = L^2(N_1) \times L^2(N_2)$. Then*

i) $V : C_c^\infty(N_1) \times C_c^\infty(N_2) \to L_q^2$ *is isometric and can be extended to an isometry* $\tilde{V} : H \to L_q^2$*, which we shall again denote by* V *in the following.*

ii) *The operator* A *is selfadjoint and its spectrum is* $\sigma(A) = [a_1, +\infty)$.

iii) $V : H \to L_q^2$ *is a spectral representation of* H *with respect to* A*. In particular,* V *is surjective.*

iv) *For* $l \in \mathbb{N}$ *the following statements are equivalent:*
   (a) $u \in D(A^l)$,
   (b) $\lambda \mapsto \lambda^l (Vu)_k(\lambda) \in L^2((a_k, +\infty), q_k),\ k = 1, 2.$

We are now interested in the Abstract Cauchy Problem

$$(\text{ACP}) : u_{tt}(t) + Au(t) = 0,\ t > 0,\ \text{with } u(0) = u_0,\ u_t(0) = 0.$$

Here, the zero initial condition for the velocity is just chosen for simplicity, as we will not deal with the general case in this contribution.

By the surjectivity of V (cf. Theorem 2.3 (ii)) there exists an initial condition $u_0 \in H$ satisfying

Condition (A): $(Vu_0)_2 \equiv 0$ and $(Vu_0)_1 \in C_c^2((a_2, \infty))$.

Remark 2.4.

i) For u_0 satisfying (A) there exist $a_2 < \lambda_{\min} < \lambda_{\max} < \infty$ such that

$$\operatorname{supp}(Vu_0)_1 \subset [\lambda_{\min}, \lambda_{\max}].$$

ii) If $u_0 \in H$ satisfies (A), then $u_0 \in D(A^\infty) = \bigcap_{l \geq 1} D(A^l)$, due to Theorem 2.3 (iv), since $\lambda \mapsto \lambda^l (Vu)_m(\lambda) \in L^2((a_m, +\infty), q_m),\ m = 1, 2$ for all $l \in \mathbb{N}$ by the compactness of $\operatorname{supp}(Vu_0)_m$.

Theorem 2.5 (Solution formula of (ACP) in a special case). *Suppose that* u_0 *satisfies Condition* (A)*. Then there exists a unique solution* u *of* (ACP) *with* $u \in C^l([0, +\infty), D(A^{m/2}))$ *for all* $l, m \in \mathbb{N}$*. For* $x \in N_2$ *we have the representation*

$$u_2(t, x) = \frac{1}{2}\big(u_+(t, x) + u_-(t, x)\big)$$

with

$$u_\pm(t, x) := \int_{\lambda_{\min}}^{\lambda_{\max}} e^{\pm i\sqrt{\lambda}t} q_1(\lambda) e^{-i\xi_2(\lambda)x} (Vu_0)_1(\lambda) d\lambda. \tag{3}$$

Proof. Since $v_0 = u_t(0) = 0$, we have for the solution of (ACP) the representation

$$u(t) = V^{-1} \cos(\sqrt{\lambda}t) V u_0.$$

(cf. for example [1, Theorem 5.1]). The expression for V^{-1} given in [2] yields the formula for $u_\pm$. □

Remark 2.6. A solution formula for arbitrary initial conditions which is valid on all branches is available in [2]. This general expression is not needed in the following.

3. L^∞-time decay

Next, we quote the result on L^∞-time decay of solutions to the problem (ACP) from [3]. Here, we only consider special initial conditions that are localized in energy in a compact interval contained in (a_2, a_2+1). For these it is possible to give very explicit estimates for all the constants that appear in an asymptotic expansion of the solution to the order $t^{1/2}$.

Theorem 3.1. *Let $0 < \alpha < \beta < 1$ and $\psi \in C_c^2((\alpha,\beta))$ with $\|\psi\|_\infty = 1$ be given. Setting $\tilde{\psi}(\lambda) := \psi(\lambda - a_2)$, we choose the initial condition $u_0 \in H$ satisfying $(Vu_0)_2 \equiv 0$ and $(Vu_0)_1 = \tilde{\psi}$. Then u_0 satisfies condition (A). Furthermore, let u_+ be defined as in Theorem 2.5.*

Then there is a constant $C(\psi,\alpha,\beta)$ independent of a_1 and a_2, such that for all $t \in \mathbb{R}^+$ and all $x \in N_2$ with

$$\sqrt{\frac{a_2+\beta}{\beta}} \le \frac{t}{x} \le \sqrt{\frac{a_2+\alpha}{\alpha}} \tag{4}$$

we have

$$\left|u_+(t,x) - H(t,x,u_0)\cdot t^{-1/2}\right| \le C(\psi,\alpha,\beta)\cdot t^{-1},$$

where

$$H(t,x,u_0) := e^{-i\varphi(p_0,t,x)}(2i\pi)^{1/2} a_2^{3/4}\ h_1(t,x)\ h_2(t,x)\ (Vu_0)_1(a_2+p_0^2)$$

with

$$\varphi(p,t,x) := \sqrt{a_2+p^2}\,t - px,\ p_0 := \sqrt{\frac{a_2x^2}{(t^2-x^2)}},$$

$$h_1(t,x) := \left(\frac{(t/x)^2}{(t/x)^2-1}\right)^{3/4},\ h_2(t,x) := \frac{\sqrt{(a_2-a_1)\left((t/x)^2-1\right)+a_1}}{\left(\sqrt{(a_2-a_1)\left((t/x)^2-1\right)+a_2}+\sqrt{a_2}\right)^2}.$$

It holds further

$$\left|H(t,x,u_0)\right| \le g(a_1,a_2,\beta) := \sqrt{2\pi}\frac{\sqrt{\beta}(a_2+\beta)^{3/4}}{\sqrt{a_2}\sqrt{a_2-a_1+\beta}} \sim \sqrt{2\pi\beta}\,a_2^{-1/4} \text{ as } a_2 \to +\infty.$$

Proof. This is contained in Equation (8) from the proof of Theorem 3.2 and in Theorem 4.1 in [3]. □

Remark 3.2.

i) Note that (4) is equivalent to

$$v_{\min} \le v(t,x) := (t/x)^2 - 1 \le v_{\max}\,.$$

where $v_{\min} := \dfrac{a_2}{\lambda_{\max}-a_2} = \dfrac{a_2}{\beta}$ and $v_{\max} := \dfrac{a_2}{\lambda_{\min}-a_2} = \dfrac{a_2}{\alpha}$.

ii) For later use we define $p_{\min} := \xi_2(\lambda_{\min})$ and $p_{\max} := \xi_2(\lambda_{\max})$.

iii) The proof of Theorem 3.1 has been given in [3]. It applies the stationary phase method in a version by L. Hörmander [10], in which partial integration is used to evaluate the influence of large parameters (here t) in the phase function of oscillating integrands. This requires that the amplitude is sufficiently regular. In fact the constant $C(\psi, \alpha, \beta)$ depends on the derivatives up to the order 2 of ψ, which explains the hypothesis of twice differentiability in condition (A). For Theorem 2.5 the twice differentiability is not necessary.

iv) The phase function in the integrand of $u_-(t, \cdot)$ has no stationary points. Therefore we conjecture that the L^∞-norm of $u_-(t, \cdot)$ decays at least as t^{-1}, much quicker than the L^∞-norm of $u_+(t, \cdot)$, which decays as $t^{-1/2}$. This seems to be reasonable, because this term corresponds to the incoming part of the solution, whose energy should asymptotically disappear in the boundary point 0 of N_2.

4. Energy flow

In this section we use the asymptotic expansion from Section 3 to estimate the outgoing solution from above and below in the cones given by the group velocities corresponding to the bounds of the energy band of the initial condition. This leads to time independent asymptotic upper (Theorem 4.5) and lower (Theorem 4.1) estimates of the L^2-norm of the solution on the space interval inside the cones.

Further an upper estimate of the L^2-norm of the outgoing solution on the whole branch N_2 is obtained using Plancherel's theorem (Theorem 4.2).

This information leads to our main result (Theorem 4.4), where we give an upper bound for the ratio of the L^2-norm on the whole branch N_2 and the L^2-norm inside the cone, which is asymptotically independent of the height of the potential step: if the higher potential a_2 tends to infinity, our upper bound tends to a constant independent of a_2. This means that a significant part of the energy stays inside cones in space-time. The last observation might lead to the idea of viewing in general the total energy as being subdivided in a (deterministic) part inside the group line cones and a (non deterministic) part outside these cones. In our case, the ratio of these two energies is quite independent of the experimental configuration, but depends mainly on the chosen energy band.

Theorem 4.1. *In the setting of Theorem* 3.1 *suppose that* $\psi(\mu) \geq m > 0$ *for* $\mu \in [\alpha', \beta']$ *with* $\alpha < \alpha' < \beta' < \beta$. *Then we have*

i) *the lower estimate for the coefficient of* $t^{-1/2}$:

$$\begin{aligned} \left|H(t, x, u_0)\right| &\geq f(a_2, \alpha, \beta) \cdot m \\ &:= \sqrt{2\pi}\, a_2^{3/4} \left(\frac{\beta}{a_2} + 1\right)^{3/4} \frac{1}{\sqrt{a_2}} \sqrt{\frac{a_2 - a_1}{\alpha} + 1} \left(\sqrt{\frac{a_2 - a_1}{\alpha} + 1} + 1\right)^{-2} m. \\ &\sim \sqrt{2\pi\alpha}\, a_2^{-1/4} m, \quad as\ a_2 \to \infty \end{aligned}$$

for all (t,x) *satisfying*

$$\sqrt{\frac{a_2+\beta'}{\beta'}} \leq \frac{t}{x} \leq \sqrt{\frac{a_2+\alpha'}{\alpha'}}\ .$$

ii) *and the lower estimate for the solution*

$$\forall \varepsilon > 0\ \exists t_0 > 0\ \forall t > t_0 \quad \left|u_+(t,x)\right| \geq \left(f(a_2,\alpha,\beta)m - \varepsilon\right)\, t^{-1/2}$$

for all (t,x) *satisfying*

$$\sqrt{\frac{a_2+\beta'}{\beta'}} \leq \frac{t}{x} \leq \sqrt{\frac{a_2+\alpha'}{\alpha'}}\ .$$

Proof. Note that it is always possible to choose the initial condition in the indicated way, thanks to the surjectivity of V, cf. Theorem 2.3 ii).

(i): Theorem 3.1 implies

$$\left|H(t,x,u_0)\right| = \sqrt{2\pi}a_2^{3/4}h_1(t,x)\left|h_2(t,x)\right| \cdot \left|(Vu_0)_1(a_2+p_0^2)\right|.$$

We estimate

$$\begin{aligned}\left|h_2(t,x)\right| &= \left|\frac{\sqrt{(a_2-a_1)((t/x)^2-1)+a_2}}{\left(\sqrt{(a_2-a_1)((t/x)^2-1)+a_2}+\sqrt{a_2}\right)^2}\right| \\ &\geq \frac{\sqrt{(a_2-a_1)v_{\max}+a_2}}{\left(\sqrt{(a_2-a_1)v_{\max}+a_2}+\sqrt{a_2}\right)^2} \\ &\geq \frac{\sqrt{(a_2-a_1)\frac{a_2}{\alpha}+a_2}}{\left(\sqrt{(a_2-a_1)\frac{a_2}{\alpha}+a_2}+\sqrt{a_2}\right)^2} \\ &= \frac{1}{\sqrt{a_2}}\frac{\sqrt{\frac{a_2-a_1}{\alpha}+1}}{\left(\sqrt{\frac{a_2-a_1}{\alpha}+1}+1\right)^2} \\ &\sim \frac{\sqrt{\alpha}}{a_2} \quad \text{as } a_2 \to \infty\end{aligned}$$

Here we used, that $b \mapsto \frac{b}{(b+c)^2}$ is decreasing for $b > c \geq 0$. Further

$$\left|h_1(t,x)\right| \geq \left(\frac{\frac{\lambda_{\max}}{\lambda_{\max}-a_2}}{\frac{\lambda_{\max}}{\lambda_{\max}-a_2}-1}\right)^{3/4} = \left(\frac{\beta}{a_2}+1\right)^{3/4} \to 1 \text{ as } a_2 \to \infty$$

This implies (i).

(ii): Using the lower triangular inequality we find that $\forall \varepsilon > 0\ \exists t_0 > 0\ \forall t > t_0$:

$$\begin{aligned}\left|u_+(t,x)\right| &\geq \left|u_+(t,x) - H(t,x,u_0)\cdot t^{-1/2} + H(t,x,u_0)\cdot t^{-1/2}\right| \\ &\geq \left|\left|H(t,x,u_0)\right|\cdot t^{-1/2} - \left|u_+(t,x) - H(t,x,u_0)\cdot t^{-1/2}\right|\right| \\ &\geq \left(f(a_2,\alpha,\beta)m - \varepsilon\right)\, t^{-1/2}\end{aligned}$$

for all (t, x) satisfying

$$\sqrt{\frac{a_2 + \beta'}{\beta'}} \leq \frac{t}{x} \leq \sqrt{\frac{a_2 + \alpha'}{\alpha'}} .$$

This shows (ii). □

Theorem 4.2. *Suppose the setting of Theorem* 3.1. *Then we have*

$$\begin{aligned} \|u_+(t, \cdot)\|_{L^2(N_2)} &\leq \frac{\sqrt{\beta}}{\sqrt{a_2 - a_1 + \alpha}\sqrt[4]{\alpha}} \\ &\sim \frac{\sqrt{\beta}}{\sqrt[4]{\alpha}} \|\psi\|_{L^2((\alpha,\beta))} a_2^{-1/2} \quad \textit{as } a_2 \to \infty. \end{aligned}$$

Proof. In the expression (3) for u_+ we substitute $p := \xi_2(\lambda)$. This yields

$$u_+(t, x) = 2 \int_{p_{\min}}^{p_{\max}} e^{i\sqrt{a_2+p^2}t} q_1(a_2 + p^2) e^{-ipx} (V u_0)_1(a_2 + p^2) p \, dp.$$

Interpreting this integral as a Fourier transform we find by the Plancherel theorem

$$\begin{aligned} \|u_+(t, \cdot)\|_{L^2(N_2)} &\leq \|u_+(t, \cdot)\|_{L^2(\mathbb{R})} \\ &= \left\|p \mapsto p q_1(a_2 + p^2)(V u_0)_1(a_2 + p^2)\chi_{[p_{\min}, p_{\max}]}(p)\right\|_{L^2(\mathbb{R})}. \end{aligned}$$

Now, we use that $\operatorname{supp}((V u_0)_1)$ is contained in the interval $[a_2 + \alpha, a_2 + \beta]$, so only the range $\sqrt{\alpha} \leq p \leq \sqrt{\beta}$ is relevant. For these values of p we find

$$p q_1(a_2 + p^2) = p \frac{\sqrt{a_2 - a_1 + p^2}}{(\sqrt{a_2 - a_1 + p^2} + p)^2} \leq \frac{p}{\sqrt{a_2 - a_1 + p^2}} \leq \frac{\sqrt{\beta}}{\sqrt{a_2 - a_1 + \alpha}}.$$

So, we have

$$\|u_+(t, \cdot)\|_{L^2(N_2)} \leq \frac{\sqrt{\beta}}{\sqrt{a_2 - a_1 + \alpha}} \left\|(V u_0)_1(a_2 + p^2)\right\|_{L^2((p_{\min}, p_{\max}))}$$

and substituting back $\lambda = a_2 + p^2$ we obtain

$$\begin{aligned} \left\|(V u_0)_1(a_2 + p^2)\right\|^2_{L^2((p_{\min}, p_{\max}))} &= \int_{\lambda_{\min}}^{\lambda_{\max}} \left|(V u_0)_1(\lambda)\right|^2 \frac{d\lambda}{\sqrt{\lambda - a_2}} \\ &= \int_{a_2+\alpha}^{a_2+\beta} |\tilde{\psi}(\lambda)|^2 \frac{d\lambda}{\sqrt{\lambda - a_2}}. \end{aligned}$$

Setting, finally, $\mu = \lambda - a_2$ we end up with

$$\begin{aligned} \|u_+(t, \cdot)\|_{L^2(N_2)} &\leq \frac{\sqrt{\beta}}{\sqrt{a_2 - a_1 + \alpha}} \left(\int_\alpha^\beta |\psi(\mu)|^2 \frac{d\mu}{\sqrt{\mu}}\right)^{1/2} \\ &\leq \frac{\sqrt{\beta}}{\sqrt{a_2 - a_1 + \alpha}\sqrt[4]{\alpha}} \|\psi\|_{L^2((\alpha,\beta))} . \end{aligned}$$

□

Remark 4.3. In the proof of the last theorem we extend $u_+(t,\cdot)$ in a natural way to a function on $\mathbb{R}$, reinterpret the integral as Fourier transform and use the Plancherel theorem. We could also obtain an upper estimate of the energy by using the fact that V is a surjective isometry instead of the Plancherel theorem. But V automatically takes into account all branches which makes that this estimate would not furnish the optimal asymptotic behavior $a_2^{-1/4}$, as a_2 tends to infinity.

Theorem 4.4. *In the setting of Theorem* 3.1 *suppose that* $\psi(\mu) \geq m > 0$ *for* $\mu \in [\alpha', \beta']$ *with* $\alpha < \alpha' < \beta' < \beta$. *Then we have*

i) *the estimate from below:* $\forall \varepsilon > 0\ \exists t_0 > 0\ \forall t > t_0$

$$\left|\left|u_+(t,\cdot)\right|\right|^2_{L^2(I_t')} \geq \left(f(a_2,\alpha,\beta)m - \varepsilon\right)^2 \left(\sqrt{\frac{\beta'}{a_2+\beta'}} - \sqrt{\frac{\alpha'}{a_2+\alpha'}}\right)$$

where

$$I_t' := \left[t\sqrt{\frac{\alpha'}{a_2+\alpha'}}\,,\ t\sqrt{\frac{\beta'}{a_2+\beta'}}\right]$$

and f *is defined in Theorem* 3.1 (i).

ii) *the estimate for the ratio of the global and local energy:* $\forall \varepsilon > 0\ \exists t_0 > 0\ \forall t > t_0$

$$\frac{\left|\left|u_+(t,\cdot)\right|\right|_{L^2(N_2)}}{\left|\left|u_+(t,\cdot)\right|\right|_{L^2(I_t')}} \leq \frac{\sqrt{\beta}\ \|\psi\|_{L^2((\alpha,\beta))}}{\sqrt{a_1-a_2+\alpha}\sqrt[4]{\alpha}}\left(f(a_2,\alpha,\beta)m - \varepsilon\right)^{-1}\left(\sqrt{\frac{\beta'}{a_2+\beta'}} - \sqrt{\frac{\alpha'}{a_2+\alpha'}}\right)^{-1/2}$$

iii) *and*

$$\limsup_{t\to\infty}\frac{\left|\left|u_+(t,\cdot)\right|\right|_{L^2(N_2)}}{\left|\left|u_+(t,\cdot)\right|\right|_{L^2(I_t')}} \leq \frac{\sqrt{\beta}\ \|\psi\|_{L^2((\alpha,\beta))}}{\sqrt{a_1-a_2+\alpha}\sqrt[4]{\alpha}}\left(f(a_2,\alpha,\beta)m\right)^{-1}\left(\sqrt{\frac{\beta'}{a_2+\beta'}} - \sqrt{\frac{\alpha'}{a_2+\alpha'}}\right)^{-1/2}$$
$$\sim (2\pi)^{-1/2}m^{-1}\beta^{1/2}\alpha^{-3/4}\left(\sqrt{\beta'}-\sqrt{\alpha'}\right)^{-1/2}\|\psi\|_{L^2((\alpha,\beta))} \text{ as } a_2 \to \infty.$$

Proof. (i): Follows from

$$\left|\left|u_+(t,\cdot)\right|\right|^2_{L^2(I_t')} \geq \left(\inf_{x\in I_t'}\left|u_+(t,\cdot)\right|\right)^2 \cdot |I_t'|$$

and Theorem 4.1 (ii).

(ii): Follows from (i) and Theorem 4.2.

(iii): Direct consequence of (ii) and Theorem 4.1 (i). □

Theorem 4.5. *Consider the setting of Theorem* 3.1.

i) $\forall \varepsilon > 0 \; \exists t_1 > 0 \; \forall t > t_1$:

$$\left|u_+(t,x)\right| \leq \left(g(a_1,a_2,\beta)+\varepsilon\right) t^{-1/2}$$

for all (t,x) *satisfying*

$$\sqrt{\frac{a_2+\beta}{\beta}} \leq \frac{t}{x} \leq \sqrt{\frac{a_2+\alpha}{\alpha}} .$$

ii) $\forall \varepsilon > 0 \; \exists t_1 > 0 \; \forall t > t_1$:

$$\left|\left|u_+(t,\cdot)\right|\right|^2_{L^2(I_t)} \leq \left(g(a_1,a_2,\beta)+\varepsilon\right)^2 \left(\sqrt{\frac{\beta}{a_2+\beta}} - \sqrt{\frac{\alpha}{a_2+\alpha}}\right).$$

iii)

$$\liminf_{t\to\infty} \left|\left|u_+(t,\cdot)\right|\right|^2_{L^2(I_t)} \leq g(a_1,a_2,\beta)^2 \left(\sqrt{\frac{\beta}{a_2+\beta}} - \sqrt{\frac{\alpha}{a_2+\alpha}}\right)$$
$$\sim 2\pi\beta(\sqrt{\beta}-\sqrt{\alpha})\; a_2^{-1} \;\; \text{as } a_2 \to \infty.$$

Proof. (i): Theorem 3.1 implies that $\forall \varepsilon > 0 \; \exists t_1 > 0 \; \forall t > t_1$:

$$\begin{aligned}\left|u_+(t,x)\right| &\leq \left|u_+(t,x) - H(t,x,u_0)t^{-1/2} + H(t,x,u_0)t^{-1/2}\right| \\ &\leq C(\psi,\alpha,\beta)t^{-1} + \left|H(t,x,u_0)\right|t^{-1/2} \\ &\leq \left(g(a_1,a_2,\beta)+\varepsilon\right) t^{-1/2}\end{aligned}$$

for (t,x) in the cone indicated there.

(ii): Follows from estimating the square of the L^2-norm against the square of the maximum of the function (using (i)) times the length of the integration interval.

(iii): Direct consequence of (ii). □

References

[1] F. Ali Mehmeti, *Spectral Theory and L^∞-time Decay Estimates for Klein-Gordon Equations on Two Half Axes with Transmission: the Tunnel Effect.* Math. Methods Appl. Sci. **17** (1994), 697–752.

[2] F. Ali Mehmeti, R. Haller-Dintelmann, V. Régnier, *Multiple tunnel effect for dispersive waves on a star-shaped network: an explicit formula for the spectral representation.* J. Evol. Equ. **12** (2012), 513–545.

[3] F. Ali Mehmeti, R. Haller-Dintelmann, V. Régnier, *The influence of the tunnel effect on L^∞-time decay.* W. Arendt, J.A. Ball, J. Behrndt, K.-H. Förster, V. Mehrmann, C. Trunk (eds): Spectral Theory, Mathematical System Theory, Evolution Equations, Differential and Difference Equations: IWOTA10; Springer, Basel; Operator Theory: Advances and Applications, **221** (2012), 11–24.

[4] F. Ali Mehmeti, V. Régnier, *Delayed reflection of the energy flow at a potential step for dispersive wave packets.* Math. Methods Appl. Sci. **27** (2004), 1145–1195.

[5] J. von Below, J.A. Lubary, *The eigenvalues of the Laplacian on locally finite networks.* Results Math. **47** (2005), no. 3-4, 199–225.

[6] Y. Daikh, *Temps de passage de paquets d'ondes de basses fréquences ou limités en bandes de fréquences par une barrière de potentiel.* Thèse de doctorat, Valenciennes, France, 2004.

[7] J.M. Deutch, F.E. Low, *Barrier Penetration and Superluminal Velocity.* Annals of Physics **228** (1993), 184–202.

[8] A. Enders, G. Nimtz, *On superluminal barrier traversal.* J. Phys. I France **2** (1992), 1693–1698.

[9] A. Haibel, G. Nimtz, *Universal relationship of time and frequency in photonic tunnelling.* Ann. Physik (Leipzig) **10** (2001), 707–712.

[10] L. Hörmander, *The Analysis of Linear Partial Differential Operators I.* Springer, 1984.

[11] V. Kostrykin, R. Schrader, *The inverse scattering problem for metric graphs and the travelling salesman problem.* Preprint, 2006 (www.arXiv.org:math.AP/0603010).

F. Ali Mehmeti and V. Régnier
Univ. Lille Nord de France
F-59000 Lille, France

and

UVHC, LAMAV, FR CNRS 2956
F-59313 Valenciennes, France
e-mail: `felix.ali-mehmeti@univ-valenciennes.fr`
`Virginie.Regnier@univ-valenciennes.fr`

R. Haller-Dintelmann
TU Darmstadt
Fachbereich Mathematik
Schloßgartenstraße 7
D-64289 Darmstadt, Germany
e-mail: `haller@mathematik.tu-darmstadt.de`

Operator Theory:
Advances and Applications, Vol. 229, 77–89

Some New Hardy-type Integral Inequalities on Cones of Monotone Functions

L.S. Arendarenko, R. Oinarov and L.-E. Persson

Dedicated to Professor Stefan Samko on the occasion of his 70th anniversary

Abstract. Some new Hardy-type inequalities with Hardy-Volterra integral operators on the cones of monotone functions are obtained. The case $1 < p \leq q < \infty$ is considered and the involved kernels satisfy conditions which are less restrictive than the classical Oinarov condition.

Mathematics Subject Classification (2010). 47G10, 47B38.

Keywords. Hardy type inequalities, boundedness, integral operators, Volterra integral operator, kernels, weighted Lebesgue spaces, maximal function, non-increasing function, non-decreasing function.

1. Introduction

Let $1 < p, q < \infty$, $\frac{1}{p} + \frac{1}{p'} = 1$, $R_+ = (0, \infty)$, v and w be nonnegative functions, such that the functions v, w, $v^{1-p'}$ and $w^{1-q'}$ are locally integrable on R_+. For a weight function v and $1 \leq p < \infty$ we define the weighted Lebesgue space $L_{p,v}(R_+)$ as the set of all measurable functions on R_+ such that

$$\|f\|_{p,v} = \left(\int_0^\infty |f(x)|^p v(x) dx \right)^{\frac{1}{p}} < \infty.$$

We consider the integral operator defined as follows:

$$Kf(x) = \int_0^x K(x,s) f(s) ds, \ x > 0 \tag{1.1}$$

where $K(x,s)$ is a nonnegative measurable kernel, and its conjugate operator

$$K^* g(s) = \int_s^\infty K(x,s) g(x) dx, \ s > 0 \tag{1.2}$$

with the same kernel.

One of the important problems in functional analysis is to find necessary and sufficient conditions for validity of the inequality

$$\left(\int_0^\infty |Kf(x)|^q\, w(x)dx\right)^{\frac{1}{q}} \leq C\left(\int_0^\infty |f(x)|^p v(x)dx\right)^{\frac{1}{p}} \tag{1.3}$$

for the operators K and K^* to hold for all functions $f \in L_{p,v}$, where the constant C does not depend on functions f, and p,q are fixed parameters. This type of problems is studied in what today is called Hardy-type inequalities (see, e.g., the books [5], [6], [7] and [10]).

In this type of research the necessity to study inequalities defined by (1.3) restricted to the cones of monotone functions has appeared. This problem has some applications in the investigation of boundedness of operators in Lorentz spaces and in the theory of embedding in Lorentz spaces. Another reason to study such inequalities is that the maximal function, which is defined by the formula $Mf(x) = \sup\limits_{x\in Q} \frac{1}{|Q|}\int_Q |f(s)|ds$, is related to the Hardy operator restricted to the cone of decreasing functions in the following precise way:

$$(Mf)^* \approx \frac{1}{x}\int_0^x f^*(s)ds, \tag{1.4}$$

where f^* is decreasing rearrangement of $|f|$. For a more extensive motivation and description of Hardy-type inequalities on the cone of monotone functions see the book [6] by A. Kufner, L. Maligranda and L.-E. Persson and the references given there, e.g., [1], [4], [8], [11]–[17]. Moreover, a new approach for this problem was considered in [3]. Concerning historical facts and further developments of the formula (1.4) (also in weighted cases) we refer to the article [2] by I. Asekritova, N. Krugljak, L. Maligranda and L.-E. Persson.

In this paper we derive necessary and sufficient conditions for boundedness of the operators K and K^* defined by (1.1) and (1.2), respectively, with kernels satisfying generalized Oinarov condition (see [9]), on the cone of monotone functions. These results are new in this generality (see the books [5], [6], [7] and the references given there).

We use the following notations and conventions: Products of type $0 \cdot \infty$ are considered to be equal to zero. We will write $A \ll B$, if $A \leq cB$ with some positive constant c, which depends only on some unessential parameters. The notation $A \approx B$ means that $A \ll B \ll A$. The symbol $\chi_E(\cdot)$ stands for the characteristic function of a set $E \subset R_+$. We also use notations $p' = \frac{p}{p-1}$, $q' = \frac{q}{q-1}$ and $V(x) = \int_0^x v(t)dt$.

The paper is organized as follows: In Section 2 we present some preliminaries for our further discussions. The main results can be found in Section 3 and the proofs of these results are given in Section 4. Finally, in Section 5 we present the corresponding results for the case of non-decreasing functions.

2. Preliminaries

At first we shall define the classes $\mathcal{O}_n^+$ and $\mathcal{O}_n^-$, $n \geq 0$, of kernels equipped with the operators K and K^* defined by (1.1) and (1.2), which were introduced in [9]. We shall write $K(\cdot,\cdot) \equiv K_n^+(\cdot,\cdot)$, if $K(\cdot,\cdot) \in \mathcal{O}_n^+$ and $K(\cdot,\cdot) \equiv K_n^-(\cdot,\cdot)$, if $K(\cdot,\cdot) \in \mathcal{O}_n^-$.

For $n \geq 0$ we define the class $\mathcal{O}_n^+$ in the following way: Let the function $K^+(x,s)$ be a non-negative and measurable function, which is defined for all $x \geq s > 0$. Moreover, the function $K^+(x,s)$ is non-decreasing in the first argument. We define the class $\mathcal{O}_0^+$ as a set of all functions of type $K_0^+(x,s) \equiv V(s)$, where the function $V(s)$ is non-negative and measurable function on the set $\{(x,s) : x \geq s \geq 0\}$.

Next we suppose that the classes $\mathcal{O}_i^+$, $i = 0,1,\ldots,n-1$, are defined and we introduce the class $\mathcal{O}_n^+$ as a set of all functions $K_n^+(x,s)$, for which there exist functions $K_i^+(x,s) \in \mathcal{O}_i^+$, $i = 0,1,\ldots,n-1$, and a number h_n such that

$$K_n^+(x,s) \leq h_n \left(\sum_{i=0}^{n-1} K_{n,i}^+(x,t) K_i^+(t,s) + K_n^+(t,s) \right)$$

for $x \geq t \geq s > 0$, where

$$K_{n,i}^+(x,t) = \inf_{0<s<t} \frac{K_n^+(x,s)}{K_i^+(t,s)}, \quad i = 0,1,\ldots,n-1.$$

It follows from the definition of the function $K_{n,i}^+$ (see, e.g., [9]) for all $n \geq 0$ that we can characterize class $\mathcal{O}_n^+$ by the formula

$$K_n^+(x,s) \approx \sum_{i=0}^{n} K_{n,i}^+(x,t) K_i^+(t,s), \quad x \geq t \geq s > 0. \tag{2.1}$$

We put $K_{n,n}^+(x,t) \equiv 1$ here.

We similarly introduce the classes $\mathcal{O}_n^-$ for $n \geq 0$. Let the function $K^-(x,s)$ be a non-negative measurable function, which is defined for all $x \geq s \geq 0$ and non-increasing in a second argument. Let the class $\mathcal{O}_0^-$ be the set of all functions $K_0^-(x,s) \equiv U(x)$, where the function $U(x)$ is a non-negative and measurable function on the set $\{(x,s) : x \geq s \geq 0\}$. Let $n \geq 1$. Suppose that the classes $\mathcal{O}_i^-$ are defined for $i = 0,1,\ldots,n-1$. We say that the function $K_n^-(x,s)$ belongs to class $\mathcal{O}_n^-$ if there exist functions $K_i^-(x,s) \in \mathcal{O}_i^-$, $i = 0,1,\ldots,n-1$, and number h_n such that the inequality

$$K_n^-(x,s) \leq h_n \left(K_n^-(x,t) + \sum_{i=0}^{n-1} K_i^-(x,t) K_{i,n}^-(t,s) \right)$$

holds for $x \geq t \geq s > 0$, where

$$K_{i,n}^-(t,s) = \inf_{t<x<\infty} \frac{K_n^-(x,s)}{K_i^-(t,s)} \quad i = 0,1,\ldots,n-1.$$

Similarly to (2.1), each class $\mathcal{O}_n^-$, $n \geq 0$ can be characterized by the equivalence

$$K_n^-(x,s) \approx \sum_{i=0}^{n} K_i^-(x,t)K_{i,n}^-(t,s), \quad x \geq t \geq s > 0, \tag{2.2}$$

where $K_{n,n}^-(t,s) \equiv 1$.

Remark 2.1. It is shown in [9] that $K_{n,i}^+(x,s)$ and $K_{i,n}^-(x,s)$ can be arbitrary non-negative functions, measurable on the set $\{(x,s) : 0 < s \leq x < \infty\}$ and satisfying the conditions (2.1) and (2.2), respectively.

We shall use the following property of the class $\mathcal{O}_n^+$:

Lemma 2.2. *Let $n \geq 0$ and $K(t,s) \in \mathcal{O}_n^+$. Then the function $\widetilde{K}(x,s) = \int_s^x K(t,s)dt$ belongs to the class $\mathcal{O}_{n+1}^+$.*

Proof. Let $y \in [s,x]$. Since $K(t,s) \in \mathcal{O}_n^+$, we can apply (2.1) for y, $s \leq y \leq t$, and obtain that

$$\begin{aligned}
\widetilde{K}(x,s) &= \int_s^x K(t,s)dt \\
&= \int_s^y K(t,s)dt + \int_y^x K(t,s)dt \\
&\approx \int_s^y K(t,s)dt + \sum_{i=0}^{n} K_i^+(y,s) \int_y^x K_{n,i}^+(t,y)dt \\
&= \widetilde{K}(y,s) + \sum_{i=0}^{n} K_i^+(y,s) \int_y^x K_{n,i}^+(t,y)dt \\
&= \widetilde{K}(y,s) + \sum_{i=0}^{n} K_i^+(y,s)\widetilde{K}_{n,i}^+(x,y),
\end{aligned}$$

where $\widetilde{K}_{n,i}^+(x,y) = \int_y^x K_{n,i}^+(t,y)$. Hence, the function $\widetilde{K}(x,s)$ belongs to $\mathcal{O}_{n+1}^+$ by definition. □

To prove the main results of the paper we use the Saywer duality principle (see (2.4)) and the criterion of boundedness of the operator $\mathbf{K}$ defined by (1.1), with kernel from the class $\mathcal{O}_n^+ \bigcup \mathcal{O}_n^-$, $n \geq 0$. The criterion has been obtained in [9] and we rewrite it in a more convenient form for us in the following theorem.

Theorem A. *Let $1 < p \leq q < \infty$ and the kernel $K(\cdot,\cdot)$ of the operator $\mathbf{K}$ defined by (1.1) belong to the class $\mathcal{O}_n^+ \bigcup \mathcal{O}_n^-$, $n \geq 0$. Then the inequality*

$$\|\mathbf{K}f\|_{q,w} \leq C\|f\|_{p,v}, \quad \forall f \in L_{p,v} \tag{2.3}$$

holds if and only if one of the following conditions is satisfied:

$$A^+ = \sup_{z>0} \left(\int_z^\infty \left(\int_0^z K^{p'}(x,s) v^{1-p'}(s) ds \right)^{\frac{q}{p'}} w(x) dx \right)^{\frac{1}{q}} < \infty,$$

$$A^- = \sup_{z>0} \left(\int_0^z \left(\int_z^\infty K^q(x,s) w(x) dx \right)^{\frac{p'}{q}} v^{1-p'}(s) ds \right)^{\frac{1}{p'}} < \infty.$$

If C is the best constant in the inequality (2.3)*, then* $C \approx A^+ \approx A^-$.

Another important result we will mention is the Saywer duality principle. In 1990 E. Sawyer [14] developed a method which makes it possible to reduce the inequality (1.3) for all non-increasing functions to an inequality of the same form, but for arbitrary functions. For this he proved a reverse Hölder inequality in the following form (nowadays called the Sawyer duality principle):

$$\sup_{0\le g\downarrow} \frac{\int_0^\infty gf}{\left(\int_0^\infty g^p v\right)^{\frac{1}{p}}} \approx \left(\int_0^\infty \left(\int_0^x f \right)^{p'} V^{-p'}(x) v(x) dx \right)^{\frac{1}{p'}} + \frac{\int_0^\infty f}{\left(\int_0^\infty v\right)^{\frac{1}{p}}}, \ f \ge 0, \quad (2.4)$$

where the notation $0 \le g\downarrow$ means that g is a non-negative, non-increasing function.

Let T be an operator defined by the rule $Tg(x) = \int_0^\infty K(x,s) g(s) ds$, where $K(x,s) \ge 0$. By using (2.4) in the case $1 < p, q < \infty$, we obtain (see, e.g., [7], [17]) that the inequality

$$\left(\int_0^\infty (Tg)^q w \right)^{\frac{1}{q}} \le C \left(\int_0^\infty g^p v \right)^{\frac{1}{p}} \quad \forall 0 \le g \downarrow$$

holds if and only is the following two inequalities hold:

$$\left(\int_0^\infty \left(\int_0^x T^* f \right)^{p'} V^{-p'}(x) v(x) dx \right)^{\frac{1}{p'}} \le C \left(\int_0^\infty f^{q'} w^{1-q'} \right)^{\frac{1}{q'}} \quad \forall f \ge 0, \quad (2.5)$$

$$\int_0^\infty T^* f \left(\int_0^\infty v \right)^{-\frac{1}{p}} \le C \left(\int_0^\infty f^{q'} w^{1-q'} \right)^{\frac{1}{q'}} \quad \forall f \ge 0, \quad (2.6)$$

where $T^* f(s) = \int_0^\infty K(x,s) f(x) dx$.

Usually it is more convenient to use the dual form of the inequality (2.5):

$$\left(\int_0^\infty \left(T \left(\int_s^\infty h \right) \right)^q w(x) dx \right)^{\frac{1}{q}} \le C \left(\int_0^\infty h^p(x) V^p(x) v^{1-p}(x) dx \right)^{\frac{1}{p}}, \quad (2.7)$$

and it follows from the duality principle in Lebesgue spaces that (2.6) is equivalent to

$$\left(\int_0^\infty \left(\int_0^\infty K(x,s) ds \right)^q w(x) dx \right)^{\frac{1}{q}} \le C \left(\int_0^\infty v(x) dx \right)^{\frac{1}{p}}. \quad (2.8)$$

3. The main results

Our main result for the operator **K** defined by (1.1) on the cone of non-increasing functions reads as follows.

Theorem 3.1. *Let $1 < p \le q < \infty$, $\int_0^\infty v < \infty$ and the kernel $K(\cdot,\cdot)$ of the operator* **K** *defined by* (1.1) *belong to the class $\mathcal{O}_n^-$, $n \ge 0$. Then the inequality*

$$\|\mathbf{K}g\|_{q,w} \le C\|g\|_{p,v}, \quad \forall 0 \le g \downarrow \tag{3.1}$$

holds if and only if $A_1 + A_2 + A_3 < \infty$ or $A_1 + A_2 + A_4 < \infty$, where

$$A_1 = \sup_{z>0}\left(\int_z^\infty V^{-p'}(t)v(t)dt\right)^{\frac{1}{p'}}\left(\int_0^z\left(\int_0^x K(x,s)ds\right)^q w(x)dx\right)^{\frac{1}{q}},$$

$$A_2 = \left(\int_0^\infty w(x)\left(\int_0^x K(x,s)ds\right)^q dx\right)^{\frac{1}{q}}\left(\int_0^\infty v(t)dt\right)^{-\frac{1}{p}},$$

$$A_3 = \sup_{z>0}\left(\int_z^\infty\left(\int_0^z\left(\int_0^t K(x,s)ds\right)^{p'} V^{-p'}(t)v(t)dt\right)^{\frac{q}{p'}} w(x)dx\right)^{\frac{1}{q}},$$

$$A_4 = \sup_{z>0}\left(\int_0^z\left(\int_z^\infty\left(\int_0^t K(x,s)ds\right)^q w(x)dx\right)^{\frac{p'}{q}} V^{-p'}(t)v(t)dt\right)^{\frac{1}{p'}}.$$

The corresponding result for the operator $\mathbf{K}^*$ defined by (1.2) reads:

Theorem 3.2. *Let $1 \le p \le q < \infty$, $\int_0^\infty v < \infty$ and the kernel $K(\cdot,\cdot)$ of the operator* $\mathbf{K}^*$ *defined by* (1.2) *belong to the class $\mathcal{O}_n^+$, $n \ge 0$. Then the inequality*

$$\|K^*g\|_{q,w} \le C\|g\|_{p,v} \quad \forall 0 \le g \downarrow \tag{3.2}$$

holds if and only if $B_1 + B_3 < \infty$ or $B_2 + B_3 < \infty$, where

$$B_1 = \sup_{z>0}\left(\int_0^z\left(\int_z^\infty\left(\int_s^x K(t,s)dt\right)^{p'} V^{-p'}(x)v(x)dx\right)^{\frac{q}{p'}} w(s)ds\right)^{\frac{1}{q}},$$

$$B_2 = \sup_{z>0}\left(\int_z^\infty\left(\int_0^z\left(\int_s^x K(t,s)dt\right)^q w(s)ds\right)^{\frac{p'}{q}} V^{-p'}(x)v(x)dx\right)^{\frac{1}{p'}},$$

$$B_3 = \left(\int_0^\infty w(s)\left(\int_s^\infty K(x,s)dx\right)^q ds\right)^{\frac{1}{q}}\left(\int_0^\infty v(t)dt\right)^{-\frac{1}{p}}.$$

In case $\int_0^\infty v = \infty$ Theorems 3.1 and 3.2 take the following forms respectively:

Theorem 3.3. *Let $1 < p \le q < \infty$, $\int_0^\infty v = \infty$ and the kernel $K(\cdot,\cdot)$ of the operator* **K** *defined by* (1.1) *belong to the class $\mathcal{O}_n^-$, $n \ge 0$. Then the inequality* (3.1) *holds* *if and only if* $A_5 + A_3 < \infty$ *or* $A_5 + A_4 < \infty$, *where*

$$A_5 = \sup_{z>0} V^{-\frac{1}{p}}(z) \left(\int_0^z \left(\int_0^x K(x,s)ds \right)^q w(x)dx \right)^{\frac{1}{q}},$$

and A_3 *and* A_4 *are defined in Theorem* 3.1.

Theorem 3.4. *Let* $1 \le p \le q < \infty$, $\int_0^\infty v = \infty$ *and the kernel* $K(\cdot,\cdot)$ *of the operator* $\mathbf{K}^*$ *defined by* (1.2) *belong to the class* $\mathcal{O}_n^+$, $n \ge 0$. *Then the inequality* (3.2) *holds if and only if* $B_1 < \infty$ *or* $B_2 < \infty$. *Here* B_1 *and* B_2 *are defined in Theorem* 3.2.

Remark 3.5. The method we used to prove Theorems 3.1 and 3.3 does not work for the other case when the kernel of the operator K defined by (1.1) belongs the classes $\mathcal{O}_n^+$, $n \ge 0$. Hence, we leave this as an open question. Similarly, we leave as an open question whether inequality (3.2) holds also when the kernel of the operator $\mathbf{K}^*$ defined by (1.2) belongs to the classes $\mathcal{O}_n^-$, $n \ge 0$.

4. Proofs

Proof of Theorem 3.1. First we reduce (3.1) to the inequalities on non-negative functions. The condition $A_2 < \infty$ follows from (2.8) with T replaced by the operator K. Hence, we need to derive necessary and sufficient conditions for the inequality

$$\begin{aligned} &\left(\int_0^\infty \left(\int_0^x K(x,s) \left(\int_s^\infty h(t)dt \right) ds \right)^q w(x)dx \right)^{\frac{1}{q}} \\ &\le C_1 \left(\int_0^\infty h^p(x) V^p(x) v^{1-p}(x)dx \right)^{\frac{1}{p}}, \quad \forall h \ge 0, \end{aligned} \tag{4.1}$$

to hold.

Let us consider the left-hand side of (4.1). Dividing the expression into two parts by x, $s \le x < \infty$, and then changing the order of integration, we obtain that

$$\begin{aligned} &\left(\int_0^\infty \left(\int_0^x K(x,s) \left(\int_s^\infty h(t)dt \right) ds \right)^q w(x)dx \right)^{\frac{1}{q}} \\ &\approx \left(\int_0^\infty \left(\int_x^\infty h(t)dt \right)^q \left(\int_0^x K(x,s)ds \right)^q w(x)dx \right)^{\frac{1}{q}} \\ &+ \left(\int_0^\infty \left(\int_0^x h(t) \int_0^t K(x,s)dsdt \right)^q w(x)dx \right)^{\frac{1}{q}}. \end{aligned}$$

Hence, we can divide (4.1) into two inequalities:

$$\left(\int_0^\infty \left(\int_0^x h(t) \int_0^t K(x,s)dsdt \right)^q w(x)dx \right)^{\frac{1}{q}} \le C_3 \left(\int_0^\infty h^p(x) V^p(x) v^{1-p}(x)dx \right)^{\frac{1}{p}} \tag{4.2}$$

and

$$\left(\int_0^\infty \left(\int_x^\infty h(t)dt\right)^q \left(\int_0^x K(x,s)ds\right)^q w(x)dx\right)^{\frac{1}{q}} \\ \leq C_4 \left(\int_0^\infty h^p(x)V^p(x)v^{1-p}(x)dx\right)^{\frac{1}{p}}.$$

The last inequality is a Hardy-type inequality, which holds (see, for example [6], [7]) if and only if

$$A_1 = \sup_{z>0} \left(\int_z^\infty V^{-p'}(t)v(t)dt\right)^{\frac{1}{p'}} \left(\int_0^z \left(\int_0^x K(x,s)ds\right)^q w(x)dx\right)^{\frac{1}{q}}$$

is finite.

Next we consider (4.2). By using the fact the kernel $K(x,s)$ belongs to the class $\mathcal{O}_n^-$, $n \geq 0$, we obtain that the inequality (4.2) is equivalent to the following $n+1$ inequalities

$$\left(\int_0^\infty \left(\int_0^x h(t)K_i(x,t)\int_0^t K_{i,n}(t,s)dsdt\right)^q w(x)dx\right)^{\frac{1}{q}} \\ \leq \left(\int_0^\infty h^p(x)V^p(x)v^{1-p}(x)dx\right)^{\frac{1}{p}}, \quad i = 0,1,\ldots,n. \tag{4.3}$$

It is easy to see that (4.3) is equivalent to

$$\left(\int_0^\infty \left(\int_0^x f(t)K_i(x,t)dt\right)^q w(x)dx\right)^{\frac{1}{q}} \\ \leq \left(\int_0^\infty f^p(x)v_i(x)dx\right)^{\frac{1}{p}}, \quad i = 0,1,\ldots,n,$$

where $v_i(x) = \left(\int_0^x K_{i,n}^-(x,s)ds\right)^{-p} V^p(x)v^{1-p}(x)$, $i = 0,1,\ldots,n$. Hence we can apply Theorem A and obtain that a necessary and sufficient condition for the validity of (4.2) is one of the following conditions:

$$A_{3,i} = \sup_{z>0}\left(\int_z^\infty \left(\int_0^z (K_i^-(x,t))^{p'} v_i^{1-p'}(t)dt\right)^{\frac{q}{p'}} w(x)dx\right)^{\frac{1}{q}} < \infty, \quad i = 0,1,\ldots,n,$$

$$A_{4,i} = \sup_{z>0}\left(\int_0^z \left(\int_z^\infty (K_i^-(x,t))^q w(x)dx\right)^{\frac{p'}{q}} v_i^{1-p'}(t)dt\right)^{\frac{1}{p'}} < \infty, \quad i = 0,1,\ldots,n.$$

By summing $A_{3,i}$ with respect to i and substituting the expression for v_i, we get that

$$\sum_{i=0}^{n} A_{3,i} = \sum_{i=0}^{n} \sup_{z>0} \left(\int_z^\infty \left(\int_0^z \left(K_i^-(x,t) \int_0^t K_{i,n}^-(t,s)ds \right)^{p'} V^{-p'}(t)v(t)dt \right)^{\frac{q}{p'}} w(x)dx \right)^{\frac{1}{q}}$$

$$\approx \sup_{z>0} \left(\int_z^\infty \left(\int_0^z \left(\int_0^t \sum_{i=0}^{n} K_i^-(x,t)K_{i,n}^-(t,s)ds \right)^{p'} V^{-p'}(t)v(t)dt \right)^{\frac{q}{p'}} w(x)dx \right)^{\frac{1}{q}}$$

$$\approx \sup_{z>0} \left(\int_z^\infty \left(\int_0^z \left(\int_0^t K(x,s)ds \right)^{p'} V^{-p'}(t)v(t)dt \right)^{\frac{q}{p'}} w(x)dx \right)^{\frac{1}{q}} = A_3.$$

Similarly we have $\sum\limits_{i=0}^{n} A_{4,i} \approx A_4$.

It follows that the inequality (4.2) hold if and only if A_3 or A_4 is finite and the proof is complete. □

Proof of Theorem 3.2. Similarly to the proof of Theorem 3.1, we need to study the following inequality:

$$\left(\int_0^\infty \left(\int_0^x Kf(t)dt \right)^{p'} \left(\int_0^x v(s)ds \right)^{-p'} v(x)dx \right)^{\frac{1}{p'}} \le C_1 \left(\int_0^\infty g^{q'}(x)w^{1-q'}(x) \right)^{\frac{1}{q'}}. \tag{4.4}$$

By using the Fubini Theorem 4.4 can be rewritten as

$$\left(\int_0^\infty \left(\int_0^x f(s) \int_s^x K(t,s)dtds \right)^{p'} \left(\int_0^x v(s)ds \right)^{-p'} v(x)dx \right)^{\frac{1}{p'}}$$
$$\le C_1 \left(\int_0^\infty g^{q'}(x)w^{1-q'}(x) \right)^{\frac{1}{q'}}.$$

Due to Lemma 2.2 the function $\widetilde{K}(x,s) = \int_s^x K(t,s)dt$ belongs to the class $\mathcal{O}_{n+1}^+$ and it follows from Theorem A that the inequality (4.4) holds if and only if one of the quantities

$$B_1 = \sup_{z>0} \left(\int_0^z \left(\int_z^\infty \left(\int_s^x K(t,s)dt \right)^{p'} \left(\int_0^x v(y)dy \right)^{-p'} v(x)dx \right)^{\frac{q}{p'}} w(s)ds \right)^{\frac{1}{q}},$$

$$B_2 = \sup_{z>0} \left(\int_z^\infty \left(\int_0^z \left(\int_s^x K(t,s)dt \right)^q w(s)ds \right)^{\frac{p'}{q}} \left(\int_0^x v(y)dy \right)^{-p'} v(x)dx \right)^{\frac{1}{p'}}$$

is finite.

The condition $B_3 < \infty$ follows from (2.8) and the proof is complete. □

Proofs of Theorem 3.3 *and Theorem* 3.4. The proofs are similar to the proofs of Theorem 3.1 and Theorem 3.2. Hence, we omit the details here.

5. The non-decreasing case

Let notation $0 \leq g \uparrow$ means that g is a non-negative and non-decreasing function. The duality principle for non-decreasing functions reads as follows (see, e.g., [4], [17]):

$$\sup_{0 \leq g \uparrow} \frac{\int_0^\infty gf}{\left(\int_0^\infty g^p v\right)^{\frac{1}{p}}} \approx \left(\int_0^\infty \left(\int_x^\infty f \right)^{p'} \left(\int_x^\infty v \right)^{-p'} v(x)dx \right)^{\frac{1}{p'}} + \frac{\int_0^\infty f}{\left(\int_0^\infty v\right)^{\frac{1}{p}}}, \quad (5.1)$$

where f is a non-negative function.

By using (5.1) and similar arguments as in our previous section, we can derive the following results for non-decreasing functions.

Theorem 5.1. *Let $1 \leq p \leq q < \infty$, $\int_0^\infty v < \infty$ and the kernel $K(\cdot,\cdot)$ of the operator* $\mathbf{K}$ *defined by* (1.1) *belong to the class $\mathcal{O}_n^-$, $n \geq 0$. Then the inequality*

$$\|\mathbf{K}g\|_{q,w} \leq C\|g\|_{p,v}, \quad \forall 0 \leq g \uparrow \quad (5.2)$$

holds if and only if one of the conditions $\widetilde{A}_1 + \widetilde{A}_3 < \infty$, $\widetilde{A}_2 + \widetilde{A}_3 < \infty$ is satisfied. Here

$$\widetilde{A}_1 = \sup_{z>0} \left(\int_z^\infty \left(\int_0^z \left(\int_s^x K(x,t)dt \right)^{p'} \left(\int_s^\infty v(y)dy \right)^{p'} v(s)ds \right)^{\frac{q}{p'}} w(x)dx \right)^{\frac{1}{q}},$$

$$\widetilde{A}_2 = \sup_{z>0} \left(\int_0^z \left(\int_z^\infty \left(\int_s^x K(x,t)dt \right)^q w(x)dx \right)^{\frac{p'}{q}} \left(\int_s^\infty v(t)dt \right)^{-p'} v(s)ds \right)^{\frac{1}{p'}},$$

$$\widetilde{A}_3 = \left(\int_0^\infty \left(\int_0^x K(x,s)ds \right)^q w(x)dx \right)^{\frac{1}{q}} \left(\int_0^\infty v(t)dt \right)^{-\frac{1}{p}}.$$

Theorem 5.2. *Let $1 \leq p \leq q < \infty$, $\int_0^\infty v = \infty$ and the kernel $K(\cdot,\cdot)$ of the operator* $\mathbf{K}$ *defined by* (1.1) *belong to the class $\mathcal{O}_{\underset{\sim}{n}}^-$, $n \geq 0$. Then the inequality* (5.2) *holds if and only if $\widetilde{A}_1 < \infty$ or $\widetilde{A}_2 < \infty$, and $\widetilde{A}_1$, $\widetilde{A}_2$ are defined in Theorem* 5.1.

Theorem 5.3. *Let $1 < p \le q < \infty$, $\int_0^\infty v < \infty$ and the kernel $K(\cdot,\cdot)$ of the operator $\mathbf{K}^*$ defined by (1.2) belong to the class $\mathcal{O}_n^+$, $n \ge 0$. Then the inequality*

$$\|\mathbf{K}^* g\|_{q,w} \le C\|g\|_{p,v}, \quad \forall 0 \le g\uparrow \tag{5.3}$$

holds if and only if one of conditions $\widetilde{B}_1 + \widetilde{B}_2 + \widetilde{B}_3 < \infty$, $\widetilde{B}_1 + \widetilde{B}_2 + \widetilde{B}_4 < \infty$ is satisfied, where

$$\widetilde{B}_1 = \sup_{z>0}\left(\int_0^z \left(\int_t^\infty v(s)ds\right)^{-p'} v(t)dt\right)^{\frac{1}{p'}} \left(\int_z^\infty \left(\int_t^\infty K(x,t)dx\right)^q w(t)dt\right)^{\frac{1}{q}},$$

$$\widetilde{B}_2 = \left(\int_0^\infty \left(\int_s^\infty K(x,s)dx\right)^q w(s)ds\right)^{\frac{1}{q}} \left(\int_0^\infty v(t)dt\right)^{-\frac{1}{p}},$$

$$\widetilde{B}_3 = \sup_{z>0}\left(\int_0^z \left(\int_z^\infty \left(\int_t^\infty K(x,s)dx\right)^{p'} \left(\int_t^\infty v(s)ds\right)^{-p'} v(t)dt\right)^{\frac{q}{p'}} w(s)ds\right)^{\frac{1}{q}},$$

$$\widetilde{B}_4 = \sup_{z>0}\left(\int_z^\infty \left(\int_0^z \left(\int_t^\infty K(x,s)dx\right)^q w(s)ds\right)^{\frac{p'}{q}} \left(\int_t^\infty v(s)ds\right)^{-p'} v(t)dt\right)^{\frac{1}{p'}}.$$

Theorem 5.4. *Let $1 < p \le q < \infty$, $\int_0^\infty v = \infty$ and the kernel $K(\cdot,\cdot)$ of the operator $\mathbf{K}^*$ defined by (1.2) belong to the class $\mathcal{O}_n^+$, $n \ge 0$. Then the inequality (5.3) holds if and only if one of conditions $\widetilde{B}_5 + \widetilde{B}_3 < \infty$, $\widetilde{B}_5 + \widetilde{B}_4 < \infty$ is satisfied, where*

$$\widetilde{B}_5 = \sup_{z>0}\left(\int_0^\infty \left(\int_s^\infty K(x,s)dx\right)^q w(s)ds\right)^{\frac{1}{q}} \left(\int_z^\infty v(t)dt\right)^{-\frac{1}{p}},$$

and $\widetilde{B}_3$, $\widetilde{B}_4$ are defined in Theorem 5.3.

Proofs. For proving Theorems 5.1–5.4 we apply the method we used in the non-increasing case. In particular, the inequality

$$\left(\int_0^\infty (Kg)^q w\right)^{\frac{1}{q}} \le C\left(\int_0^\infty g^p v\right)^{\frac{1}{p}} \quad \forall 0 \le g\uparrow$$

is equivalent to the following two inequalities

$$\left(\int_0^\infty \left(\int_x^\infty K^* f\right)^{p'} \left(\int_x^\infty v\right)^{p'} v(x)dx\right)^{\frac{1}{p'}} \le C\left(\int_0^\infty f^{q'} w^{1-q'}\right)^{\frac{1}{q'}}, \quad \forall f \ge 0,$$

$$\int_0^\infty K^* f \left(\int_0^\infty v\right)^{-\frac{1}{p}} \le C\left(\int_0^\infty f^{q'} w^{1-q'}\right)^{\frac{1}{q'}}, \quad \forall f \ge 0.$$

Also all other arguments and formulas from proofs of Theorems 3.1–3.4 can be modified in a similar way, so we omit the details.

Acknowledgment

The authors thank both referees for some generous advices, which have improved the final form of this paper.

References

[1] M. Ariño and B. Muckenhoupt, *Maximal functions on classical Lorentz spaces and Hardy's inequality with weights for non-increasing functions.* Trans. Amer. Math. Soc. **320** (1990), 727–735.

[2] I. Asekritova, N. Krugljak, L. Maligranda and L.-E. Persson, *Distribution and rearrangement estimates of the maximal function and interpolation.* Studia Math **124** (1997), no. 2, 107–132.

[3] A. Gogatishvili, M. Johansson, C. A. Okpoti and L.-E. Persson, *Characterisation of embeddings in Lorentz spaces.* Bull. Austral. Math. Soc. **76** (2007), no. 1, 69–92.

[4] H.P. Heinig and V.D. Stepanov, *Weighted Hardy inequalities for increasing functions.* Canad. J. Math. **45** (1993), no. 1, 104–116.

[5] V. Kokilashvili, A. Meskhi and L.-E. Persson, *Weighted norm inequalities for integral transforms with product kernels.* Nova Science Publishers Inc. New York, 2010.

[6] A. Kufner, L. Maligranda and L.-E. Persson, *The Hardy inequality. About its history and some related results.* Vydavatelský Servis, Plseň, 2007.

[7] A. Kufner and L.-E. Persson, *Weighted inequalities of Hardy type.* World Scientific, New Jersey, 2003.

[8] C.J. Neugebauer, *Weights for the Hardy operator on non-decreasing functions.* Int. J. Pure Appl. Math. **60** (2010), no. 1, 71–82.

[9] R. Oinarov, *Boundedness and compactness of Volterra-type integral operators.* Sibirsk. Mat. Zh. **48** (2007), no. 5, 1100–1115; English transl. in Siberian Math J. **48** (2007), no. 5, 884–896.

[10] B. Opic and A. Kufner, *Hardy-type inequalities.* Pitman Research Notes in Mathematics, Series 219, Longman Science and Technology, Harlow, 1990.

[11] L.-E. Persson, O.V. Popova and V.D. Stepanov, *Two-sided Hardy-type inequalities for monotone functions.* Doklady Mathematics **80** (2009), no. 3, 814–817.

[12] L.-E. Persson, O.V. Popova and V.D. Stepanov, *Two-sided Hardy-type inequalities for monotone functions.* Complex Variables and Elliptic Equations **55** (2010), no. 8-10, 973–989.

[13] L.-E. Persson, V.D. Stepanov and E.P. Ushakova, *Equivalence of Hardy-type inequalities with general measures on the cones of non-negative respective non-increasing functions.* Proc. Amer. Math. Soc. **134** (2006), 2363–2372.

[14] E. Sawyer, *Boundedness of classical operators on classical Lorentz spaces.* Studia Math. **96** (1990), 145–158.

[15] G. Sinnamon, *Hardy's inequality and monotonocity.* Function Spaces and Nonlinear Analysis (eds.: P. Drábec and J. Rákosnik), Mathematical Institute of the Academy of Sciences of the Czech Republic, Prague, 2005, 292–310.

[16] V.D. Stepanov, *The weighted Hardy's inequality for non-increasing functions.* Trans. Amer. Math. Soc. **338** (1993), 173–186.

[17] V.D. Stepanov, *Integral operators on the cone of monotone functions.* J. London Math. Soc. **48** (1993), 465–487.

L.S. Arendarenko and R. Oinarov
Department of Fundamental and Applied Mathematics
Eurasian National University
Munaitpasov str. 4
Astana, 010008, Kazakhstan
e-mail: `arendarenko_l@mail.ru`
`o_ryskul@mail.ru`

L.-E. Persson
Department of Engineering Sciences and Mathematics
Luleå University of Technology
SE-971 87 Luleå, Sweden

and

Narvik University College
Postboks 385
N-8505 Narvik, Norway
e-mail: `larserik@ltu.se`

Operator Theory:
Advances and Applications, Vol. 229, 91–99

On a Boundary Value Problem for a Class of Generalized Analytic Functions

P. Berglez and T.T. Luong

Abstract. For solutions of the Bers-Vekua equation $Dv := v_{\bar{z}} - c(z,\bar{z})\,\bar{v} = 0$ defined in a domain $\mathbb{D} \subset \mathbb{C}$ we consider Riemann-Hilbert type boundary conditions.

In the case of the existence of certain differential operators with which all the solutions of $Dv = 0$ defined in $\mathbb{D}$ can be generated from a function f holomorphic in $\mathbb{D}$ the boundary value problem is reduced to a Goursat problem for f in essence. For certain classes of coefficients c and domains $\mathbb{D}$ we show how this problem can be solved explicitly.

For the poly-pseudoanalytic functions obeying the differential equation $D^n v = 0, n \in \mathbb{N}$, we investigate an appropriate boundary value problem and show the equivalence of this problem to n boundary value problems for generalized analytic functions.

Mathematics Subject Classification (2010). Primary 30G20; Secondary 35C50, 35F15.

Keywords. Boundary value problems, generalized analytic functions, differential operators.

1. Introduction

In this paper we consider boundary value problems for generalized analytic functions governed by the so-called Bers-Vekua equation (see, e.g., [8], [18])

$$v_{\bar{z}} = c(z,\bar{z})\,\bar{v} \tag{1.1}$$

with $z = x + iy$. If the real variables x and y are continued into a complex domain we obtain functions of the two complex variables $z = x + iy$ and $\zeta = x - iy$ and equation (1.1) takes the form

$$w_\zeta = c(z,\zeta)\, w^* \tag{1.2}$$

where $w^*(\zeta, z)$ is the conjugate function to $w(z,\zeta)$. For $\zeta = \bar{z}$ equations (1.1) and (1.2) are equivalent. Thus every solution $w(z,\zeta)$ of (1.2) holomorphic in $\mathbb{D} \times \bar{\mathbb{D}}$, where $\mathbb{D}$ is a fundamental domain of (1.1) in the sense of I.N. Vekua [19] and $\bar{\mathbb{D}}$ denotes the mirror image of $\mathbb{D}$ with respect to the real axis, is a solution of (1.1) defined in $\mathbb{D}$ if we use $\zeta = \bar{z}$.

We start with a differential equation of the form (1.2) for which certain differential operators of Bauer-type exist such that all the solutions w of (1.2) may be represented as the image of a so-called generating function holomorphic in $\mathbb{D}$ under these operators. A necessary and sufficient condition on the coefficient c under which such operators exist can be found in [6]. Here we present an additional condition on this generating function such that it is determined uniquely by the solution w. Representations for generalized analytic functions using these Bauer operators are discussed in [2].

In the present paper we consider a particular Bers-Vekua equation of the form $v_{\bar{z}} = \dfrac{m}{1 - z\bar{z}}\,\bar{v}$, $m \in \mathbb{N}$, for the solutions of which such a representation is available. This differential equation can be found in the Ising field theory on a pseudosphere [10], in the discussion of Coulomb systems living on a surface of constant negative curvature [12], or in the investigation of the tau functions for the Dirac operator in the Poincaré disk [16]. For recent investigations on the application of pseudo-analytic functions to physical problems see, e.g., [13] and [9]. Biquaternionic Bers-Vekua type equations arising from the factorization of elliptic operators are studied in [14].

In the second part we consider a boundary value problem for these generalized analytic functions which may be represented by such Bauer operators. We show that this problem is equivalent to a certain ordinary differential equation for the generating function defined on the boundary of the domain considered. Once this function is determined on the boundary we can express it in the whole domain considered by a Cauchy integral. In the case of the particular Bers-Vekua equation (1.1) with $c = m/(1 - z\bar{z}), m \in \mathbb{N}$, we show how to solve the differential equation occurring there explicitly using Fourier expansions for the functions involved. Boundary value problems for generalized analytic functions with a singular point are investigated in [17] where further references on this topic are given.

Finally we consider solutions of the iterated Bers-Vekua equation which are called poly-pseudoanalytic functions. A decomposition theorem for these functions was presented in [7] whereas in [15] polynomial generalized Bers-Vekua equations were treated. These poly-pseudoanalytic functions can be considered as a generalization of the polyanalytic functions investigated in detail in [1]. For these functions we formulate a suitable boundary value problem and show how it can be solved.

2. Differential operators for pseudoanalytic functions

Under certain conditions on the coefficient c analytic in $\mathbb{D} \times \bar{\mathbb{D}}$ for the differential equation

$$w_\zeta = c\,w^* \tag{2.1}$$

there exist differential operators X_m and X^*_{m-1} of the form

$$X_m = \sum_{k=0}^{m} a_k(z,\zeta)\,\frac{\partial^k}{\partial z^k}\,, \quad X^*_{m-1} = \sum_{k=0}^{m-1} b_k(z,\zeta)\,\frac{\partial^k}{\partial \zeta^k}\,,\ m \in \mathbb{N}$$

such that all the solutions of (2.1) analytic in $\mathbb{D} \times \bar{\mathbb{D}}$ can be given by the Bauer operator B as

$$w(z,\zeta) = B\,g(z) := X_m\,g(z) + X^*_{m-1}\,g^*(\zeta) \tag{2.2}$$

where g is a function holomorphic in $\mathbb{D}$ (see [6]). For particular Bers-Vekua equations for which such Bauer-operators B exist see [2], [4] and [5].

On the other hand after I.N. Vekua [18] all the solutions of (2.1) analytic in $\mathbb{D} \times \bar{\mathbb{D}}$ can be represented using an integral operator I in the following way

$$w(z,\zeta) = I\,\Phi(z) := \Phi(z) + \int_{z_0}^{z} \Gamma_1(z,\zeta,t,\zeta_0)\,\Phi(t)\,dt + \int_{\zeta_0}^{\zeta} \Gamma_2(z,\zeta,z_0,\tau)\,\Phi^*(\tau)\,d\tau \tag{2.3}$$

with $z_0 \in \mathbb{D}$ and $\zeta_0 \in \bar{\mathbb{D}}$ where the function Φ is holomorphic in $\mathbb{D}$ and determined uniquely by

$$\Phi(z) = w(z,\zeta_0). \tag{2.4}$$

Here Γ_1 and Γ_2 denote the Vekua resolvents which can be calculated using the Bauer operators in an explicit way (see [6]). In view of the representation (2.2) we have

$$w(z,\zeta_0) = \sum_{k=0}^{m} a_k(z,\zeta_0)\,g^{(k)}(z) + \sum_{k=0}^{m-1} b_k(z,\zeta_0)\,(g^*)^{(k)}(\zeta_0).$$

Now let us assume that the origin is an element of $\mathbb{D}$ and that the function g obeys the conditions

$$g(0) = g'(0) = \cdots = g^{(m-1)}(0) = 0. \tag{2.5}$$

We denote by $H_{\mathbb{D}}$ the set of functions holomorphic in $\mathbb{D}$ and by $H_{\mathbb{D}}(m,0)$ the subset of $H_{\mathbb{D}}$ with the property (2.5). Setting $\zeta_0 = 0$ we can write $\Phi(z) = w(z,0)$ and we have

$$\Phi(z) = \sum_{k=0}^{m} a_k(z,0)\,g^{(k)}(z)$$

which together with the conditions in (2.5) determines the function g in a unique way once the function Φ is given.

Theorem 2.1. *Let the coefficient c in (2.1) be analytic in $\mathbb{D} \times \bar{\mathbb{D}}$ and let $\mathbb{D}$ contain the origin. Assume further the existence of a differential operator B of Bauer-type in the form (2.2) such that for g holomorphic in $\mathbb{D}$ the expression*

$$w(z,\zeta) = B\,g(z) \tag{2.6}$$

is a solution of (2.1) defined in $\mathbb{D} \times \bar{\mathbb{D}}$. Then for each solution w of (2.1) defined in $\mathbb{D}$ there exists a unique function $g \in H_{\mathbb{D}}(m,0)$ such that for w the representation (2.6) holds.

Now let us consider the particular differential equation

$$v_{\bar{z}} = \frac{m}{1 - z\bar{z}}\,\bar{v}\,, \; m \in \mathbb{N}. \tag{2.7}$$

Each disk $K_R := \{z \,\big|\, |z| < R < 1\}$ is a fundamental domain of eq. (2.7) for which we can prove that the expression

$$\begin{aligned} v &= \hat{X}_m\, g + \overline{\hat{X}^*_{m-1}\, g} \\ &:= \sum_{k=0}^{m} m\, A_k^m \chi^{m-k}\, g^{(k)}(z) + \sum_{k=0}^{m-1} (m-k) A_k^m \frac{\bar{\chi}^{m-k-1}}{1 - z\bar{z}}\, \overline{g^{(k)}(z)} \end{aligned} \tag{2.8}$$

with $A_k^m = \dfrac{(2m-k-1)!}{k!\,(m-k)!}$ and $\chi = \dfrac{\bar{z}}{1 - z\bar{z}}$ is a solution in K_R for all $g \in H_{K_R}$. For another representation of the solutions of (2.7) see [3] where the corresponding Vekua resolvents are given – even in the case $m > 0$ – also.

3. Boundary value problems and Bauer operators

We consider the boundary value problem

$$v_{\bar{z}} = c\,\bar{v} \quad \text{in} \quad \mathbb{D} \tag{3.1}$$

$$\operatorname{Re}(v) = \Psi \quad \text{on} \quad \partial\mathbb{D} \tag{3.2}$$

where c is an arbitrary analytic function defined in $\mathbb{D}$ and Ψ is Hölder-continuous on $\partial\mathbb{D}$. I.N. Vekua [19] presented theorems concerning the existence and uniqueness of the solution of this problem. He proved that this boundary value problem is equivalent to a singular integral equation for a certain density function the kernel of which is depending on the coefficient c.

In the following we will show that under the assumption that there exists a representation of the solution of (3.1) using a Bauer-type operator in the form

$$v(z, \bar{z}) = [B\, g(z)]_{\zeta = \bar{z}} \tag{3.3}$$

with B from (2.2) and g holomorphic in $\mathbb{D}$ this boundary value problem can be solved explicitly in a direct way. Since according to Theorem 2.1 the generating function $g \in H_{\mathbb{D}}(m, 0)$ is determined uniquely by the solution v we can state that solving the boundary value problem (3.1)–(3.2) is equivalent to finding the suitable generating function g.

Now the boundary condition (3.2) in connection with the representation (3.3) for v leads to the differential equation

$$\operatorname{Re}\left\{ \sum_{k=0}^{m} a_k(\xi)\, g^{(k)}(\xi) + \sum_{k=0}^{m-1} b_k(\xi)\, \overline{g^{(k)}(\xi)} \right\} = \Psi(\xi) \tag{3.4}$$

where $\xi \in \partial\mathbb{D}$ and $a_k(\xi) := a_k(z, \bar{z})|_{\partial\mathbb{D}}, b_k(\xi) := b_k(z, \bar{z})|_{\partial\mathbb{D}}$ is used.

With respect to the condition $g \in H_{\mathbb{D}}(m,0)$ we use for g the expansion $g(z) = \sum_{j=m}^{\infty} \gamma_j \, z^j, \gamma_j \in \mathbb{C}$, and in particular

$$g(\xi) = \sum_{j=m}^{\infty} \gamma_j \, \xi^j \quad \text{on} \quad \partial\mathbb{D}. \tag{3.5}$$

With $g(\xi)$ in the form (3.5) the boundary condition (3.4) can be written as

$$\mathrm{Re}\left\{ \sum_{k=0}^{m} a_k(\xi) \sum_{j=m}^{\infty} \frac{j!}{(j-k)!} \gamma_j \xi^{j-k} + \sum_{k=0}^{m-1} b_k(\xi) \sum_{j=m}^{\infty} \frac{j!}{(j-k)!} \bar{\gamma}_j \bar{\xi}^{j-k} \right\} = \Psi(\xi).$$

Once we have determined $g(\xi)$ the function $g \in H_{\mathbb{D}}$ can be calculated by the Cauchy integral

$$g(z) = \frac{1}{2\pi i} \int_{\partial\mathbb{D}} \frac{g(\tau)}{\tau - z} \, d\tau.$$

In particular for the differential equation (2.7) where we have $c = m/(1 - z\bar{z})$ there exists such a Bauer-type operator given by (2.8). Since the coefficients a_k and b_k are known explicitly here, we can solve the differential equation (3.4) for the function g in the following way. The boundary condition $\mathrm{Re}(v) = \Psi$ on ∂K_R leads to

$$\mathrm{Re}\left\{ \sum_{k=0}^{m} \frac{A_k^m}{(1-\xi\bar{\xi})^{m-k}} \times \left[m\,\bar{\xi}^{m-k} \sum_{j=m}^{\infty} \frac{j!\gamma_j \xi^{j-k}}{(j-k)!} + (m-k)\,\xi^{m-k-1} \sum_{j=m}^{\infty} \frac{j!\bar{\gamma}_j \bar{\xi}^{j-k}}{(j-k)!} \right] \right\} = \Psi(\xi).$$

Introducing the real parameter $t \in [0, 2\pi]$ by $\xi = R\,e^{it} \in \partial K_R$ we have

$$\mathrm{Re}\left\{ \sum_{j=m}^{\infty} c_j \, \gamma_j \, R^{j-m} \, e^{i(j-m)t} + \sum_{j=m}^{\infty} d_j \, \bar{\gamma}_j \, R^{j-m+1} \, e^{-i(j-m+1)t} \right\} = \Psi(\xi)$$

with

$$c_j = \sum_{k=0}^{m} m A_k^m \frac{j!}{(j-k)!} \frac{R^{2(m-k)}}{(1-R^2)^{m-k}} > 0$$

$$d_j = \sum_{k=0}^{m-1} (m-k) A_k^m \frac{j!}{(j-k)!} \frac{R^{2(m-k-1)}}{(1-R^2)^{m-k}} > 0.$$

Now we use $\gamma_j = \alpha_j + i\,\beta_j, \alpha_j, \beta_j \in \mathbb{R}\,, j \geq m$, and $e^{i\sigma} = \cos\sigma + i\,\sin\sigma$ for $\sigma \in \mathbb{R}$ with which we get

$$\begin{aligned}\operatorname{Re}(v)\big|_{\partial K_R} &= c_m\,\alpha_m + \sum_{j=1}^{\infty} \big(c_{m+j}\alpha_{m+j} + d_{m+j-1}\alpha_{m+j-1}\big) R^j\,\cos(jt)\\ &\quad - \sum_{j=1}^{\infty} \big(c_{m+j}\beta_{m+j} + d_{m+j-1}\beta_{m+j-1}\big) R^j\,\sin(jt).\end{aligned}$$

Now the boundary function Ψ is assumed to possess a uniformly convergent Fourier series of the form

$$\Psi(t) = \varphi_0 + \sum_{j=1}^{\infty} \big(\varphi_j\,\cos(jt) + \psi_j\,\sin(jt)\big). \tag{3.6}$$

Comparing the last two expressions we are led to the following linear system for the coefficients α_j and β_j

$$\begin{aligned} c_m\,\alpha_m &= \varphi_0\\ \big(c_{m+j}\alpha_{m+j} + d_{m+j-1}\alpha_{m+j-1}\big) R^j &= \varphi_j\,, \quad j = 1,2,\dots\\ -\big(c_{m+j}\beta_{m+j} + d_{m+j-1}\beta_{m+j-1}\big) R^j &= \psi_j\,, \quad j = 1,2,\dots\end{aligned}$$

Here $\beta_m \in \mathbb{R}$ can be chosen arbitrarily and then the remaining coefficients can be calculated recursively in a unique way as

$$\begin{aligned}\alpha_m &= \frac{\varphi_0}{c_m}\\ \alpha_{m+j} &= \frac{\varphi_j - d_{m+j-1}\,R^j\,\alpha_{m+j-1}}{c_{m+j}\,R^j}\,, \quad j = 1,2,\dots\\ \beta_{m+j} &= -\frac{\psi_j + d_{m+j-1}\,R^j\,\beta_{m+j-1}}{c_{m+j}\,R^j}\,, \quad j = 1,2,\dots\end{aligned} \tag{3.7}$$

Summarizing we have the following

Theorem 3.1. *The boundary value problem*

$$v_{\bar z} = \frac{m}{1 - z\bar z}\,\bar v \quad \textit{in} \quad K_R := \{z \mid |z| < R, 0 < R < 1\}$$

$$\operatorname{Re}(v) = \Psi \quad \textit{on} \quad \partial K_R := \{z \mid |z| = R\}$$

with Ψ in the form (3.6) *has the solution*

$$v = \hat X_m\,g + \overline{\hat X^*_{m-1}\,g}$$

where the differential operators $\hat X_m$ and $\hat X^*_{m-1}$ are given by (2.8) *and the generating function g can be calculated by*

$$g(z) = \frac{1}{2\pi i}\int_{\partial K_R} \frac{\hat g(\tau)}{\tau - z}\,d\tau$$

using

$$\hat{g}(t) = g(R\,e^{it}) = \sum_{j=m}^{\infty} (\alpha_j + i\beta_j)\,R^j\,e^{ijt}.$$

Here $\beta_m \in \mathbb{R}$ *can be chosen arbitrarily and the coefficients* α_j, $j \geq m$, *and* β_j, $j \geq m+1$, *are given recursively by* (3.7).

The differential equation for the complex potentials w for equation (2.7) in the sense of I.N. Vekua [18] is given by

$$w_{z\bar{z}} - \frac{\bar{z}}{1 - z\bar{z}}\,w_{\bar{z}} - \frac{m^2}{(1-z\bar{z})^2}\,w = 0.$$

This is a particular case of the elliptic equation

$$w_{z\bar{z}} - \frac{(n-m)\,\bar{z}}{1 - z\bar{z}}\,w_{\bar{z}} - \frac{n(m+1)}{(1-z\bar{z})^2}\,w = 0 \tag{3.8}$$

which for $m, n \in \mathbb{N}$ has been investigated in [2]. In the case $m = n \in \mathbb{N}$ this equation reduces to the Bauer-Peschl equation

$$w_{z\bar{z}} - \frac{n(n+1)}{(1-z\bar{z})^2}\,w = 0$$

for which boundary value problems were solved in [11]. Thus we hope that boundary value problems for (3.8) can be treated in a similar manner.

4. Poly-pseudoanalytic functions

Now we consider the iterated Bers-Vekua equation

$$D^n\,v = 0\,,\; n \in \mathbb{N} \quad \text{with} \quad D\,v := v_{\bar{z}} - c\,\bar{v} \tag{4.1}$$

with c analytic in $\mathbb{D}$ where $\mathbb{D}$ denotes a fundamental domain of $Dv = 0$. If $c \equiv 0$ equation (4.1) represents the differential equation for the polyanalytic functions which were investigated well in the past (see, e.g., [1]). Since after L. Bers [8] the solutions of $Dv = 0$ are called pseudoanalytic functions also, the solutions of (4.1) are named poly-pseudoanalytic functions. From [7] we have the following

Theorem 4.1.

1. *Let the functions* $v_k, k = 0, \ldots, n-1$, *be pseudoanalytic in* $\mathbb{D}$ *which means that they are solutions of the Bers-Vekua equation*

$$Dv_k = 0\,, k = 0, \ldots, n-1 \tag{4.2}$$

with D *from* (4.1). *Then the function* v *according to*

$$v = v_0 + \eta v_1 + \cdots + \eta^{n-1} v_{n-1}\,, \quad \eta = z + \bar{z} \tag{4.3}$$

represents a solution of (4.1) *defined in* $\mathbb{D}$.

2. *For each solution* v *of* (4.1) *defined in* $\mathbb{D}$ *there exist functions* $v_0, \ldots, v_{n-1}$, *pseudoanalytic in* $\mathbb{D}$ *such that* v *is given by* (4.3).

3. *For each solution* v *of* (4.1) *in the form* (4.3) *the functions* v_k *are determined uniquely by*

$$v_k = \frac{1}{k!} \sum_{j=0}^{n-1-k} \frac{(-1)^j}{j!} \eta^j D^{k+j} v \,. \tag{4.4}$$

Now we can state the following boundary value problem for the poly-pseudoanalytic functions. We look for functions v defined in $\mathbb{D} \cup \partial\mathbb{D}$ such that

$$D^n v = 0 \quad \text{in} \quad \mathbb{D}\,, \quad n \in \mathbb{N} \tag{4.5}$$

$$\operatorname{Re}\left(D^j v\right) = \Psi_j,\ j = 0, 1, \ldots, n-1, \quad \text{on} \quad \partial\mathbb{D}. \tag{4.6}$$

Since the functions $v_k, k = 0, \ldots, n-1$, in the decomposition (4.3) of the poly-pseudoanalytic function v are determined uniquely by v according to (4.4), the boundary conditions (4.6) for the function v lead to the following boundary conditions for the corresponding pseudoanalytic functions v_k:

$$\operatorname{Re}(v_k) = \frac{1}{k!} \sum_{j=0}^{n-1-k} \frac{(-1)^j}{j!} \tilde{\eta}^j \Psi_{k+j}\,, k = 0, \ldots, n-1, \text{ on } \partial\mathbb{D} \tag{4.7}$$

with $\tilde{\eta} = (z + \bar{z})\big|_{\partial\mathbb{D}}$. So we can prove the following

Theorem 4.2. *The boundary value problem* (4.5), (4.6) *for the poly-pseudoanalytic function* v *is equivalent to the system of boundary value problems for the pseudoanalytic functions* $v_k, k = 0, \ldots, n-1$, *consisting of the differential equations* (4.2) *defined in* $\mathbb{D}$ *and the boundary conditions* (4.7). *Its solution can be given in the form* (4.3) *where the functions* v_k *are the solutions of the boundary value problems* (4.2), (4.7).

Acknowledgement

The second author was supported by the Vietnam Ministry of Education and Training (Project No. 322).

References

[1] M.B. Balk, *Polyanalytic Functions.* Mathematical Research **63**, Akademie-Verlag, 1991.

[2] K.W. Bauer, S. Ruscheweyh, *Differential Operators for Partial Differential Equations.* Lecture Notes in Math. **792**, Springer, 1980.

[3] K.W. Bauer, *Bestimmung und Anwendung von Vekua-Resolventen.* Monatsh. Math. **85** (1977), 89–97.

[4] K.W. Bauer, *Differentialoperatoren bei verallgemeinerten Euler-Gleichungen.* Ber. Math.-Statist. Sekt. Forsch. Graz, **121** (1979), 1–17.

[5] K.W. Bauer, *Determination and application of Vekua resolvents.* In: R. Conti (ed.), Recent advances in differential equations, 37–43, Academic Press, 1981.

[6] P. Berglez, *Differentialoperatoren bei partiellen Differentialgleichungen.* Ann. Mat. Pura Appl. **143** (1984), 155–185.

[7] P. Berglez, *On the solutions of the iterated Bers-Vekua equation.* In: H. Florian et al. (eds.), Functional-Analytic and Complex Methods, their Interactions, and Applications to Partial Differential Equations, 266–275, World Scientific, 2001.

[8] L. Bers, *Theory of Pseudo-Analytic Functions.* New York University, 1953.

[9] R. Castillo Perez, V.V. Kravchenko, R. Reséndiz Vázquez, *Solution of boundary and eigenvalue problems for second-order elliptic operators in the plane using pseudoanalytic formal powers.* Math. Methods Appl. Sci. **34** (2011), 455–468.

[10] B. Doyon, P. Fonseca, *Ising field theory on a pseudosphere.* J. Stat. Mech. Theory Exp. P07002 (2004), arXiv:hep-th/0404136v1.

[11] R. Heersink, *Über Lösungsdarstellungen und funktionentheoretische Methoden bei elliptischen Differentialgleichungen.* Ber. Math.-Statist. Sekt. Forsch. Graz **67** (1976), 1–79.

[12] B. Jancovici, G. Téllez, *Two-dimensional Coulomb systems on a surface of constant negative curvature.* J. Stat. Phys. **91** (1998), 953–977.

[13] V.V. Kravchenko, *Applied Pseudoanalytic Function Theory*, Birkhäuser, 2009.

[14] V.V. Kravchenko, S. Tremblay, *Spatial pseudoanalytic functions arising from the factorization of linear second-order elliptic operators.* Math. Methods Appl. Sci. **34** (2011), 1999–2010.

[15] M. Ku, D. Wang, L. Dong, *Solutions to polynomial generalized Bers-Vekua equations in Clifford analysis.* Complex Anal. Oper. Theory **6** (2012), 407–424.

[16] J. Palmer, M. Beatty, C.A.Tracy, *Tau functions for the Dirac operator on the Poincaré disk.* Commun. Math. Phys. **165** (1994), 97–173.

[17] A.Y. Timofeev, *Dirichlet problem for generalized Cauchy-Riemann systems with a singular point.* In: G. Giorgadze (ed.), Recent Developments in Generalized Analytic Functions and Their Applications: Proceedings of the International Conference on Generalized Analytic Functions and Their Applications, Tbilisi, 2011.

[18] I.N. Vekua, *Generalized Analytic Functions.* Pergamon, 1962.

[19] I.N. Vekua, *New methods for solving elliptic equations.* North-Holland, 1967.

P. Berglez and T.T. Luong
Department of Mathematics
Graz University of Technology
A-8010 Graz, Austria
e-mail: `berglez@tugraz.at`
`luongtuyetdhxd@gmail.com`

Operator Theory:
Advances and Applications, Vol. 229, 101–122

The Factorization Problem: Some Known Results and Open Questions

Albrecht Böttcher and Ilya M. Spitkovsky

For Stefan Samko on his 70th birthday

Abstract. This is a concise survey of some results and open problems concerning Wiener-Hopf factorization and almost periodic factorization of matrix functions. Several classes of discontinuous matrix functions are considered. Also sketched is the abstract framework which unifies the two types of factorization.

Mathematics Subject Classification (2010). Primary 47A68; Secondary 30E25, 43A75, 45F15, 47B35 .

Keywords. Wiener-Hopf factorization, almost periodic factorization, matrix function.

1. Introduction

The factorization of matrix functions considered here has its roots in the Riemann-Hilbert boundary value problem. The latter consists in finding n-vector functions $\phi^{\pm}$ which are analytic in the domains $D^{\pm}$ into which the extended complex plane $\mathbb{C} \cup \{\infty\}$ is partitioned by a simple closed rectifiable curve Γ and whose boundary values on Γ satisfy the equation

$$\phi^{+} = G\phi^{-} + g. \tag{1.1}$$

Here G and g are given $n \times n$ matrix and n-vector functions, respectively, both residing on Γ.

Problem (1.1) can be recast in terms of the singular integral operator $R_G := P_+ - GP_-$, where $P_{\pm} = \frac{1}{2}(I \pm S)$ are the projections associated with the Cauchy singular integral operator S defined by

$$(S\phi)(t) = \frac{1}{\pi i} \int_{\Gamma} \phi(\tau) \frac{d\tau}{\tau - t}, \quad t \in \Gamma.$$

In the case where Γ is the unit circle $\mathbb{T}$, problem (1.1) is equivalent to an equation with block Toeplitz operators, and if $\Gamma = \mathbb{R}$, then problem (1.1) amounts to a system of Wiener-Hopf integral equations.

In all these settings, appropriate factorizations of the matrix function G play a key role. They may lead to a solution of the equation in "closed form" or may at least help to understand basic properties of the equation. We here focus our attention on Wiener-Hopf factorization and on almost periodic factorization, embarking on aspects which unify these factorizations in the final section.

The topic we have chosen for this survey is too large to be covered by a paper of reasonable length. We therefore had to make a selection. The philosophy behind our selection is that, except for the case of rational matrix functions, there is so far no universal strategy to determine when a matrix function is factorable and all the more no general method to obtain the factors. Consequently, one has to treat each matrix function as an individual and to try and take advantage of the peculiarities of the individual in order to get something useful. These peculiarities are smoothness and discontinuities on the one hand and additional structure (such as triangularity) on the other.

Wiener-Hopf factorization makes sense for matrix functions with entries in L^∞ and is hence confronting us with plenty of types of discontinuities. Therefore, in Section 2 we focus our attention on existence results for Wiener-Hopf factorization in dependence on solely the lack of smoothness or on the discontinuities. We deliberately decided not to include results based on additional structural properties, although in recent years notable progress has been made in proving the existence of a factorization and in constructing it for several special classes of (mostly 2×2) matrix functions, for example in [20, 22, 27, 28, 34, 35, 36, 43], to name a few. These results definitely deserve a separate survey.

On the other side, almost periodic matrix functions have a fixed type of discontinuities, and we thought it was worthwhile to pick this class in order to illustrate the variety of results on AP factorization in dependence on additional structure. Section 3 deals with triangular 2×2 almost periodic matrix functions, and the structure comes from the almost periodic polynomial which constitutes the $2, 1$ entry.

The objective of the final Section 4 is to discuss the two types of factorization from a unified point of view. We there show that certain common aspects of the factorizations can be understood and explained by having recourse to a few basic notions of abstract harmonic analysis.

2. Wiener-Hopf factorization

For technical reasons it will be convenient to suppose throughout this section that $\infty \in D^-$ (implying in particular that Γ is bounded) and $0 \in D^+$. There is no restriction of generality in supposing so, since a simple fractional linear transformation can be employed to enforce these conditions if they are not met right from the start.

By a *Wiener-Hopf factorization* (or simply a *factorization*) of G we understand a representation

$$G = G_+ \Lambda G_-^{-1} \tag{2.1}$$

almost everywhere on Γ, where Λ is a diagonal matrix function whose diagonal entries are of the form t^{κ_j} $(t \in \Gamma)$ with $\kappa_j \in \mathbb{Z}$ and where $G_\pm$ are matrix functions which can be continued to analytic and invertible matrix functions in $D_\pm$ and which are subject to further restrictions, depending on the specific type of the factorization. For example, such a further restriction could be that $G_\pm$ and $G_\pm^{-1}$ are required to belong to certain Lebesgue spaces on Γ. Different versions of such factorizations have been introduced, employed, and developed in pioneering works by many mathematicians, including Plemelj, Wiener and Hopf, Gakhov, Muskhelishvili and Vekua, Khvedelidze, Simonenko, Widom, Gohberg and Krein. The monographs [29, 41, 57, 61, 60] contain historical notes. We remark that one could replace (2.1) by $G = G_+\Lambda G_-$, which, however, causes a little change when passing to L_p factorization. We also want to mention that in part of the literature, especially in the context of Toeplitz operators, which are defined as $T_G := P_+G \,|\, \operatorname{Ran} P_+$, one works with (2.1) replaced by $G = G_-\Lambda G_+$.

There exist various settings of problem (1.1), differing by the conditions on G and the boundary behavior of $\phi^\pm$, and for each such setting there is a corresponding version of the factorization (2.1), varying by the boundary behavior of $G_\pm$ and their inverses. In each case there is an intrinsic relation between the solvability properties of (1.1) (equivalently, of the operator R_G) and factorability of G. We will first address the situation of *Banach algebras of continuous functions.*

Denote by $C(\Gamma)$ the Banach algebra of all continuous functions on Γ with the uniform norm. Let $\mathcal{C}$ be a subalgebra of $C(\Gamma)$ containing the identity and being a Banach algebra on its own with respect to some other norm, $\|.\|_{\mathcal{C}}$. Assume that $\mathcal{C}$ is continuously embedded in $C(\Gamma)$. The set $\mathcal{C}_\pm$ of all functions in $\mathcal{C}$ having continuous extensions to $D^\pm \cup \Gamma$ which are analytic in $D^\pm$ is a closed subalgebra of $\mathcal{C}$. The intersection $\mathcal{C}_+ \cap \mathcal{C}_-$ consists of constants only. The algebra $\mathcal{C}$ is called *decomposing* if it coincides with the sum $\mathcal{C}_+ + \mathcal{C}_-$. We will suppose that $\mathcal{C}$ is *inverse closed* (that is, if $f \in \mathcal{C}$ is invertible as an element of $C(\Gamma)$, its inverse automatically lies in $\mathcal{C}$ as well) and contains the set $\mathcal{R}$ of all rational functions with poles outside of Γ. All such algebras, decomposable or not, will be called C-algebras for short.

A *factorization in* $\mathcal{C}$ is a representation (2.1) in which $G_+^{\pm 1}$ and $G_-^{\pm 1}$ entrywise belong to $\mathcal{C}_+$ and $\mathcal{C}_-$, respectively. Of course, the invertibility of $\det G$ in $\mathcal{C}$ is necessary for G to admit such a factorization.

In the scalar setting ($n = 1$, so that $\det G = G$) this necessary condition is also sufficient if and only if $\mathcal{C}$ is decomposing. The invertibility of G is then also necessary and sufficient for the operator R_G to be Fredholm, in which case the index of R_G coincides with the winding number of G. For all specific algebras $\mathcal{C}$ studied in the literature, these equivalencies survive the transition to the matrix case

$(n > 1)$. This is true, in particular, for $\mathcal{R}$-*algebras* [38] (that is, algebras $\mathcal{C}$ in which $\mathcal{R}$ is dense with respect to $\|.\|_{\mathcal{C}}$), and in the classical case of Hölder-continuous functions on Lyapunov curves (see, e.g., [61, 85]). However, in the general setting it has only been proven [18, 19, 42] that factorability in $\mathcal{C}$ of every invertible G is equivalent to the decomposability of $\mathcal{C}$ and the Fredholmness of *all* operators from a certain family, including all R_G with invertible $\det G$. Factorability of a particular G easily implies the Fredholmness of the respective R_G, but whether or not the converse is true (and is equivalent to merely the invertibility of $\det G$) in any C-algebra, is not known. This remains an open question.

The algebra $C(\Gamma)$ is not decomposing (if it were, the factorization theory in C-algebras discussed above would be of no practical interest). Therefore, invertible continuous matrix (and even scalar) functions do in general not admit factorization (2.1) with continuous $G_\pm$. In order to embrace this situation fully, the boundary conditions on $G_\pm$ should be relaxed. This is how the concept of L_p factorization emerges.

To this end, recall the notion of the *Smirnov classes* $E_p^\pm$ which by definition consist of functions ψ such that $(\psi\circ\omega_\pm)\sqrt[p]{\omega'_\pm}$ belongs to the Hardy class H_p in the unit disk $\mathbb{D}$. Here $\omega_\pm$ is a conformal mapping of $\mathbb{D}$ onto $D^\pm$ and $0 < p < \infty$. The class $E_\infty^\pm$ consists of all functions which are analytic and bounded in $D^\pm$. We will identify functions from $E_p^\pm$ with their trace on Γ; the latter is a function in $L_p := L_p(\Gamma)$ and defines its analytic extension into $D^\pm$ uniquely. With this convention in mind, $E_p^+\cap E_p^-$ $(p\ge 1)$ consists of constant functions only.

Let $1 < p < \infty$ and put $q = p/(p-1)$. A Wiener-Hopf factorization (2.1) is called an L_p *factorization* of G if entry-wise $G_\pm \in E_p^\pm$, $G_\pm^{-1}\in E_q^\pm$ and if the operator $G_-\Lambda^{-1}P_-G_+^{-1}$ (a priori just closed and densely defined) is bounded on $L_p^n(\Gamma)$. (When replacing (2.1) by $G = G_+\Lambda G_-$, one has to require that $G_+\in E_p^+$, $G_-\in E_q^-$, $G_+^{-1}\in E_q^+$, $G_-^{-1}\in E_p^-$, and that $G_-^{-1}\Lambda P_- G_+^{-1}$ is bounded.) Having an L_p factorization of G allows us to solve problem (1.1) and to understand the Fredholm properties of the operator R_G: for example, a pioneering result by Simonenko [80] says that if G is entry-wise in $L_\infty := L_\infty(\Gamma)$, then R_G is Fredholm on $L_p^n(\Gamma)$ if and only if G admits an L_p factorization.

We remark that L_p factorization is frequently also referred to as generalized or Φ factorization in L_p. After preliminary work by Khvedelidze, this type of factorization was independently introduced by Simonenko [78, 79] and Widom [86, 87] already around 1960.

An L_p factorization can possibly exist only if $\det G$ is invertible in L_∞ and the operator S is bounded on $L_p(\Gamma)$, that is, when Γ is an *Ahlfors-David-Carleson curve*:

$$\sup_{t\in\Gamma}\sup_{\epsilon>0}|\Gamma(t,\epsilon)|/\epsilon < \infty,$$

where $|\Gamma(t,\epsilon)|$ stands for the length of the portion $\Gamma(t,\epsilon)$ of Γ located in the disk of radius ϵ centered at t.

It is also natural to consider the weighted L_p setting, with weights ρ for which S is bounded on

$$L_{p,\rho} := L_{p,\rho}(\Gamma) := \{f \colon \rho f \in L_p(\Gamma)\}, \quad \|f\|_{L_{p,\rho}} = \|\rho f\|_{L_p}.$$

These are exactly the weights satisfying the A_p condition

$$\sup_{t\in\Gamma}\sup_{\epsilon>0} \frac{1}{\epsilon} \left(\int_{\Gamma(t,\epsilon)} \rho(t)^p \, |dt| \right)^{1/p} \left(\int_{\Gamma(t,\epsilon)} \rho(t)^{-q} \, |dt| \right)^{1/q} < \infty,$$

implying in particular that $\rho \in L_p$, $\rho^{-1} \in L_q$. It follows that the weight ρ is the trace of functions $\rho_\pm \in E_p^\pm$ such that $\rho_\pm^{-1} \in E_q^\pm$. We put $E_{r,\rho}^\pm = \{f \colon \rho_\pm f \in E_r^\pm\}$ and introduce the notion of $L_{p,\rho}$ *factorization* by changing $E_p^\pm$, $E_q^\pm$, L_p to their weighted analogues $E_{p,\rho}^\pm$, $E_{q,\rho}^\pm$, $L_{p,\rho}$ in the definition of the L_p factorization above.

As it happens, the invertibility of $\det G$ is not only necessary but also sufficient for the L_p factorability of $G \in C^{n\times n}(\Gamma)$. Moreover, if this condition is satisfied, then the factorization is the same for all $p \in (1, \infty)$ and for all weights $\rho \in A_p$. In this generality, the result can be found in [10], where its long history is described in full detail; see also [39]. This settles the question of the existence of an $L_{p,\rho}$ factorization in the case of continuous matrix functions.

The situation is more complicated (and therefore interesting) in the discontinuous case. The invertibility of $\det G$, while still necessary, is not sufficient anymore for G to be factorable, and various additional conditions (depending on the nature of discontinuities) emerge.

The algebra $E_\infty^\pm + C$. This algebra is defined as the algebraic sum of $E_\infty^\pm$ and $C(\Gamma)$. It turns out that this sum is a closed subspace (and therefore a closed subalgebra) of L_∞. This algebra is not inverse closed: a function $G \in E_\infty^\pm + C$ is invertible in $E_\infty^\pm + C$ if and only if its harmonic extension to $D^\pm$ is bounded away from zero in some neighborhood of Γ, which is a more restrictive requirement than sole invertibility in L_∞. The latter amounts to G being bounded away from zero on Γ only.

A matrix function G with entries in $E_\infty^\pm + C$ admits an $L_{p,\rho}$ factorization if and only if $\det G$ is invertible in the algebra. As it was the case with continuous matrix functions, there is no dependency on p or ρ. For L_2 factorization on $\mathbb{T}$, this result goes back to Douglas [32]; further references and comments can be found in [39, 57].

The algebra PC. This algebra consists of all functions on Γ which have essential one-sided limits at each point of Γ. For Lyapunov curves Γ and power weights $\rho(t) = \prod |t - t_j|^{\beta_j}$, the $L_{p,\rho}$ factorization theory of matrix functions with entries in PC starts with Widom's paper [87] and was elaborated to a comprehensive body of knowledge by Gohberg and Krupnik (see [41, 39] for the full history, detailed statements, and references). The key point here is that the factorization depends on p and ρ. Namely, for $G \in PC^{n\times n}$ to admit an $L_{p,\rho}$ factorization it is necessary and sufficient that

$$\frac{1}{2\pi}\omega \neq \frac{1}{p} + \beta(t). \tag{2.2}$$

Here ω stands for all possible values of the arguments of the eigenvalues $\lambda_j(t)$ of $G(t-0)^{-1}G(t+0)$ with $t \in \Gamma$, while $\beta(t) = \beta_j$ if $t = t_j$ is a node of the power weight ρ and $\beta(t) = 0$ otherwise.

It was discovered later by Yu.I. Karlovich and the authors that for PC matrix functions a qualitative change occurs when passing to Ahlfors-David-Carleson curves and A_p weights; see [84] and the monograph [10]. Namely, given such a curve Γ and such a weight ρ, one may associate with every point $t \in \Gamma$ two real-valued functions $\alpha_t \le \beta_t$ on $\mathbb{R}$, one concave and one convex and both with linear asymptotics at $\pm\infty$, so that $G \in PC^{n\times n}$ is $L_{p,\rho}$ factorable if and only if

$$\frac{1}{2\pi}\arg\lambda_j(t) \neq \frac{1}{p} + \theta\alpha_t(|\lambda_j(t)|) + (1-\theta)\beta_t(|\lambda_j(t)|), \quad \theta \in [0,1]. \tag{2.3}$$

For Lyapunov and some other curves, α_t and β_t are constant, so that (2.3) simplifies to $\arg\lambda_j(t)$ being forbidden from some closed interval, which is just the case covered in [84]. For certain weights (called "power-like") we also have $\alpha_t = \beta_t$, and in these cases the interval collapses to a single point, as it was the case in the Widom-Gohberg-Krupnik setting.

The algebra PQC. Let now $\Gamma = \mathbb{T}$. By definition, QC is the intersection of $H^+_\infty + C$ with $H^-_\infty + C$. Hence QC is a C^*-algebra. Its maximal ideal space $\mathcal{M}(QC)$ consists of the fibers $\mathcal{M}_t(QC)$, $t \in \mathbb{T}$, where $\xi \in \mathcal{M}_t(QC)$ if and only if $\xi(a) = a(t)$ for all $a \in C(\mathbb{T})$. The smallest closed subalgebra of L_∞ containing QC and PC is PQC, the algebra of *piecewise quasicontinuous functions.* This is also a C^*-algebra, and Sarason [76] discovered that the fibers $\mathcal{M}_t(QC)$ refine further into $\mathcal{M}_t^-(QC) \cup \mathcal{M}_t^+(QC) \cup \mathcal{M}_t^0(QC)$. For $\xi \in \mathcal{M}_t^\pm(QC)$ the fiber $\mathcal{M}_\xi(PQC)$ consists of just one element, which assumes the value $a(t\pm 0)$ on functions $a \in PC$. Denote this element by $\xi_\pm$. On the other hand, for $\xi \in \mathcal{M}_t^0(QC)$ the fiber $\mathcal{M}_\xi(PQC)$ is a doubleton, denoted by $\{\xi_+, \xi_-\}$, where again $\xi_\pm(a) = a(t\pm 0)$ for $a \in PC$.

A factorability criterion in L_2 for scalar PQC functions (in terms of the related Toeplitz operators) was given by Sarason [76]. It was extended to arbitrary p and power weights in [15] and states that a scalar function $G \in PQC$ is $L_{p,\rho}$ factorable if and only if condition (2.2) is satisfied for ω covering the set of arguments of $\xi_+(G)/\xi_-(G)$ for all $t \in \mathbb{T}$, $\xi \in \mathcal{M}_t^0(QC)$. The matrix version of this result also holds; see [13].

Since the results for PC matrices are now available in the setting of general curves and weights, it would be natural to bring the PQC setting to the same level, or at least to tackle the case of arbitrary A_p weights while $\Gamma = \mathbb{T}$. Both versions presently remain open problems.

The PQS setting. Recall that the *field of values* (also known as the numerical range or the Hausdorff set) of a given $n \times n$ matrix A is defined as $F(A) = \{\langle Ax, x\rangle \colon \|x\| = 1\}$. Since $F(A)$ is a compact and convex subset of $\mathbb{C}$, it is contained in some sector S with vertex at the origin and of opening strictly less than π as long as it does not contain the origin. In this case, A is called *sectorial.* Let again $\Gamma = \mathbb{T}$ and ρ be a power weight. A matrix function G is by definition *(p, ρ)-sectorial over QC* if for all $t \in \mathbb{T}$, $\xi \in \mathcal{M}_t(QC)$ there exist constant invertible matrices B, C such that for some γ from the interval with endpoints $-\beta(t)$ and $-\beta(t) + (p-2)/p$ and a sector $\mathcal{S}$ with vertex at the origin and of opening strictly less than $2\pi/\max\{p, q\}$,

$$F(BG(x)C) \subset \begin{cases} \mathcal{S} & \text{for } x \in \mathcal{M}_{\xi_-}(L_\infty), \\ e^{2\pi i\gamma}\mathcal{S} & \text{for } x \in \mathcal{M}_{\xi_+}(L_\infty). \end{cases}$$

It was shown in [14] that a matrix function is $L_{p,\rho}$ factorable whenever it is (p, ρ)-sectorial over QC. This result covers in particular the case of *piecewise sectorial* matrix functions considered in [82] and unifies the PQC results with the classical factorability results for matrix functions satisfying the sectoriality condition $F(G(t)) \subset \mathcal{S}$ a.e. on $\mathbb{T}$. Transition to arbitrary A_p weights is a standing question.

We note that a further natural step, namely passage to arbitrary Ahlfors-David-Carleson curves, represents a potentially bigger challenge here than in the purely PQC setting. Indeed, condition (2.3) suggests that even in the scalar case the factorability condition then depends on the absolute values of $G(t)$ and not only on the behavior of its argument. Consequently, changes are called for in the notion of sectoriality itself. Perhaps $\mathcal{S}$ should be substituted by a different shape, determined by the geometry of Γ, the behavior of ρ, and their interplay.

Another open sectoriality-related question is as follows. For smooth Γ the already mentioned results imply that any matrix function G of the form $G = X \cdot H$, where X is invertible in $(E_\infty^+ + C)^{n \times n}$ and $F(H(t)) \subset \mathcal{S}$ a.e. on Γ for some sector $\mathcal{S}$ with opening smaller than $2\pi/\max\{p, q\}$, is L_p factorable. Due to the symmetry of the condition imposed, it is actually L_r factorable for all r between p and q. The converse is also true if either $p = 2$ [64, 66] or $n = 1$ [56]. But there are no counterexamples known without these additional conditions. Proving the converse in general is another open problem.

The algebra SAP. This algebra, consisting of the *semi-almost periodic functions,* is of particular interest; see Chapter 4 of [12]. Here things are simpler under the assumption that $\Gamma = \mathbb{R}$, although this does not fit with our standing requirement that Γ be bounded. The notion of $L_{p,\rho}$ factorization requires some mild modifications when replacing Γ with $\mathbb{R}$; see, for example, page 306 of [12]. We introduce the abbreviation $e_\lambda(x) := e^{i\lambda x}$ $(x \in \mathbb{R})$. Almost periodic polynomials are, by definition, finite linear combinations over $\mathbb{C}$ of the functions e_λ, $\lambda \in \mathbb{R}$. They form a (non-closed) subalgebra APP of $L_\infty(\mathbb{R})$. Its uniform closure is the classical Bohr algebra AP, while the closure under the stronger Wiener norm $\left\|\sum c_j e_{\lambda_j}\right\| := \sum |c_j|$ is the algebra $APW (\subset AP)$.

For every $f \in AP$ there exists the *Bohr mean value*

$$\mathbf{M}(f) = \lim_{T\to\infty} \frac{1}{2T} \int_{-T}^{T} f(x)\,dx$$

and therefore also the *Bohr-Fourier coefficients* $\widehat{f}(\lambda) := \mathbf{M}(e_{-\lambda}f)$. The latter differ from zero only for at most countably many values of $\lambda \in \mathbb{R}$; the set of all such λ constitutes the *Bohr-Fourier spectrum* $\Omega(f)$ of f.

The C^*-algebra SAP is defined as the subalgebra of L_∞ generated by AP and the algebra $C(\overline{\mathbb{R}})$ of all functions which are continuous on the two-point compactification of $\mathbb{R}$. Introduced by Sarason [77], it actually consists of all continuous functions f on $\mathbb{R}$ for which there exist $f_\pm \in AP$ such that $\lim_{x\to\pm\infty}(f(x)-f_\pm(x)) = 0$. These $f_\pm$ are defined by f uniquely, and we will call them the *AP-representatives of f at $\pm\infty$*. For vector or matrix functions F, their Bohr-Fourier coefficients (if $F \in AP^{n\times n}$) or AP representatives (if $F \in SAP^{n\times n}$) are understood entry-wise.

The factorization theory for matrix functions with entries in SAP was developed by Yu.I. Karlovich and one of the authors (starting with [48, 50]), and a systematic exposition can be found in [12]. The main result can be stated as follows: $G \in SAP^{n\times n}$ is L_p factorable if and only if (i) $G(t)$ is invertible for all $t \in \mathbb{R}$, (ii) the AP representatives $G_\pm$ of G admit *canonical* L_p factorizations, that is, representations of type (2.1) with missing middle factor, and (iii) for the *geometric means* $\mathbf{d}(G_\pm)$ of $G_\pm$, the arguments of all the eigenvalues of $\mathbf{d}(G_-)^{-1}\mathbf{d}(G_+)$ differ from $2\pi/p$. The definition of the geometric mean will be given in the next paragraph.

Note that for $F \in APW^{n\times n}$ a canonical L_p factorization $F = F_+F_-^{-1}$, if it exists, automatically has the property that the entries of $F_\pm$ and $F_\pm^{-1}$ belong to $APW^\pm$, the subalgebras of APW consisting of the functions whose Bohr-Fourier spectrum is contained in $\mathbb{R}_\pm$. In this situation the geometric mean $\mathbf{d}(F)$ is defined as $\mathbf{M}(F_+)\mathbf{M}(F_-^{-1})$. It can be shown that $\mathbf{d}(.)$, defined as described, can then be extended by continuity to the set of all canonically L_p factorable AP matrix functions.

Matrix functions $G \in AP^{n\times n}$ can be thought of as belonging to $SAP^{n\times n}$ and having AP representatives at $\pm\infty$ coinciding with G itself. Consequently, a byproduct of the criterion stated above is that if an AP matrix function is L_p factorable, then the factorization is automatically canonical. We will again embark on the factorability of AP matrix functions in Section 3.

The factorability criterion for SAP matrix functions generalizes immediately to the case of weighted spaces $L_{p,\rho}$ with weights having power behavior at ∞. The case of A_p weights was tackled in [11], but only under the a priori condition that the AP representatives of G are factorable. Lifting this condition remains an open problem.

The algebra $[SO, SAP]$. A continuous and bounded function f on $\mathbb{R}$ is said to be *slowly oscillating* if

$$\lim_{x\to+\infty} \sup\{|f(t)-f(\tau)| : t,\tau \in [-2x,x] \text{ or } t,\tau \in [x,2x]\} = 0.$$

The set of all slowly oscillating functions forms a C^*-subalgebra of QC, denoted by SO, and the notation $[SO, SAP]$ stands for the smallest closed subalgebra of $L_\infty(\mathbb{R})$ containing SO and SAP. It is crucial that SO and SAP are *asymptotically* *independent* algebras, that is, the fiber $\mathcal{M}_\infty([SO, SAP])$ is naturally isomorphic to $\mathcal{M}_\infty(SO)\times\mathcal{M}_\infty(SAP)$. Using this favorable circumstance, a factorability criterion was established [7]. This criterion says that a matrix function G with entries in $[SO, SAP]$ is L_p factorable if and only if G is invertible on $\mathbb{R}$ and, for every $\xi \in \mathcal{M}_\infty(SO)$, the restriction $G_\xi(\in SAP^{n\times n})$ under the above-mentioned natural isomorphism satisfies the SAP factorability criterion.

Extension of this result to A_p weights is of course an open question, since it is open even for SAP. It would also be interesting to consider (even in the unweighted case) the algebra generated by PQC and AP, which contains SAP and therefore $[SO, SAP]$.

Variable Lebesgue spaces. Let (X, μ) be a measure space and $p\colon X \to [1,\infty]$ be a measurable a.e. finite function on X. The space $L^{p(\cdot)}(X)$ consists of all (equivalence classes of) measurable functions f on X for which

$$I_p(f/\lambda) := \int_X |f(x)/\lambda)|^{p(x)}\, d\mu(x) < \infty$$

for some $\lambda > 0$. The infimum of all λ for which $I_p(f/\lambda) \le 1$ is by definition the norm of f in $L^{p(\cdot)}(X)$. Of course, for constant p the space $L^{p(\cdot)}(X)$ turns into the standard Lebesgue space $L^p(X)$. We will always suppose that $1 < p_- = \operatorname{ess\,inf} p(x)$, $p_+ = \operatorname{ess\,sup} p(x) < \infty$, under which conditions $L^{p(\cdot)}(X)$ is a reflexive Banach space.

A Fredholm criterion for singular integral operators with PC coefficients in the spaces $L^{p(\cdot)}(\Gamma)$ on sufficiently smooth curves (Lyapunov or Radon curves without cusps) and with p satisfying

$$|p(x) - p(y)| \le \frac{C}{-\log|x-y|}, \quad |x-y| \le \frac{1}{2}, \tag{2.4}$$

was obtained by Kokilashvili and Samko in [55]. Of course, this criterion can be recast as an $L^{p(\cdot)}(\Gamma)$ factorability criterion for PC matrix functions. Necessary factorability conditions for PC matrix functions were established by A.Yu. Karlovich in [44] in the case of arbitrary curves Γ and weights ρ such that the operator S is bounded on $L^{p(\cdot)}(\Gamma, \rho)$ and p still satisfies (2.4). These necessary conditions can formally be obtained from (2.3) by changing the constant p to $p(t)$, and it has been conjectured that these conditions are also sufficient. Their sufficiency was indeed established in the same paper [44] in the case of power weights and the curves allowed in [55], while Ahlfors-David-Carleson curves and some special weights (including power ones) were treated in [45]. The general case remains unsolved.

Rabinovich and Samko [67] were able to settle the case where the matrix function G and the weight ρ have only a finite set of slowly oscillating singularities and Γ is an Ahlfors-David-Carleson curve which is slowly oscillating at these singularities. The generalization of Simonenko's localization principle to variable exponent spaces played a crucial role there.

Consideration of other classes of (matrix) functions, such as SAP in particular, remains an open problem for variable Lebesgue spaces. A first step in this direction was made in [46], where it was shown that for the factorability of a SAP matrix function G in $L^{p(\cdot)}(\mathbb{R})$ it is necessary that its almost periodic representatives are L_r factorable for *some* r, but anything more is not yet known.

3. *AP* factorization

As mentioned earlier, AP matrix functions can admit only canonical L_p (Wiener-Hopf) factorizations. On the other hand, any invertible *scalar* APW function g can be represented as

$$g = g_+ e_\mu g_-,$$

with $g_\pm$ invertible in $APW^\pm$ and a (uniquely defined) $\mu \in \mathbb{R}$ equal the so-called *mean motion* of g. Comparing this representation with (2.1), it seems natural in the matrix case to introduce the *APW factorization* as a representation

$$G = G_+ \Lambda G_-, \tag{3.1}$$

where $G_\pm$ are now matrix functions invertible in $(APW^\pm)^{n\times n}$, while the non-zero entries of the diagonal factor Λ are all of the form e_{μ_j}. (Writing the factorization in the form $G_+\Lambda G_-$ instead of $G_+\Lambda G_-^{-1}$ or $G_-\Lambda G_+$ is a matter of tradition.) We remark that there are deeper reasons for choosing the new form of the diagonal entries in Λ; more on that will be said in Section 4. It also makes sense to introduce the concept of *AP factorization*, where $G_\pm$ are required to be invertible in $(AP^\pm)^{n\times n}$, with $AP^\pm$ standing for the class of all AP functions with the Bohr-Fourier spectra in $\mathbb{R}_\pm$.

Many properties of AP factorization mimic those of the Wiener-Hopf factorization. Some of them, such as the description of all factorizations of a given matrix function (provided that it is factorable), in particular invariance of the *partial AP indices* μ_j up to permutations, are even obtainable along the same lines. Others, though looking innocently similar to those of Wiener-Hopf factorization (e.g., APW factorability of locally sectorial APW matrix functions) require much more involved proofs. But more importantly, there is a striking difference which AP and APW factorization in the matrix case exhibits in comparison with both such factorization in the scalar case and the traditional Wiener-Hopf factorization. Namely, starting with $n = 2$ not all invertible AP, APW, or even APP matrix functions are AP factorable! (Note that, in contrast to this, it is known since [40] that every invertible matrix function with entries in the Wiener algebra W is always Wiener-Hopf factorable in W.) This phenomenon manifests itself already in the case of triangular matrix functions, such as

$$G_{f,\lambda} = \begin{bmatrix} e_\lambda & 0 \\ f & e_{-\lambda} \end{bmatrix}. \tag{3.2}$$

These particular matrix functions arise in applications to convolution type equations on finite intervals, λ being the length of the interval and f being related to AP representatives of the Fourier transform of the equations' kernel. Consequently, a lot of effort was put into the factorization problem for the matrices

(3.2) with $f \in APW$. It is easy to show that without loss of generality one may assume that $\Omega(f) \subset (-\lambda, \lambda)$, and this condition will be silently imposed. The main results presently known are as follows.

Commensurable case. If the distances between the points of $\Omega(f)$ are commensurable, that is, $\Omega(f) \subset -\nu + h\mathbb{Z}$ for some $\nu, h \in \mathbb{R}$, then (3.2) is APW factorable, and the factorization can be constructed explicitly ([51, 54]; see also [12, Section 14.4]). This case covers in particular all *binomial* f, treated earlier in [50].

Big gap case. Suppose there exist $-\nu, \alpha \in [-\lambda, \lambda]$ belonging to the closure of $\Omega(f) \cup \{-\lambda, \lambda\}$ and such that $\alpha + \nu \geq \lambda$, $(-\nu, \alpha) \cap \Omega(f) = \emptyset$. Then $G_{f,\lambda}$ is APW factorable and the partial AP indices equal $\pm(\alpha + \nu - \lambda)$ if $-\nu, \alpha \in \Omega(f) \cup \{-\lambda, \lambda\}$, whereas $G_{f,\lambda}$ is not AP factorable otherwise. Thus, in this case the factorization of $G_{f,\lambda}$ (if it exists) is canonical if and only if the gap width is exactly λ.

Note that the "big gap" condition holds with $\alpha = \lambda$ ($\nu = \lambda$) if and only if $\Omega(f)$ lies to the left (respectively, to the right) of the origin, in which case $-\nu$ (respectively, α) is the point in the closure of $\Omega(f)$ which is closest to the origin, and the partial AP indices (provided that the factorization exists) are by absolute value equal to this distance. This is the so-called *one-sided case.*

For $f \in APP$, the condition $-\nu, \alpha \in \Omega(f) \cup \{-\lambda, \lambda\}$ holds automatically, and the APW factorability of $G_{f,\lambda}$ in this case was established in [50] (one-sided case) and in [65] (when either $(-\lambda, -\nu)$ or (α, λ) was disjoint with $\Omega(f)$). These results were extended to $f \in APW$ in [69], while the sufficiency part in its full generality was obtained in [24]. Finally, the necessity is in [21].

Special trinomials. Consider f for which $\Omega(f)$ consists of exactly three points:

$$f = c_{-1}e_{-\nu} + c_0 e_{\mu} + c_1 e_{\alpha}, \quad -\lambda < -\nu < \mu < \alpha < \lambda. \tag{3.3}$$

We may without loss of generality suppose that $\nu, \alpha > 0$ since otherwise this would be a one-sided case. Also, let $(\alpha + \nu)/(\mu + \nu)$ be irrational, to avoid commensurability. Herewith what is known about this situation.

Matrix functions of the form (3.2) with f given by (3.3) are APW factorable provided that $\alpha + \nu + |\mu| \geq \lambda$, unless

$$\alpha + \nu = \lambda, \ \mu = 0, \text{ and } |c_{-1}|^{\alpha} \cdot |c_1|^{\nu} = |c_0|^{\lambda}. \tag{3.4}$$

In the exceptional case (3.4), $G_{f,\lambda}$ is not AP factorable, in spite of being an AP polynomial – historically the first example of this nature.

These results are from [52, 65] and can also be found in [12, Chapter 15]. These publications contain recursive procedures for the factorization construction, terminating in finitely many steps if $\alpha + \nu > \lambda$, $\mu \neq 0$. If the first two equalities in (3.4) hold while the last one fails, the factorization of $G_{f,\lambda}$ is canonical but the recursion does not terminate. Explicit formulas for the factorization in this case were obtained via an alternative approach (involving a corona problem) [6]; see also [12, Chapter 23]. In the "borderline" cases $\alpha + \nu = \lambda$, $\mu \neq 0$ and $\alpha + \nu > \lambda$, $\mu = 0$, the factorization of $G_{f,\lambda}$ is canonical, and explicit (not just recursive) formulas for it are in [27] for the former borderline case and in [53] for the latter. Paper [53] also contains explicit formulas for the former borderline case which differ in their form from those of [27].

For $\mu = 0$, the restriction $\alpha+\nu \geq \lambda$ was further relaxed to $\alpha+\nu+\max\{\alpha,\nu\} \geq \lambda$, and the respective factorability criteria were established in [68]. As in the case $\alpha+\nu = \lambda$, $\mu = 0$, the factorization is canonical whenever it exists. The explicit factorization formulas for this setting are contained in [25].

There are other $\{-\nu, \mu, \alpha\}$ configurations for which the factorability problem is settled. We mention, for example, the case $\mu/\nu \in \mathbb{N}$, $2\alpha+\nu \geq \lambda$ [68], or

$$\mu \geq \alpha\frac{n-1}{n}, \qquad \frac{\lambda}{\alpha+\nu} \geq n-1, \text{ where } n = \left\lceil \frac{\lambda}{\mu+\nu} \right\rceil$$

(in the latter case actually the factorability persists when any number of additional exponents is allowed between μ and α) [24]. Nevertheless, the factorability problem for the matrices $G_{f,\lambda}$ with *arbitrary* trinomials (3.3), ten years after it was characterized as "the biggest mystery of the AP factorization theory" [12, Notes to Chapter 15], still remains open.

Shifted grids. Consider the matrix function (3.2) with

$$f = \sum_{j=0}^{n}(c_j + d_j e_{-\lambda})e_{\mu+jh} \tag{3.5}$$

for some $\mu, h > 0$ and $n < (\lambda-\mu)/h$. Let us also suppose that λ/h is irrational, again to avoid commensurability. If $\mu = 0$ and $c_j = d_j$, it was shown in [47] that $G_{f,\lambda}$ admits a canonical AP factorization if and only if the polynomial $\sum_{j=0}^{n} c_j t^j$ has no zeros on the unit circle. On the other hand, for $n = 1$ and under no additional restrictions on μ and the coefficients, $G_{f,\lambda}$ is canonically AP factorable if and only if

$$|c_1|^{\lambda-\mu}\,|d_1|^{\mu} \neq |c_2|^{\lambda-\mu-h}\,|d_1|^{\mu+h}. \tag{3.6}$$

This result (in different terms) was established in [3], while the factorization formulas for this case were obtained in [5]. In [68] it was shown in addition that $G_{f,\lambda}$ is not AP factorable at all when (3.6) fails.

Apparently, there should exist a unifying result, that is, a factorability criterion for matrices (3.2) with f given by (3.5) with no additional restrictions on μ, h, and the coefficients. Finding this result is yet another challenge.

There are not that many approaches to the AP factorization problem of matrix functions, even triangular ones. Many of the earlier results listed above were originally obtained by using the division algorithm introduced in [49], explained in detail in [51], [12, Section 14.5], and also used in [37]. Then, in [8], the so-called *Portuguese transformation* was developed. A recent criterion [21] states that a 2×2 (not necessarily triangular) APW matrix function G admits an APW factorization if and only if there exist vector functions $\psi^{\pm}$ with components in $APW^{\pm}$ such that $e_{-\kappa/2}G\psi^{-} = \psi^{+}$, where κ is the mean motion of $\det G$, and the components of ψ^{-} ($e_{-\delta}\psi^{+}$ for some $\delta \geq 0$) satisfy corona conditions in the lower (respectively, upper) half-plane. This δ, if it exists, is defined uniquely, and the partial AP indices of

G equal $\pm\delta + \kappa/2$. So, the existence (or non-existence) of an AP factorization of G can in principle be established if at least one non-zero solution to the respective homogenous Riemann-Hilbert problem is available. Sometimes these solutions can be constructed via considering kernels of asymmetric Toeplitz operators with scalar symbols. This approach looks promising, and some its implementations are contained in paper [23].

In all the cases described above, matrix functions which are not AP factorable can be approximated by AP factorable ones. This phenomenon actually persists in all the cases for which factorability criteria are presently known. It was therefore a natural guess that the set of AP factorable matrix functions is dense in the set of the invertible matrix functions in $AP^{n\times n}$. However, this is not true: it follows from a more general result in [17] that for every $n \geq 2$ infinitely many pathwise connected components of the set of all invertible AP or APW matrix functions of order n do not contain any AP factorable matrices. Moreover, these components actually do not contain any matrix functions from the closed subgroup generated by the AP factorable matrices.

The desired denseness result might hold nevertheless for matrix functions of the form (3.2). This still remains to be settled.

Equivalence. Let us say that two APW matrix functions G_1 and G_2 of size $n\times n$ are *equivalent* if there exist $n\times n$ matrix functions $X_\pm$ which are invertible in $(APW^\pm)^{n\times n}$ such that $G_1 = X_+G_2X_-$. Of course, this is indeed an equivalence relation, and an invertible matrix function G is APW factorable if and only if it is equivalent to a diagonal matrix function whose nonzero entries are of the form e_{μ_j} with $\mu_j \in \mathbb{R}$. The numbers μ_j are then defined uniquely up to a permutation, and they determine the equivalence class of G.

As we have seen, starting with $n = 2$ not all invertible APW matrix functions of order n are AP factorable. Consequently, there are equivalence classes free of diagonal representatives. What are their canonical representatives? How to describe (all of) them? Does every such class contains triangular matrix functions, or in other words, is any invertible APW matrix function triangularizable? All these questions are open. Even for $n = 2$!

4. Factorization on compact abelian groups

One can consider Wiener-Hopf and AP factorizations as concrete instances of more general factorizations in different ways. One possibility is based on passing from matrix functions G to operators generated by these matrix functions and then studying appropriate factorizations of the operators. This approach leads to the notion of abstract Wiener-Hopf factorization. The factors are in general no longer matrix functions but operators with special properties, which are related to the invariance of certain subspaces. We will not embark on this topic here and refer the reader to the papers [31, 63] and the book [81] for good introductions into this strategy.

Alternatively, it is beneficial to invoke concepts of abstract harmonic analysis in order to understand the unifying idea behind AP (APW) factorization and Wiener-Hopf factorization in $C(\mathbb{T})$ (respectively, the Wiener algebra $W(\mathbb{T})$). The purpose of this section is to introduce the reader into the latter point of view.

Let $\mathcal{G}$ be a (multiplicative) connected compact abelian group and let Γ be its (additive) character group. The Γ used in this section should not be confused with the curve Γ occurring in Sections 1 and 2. Recall that Γ consists of the continuous homomorphisms of $\mathcal{G}$ into the group $\mathbb{T}$ of unimodular complex numbers. Since $\mathcal{G}$ is compact, Γ is discrete (in the natural topology as a dual locally compact abelian group) [75, Theorem 1.2.5], and since $\mathcal{G}$ is connected, Γ is torsion free [75, Theorem 2.5.6]. By duality, $\mathcal{G}$ is the character group of Γ. Note that the character group of every torsion free abelian group with the discrete topology is connected and compact [75, Theorems 1.2.5, 2.5.6].

It is well known [75] that, because $\mathcal{G}$ is connected, Γ can be made into a linearly ordered group. So let $\preceq$ be a fixed linear order such that $(\Gamma, \preceq)$ is an ordered group. Let

$$\Gamma_+ = \{x \in \Gamma : x \succeq 0\}, \quad \Gamma_- = \{x \in \Gamma : x \preceq 0\}.$$

Widely used standard examples of Γ are $\mathbb{Z}$ (the group of integers), $\mathbb{Q}$ (the group of rationals with the discrete topology), $\mathbb{R}$ (the group of reals with the discrete topology), and $\mathbb{Z}^k$, $\mathbb{R}^k$ ($k \in \mathbb{Z}$) with lexicographic or other ordering.

Let $C(\mathcal{G})$ be the unital Banach algebra of (complex-valued) continuous functions on $\mathcal{G}$ (in the uniform topology), and let $P(\mathcal{G})$ be the (non-closed) subalgebra of $C(\mathcal{G})$ of all finite linear combinations of functions $\langle j, \cdot \rangle$, $j \in \Gamma$, where $\langle j, g \rangle$ stands for the action of the character $j \in \Gamma$ on the group element $g \in \mathcal{G}$ (thus, $\langle j, g \rangle \in \mathbb{T}$). Note that $P(\mathcal{G})$ is dense in $C(\mathcal{G})$ (this fact is a corollary of the Stone-Weierstrass theorem). For

$$a = \sum_{k=1}^{m} a_{j_k} \langle j_k, . \rangle \in P(\mathcal{G})$$

with distinct $j_1, \dots, j_k \in \Gamma$ and $a_{j_k} \neq 0$ for $k = 1, 2, \dots, m$, the *Bohr-Fourier spectrum* is defined as the finite set $\sigma(a) := \{j_1, \dots, j_k\}$. The notion of the Bohr-Fourier spectrum is extended from functions in $P(\mathcal{G})$ to $C(\mathcal{G})$ by continuity; indeed, since the Bohr-Fourier coefficients are continuous in the uniform topology, we can use approximations of a given element in $C(\mathcal{G})$ by elements of $P(\mathcal{G})$. The Bohr-Fourier spectrum of $A = [a_{ij}]_{i,j-1}^{n} \in C(\mathcal{G})^{n \times n}$ is, by definition, the union of the Bohr-Fourier spectra of the a_{ij}'s.

For $\mathcal{G} = \mathbb{T}$ we have $\Gamma = \mathbb{Z}$, the characters are $j(t) = t^j$, and the Bohr-Fourier coefficients and spectra are simply the classical Fourier objects. On the other hand, $\Gamma = \mathbb{R}$ (with the discrete topology) corresponds to $\mathcal{G}$ being the Bohr compactification of $\mathbb{R}$ (with its standard topology). In this case the characters are e_λ, we have $C(\mathcal{G}) = AP$, and the Bohr-Fourier coefficients and spectra go into their namesakes for AP functions introduced earlier.

As in the AP case, the Bohr-Fourier spectra of elements of $C(\mathcal{G})$ are at most countable also for general $\mathcal{G}$; a proof for the case $\Gamma = \mathbb{R}$ can be found, for example, in [30, Theorem 1.15], and this proof can be easily extended to general connected compact abelian groups $\mathcal{G}$.

We say that a unital Banach algebra $\mathcal{B} \subseteq C(\mathcal{G})$ is *admissible* if

(1) $P(\mathcal{G})$ is dense in $\mathcal{B}$ and

(2) $\mathcal{B}$ is inverse closed, i.e., if $X \in \mathcal{B} \cap GL(C(\mathcal{G}))$ implies $X \in GL(\mathcal{B})$.

Important examples of admissible algebras are $C(\mathcal{G})$ itself and the *Wiener algebra* $W(\mathcal{G})$, which, by definition, consists of all functions a on $\mathcal{G}$ of the form

$$a(g) = \sum_{j \in \Gamma} a_j \langle j, g \rangle, \qquad g \in \mathcal{G}, \tag{4.1}$$

where $a_j \in \mathbb{C}$ and $\sum_{j \in \Gamma} |a_j| < \infty$. The norm in $W(\mathcal{G})$ is defined by

$$\|a\|_1 = \sum_{j \in \Gamma} |a_j|.$$

The inverse closedness of $W(\mathcal{G})$ follows from the Bochner-Phillips theorem [9] (a generalization of Wiener's classical theorem for the case when $\mathcal{G} = \mathbb{T}$).

Other examples of admissible algebras are weighted Wiener algebras. A function $\nu : \Gamma \to [1, \infty)$ is called a *weight* if

$$\nu(\gamma_1 + \gamma_2) \le \nu(\gamma_1)\nu(\gamma_2)$$

for all $\gamma_1, \gamma_2 \in \Gamma$ and

$$\lim_{m \to \infty} m^{-1} \log(\nu(m\gamma)) = 0$$

for every $\gamma \in \Gamma$. The weighted Wiener algebra $W_\nu(\mathcal{G})$ consists of all functions a on $\mathcal{G}$ of the form (4.1) with

$$\sum_{j \in \Gamma} \nu(j)|a_j| < \infty$$

and with the norm

$$\|a\|_\nu = \sum_{j \in \Gamma} \nu(j)|a_j|.$$

One can show that $W_\nu(\mathcal{G})$ is indeed an inverse closed unital Banach algebra; see [4] for inverse closedness.

For an admissible algebra $\mathcal{B}$, we denote by $\mathcal{B}_\pm$ the closed unital subalgebra of $\mathcal{B}$ formed by the elements of $\mathcal{B}$ with Bohr-Fourier spectrum in $\Gamma_\pm$.

We are now ready to introduce the concept of factorization in the connected compact abelian group setting. Let $\mathcal{B}$ be an admissible algebra, and let $A \in \mathcal{B}^{n \times n}$. A representation of the form

$$A(g) = A_+(g) \left(\operatorname{diag}\left(\langle j_1, g \rangle, \dots, \langle j_n, g \rangle\right)\right) A_-(g), \quad g \in \mathcal{G}, \tag{4.2}$$

where $A_\pm, A_\pm^{-1} \in \mathcal{B}_\pm^{n \times n}$ and $j_1, \dots, j_n \in \Gamma$, is called a (left) *$\mathcal{B}$-factorization* of A (with respect to the order $\preceq$). One can prove that the elements $j_1, \dots, j_n$ in (4.2) are uniquely determined by A, up to a permutation.

When $\mathcal{G} = \mathbb{T}$ and $\mathcal{B}$ is $C(\mathbb{T})$ or $W(\mathbb{T})$, we thus end up with the Wiener-Hopf factorization in the respective algebra of continuous functions as in Section 2. When $\mathcal{G}$ is the group $\mathbb{R}$ with the discrete topology, the AP and APW factorizations from Section 3 emerge. Mimicking the terminology from these situations, we call $j_1, \dots, j_n$ from (4.2) the *partial indices* of A. The sum $j_1 + \cdots + j_n$ is the *total* index of A. For $n = 1$, the only partial index of A (therefore coinciding with its total index) is simply called the *index* of A.

For $\mathcal{G} = \mathbb{R}^k$ the notion of $\mathcal{B}$-factorization yields AP and APW factorizations in several variables, with the partial indices being k-tuples of real numbers. We refer to these factorizations as AP^k and APW^k factorizations. In the form presented here, they were introduced in [72] and further considered in [73, 74]. We remark that $W(\mathcal{G})$-factorization appeared already in [1] in the scalar case and in [2] in the matrix case. In its full generality, factorization on compact abelian groups was treated starting with [59, 58]; see [33] for the factorization in weighted Wiener algebras.

Among other things, papers [2, 72, 73] establish the equivalence between the existence of a canonical factorization in the appropriate setting and the invertibility of respective Toeplitz operators. One question raised in [72, 73] concerns the description of the algebras for which Fredholmness and invertibility of Toeplitz operators with matrix symbols occur only simultaneously (for scalar symbols the answer is known from [1]). That question is prompted by the observation that this equivalence does not hold for $W(\mathbb{T})$ but that it does for APW. The latter result was extended to APW^k in [73], but for algebras generated by subgroups of $\mathbb{R}^k$ the question remains open. In the abstract abelian groups setting it was brought up again in [16].

In [74], for hermitian matrix function admitting a (not necessarily canonical) AP^k factorization, a special form of this factorization was derived. The $C(\mathbb{T})$ version of this result goes back to [62], see also [29, Chapter V], while the AP version is in [83]. This result surely holds in the $\mathcal{B}$-factorization setting but is yet to be recorded.

In the same paper [74], the behavior of the partial indices under small perturbations is discussed. An order $\succ$ on the n-tuples with $\mathbb{R}^k$ entries is introduced such that if $\lambda \succ \mu$ and A is an AP^k factorable matrix function with the set λ of partial indices, then in every neighborhood of A there are AP^k factorable matrix functions with the set μ of partial indices. This fact, along with its converse, was established in [50] for $k = 1$, and it is of course an AP version of the Gohberg-Krein [40] description for the case $\mathcal{G} = \mathbb{T}$. However, for $k > 1$ the converse is known to hold only for n-tuples λ which are minimal in the sense of the order $\succ$ (this result, though incomplete, allows us to prove that the minimal n-tuples are the only stable ones under small perturbations). It is conjectured in [74] that the converse holds in general, but the problem is presently open.

We mention also the following result of [70]: whenever a matrix function A is $\mathcal{B}$-factorable for some Banach algebra $\mathcal{B}$ associated with the character group Γ while the Bohr-Fourier spectrum of A lies in a subgroup Γ', it can be arranged that the Bohr-Fourier spectra of all factors lie in Γ'. In particular, the partial indices of A must be from Γ'. This result is rather obvious for $\Gamma = \mathbb{Z}$ but was posed as an open question (even for $\mathcal{B} = AP$) in [74], where it was settled for matrix functions admitting a canonical AP^k factorization. The case of canonical AP factorization was settled earlier in [6] and independently in [71].

Finally, the non-density of the set of factorable matrix functions in the set of all invertible matrix functions, mentioned in Section 3 for the AP setting, was actually proved in [17] in a more general framework. To be more specific, the result holds for any admissible Banach algebra $\mathcal{B}$ such that the character group Γ contains a copy of $\mathbb{Z}^3$. On the other hand, if Γ is isomorphic to a subgroup of $\mathbb{Q}$, then the set of $\mathcal{B}$-factorable matrix functions is dense. It is not clear what happens in the intermediate cases, in particular for Γ isomorphic to $\mathbb{Z}^2$.

The paramount problem, of course, is that of finding non-trivial conditions for the existence of a $\mathcal{B}$-factorization. For triangular matrix functions (even in the case when the order on Γ is not archimedean), some results in this direction can be found in [59, 58], but the bulk of the work definitely lies ahead.

References

[1] V. Adukov, *Wiener-Hopf operators on a subsemigroup of a discrete torsion free abelian group*, Integral Equations and Operator Theory **16** (1993), 305–332.

[2] ———, *On invertibility of matrix Wiener-Hopf operator on discrete linearly ordered abelian group*, Integral Equations and Operator Theory **23** (1995), 373–386.

[3] S. Avdonin, A. Bulanova, and W. Moran, *Construction of sampling and interpolating sequences for multi-band signals. The two-band case*, Int. J. Appl. Math. Comput. Sci. **17** (2007), 143–156.

[4] R. Balan and I. Krishtal, *An almost periodic noncommutative Wiener's lemma*, J. Math. Anal. Appl. **370** (2010), 339–349.

[5] M.A. Bastos, A. Bravo, Yu.I. Karlovich, and I.M. Spitkovsky, *Constructive factorization of some almost periodic triangular matrix functions with a quadrinomial off diagonal entry*, J. Math. Anal. Appl. **376** (2011), 625–640.

[6] M.A. Bastos, Yu.I. Karlovich, A.F. dos Santos, and P.M. Tishin, *The corona theorem and the canonical factorization of triangular AP-matrix functions – Effective criteria and explicit formulas*, J. Math. Anal. Appl. **223** (1998), 523–550.

[7] M.A. Bastos, Yu.I. Karlovich, and B. Silbermann, *Toeplitz operators with symbols generated by slowly oscillating and semi-almost periodic matrix functions*, Proc. London Math. Soc. **89** (2004), 697–737.

[8] M.A. Bastos, Yu.I. Karlovich, I.M. Spitkovsky, and P.M. Tishin, *On a new algorithm for almost periodic factorization*, Oper. Theory Adv. Appl. **103** (1998), 53–74.

[9] S. Bochner and R.S. Phillips, *Absolutely convergent Fourier expansion for non-commutative normed rings*, Ann. of Math. **43** (1942), 409–418.

[10] A. Böttcher and Yu.I. Karlovich, *Carleson curves, Muckenhoupt weights, and Toeplitz operators*, Birkhäuser Verlag, Basel and Boston 1997.

[11] A. Böttcher, Yu.I. Karlovich, and I.M. Spitkovsky, *Toeplitz operators with semi-almost periodic symbols on spaces with Muckenhoupt weight*, Integral Equations and Operator Theory **18** (1994), 261–276.

[12] ———, *Convolution operators and factorization of almost periodic matrix functions*, Birkhäuser Verlag, Basel and Boston 2002.

[13] A. Böttcher, S. Roch, B. Silbermann, and I. Spitkovsky, *A Gohberg-Krupnik-Sarason symbol calculus for algebras of Toeplitz, Hankel, Cauchy, and Carleman operators*, Oper. Theory Adv. Appl. **48** (1990), 189–234.

[14] A. Böttcher, B. Silbermann, and I.M. Spitkovskiĭ, *Toeplitz operators with piecewise quasisectorial symbols*, Bull. London Math. Soc. **22** (1990), 281–286.

[15] A. Böttcher and I. Spitkovsky, *Toeplitz operators with PQC symbols on weighted Hardy spaces*, J. Functional Analysis **97** (1991), 194–214.

[16] A. Brudnyi, L. Rodman, and I.M. Spitkovsky, *Factorization versus invertibility of matrix functions on compact abelian groups*, Oper. Theory Adv. Appl. **218** (2012), 225–240.

[17] ———, *Non-denseness of factorable matrix functions*, J. Functional Analysis **261** (2011), 1969–1991.

[18] M.S. Budjanu and I.C. Gohberg, *General theorems on the factorization of matrix-valued functions. I. Fundamental theorems*, Amer. Math. Soc. Transl. **102** (1973), 1–14.

[19] ———, *General theorems on the factorization of matrix-valued functions. II. Some tests and their consequences*, Amer. Math. Soc. Transl. **102** (1973), 15–26.

[20] M.C. Câmara and C. Diogo, *Invertibility of Toeplitz operators and corona conditions in a strip*, J. Math. Anal. Appl. **342** (2008), 1297–1317.

[21] M.C. Câmara, C. Diogo, and Yu.I. Karlovich, and I.M. Spitkovsky, *Factorizations, Riemann-Hilbert problems and the corona theorem*, arXiv:1103.1935v1 [math.FA] (2011), 1–32, accepted by the Journal of the London Mathematical Society.

[22] M.C. Câmara, C. Diogo, and L. Rodman, *Fredholmness of Toeplitz operators and corona problems*, J. Functional Analysis **259** (2010), 1273–1299.

[23] M.C. Câmara, Yu.I. Karlovich, and I.M. Spitkovsky, *Kernels of asymmetric Toeplitz operators and applications to almost periodic factorization*, Complex Analysis and Operator Theory, to appear.

[24] ———, *Almost periodic factorization of some triangular matrix functions*, Oper. Theory Adv. Appl. **190** (2009), 171–190.

[25] ———, *Constructive almost periodic factorization of some triangular matrix functions*, J. Math. Anal. Appl. **367** (2010), 416–433.

[26] M.C. Câmara and M.C. Martins, *Explicit almost periodic factorization for a class of triangular matrix functions*. J. Anal. Math. **103** (2007), 221–260.

[27] M.C. Câmara, A.F. dos Santos, and M.C. Martins, *A new approach to factorization of a class of almost-periodic triangular symbols and related Riemann-Hilbert problems*, J. Functional Analysis **235** (2006), 559–592.

[28] M.C. Câmara, A.F. dos Santos, and P.F. dos Santos, *Matrix Riemann-Hilbert problems and factorization on Riemann surfaces*, J. Functional Analysis **255** (2008), 228–254.

[29] K.F. Clancey and I. Gohberg, *Factorization of matrix functions and singular integral operators*, Birkhäuser Verlag, Basel and Boston 1981.

[30] C. Corduneanu, *Almost periodic functions*, J. Wiley & Sons, 1968.

[31] A. Devinatz and M. Shinbrot, *General Wiener-Hopf operators*, Trans. Amer. Math. Soc. **145** (1969), 467–494.

[32] R.G. Douglas, *Banach algebra techniques in operator theory*, Academic Press, New York 1972.

[33] T. Ehrhardt, C.V.M. van der Mee, L. Rodman, and I.M. Spitkovsky, *Factorizations in weighted Wiener algebras on ordered abelian groups*, Integral Equations and Operator Theory **58** (2007), 65–86.

[34] T. Ehrhardt and F.-O. Speck, *Transformation techniques towards the factorization of non-rational* 2×2 *matrix functions*, Linear Algebra Appl. **353** (2002), 53–90.

[35] T. Ehrhardt and I.M. Spitkovsky, *Factorization of piecewise-constant matrix functions, and systems of linear differential equations*, St. Petersburg Math. J. **13** (2002), 939–991.

[36] I. Feldman, I. Gohberg, and N. Krupnik, *An explicit factorization algorithm*, Integral Equations and Operator Theory **49** (2004), 149–164.

[37] M. Gelfand and I.M. Spitkovsky, *Almost periodic factorization: Applicability of the division algorithm*, Voronezh Winter Mathematical Schools (P. Kuchment and V. Lin, eds.), vol. 184, Advances in the Mathematical Sciences, no. 2, Amer. Math. Soc., 1998, pp. 97–109.

[38] I. Gohberg, *The factorization problem in normed rings, functions of isometric and symmetric operators, and singular integral equations*, Uspehi Mat. Nauk **19** (1964), 71–124 (in Russian).

[39] I. Gohberg, M.A. Kaashoek, and I.M. Spitkovsky, *An overview of matrix factorization theory and operator applications*, Oper. Theory Adv. Appl. **141** (2003), 1–102.

[40] I. Gohberg and M.G. Krein, *Systems of integral equations on a half-line with kernel depending upon the difference of the arguments*, Amer. Math. Soc. Transl. **14** (1960), 217–287.

[41] I. Gohberg and N. Krupnik, *One-dimensional linear singular integral equations*, Vols. 1 and 2, Birkhäuser Verlag, Basel and Boston 1992.

[42] G. Heinig and B. Silbermann, *Factorization of matrix functions in algebras of bounded functions*, Oper. Theory Adv. Appl. **14** (1984), 157–177.

[43] G. Janashia, E. Lagvilava, and L. Ephremidze, *A new method of matrix spectral factorization*, IEEE Trans. Inform. Theory **57** (2011), 2318–2326.

[44] A.Yu. Karlovich, *Fredholmness of singular integral operators with piecewise continuous coefficients on weighted Banach function spaces*, J. Integral Equations Appl. **15** (2003), 263–320.

[45] ———, *Singular integral operators on variable Lebesgue spaces over arbitrary Carleson curves*, Oper. Theory Adv. Appl. **202** (2010), 321–336.

[46] A.Yu. Karlovich and I.M. Spitkovsky, *On singular integral operators with semi-almost periodic coefficients on variable Lebesgue spaces*, J. Math. Anal. Appl. **384** (2011), 706–725.

[47] Yu.I. Karlovich, *Approximation approach to canonical APW factorability*, Izv. Vuzov., Sev.-Kavk. Region, 2005, pp. 143–151 (in Russian).

[48] Yu.I. Karlovich and I.M. Spitkovsky, *On the Noether property for certain singular integral operators with matrix coefficients of class SAP and the systems of convolution equations on a finite interval connected with them*, Soviet Math. Dokl. **27** (1983), 358–363.

[49] ———, *Factorization of almost periodic matrix functions and Fredholm theory of Toeplitz operators with semi almost periodic matrix symbols*, Linear and Complex Analysis Problem Book: 199 research problems, Lecture Notes in Mathematics **1043** (1984), 279–282.

[50] ———, *Factorization of almost periodic matrix-valued functions and the Noether theory for certain classes of equations of convolution type*, Mathematics of the USSR, Izvestiya **34** (1990), 281–316.

[51] ———, *Almost periodic factorization: An analogue of Chebotarev's algorithm*, Contemporary Math. **189** (1995), 327–352.

[52] ———, *Factorization of almost periodic matrix functions*, J. Math. Anal. Appl. **193** (1995), 209–232.

[53] ———, *Almost periodic polynomial factorization of some triangular matrix functions*, Oper. Theory Adv. Appl. **202** (2010), 337–354.

[54] Yu.I. Karlovich, I.M. Spitkovsky, and R.A. Walker, *Almost periodic factorization of block triangular matrix functions revisited*, Linear Algebra Appl. **293** (1999), 199–232.

[55] V. Kokilashvili and S. Samko, *Singular integrals in weighted Lebesgue spaces with variable exponent*, Georgian Math. J. **10** (2003), 145–156.

[56] N.Ya. Krupnik, *A criterion for singular integral operators with measurable coefficients to be Noetherian*, Sakharth. SSR Mecn. Akad. Moambe **80** (1975), 533–536.

[57] G.S. Litvinchuk and I.M. Spitkovsky, *Factorization of measurable matrix functions*, Birkhäuser Verlag, Basel and Boston 1987.

[58] C.V.M. van der Mee, L. Rodman, and I.M. Spitkovsky, *Factorization of block triangular matrix functions with off diagonal binomials*, Oper. Theory Adv. Appl. **160** (2005), 423–437.

[59] C.V.M. van der Mee, L. Rodman, I.M. Spitkovsky, and H.J. Woerdeman, *Factorization of block triangular matrix functions in Wiener algebras on ordered abelian groups*, Oper. Theory Adv. Appl. **149** (2004), 441–465.

[60] S.G. Mikhlin and S. Prössdorf, *Singular integral operators*, Springer-Verlag, Berlin 1986.

[61] N.I. Muskhelishvili, *Singular integral equations*, Dover Publications Inc., New York 1992.

[62] A.M. Nikolaichuk and I.M. Spitkovsky, *The Riemann boundary value problem with a Hermitian matrix*, Soviet Math. Dokl. **16** (1975), 533–536.

[63] V. Pellegrini, *General Wiener-Hopf operators and the numerical range of an operator*, Proc. Amer. Math. Soc. **38** (1973), 141–146.

[64] H.R. Pousson, *Systems of Toeplitz operators on H^2*, Proc. Amer. Math. Soc. **19** (1968), 603–608.

[65] D. Quint, L. Rodman, and I.M. Spitkovsky, *New cases of almost periodic factorization of triangular matrix functions*, Michigan Math. J. **45** (1998), 73–102.

[66] M. Rabindranathan, *On the inversion of Toeplitz operators*, J. Math. Mech. **19** (1969/1970), 195–206.

[67] V. Rabinovich and S. Samko, *Pseudodifferential operators approach to singular integral operators in weighted variable exponent Lebesgue spaces on Carleson curves*, Integral Equations and Operator Theory **69** (2011), 405–444.

[68] A. Rastogi, L. Rodman, and I.M. Spitkovsky, *Almost periodic factorization of* 2×2 *matrix functions: New cases of off diagonal spectrum*, **202** (2010), 469–487.

[69] L. Rodman and I.M. Spitkovsky, *Almost periodic factorization and corona theorem*, Indiana Univ. Math. J. **47** (1998), 1243–1256.

[70] ———, *Factorization of matrix functions with subgroup supported Fourier coefficients*, J. Math. Anal. Appl. **323** (2006), 604–613.

[71] L. Rodman, I.M. Spitkovsky, and H.J. Woerdeman, *Carathéodory-Toeplitz and Nehari problems for matrix-valued almost periodic functions*, Trans. Amer. Math. Soc. **350** (1998), 2185–2227.

[72] ———, *Factorization of almost periodic matrix functions of several variables and Toeplitz operators*, Oper Theory. Adv. Appl. **122** (2001), 385–416.

[73] ———, *Fredholmness and invertibility of Toeplitz operators with matrix almost periodic symbols*, Proc. Amer. Math. Soc. **130** (2002), 1365–1370.

[74] ———, *Noncanonical factorization of almost periodic multivariable matrix functions*, Oper. Theory Adv. Appl. **142** (2003), 311–344.

[75] W. Rudin, *Fourier analysis on groups*, John Wiley & Sons Inc., New York 1990.

[76] D. Sarason, *Toeplitz operators with piecewise quasicontinuous symbols*, Indiana Univ. Math. J. **26** (1977), 817–838.

[77] ———, *Toeplitz operators with semi-almost periodic symbols*, Duke Math. J. **44** (1977), no. 2, 357–364.

[78] I.B. Simonenko, *Riemann's boundary problem with a measurable coefficient*, Soviet Math. Dokl. **1** (1960), 1295–1298.

[79] ———, *The Riemann boundary value problem for n pairs of functions with measurable coefficients and its application to the investigation of singular integrals in the spaces* L^p *with weight*, Izv. Akad. Nauk SSSR. Ser. Mat. **28** (1964), 277–306.

[80] ———, *Some general questions of the theory of the Riemann boundary value problem*, Izv. Akad. Nauk SSSR. Ser. Mat. **2** (1968), 1091–1099.

[81] F.-O. Speck, *General Wiener-Hopf factorization methods*, Pitman, Boston 1985.

[82] I.M. Spitkovskiĭ, *Singular integral operators with a piecewise sectorial matrix symbol*, Soobshch. Akad. Nauk Gruzin. SSR **121** (1986), 249–252.

[83] I.M. Spitkovsky, *On the factorization of almost periodic matrix functions*, Math. Notes **45** (1989), no. 5–6, 482–488.

[84] ———, *Singular integral operators with PC symbols on the spaces with general weights*, J. Functional Analysis **105** (1992), 129–143.

[85] N.P. Vekua, *Systems of singular integral equations*, P. Noordhoff Ltd., Groningen 1967.

[86] H. Widom, *Inversion of Toeplitz matrices. III*, Notices Amer. Math. Soc. **7** (1960), 63.

[87] ———, *Singular integral equations in L_p*, Trans. Amer. Math. Soc. **97** (1960), 131–160.

Albrecht Böttcher
Fakultät für Mathematik
TU Chemnitz
D-09107 Chemnitz, Germany
e-mail: aboettch@mathematik.tu-chemntiz.de

Ilya M. Spitkovsky
Department of Mathematics
College of William and Mary
Williamsburg, VA 23187, USA
e-mail: ilya@math.wm.edu, imspitkovsky@gmail.com

Operator Theory:
Advances and Applications, Vol. 229, 123–137

A Class of Sub-elliptic Equations on the Heisenberg Group and Related Interpolation Inequalities

Jianqing Chen and Eugénio M. Rocha

To Professor Stefan Samko on the occasion of his 70th birthday

Abstract. We firstly prove the existence of least energy solutions to a class of sub-elliptic equations on the Heisenberg group. Then we use this least energy solution to give a sharp estimate to the smallest positive constant in the Gagliardo-Nirenberg inequality on the Heisenberg group. Finally we point out some extensions to the quasilinear sub-elliptic case.

Mathematics Subject Classification (2010). Primary 35J20; Secondary 35H05.

Keywords. Sub-elliptic equations, least energy solutions, interpolation inequalities.

1. Introduction

The well-known Gagliardo-Nirenberg inequality says: there is a positive constant C such that

$$\int_{\mathbb{R}^N} |u|^q dx \leq C \left(\int_{\mathbb{R}^N} |\nabla u|^2 dx \right)^{\frac{N(q-2)}{4}} \left(\int_{\mathbb{R}^N} |u|^2 dx \right)^{\frac{2q-N(q-2)}{4}} \tag{1.1}$$

holds for all $u \in W^{1,2}(\mathbb{R}^N)$, where $2 < q < 2^*$ and $2^* = 2N/(N-2)$ for $N \geq 3$ and $2^* = +\infty$ for $N = 2$. Denoted by C_{GN} the smallest positive constant C such

This work was supported by *FEDER* funds through *COMPETE* – Operational Programme Factors of Competitiveness ("Programa Operacional Factores de Competitividade") and by Portuguese funds through the *Center for Research and Development in Mathematics and Applications* (University of Aveiro) and the Portuguese Foundation for Science and Technology ("FCT – Fundação para a Ciência e a Tecnologia"), within project PEst-C/MAT/UI4106/2011 with COMPETE number FCOMP-01-0124-FEDER-022690.

that (1.1) holds. Weinstein [17] established a sharp estimate of C_{GN} by the least energy solution of the following semilinear elliptic equation

$$-\Delta u + u = |u|^{q-2}u, \quad u \in W^{1,2}(\mathbb{R}^N). \tag{1.2}$$

These results have been used extensively in the study of nonlinear Schrödinger equations, see, e.g., [12, 13]. Further applications of these results can be expected.

In this paper, we are concerned with a class of sub-elliptic equations on the Heisenberg group and related Sobolev inequalities and interpolation inequalities. The whole paper contains two parts. In the first part, we focus on semilinear equations and explain our strategy in detail via a simple model. In the second part, we point out some extensions to quasilinear equations. Let us begin with some definitions and some useful results. The Heisenberg group $\mathbb{H}^N$, whose points will be denoted by $\xi = (x, y, t)$, is identified with the Lie group $(\mathbb{R}^{2N+1}, \circ)$ with composition law defined by

$$\xi \circ \xi' = (x + x', y + y', t + t' + 2(\langle x', y\rangle - \langle x, y'\rangle)), \tag{1.3}$$

where $\langle\cdot,\cdot\rangle$ denotes the inner product in $\mathbb{R}^N$. A family of dilation on $\mathbb{H}^N$ is defined as

$$\delta_\tau(x, y, t) = (\tau x, \tau y, \tau^2 t), \quad \tau > 0. \tag{1.4}$$

The homogeneous dimension with respect to the dilation is $Q = 2N + 2$. Left invariant vector fields on the Heisenberg group have the form, e.g.,

$$X_j = \frac{\partial}{\partial x_j} + 2y_j\frac{\partial}{\partial t}, \quad Y_j = \frac{\partial}{\partial y_j} - 2x_j\frac{\partial}{\partial t}, \quad j = 1, 2, \dots, N. \tag{1.5}$$

We denote the horizontal gradient by $\nabla_H = (X_1, \dots, X_N, Y_1, \dots, Y_N)$ and write

$$\operatorname{div}_H(\nu_1, \nu_2, \dots, \nu_{2N}) = \sum_{j=1}^{N}(X_j\nu_j + Y_j\nu_{N+j}).$$

In this way, the sub-Laplacian Δ_H is expressed by

$$\Delta_H := \sum_{j=1}^{N}(X_j^2 + Y_j^2) = \operatorname{div}_H(\nabla_H).$$

There are several results in the literature about the existence of solutions to semilinear elliptic equations on the Heisenberg group. We refer the interested readers to [1, 2, 3, 6, 7, 10, 14, 15] and the references therein.

Our goal here is to give a sharp estimate to the smallest positive constant in the following inequality. Let $2 < p < 2 + \frac{2}{N}$. Then there is a positive constant C such that

$$\int_{\mathbb{H}^N} |u|^p d\xi \le C\left(\int_{\mathbb{H}^N} |\nabla_H u|^2 d\xi\right)^{\frac{Q(p-2)}{4}} \left(\int_{\mathbb{H}^N} |u|^2 d\xi\right)^{\frac{2p-Q(p-2)}{4}} \tag{1.6}$$

for all $u \in S_0^{1,2}(\mathbb{H}^N)$, where $S_0^{1,2}(\mathbb{H}^N)$ is the closure of $C_0^\infty(\mathbb{H}^N)$ under the norm

$$\|u\| := \left(\int_{\mathbb{H}^N} \left(|\nabla_H u|^2 + |u|^2\right) d\xi\right)^{1/2}.$$

Inequality (1.6) is the Heisenberg group counterpart of the usual Gagliardo-Nirenberg inequality. Denoted by C_H the smallest positive constant C such that (1.6) holds. In order to estimate C_H, we consider firstly the following sub-elliptic equation with pure power nonlinearity

$$-\triangle_H u(\xi) + u(\xi) = |u(\xi)|^{p-2}u(\xi), \quad u \in S_0^{1,2}(\mathbb{H}^N), \tag{1.7}$$

where $2 < p < 2 + \frac{2}{N}$. A function $u \in S_0^{1,2}(\mathbb{H}^N)$ is said to be a solution of (1.7) if and only if for any $\psi \in S_0^{1,2}(\mathbb{H}^N)$, there holds

$$\int_{\mathbb{H}^N} \left(\nabla_H u \nabla_H \psi + u\psi - |u|^{p-2}u\psi\right) d\xi = 0. \tag{1.8}$$

On $S_0^{1,2}(\mathbb{H}^N)$, we define the functionals

$$L(u) = \frac{1}{2}\int_{\mathbb{H}^N} |\nabla_H u|^2 d\xi + \frac{1}{2}\int_{\mathbb{H}^N} |u|^2 d\xi - \frac{1}{p}\int_{\mathbb{H}^N} |u|^p d\xi$$

and

$$I(u) = \int_{\mathbb{H}^N} \left(|\nabla_H u|^2 + |u|^2 - |u|^p\right) d\xi.$$

Denote the Nehari set by $\mathcal{N} = \left\{u \in S_0^{1,2}(\mathbb{H}^N)\backslash\{0\} \ : \ I(u) = 0\right\}$ and set

$$d = \inf\{L(u) \ : \ u \in \mathcal{N}\}. \tag{1.9}$$

Definition 1.1. Let Γ be the set of the solutions of (1.7). Namely,

$$\Gamma = \{\phi \in S_0^{1,2}(\mathbb{H}^N) \ : \ L'(\phi) = 0 \text{ and } \phi \neq 0\}.$$

Let $\mathcal{G}$ be the set of least energy solutions of (1.7), that is,

$$\mathcal{G} = \{u \in \Gamma \ : \ L(u) \leq L(v) \text{ for any } \ v \in \Gamma\}.$$

Our main results are the following two theorems.

Theorem 1.2. *If $2 < p < 2+\frac{2}{N}$, then (1.7) has a least energy solution $\phi \in S_0^{1,2}(\mathbb{H}^N)$ and $d = L(\phi)$.*

Theorem 1.3. *Let ϕ be a least energy solution of (1.7). Then the smallest positive constant C_H can be characterized by*

$$\begin{aligned} C_H^{-1} &= \frac{2p - Q(p-2)}{2p}\left(\frac{Q(p-2)}{2p - Q(p-2)}\right)^{\frac{Q(p-2)}{4}} \|\phi\|_{L^2}^{p-2} \\ &= \frac{2p - Q(p-2)}{2p}\left(\frac{Q(p-2)}{2p - Q(p-2)}\right)^{\frac{Q(p-2)}{4}} \left(\frac{2p - Q(p-2)}{p-2}d\right)^{\frac{p-2}{2}}, \end{aligned} \tag{1.10}$$

where d is defined in (1.9).

Remark 1.4. We do not know if ϕ is a unique least energy solution of (1.7). But the second equality in (1.10) implies that C_H is independent of the choice of ϕ. We also point out that if the homogeneous dimension Q is replaced by spatial dimension N, then this constant is consistent with the result of Weinstein [17] in euclidean case.

Before stating the main idea for estimating C_H, we recall that Weinstein [17] has managed to give a sharp estimate of C_{GN} by solving directly the following minimization problem

$$C_{GN}^{-1} = \inf_{u \in W^{1,2}(\mathbb{R}^N)\setminus\{0\}} \left(\int_{\mathbb{R}^N} |\nabla u|^2 dx\right)^{\frac{N(q-2)}{4}} \left(\int_{\mathbb{R}^N} |u|^2 dx\right)^{\frac{2q-N(q-2)}{4}} \left(\int_{\mathbb{R}^N} |u|^q dx\right)^{-1}.$$

Key elements of the approach in [17] contain three aspects:

(i) select a minimizing sequence $(u_n)_n$;
(ii) use symmetric rearrangement and then select a new minimizing sequence which is contained in $W_r^{1,2}(\mathbb{R}^N)$, where $W_r^{1,2}(\mathbb{R}^N) = \{u \in W^{1,2}(\mathbb{R}^N) : u(x) = u(|x|)\}$;
(iii) use the fact that $W_r^{1,2}(\mathbb{R}^N) \hookrightarrow L^q(\mathbb{R}^N)$ is compact for $N \geq 2,\ 2 < q < 2^*$ to get a minimizer.

Note that when $N = 1$, one may not have compact embedding from $W_r^{1,2}(\mathbb{R})$ to $L^q(\mathbb{R})$. For the sharp estimate of C_H in (1.6), one may try to estimate C_H by studying the a similar minimization problem as done in [17]. However, due to the structure of Heisenberg group $\mathbb{H}^N$, we do not know if a function $u(\xi)$ can be symmetrized by rearrangement. Hence we can not study the minimization problem by restricting the discussion on the function space of radially symmetric functions. Since we do not have similar compact embedding as those in the Sobolev space $W^{1,2}(\mathbb{R}^N)$. Hence it seems that the method in [17] can not be used to prove Theorem 1.3. Our idea of proving Theorem 1.3 is based on three steps. Firstly, we study the existence of a least energy solution ϕ of (1.7) and some properties of this solution; secondly, we estimate the smallest positive constant C_H both from above and below, instead of solving a minimization problem; finally combining these estimates and Theorem 1.2, we prove Theorem 1.3.

This paper is organized as follows. In Section 2, we prove Theorem 1.2. In Section 3, we use the results obtained in Section 2 to prove Theorem 1.3. In Section 4, we give some remarks. We also compare C_H with another constant C_S, where C_S is the smallest of all positive constants C such that

$$\left(\int_{\mathbb{H}^N} |u|^p d\xi\right)^{\frac{2}{p}} \leq C \int_{\mathbb{H}^N} \left(|\nabla_H u|^2 + |u|^2\right) d\xi \tag{1.11}$$

holds for any $u \in S_0^{1,2}(\mathbb{H}^N)$, see [8]. In Section 5, we extend Theorem 1.2 and Theorem 1.3 to a class of quasilinear sub-elliptic equation on the Heisenberg group. The main results are Theorem 5.1 and Theorem 5.3.

Notations: Throughout this paper all integrals are taken over $\mathbb{H}^N$ unless stated otherwise. $\langle \cdot, \cdot \rangle_H$ denotes the dual product between $S_0^{1,2}(\mathbb{H}^N)$ and its dual space. The norm in $L^s(\mathbb{H}^N)$ is denoted by $\|\cdot\|_{L^s}$. Positive constants are denoted by C or C_j $(j = 1,\ 2,\ \cdots)$, whose value may be different at different places.

2. Existence of least energy solutions of (1.7)

This section is concerned with the proof of Theorem 1.2. We always assume that $2 < p < 2 + \frac{2}{N}$. The following lemmas will be useful in what follows.

Lemma 2.1. *For any $u \in S_0^{1,2}(\mathbb{H}^N)\backslash\{0\}$, there is a unique $\theta_u > 0$ such that $\theta_u u \in \mathcal{N}$. Moreover, if $I(u) < 0$, then $0 < \theta_u < 1$.*

Proof. For any $u \in S_0^{1,2}(\mathbb{H}^N)\backslash\{0\}$, one can deduce from direct computation that there is

$$\theta_u = \|u\|^{\frac{2}{p-2}} \|u\|_{L^p}^{-\frac{p}{p-2}}$$

such that $\theta_u u \in \mathcal{N}$. It is easy to see that this θ_u is unique. From above and the expression of $I(u) = \|u\|^2 - \|u\|_{L^p}^p$, we know that if $I(u) < 0$, i.e., $\|u\|^2 < \|u\|_{L^p}^p$, then $0 < \theta_u < 1$. □

Lemma 2.2. *There is $\rho > 0$ such that for all $u \in \mathcal{N}$, $\|u\| \geq \rho$.*

Proof. For any $u \in \mathcal{N}$, we have from inequality (1.6) that

$$\|u\|^2 = \|u\|_{L^p}^p \leq C\|\nabla_H u\|_{L^2}^{\frac{Q(p-2)}{2}} \|u\|_{L^2}^{\frac{2p-Q(p-2)}{2}} \leq C\|u\|^{\frac{Q(p-2)}{2}} \|u\|^{\frac{2p-Q(p-2)}{2}} \leq C\|u\|^p,$$

which implies that $\|u\|^{p-2} \geq C^{-1}$. Choosing $\rho = C^{-\frac{1}{p-2}} > 0$, we get that $\|u\| \geq \rho$ for all $u \in \mathcal{N}$. □

Lemma 2.3. *Let $2 < p < 2 + \frac{2}{N}$ and $\Omega \subset \mathbb{H}^N$ be a smooth bounded domain. Then the embedding $S_0^{1,2}(\Omega) \hookrightarrow L^p(\Omega)$ is compact.*

Proof. Let $(u_n)_{n\in\mathbb{N}} \subset S_0^{1,2}(\Omega)$ be a bounded sequence. We may assume that $u_n \rightharpoonup u$ weakly in $S_0^{1,2}(\Omega)$. Using inequality (1.6), we deduce that

$$\begin{aligned} \|u_n - u\|_{L^p}^p &\leq C\|\nabla_H u_n - \nabla_H u\|_{L^2}^{\frac{Q(p-2)}{2}} \|u_n - u\|_{L^2}^{\frac{2p-Q(p-2)}{2}} \\ &\leq C\left(\|u_n\| + \|u\|\right)^{\frac{Q(p-2)}{2}} \|u_n - u\|_{L^2}^{\frac{2p-Q(p-2)}{2}}. \end{aligned}$$

Note that $S_0^{1,2}(\Omega) \hookrightarrow L^2(\Omega)$ is compact [7] (see also [14]). The boundedness of $(u_n)_{n\in\mathbb{N}}$ in $S_0^{1,2}(\Omega)$ implies that $\|u_n - u\|_{L^p} \to 0$ as $n \to \infty$. Hence the embedding $S_0^{1,2}(\Omega) \hookrightarrow L^p(\Omega)$ is compact. □

Lemma 2.4. *If $v \in \mathcal{N}$ and $L(v) = d$, then v is a least energy solution of* (1.7).

Proof. Since v is a minimizer of the minimum d, we obtain from the Lagrange multiplier rule that there is $\theta \in \mathbb{R}$ such that for any $\psi \in S_0^{1,2}(\mathbb{H}^N)$,

$$\langle L'(v), \psi\rangle_H = \theta\langle I'(v), \psi\rangle_H.$$

Note that

$$\langle I'(v), v\rangle_H = 2\|v\|^2 - p\int |v|^p d\xi = (2-p)\int |v|^p d\xi < 0$$

and $\langle L'(v), v\rangle_H = I(v) = 0$. We get that $\theta = 0$. Hence $L'(v) = 0$. From Definition 1.1, one sees easily that v is a least energy solution of (1.7). □

Proof of Theorem 1.2. Let $(v_n)_n \subset \mathcal{N}$ be a minimizing sequence. Then we deduce from Ekeland variational principle that there is a sequence $(u_n)_n \subset \mathcal{N}$ such that

$$L(u_n) \to d \quad \text{and} \quad L'(u_n) \to 0.$$

Using the Sobolev inequality and Lemma 2.2, we know that there are two positive constants M_1 and M_2 such that

$$M_1 \le \|u_n\| \le M_2.$$

Combining this with the fact that $\|u_n\|^2 = \int_{\mathbb{H}^N} |u_n|^p d\xi$, we have a positive constant M_3 such that

$$\limsup_{n\to\infty} \int_{\mathbb{H}^N} |u_n|^p d\xi \ge M_3 > 0. \tag{2.1}$$

Using concentration compactness lemma (originally due to [11], a recent version due to [16]), one has that if

$$\lim_{n\to\infty} \sup_{\eta\in\mathbb{H}^N} \int_{B(\eta,r)} |u_n(\xi)|^p d\xi = 0$$

for some $r > 0$, where $B(\eta, r)$ is a ball of $\mathbb{H}^N$ centered at η with radius r in the Heisenberg distance, then $u_n \to 0$ in $L^s(\mathbb{H}^N)$ for any $2 < s < 2 + \frac{2}{N}$. Therefore (2.1) implies that there exists $M_4 > 0$ and $r > 1$ such that

$$\liminf_{n\to\infty} \sup_{\eta\in\mathbb{H}^N} \int_{B(\eta,r)} |u_n(\xi)|^p d\xi \ge M_4 > 0. \tag{2.2}$$

We hence may assume that there are $\tilde{\xi}^n \in \mathbb{H}^N$ such that

$$\liminf_{n\to\infty} \int_{B(\tilde{\xi}^n,r)} |u_n(\xi)|^p d\xi \ge \frac{M_4}{2} > 0. \tag{2.3}$$

Since for any $\tilde{\xi} \in \mathbb{H}^N$, we know from direct computation that

$$L(u_n(\xi\circ\tilde{\xi})) = L(u_n(\xi)) \quad \text{and} \quad I(u_n(\xi\circ\tilde{\xi})) = I(u_n(\xi)).$$

Define $w_n(\xi) := u_n(\xi\circ\tilde{\xi}^n)$. We know that $L(w_n) = L(u_n)$ and $I(w_n) = I(u_n)$, $(w_n)_n$ is bounded in $S_0^{1,2}(\mathbb{H}^N)$ and satisfies

$$\liminf_{n\to\infty} \int_{B(0,r)} |w_n(\xi)|^p d\xi \ge \frac{M_4}{2} > 0. \tag{2.4}$$

Passing to a subsequence we may assume $w_n \rightharpoonup \phi$ weakly in $S_0^{1,2}(\mathbb{H}^N)$. Lemma 2.3 implies that $w_n \to \phi$ strongly in $L^p_{\text{loc}}(\mathbb{H}^N)$. Combining this with (2.4), we know that $\phi \neq 0$.

Next we will prove that $w_n \to \phi$ strongly in $S_0^{1,2}(\mathbb{H}^N)$. Indeed, if $I(\phi) < 0$, we get from Lemma 2.1 that there is a $0 < \theta_\phi < 1$ such that $\theta_\phi \phi \in \mathcal{N}$. Therefore using the Fatou lemma and $I(w_n) = 0$, we get that

$$\begin{aligned} d + o(1) = L(w_n) &= \left(\frac{1}{2} - \frac{1}{p}\right) \int |w_n|^p d\xi \geq \left(\frac{1}{2} - \frac{1}{p}\right) \int |\phi|^p d\xi + o(1) \\ &= \left(\frac{1}{2} - \frac{1}{p}\right) \theta_\phi^{-p} \int |\theta_\phi \phi|^p d\xi + o(1) = \theta_\phi^{-p} L(\theta_\phi \phi) + o(1). \end{aligned} \tag{2.5}$$

We now deduce from $0 < \theta_\phi < 1$ that $d > L(\theta_\phi \phi)$, which is a contradiction because of $\theta_\phi \phi \in \mathcal{N}$.

If $I(\phi) > 0$, then from the Brezis-Lieb lemma [5], one has that

$$0 = I(w_n) = I(\phi) + I(\psi_n) + o(1),$$

where $\psi_n = w_n - \phi$. Combining this with $I(\phi) > 0$, one deduces that

$$\limsup_{n\to\infty} I(\psi_n) < 0. \tag{2.6}$$

By Lemma 2.1, we have $\theta_n := \theta_{\psi_n}$ such that $\theta_n \psi_n \in \mathcal{N}$. Moreover, we claim that $\limsup_{n\to\infty} \theta_n \in (0,1)$. In fact if $\limsup_{n\to\infty} \theta_n = 1$, then there is a subsequence $(\theta_{n_j})_{j\in\mathbb{N}}$ such that $\lim_{j\to\infty} \theta_{n_j} = 1$. Thus from $\theta_{n_j}\psi_{n_j} \in \mathcal{N}$, one deduces that $I(\psi_{n_j}) = I(\theta_{n_j}\psi_{n_j}) + o(1) = o(1)$. This contradicts (2.6). Therefore $\limsup_{n\to\infty} \theta_n \in (0,1)$. Since

$$\begin{aligned} d + o(1) = L(w_n) &= \left(\frac{1}{2} - \frac{1}{p}\right) \int |w_n|^p d\xi \geq \left(\frac{1}{2} - \frac{1}{p}\right) \int |\psi_n|^p d\xi \\ &= \left(\frac{1}{2} - \frac{1}{p}\right) \theta_n^{-p} \int |\theta_n \psi_n|^p d\xi + o(1) = \theta_n^{-p} L(\theta_n \psi_n) + o(1), \end{aligned} \tag{2.7}$$

one deduces from $\limsup_{n\to\infty} \theta_n \in (0,1)$ that $d > L(\theta_n\psi_n)$, which is a contradiction because of $\theta_n \psi_n \in \mathcal{N}$.

Therefore $I(\phi) = 0$. Next we claim that $\psi_n \to 0$ in $S_0^{1,2}(\mathbb{H}^N)$. In fact if this is not true, i.e., $\|\psi_n\| \not\to 0$ as $n \to \infty$, then we have two cases:

(i) if $\int |\psi_n|^p d\xi \not\to 0$ as $n \to \infty$, we have from

$$0 = I(w_n) = I(\phi) + I(\psi_n) + o(1) = I(\psi_n) + o(1)$$

and the Brezis-Lieb lemma that

$$d + o(1) = L(w_n) = L(\phi) + L(\psi_n) + o(1) \geq d + d + o(1),$$

which is a contradiction;

(ii) if $\int |\psi_n|^p d\xi \to 0$ as $n \to \infty$, we have that

$$d + o(1) = L(w_n) = L(\phi) + \frac{1}{2}\|\psi_n\|^2 + o(1) \geq d + \frac{1}{2}\|\psi_n\|^2 + o(1) > d,$$

which is also a contradiction. Therefore we deduce that $w_n \to \phi$ in $S_0^{1,2}(\mathbb{H}^N)$. And ϕ is a minimizer of the minimum d. Lemma 2.3 implies that ϕ is a least energy solution of (1.7). The proof is complete. □

3. Sharp estimate of C_H

Throughout this section, we also assume $2 < p < 2 + \frac{2}{N}$. The aim of this section is to give a sharp estimate of C_H. Firstly, we give some properties of the least energy solutions of (1.7) which will be useful in what follows.

Lemma 3.1. *Let ϕ be a least energy solution of* (1.7). *Then*

$$\int |\nabla_H \phi|^2 d\xi = \frac{Q(p-2)}{2p - Q(p-2)} \int |\phi|^2 d\xi \text{ and } \int |\phi|^p d\xi = \frac{2p}{2p - Q(p-2)} \int |\phi|^2 d\xi.$$

Proof. Since ϕ is a least energy solution of (1.7), one has

$$\int |\nabla_H \phi|^2 d\xi + \int |\phi|^2 d\xi = \int |\phi|^p d\xi.$$

Note that for $\lambda > 0$ and $\tilde{\phi}_\lambda(\xi) := \lambda^{\frac{Q}{2}} \phi(\delta_\lambda(\xi))$, one has that

$$0 = \frac{\partial}{\partial \lambda} L(\tilde{\phi}_\lambda)\bigg|_{\lambda=1} = \int |\nabla_H \phi|^2 d\xi - \frac{Q(p-2)}{2p} \int |\phi|^p d\xi.$$

We obtain the conclusion from the above two equations. □

Next, for $\rho > 0$, set

$$T_\rho = \inf \left\{ \|u\|^2 \; : \; u \in S_0^{1,2}(\mathbb{H}^N) \text{ and } \int |u|^p d\xi = \rho \right\}.$$

Then we have the following lemma.

Lemma 3.2. *If ϕ is a minimizer obtained in Theorem* 1.2, *then ϕ is a minimizer of T_{ρ_0} with $\rho_0 = \int |\phi|^p d\xi$.*

Proof. Clearly $\|\phi\|^2 \geq T_{\rho_0}$. Now for any $u \in S_0^{1,2}(\mathbb{H}^N)$ satisfying $\int |u|^p d\xi = \int |\phi|^p d\xi$, there is a unique

$$\lambda_0 = \|u\|^{\frac{2}{p-2}} \left(\int |u|^p d\xi \right)^{-\frac{1}{p-2}}$$

such that $I(\lambda_0 u) = 0$. Since $\lambda_0 u \neq 0$ and ϕ achieves the minimum d, we have that

$$\begin{aligned} \left(\frac{1}{2} - \frac{1}{p}\right) \|\phi\|^2 = L(\phi) \leq L(\lambda_0 u) &= \left(\frac{1}{2} - \frac{1}{p}\right) \lambda_0^2 \|u\|^2 \\ &= \left(\frac{1}{2} - \frac{1}{p}\right) \|u\|^{\frac{4}{p-2}} \left(\int |u|^p d\xi \right)^{-\frac{2}{p-2}} \|u\|^2. \end{aligned}$$

It is deduced from $\int |u|^p d\xi = \int |\phi|^p d\xi$ and $\int |\phi|^p d\xi = \|\phi\|^2$ that $\|\phi\|^2 \leq \|u\|^2$. Since u is chosen arbitrarily, we get that $T_{\rho_0} \geq \|\phi\|^2$. In sum, $T_{\rho_0} = \|\phi\|^2$ and hence ϕ is a minimizer of T_{ρ_0}. □

Now we are in a position to give a sharp estimate of C_H. Denote

$$J(u)=\left(\int_{\mathbb{H}^N}|\nabla_H u|^2d\xi\right)^{\frac{Q(p-2)}{4}}\left(\int_{\mathbb{H}^N}|u|^2d\xi\right)^{\frac{2p-Q(p-2)}{4}}\left(\int_{\mathbb{H}^N}|u|^pd\xi\right)^{-1}.$$

Then the sharp estimate of C_H can be estimated by studying the following minimization problem

$$C_H^{-1}=\inf\left\{J(u)\ :\ u\in S_0^{1,2}(\mathbb{H}^N)\backslash\{0\}\right\}.$$

Proof of Theorem 1.3. The proof is divided into three steps.

Step 1. Note that $\phi\neq 0$ and $\phi\in S_0^{1,2}(\mathbb{H}^N)$. We obtain from Lemma 3.1 that

$$\begin{aligned}J(\phi)&=\left(\int_{\mathbb{H}^N}|\nabla_H\phi|^2d\xi\right)^{\frac{Q(p-2)}{4}}\left(\int_{\mathbb{H}^N}|\phi|^2d\xi\right)^{\frac{2p-Q(p-2)}{4}}\left(\int_{\mathbb{H}^N}|\phi|^pd\xi\right)^{-1}\\&=\left(\frac{Q(p-2)}{2p-Q(p-2)}\int|\phi|^2d\xi\right)^{\frac{Q(p-2)}{4}}\left(\int|\phi|^2d\xi\right)^{\frac{2p-Q(p-2)}{4}}\\&\quad\times\left(\frac{2p}{2p-Q(p-2)}\int|\phi|^2d\xi\right)^{-1}\\&=\frac{2p-Q(p-2)}{2p}\left(\frac{Q(p-2)}{2p-Q(p-2)}\right)^{\frac{Q(p-2)}{4}}\left(\int|\phi|^2d\xi\right)^{\frac{p-2}{2}}.\end{aligned}$$

This implies that

$$C_H^{-1}\leq\frac{2p-Q(p-2)}{2p}\left(\frac{Q(p-2)}{2p-Q(p-2)}\right)^{\frac{Q(p-2)}{4}}\left(\int|\phi|^2d\xi\right)^{\frac{p-2}{2}}.$$

Step 2. For any $u\in S_0^{1,2}(\mathbb{H}^N)\backslash\{0\}$, we define $w(\xi):=\lambda u(\delta_\mu(\xi))=\lambda u(\mu x,\mu y,\mu^2t)$, where λ and μ are positive parameters which will be determined later. By direct calculations, one gets firstly that

$$\int|\nabla_H w(\xi)|^2d\xi=\lambda^2\mu^{2-Q}\int|\nabla_H u(\xi)|^2d\xi,$$

$$\int|w(\xi)|^2d\xi=\lambda^2\mu^{-Q}\int|u(\xi)|^2d\xi\qquad\text{and}$$

$$\int|w(\xi)|^pd\xi=\lambda^p\mu^{-Q}\int|u(\xi)|^pd\xi.$$

Choose

$$\lambda^2\mu^{-Q}\int|u(\xi)|^2d\xi:=\int|\phi(\xi)|^2d\xi$$

and

$$\lambda^p\mu^{-Q}\int|u(\xi)|^pd\xi:=\int|\phi(\xi)|^pd\xi=\frac{2p}{2p-Q(p-2)}\int|\phi(\xi)|^2d\xi.$$

From the above two equalities, one deduces that

$$\lambda^2 = \left(\frac{2p}{2p-Q(p-2)}\right)^{\frac{2}{p-2}} \left(\int |u|^2 d\xi\right)^{\frac{2}{p-2}} \left(\int |u|^p d\xi\right)^{-\frac{2}{p-2}}$$

and

$$\mu^{2-Q} = \left(\frac{2p}{2p-Q(p-2)}\right)^{\frac{2}{p-2}\cdot\frac{2-Q}{Q}} \left(\int |u|^2 d\xi\right)^{\frac{p}{p-2}\cdot\frac{2-Q}{Q}} \times \left(\int |u|^p d\xi\right)^{-\frac{2}{p-2}\cdot\frac{2-Q}{Q}} \left(\int |\phi|^2 d\xi\right)^{-\frac{2-Q}{Q}}.$$

Secondly, from the choice of λ and μ, we have that $\int |w|^2 d\xi = \int |\phi|^2 d\xi$ and $\int |w|^p d\xi = \int |\phi|^p d\xi$. Using the fact that ϕ is a minimizer of T_{ρ_0} with $\rho_0 = \int |\phi|^p d\xi$, one gets that

$$\int |\nabla_H w|^2 d\xi \geq \int |\nabla_H \phi|^2 d\xi.$$

Hence we obtain that

$$\lambda^2 \mu^{2-Q} \int |\nabla_H u|^2 d\xi \geq \int |\nabla_H \phi|^2 d\xi = \frac{Q(p-2)}{2p-Q(p-2)} \int |\phi|^2 d\xi.$$

Therefore

$$\left(\frac{2p}{2p-Q(p-2)}\right)^{\frac{2}{p-2}(1+\frac{2-Q}{Q})} \left(\int |u|^2 d\xi\right)^{\frac{2}{p-2}+\frac{p}{p-2}\cdot\frac{2-Q}{Q}} \left(\int |u|^p d\xi\right)^{-\frac{2}{p-2}(1+\frac{2-Q}{Q})} \times \left(\int |\phi|^2 d\xi\right)^{-\frac{2-Q}{Q}} \int |\nabla_H u|^2 d\xi \geq \frac{Q(p-2)}{2p-Q(p-2)} \int |\phi|^2 d\xi.$$

It is now deduced by direct calculations that

$$J(u) \geq \frac{2p-Q(p-2)}{2p} \left(\frac{Q(p-2)}{2p-Q(p-2)}\right)^{\frac{Q(p-2)}{4}} \left(\int |\phi|^2 d\xi\right)^{\frac{p-2}{2}}.$$

Since u is chosen arbitrarily, we get that

$$C_H^{-1} \geq \frac{2p-Q(p-2)}{2p} \left(\frac{Q(p-2)}{2p-Q(p-2)}\right)^{\frac{Q(p-2)}{4}} \left(\int |\phi|^2 d\xi\right)^{\frac{p-2}{2}}.$$

From *Step* 1 and *Step* 2, one infers immediately the first equality of (1.10).

Step 3. By the fact of

$$d = L(\phi) = \frac{1}{2}\|\phi\|^2 - \frac{1}{p}\|\phi\|_{L^p}^p = \left(\frac{1}{2} - \frac{1}{p}\right)\|\phi\|_{L^p}^p = \frac{p-2}{2p-Q(p-2)} \int |\phi|^2 d\xi,$$

one deduces that

$$\int |\phi|^2 d\xi = \frac{2p-Q(p-2)}{p-2} d.$$

Combining this with the first equality of (1.10), we get the second equality of (1.10). The proof of Theorem 1.3 is complete. □

4. Further remarks on C_H

In this section, we give some remarks on the sharp estimate of C_H. Keep the following characterization of C_H in mind,

$$C_H^{-\frac{2}{p}} = \inf_{u \in S_0^{1,2}(\mathbb{H}^N)\setminus\{0\}} \frac{\left(\int |\nabla_H u|^2 d\xi\right)^{\frac{Q(p-2)}{2p}} \left(\int |u|^2 d\xi\right)^{\frac{2p-Q(p-2)}{2p}}}{\left(\int |u|^p d\xi\right)^{\frac{2}{p}}}. \tag{4.1}$$

We try to compare C_H with another smallest constant C_S in the usual Sobolev inequality on $S_0^{1,2}(\mathbb{H}^N)$. In here C_S is the smallest one of all positive constants C such that

$$\left(\int |u|^p d\xi\right)^{\frac{2}{p}} \leq C\left(\int |\nabla_H u|^2 d\xi + \int |u|^2 d\xi\right) \tag{4.2}$$

holds for any $u \in S_0^{1,2}(\mathbb{H}^N)$. Then C_S can be characterized as the following minimization problem

$$C_S^{-1} = \inf_{u \in S_0^{1,2}(\mathbb{H}^N)\setminus\{0\}} \frac{\int \left(|\nabla_H u|^2 + |u|^2\right) d\xi}{\left(\int |u|^p d\xi\right)^{\frac{2}{p}}}. \tag{4.3}$$

Our aim here is to compare $C_H^{-\frac{2}{p}}$ with C_S^{-1}. Before doing this, we give a sharp estimate of C_S.

Theorem 4.1. *Let $2 < p < 2 + \frac{2}{N}$. Then we have*

$$C_S^{-1} = \left(\frac{2p}{2p - Q(p-2)} \int |\phi|^2 d\xi\right)^{1-\frac{2}{p}}.$$

Proof. In the first place, since ϕ is a least energy solution of (1.7), we deduce from Lemma 3.1 that

$$\frac{\int \left(|\nabla_H \phi|^2 + |\phi|^2\right) d\xi}{\left(\int |\phi|^p d\xi\right)^{\frac{2}{p}}} = \left(\int |\phi|^p d\xi\right)^{\frac{p-2}{p}} = \left(\frac{2p}{2p - Q(p-2)} \int |\phi|^2 d\xi\right)^{\frac{p-2}{p}}.$$

This implies that

$$C_S^{-1} \leq \left(\frac{2p}{2p - Q(p-2)} \int |\phi|^2 d\xi\right)^{\frac{p-2}{p}}.$$

In the second place, for any $u \in S_0^{1,2}(\mathbb{H}^N)\setminus\{0\}$, $\tilde{u}(\xi) := \|\phi\|_{L^p}\|u\|_{L^p}^{-1} u(\xi)$ is such that $\int |\tilde{u}(\xi)|^p d\xi = \int |\phi(\xi)|^p d\xi$. Therefore we have from Lemma 3.2 that

$$\int \left(|\nabla_H \tilde{u}|^2 + |\tilde{u}|^2\right) d\xi \geq \int \left(|\nabla_H \phi|^2 + |\phi|^2\right) d\xi = \frac{2p}{2p - Q(p-2)} \int |\phi|^2 d\xi.$$

It is deduced from direct computations that

$$\frac{\int (|\nabla_H u|^2 + |u|^2)\, d\xi}{\left(\int |u|^p d\xi\right)^{\frac{2}{p}}} \geq \frac{2p}{2p - Q(p-2)} \int |\phi|^2 d\xi \left(\int |\phi|^p d\xi\right)^{-\frac{2}{p}}$$
$$= \left(\frac{2p}{2p - Q(p-2)} \int |\phi|^2 d\xi\right)^{1-\frac{2}{p}}.$$

Since u is chosen arbitrarily, we have that

$$C_S^{-1} \geq \left(\frac{2p}{2p - Q(p-2)} \int |\phi|^2 d\xi\right)^{1-\frac{2}{p}}.$$

Theorem 4.1 follows immediately from the above two steps. □

Remark 4.2. By the Hölder inequality, Theorem 1.2 and Theorem 4.1, we know that $(C_H)^{-\frac{2}{p}} < C_S^{-1}$.

5. Extensions to quasilinear equations

In this section, we will extend Theorem 1.2 and Theorem 1.3 to a class of quasilinear equations. For $m > 1$, define the sub-m-Laplacian $\triangle_{H,m}$ as

$$\triangle_{H,m} u := \mathrm{div}_H (|\nabla_H u|^{m-2} \nabla_H u).$$

Let $1 < m < Q := 2N + 2$ and $m < p < m_* := \frac{mQ}{Q-m}$. We consider the quasilinear equation

$$-\triangle_{H,m} u + |u|^{m-2} u = |u|^{p-2} u, \quad u \in S_0^{1,m}(\mathbb{H}^N), \tag{5.1}$$

where $u \equiv u(\xi)$, $\xi \in \mathbb{H}^N$ and $S_0^{1,m}(\mathbb{H}^N)$ is the closure of $C_0^\infty(\mathbb{H}^N)$ under the norm $\|u\|_m := \left(\int (|\nabla_H u|^m + |u|^m)\, d\xi\right)^{1/m}$. On $S_0^{1,m}(\mathbb{H}^N)$, we define the functionals

$$L_1(u) = \frac{1}{2} \int |\nabla_H u|^m d\xi + \frac{1}{2} \int |u|^m d\xi - \frac{1}{p} \int |u|^p d\xi$$

and

$$I_1(u) = \int (|\nabla_H u|^m + |u|^m - |u|^p)\, d\xi.$$

Denote the Nehari set $\mathcal{N}_1 = \left\{u \in S_0^{1,m}(\mathbb{H}^N) \backslash \{0\} \ : \ I_1(u) = 0\right\}$ and define

$$D_1 = \inf\{L_1(u) \ : \ u \in \mathcal{N}_1\}. \tag{5.2}$$

Theorem 5.1. *If $m < p < m_*$, then* (5.1) *has a least energy solution $\Phi \in S_0^{1,m}(\mathbb{H}^N)$ and $D_1 = L_1(\Phi)$.*

Remark 5.2. The proof of Theorem 5.1 follows the same line as the proof of Theorem 1.2. The only difference is that in the proof of Theorem 1.2, we have used the fact that $S_0^{1,2}(\Omega) \hookrightarrow L^p(\Omega)$ is compact for $2 < p < 2 + \frac{2}{N}$ and smooth bounded domain $\Omega \subset \mathbb{H}^N$. While in the proof of Theorem 5.1, one may use the fact that $S_0^{1,m}(\Omega) \hookrightarrow L^p(\Omega)$ is compact for $m < p < m_*$ and smooth bounded domain $\Omega \subset \mathbb{H}^N$, see [8].

The following inequality is an extension of (1.6). Let $m < p < m_*$, there is a positive constant C such that

$$\int |u|^p d\xi \leq C \left(\int |\nabla_H u|^m d\xi \right)^{\frac{Q(p-m)}{m^2}} \left(\int |u|^m d\xi \right)^{\frac{mp-Q(p-m)}{m^2}} \tag{5.3}$$

holds for all $u \in S_0^{1,m}(\mathbb{H}^N)$. Denoted by $C_{H,m}$ the smallest one of all positive constants C such that (5.3) holds. Then we have the following sharp estimate of $C_{H,m}$.

Theorem 5.3. *Let $m < p < m_*$ and Φ be a least energy solution of* (5.1). *Then the smallest positive constant $C_{H,m}$ can be characterized by*

$$\begin{aligned} C_{H,m}^{-1} &= \frac{mp - Q(p-m)}{mp} \left(\frac{Q(p-m)}{mp - Q(p-m)} \right)^{\frac{Q(p-m)}{m^2}} \|\Phi\|_{L^m}^{p-m} \qquad (5.4) \\ &= \frac{mp - Q(p-m)}{mp} \left(\frac{Q(p-m)}{mp - Q(p-m)} \right)^{\frac{Q(p-m)}{m^2}} \left(\frac{mp - Q(p-m)}{p-m} D_1 \right)^{\frac{p-m}{m}}, \end{aligned}$$

where D_1 is defined in (5.2).

Remark 5.4. The proof of Theorem 5.3 follows the same line as the proof of Theorem 1.3 with the help of the following two lemmas.

Lemma 5.5. *Let Φ be a least energy solution of* (5.1). *Then*

$$\int |\nabla_H \Phi|^m d\xi = \frac{Q(p-m)}{mp - Q(p-m)} \int |\Phi|^m d\xi$$

and

$$\int |\Phi|^p d\xi = \frac{mp}{mp - Q(p-m)} \int |\Phi|^m d\xi.$$

Next, for $\rho > 0$, set

$$T_{\rho,m} = \inf \left\{ \|u\|_m^m \; : \; u \in S_0^{1,m}(\mathbb{H}^N) \text{ and } \int |u|^p d\xi = \rho \right\}.$$

Then we have the following lemma.

Lemma 5.6. *Let Φ be the minimizer obtained in Theorem* 5.1. *Then Φ is a minimizer of $T_{\rho_0,m}$ with $\rho_0 = \int |\Phi|^p d\xi$.*

Remark 5.7. The proofs of Lemma 5.5 and Lemma 5.6 are similar to the proofs of Lemma 3.1 and Lemma 3.2.

Remark 5.8. One may also compare $C_{H,m}$ with another sharp constant $C_{S,m}$, where $C_{S,m}$ is the smallest one of all positive constant C such that

$$\left(\int |u|^p d\xi \right)^{\frac{m}{p}} \leq C \int \left(|\nabla_H u|^m + |u|^m \right) d\xi$$

holds for all $u \in S_0^{1,m}(\mathbb{H}^N)$. We leave all these to the interested readers.

Remark 5.9. One referee pointed out that the results in this paper can be extended to any sub-Laplacian on a stratified Lie group. We believe that this is interesting and can be a problem for further study. We appreciate the referees for important comments as well as pointing out references [1, 4, 7, 10] and more details about the proof of Lemma 2.3.

Appendix

In this appendix, we give a proof of inequality (1.6) for the reader's convenience.

Proof of (old 1.6). Note that $2_* = 2Q/(Q-2)$. We obtain from Hölder inequality that

$$\int |u|^p d\xi = \int |u|^{ps}|u|^{p(1-s)} d\xi \le \left(\int |u|^{2_*} d\xi\right)^{\frac{ps}{2_*}} \left(\int |u|^2 d\xi\right)^{\frac{p(1-s)}{2}},$$

where $\frac{ps}{2_*} + \frac{p(1-s)}{2} = 1$. It is deduced from simple computation that $s = \frac{Q(p-2)}{2p}$. Hence

$$\int |u|^p d\xi \le \left(\int |u|^{2_*} d\xi\right)^{\frac{(Q-2)(p-2)}{4}} \left(\int |u|^2 d\xi\right)^{\frac{2p-Q(p-2)}{4}}.$$

Combining this with the fact that $\|u\|_{L^{2_*}} \le C\|\nabla_H u\|_{L^2}$ for some positive constant C, we get inequality (1.6). The proof is complete. □

References

[1] S. Biagini, *Positive solutions for a semilinear equation on the Heisenberg group.* Boll. Un. Mat. Ital. B **(7)9** (1995), 883–900.

[2] I. Birindelli, I. Capuzzo Dolcetta and A. Cutri, *Indefinite semi-linear equations on the Heisenberg group: a priori bounds and existence.* Comm. Partial Differential Equations **23** (1998), 1123–1157.

[3] I. Birindelli, A. Cutri, *A semi-linear problem for the Heisenberg Laplacian.* Rend. Sem. Mat. Univ. Padova **94** (1995), 137–153.

[4] A. Bonfiglioli, E. Lanconelli and F. Uguzzoni, *Stratified Lie groups and potential theory for their sub-Laplacians.* Springer Monographs in Mathematics. Springer, Berlin, 2007.

[5] H. Brezis, E. Lieb, *A relation between pointwise convergence of functions and convergence of functionals.* Proc. Amer. Math. Soc. **88** (1983), 486–490.

[6] G. Citti and F. Uguzzoni, *Critical semilinear equations on the Heisenberg group: the effect of topology of the domain.* Nonli. Anal. **46** (2001), 399–417.

[7] N. Garofalo and E. Lanconelli, *Existence and non existence results for semilinear equations on the Heisenberg group.* Indiana Univ. Math. J. **41** (1992), 35–38.

[8] N. Garofalo and D. Nhieu, *Isoperimetric and Sobolev inequalities for Carnot-Carathéodory spaces and the existence of minimal surfaces.* Comm. Pure Appl. Math. **XLIX** (1996), 1081–1144.

[9] T. Horiuchi, *Best constant in weighted Sobolev inequality with weights being powers of distance from the origin.* J. Inequal. Appl. **1** (1997), 275–292.

[10] E. Lanconelli and F. Uguzzoni, *Non-existence results for semilinear Kohn-Laplace equations in unbounded domains.* Comm. Partial Differential Equations, **25** (2000), 1703–1739.

[11] P.L. Lions, *The concentration-compactness principle in the calculus of variations, The locally compact case, Part* 1 *and* 2. Ann. I.H., Anal. Nonlin. **1** (1984), 109–145 and 223–283.

[12] F. Merle and P. Raphael, *Profiles and quantization of the blow up mass for critical nonlinear Schrödinger equation.* Comm. Math. Phys. **253** (2005), 675–704.

[13] F. Merle and P. Raphael, *On universality of blow-up profile for L^2 critical nonlinear Schrödinger equation.* Invent. Math. **156** (2004), 565–672.

[14] H. Mokrani, *Semi-linear sub-elliptic equations on the Heisenberg group with a singular potential.* Comm. Pure Appl. Anal. **8** (2009), 1619–1636.

[15] F. Uguzzoni, *A non-existence theorem for a semilinear Dirichlet problem involving critical exponent on half-spaces of the Heisenberg group.* Nonlinear Differential Equations Appl. **6** (1999), 191–206.

[16] I. Schindler and K. Tintarev, *An abstract version of the concentration compactness principle.* Rev. Mat. Complut. **15** (2002), 417–436.

[17] M.I. Weinstein, *Nonlinear Schrödinger equations and sharp interpolation estimates.* Comm. Math. Phys. **87** (1983), 567–576.

Jianqing Chen and Eugénio M. Rocha
Center for Research and Development
in Mathematics and Applications
Department of Mathematics
University of Aveiro
P-3810-193 Aveiro, Portugal
e-mail: `jchen@ua.pt`
`eugenio@ua.pt`

Operator Theory:
Advances and Applications, Vol. 229, 139–155

New Types of Solutions of Non-linear Fractional Differential Equations

Mark Edelman and Laura Anna Taieb

Abstract. Using the Riemann-Liouville and Caputo Fractional Standard Maps (FSM) and the Fractional Dissipative Standard Map (FDSM) as examples, we investigate types of solutions of non-linear fractional differential equations. They include periodic sinks, attracting slow diverging trajectories (ASDT), attracting accelerator mode trajectories (AMT), chaotic attractors, and cascade of bifurcations type trajectories (CBTT). New features discovered include attractors which overlap, trajectories which intersect, and CBTTs.

Mathematics Subject Classification (2010). Primary 47Gxx; Secondary 47Hxx.

Keywords. Discrete map, attractor, fractional dynamical system, map with memory, stability.

1. Introduction

In recent years fractional calculus (FC) and fractional differential equations (FDE) became very popular in many areas of science. The books dedicated to the applications of FDEs published in 2010–2011 include [1, 2, 3] in physics in general, [4, 5] in modeling and control, [6] in viscoelasticity, [7] in systems with long-range interaction. A good review of applications of FC to chaos in Hamiltonian Systems is given in [8]. Because fractional derivatives are integro-differential operators, they are used to describe systems distributed in time and/or space: systems with long range interaction [7, 9, 10, 11, 12, 13], non-Markovian systems with memory ([14] Ch.10, [15, 16, 17, 18, 19]), fractal media [20], etc. Biological systems are probably the best examples of systems with memory. As it has been shown recently [21, 22], even processing of external stimuli by individual neurons can be described by fractional differentiation. In some cases [23, 24, 25, 26] FDEs are equivalent to the Volterra integral equations of the second kind. This kind of equations (non necessarily FDEs) is used in nonlinear viscoelasticity (see for example [27, 28])

This work was completed with the support of Yeshiva University, Courant Institute of Mathematical Sciences, and DOE Grant DE-FG0286ER53223.

and in biology (for applications to mathematical models in population biology and epidemiology see [29, 30]).

As in regular dynamics (in application to various areas in science), the nonlinearity plays a significant role in fractional dynamics. Chaos and order in nonlinear systems with long range interactions were considered in [7, 10, 11, 12], nonlinear FDEs in application to the control were reviewed in books [4, 5], nonlinear fractional reaction-diffusion systems considered in [31, 32, 33, 34]. Corresponding to the fact that in physical systems the transition from integer order time derivatives to fractional (of a lesser order) introduces additional damping similar in appearance to additional friction [14, 35], phase spaces of the systems with fractional time derivatives demonstrate different kinds of structures similar to the attractors of the dissipative dynamical systems [4, 35, 36].

As in the case of the systems which can be described by regular differential equations, general properties of systems described by FDEs can be demonstrated on the examples of maps which can be derived by integrating the FDEs (if it is possible) over a period of perturbation. Equations of some fractional maps (FM), which include among others fractional standard maps (FSM) and fractional dissipative standard maps (FDSM), were derived in a series of recent publications [7, 37, 38, 39, 40, 26] from the corresponding FDEs.

Two of the most studied examples in the regular case are the Chirikov standard map (SM) [41] and the Zaslavsky dissipative standard map (DSM) [42, 43]. The SM provides the simplest model of the universal generic area preserving map and the description of many physical systems and effects (Fermi acceleration, comet dynamics, etc.) can be reduced to the studying of the SM. The topics examined include fixed points, elementary structures of islands and a chaotic sea, and fractional kinetics [41, 44, 45]. Different properties of the DSM were discussed in [46, 47, 48, 49, 50, 51], and a rigorous proof of the existence of chaotic attractor in such type of systems was obtained in [52, 53]. It was also proved in [52, 53] that the system of this type exhibits quasi-periodic attractors, periodic sinks, transient chaos with the dynamics attracting to the sink, and the existence of the SRB measure for the purely chaotic dynamics. All this stimulated the study of the FSMs [38, 54] and FDSM [39] in hope to reveal the most general properties of the fractional dynamics.

Two-dimensional maps investigated in [38, 39, 54] were derived from the FDEs (see Section 2) and, as a result, they are discrete maps with memory in which the present state of evolution depends on all past states. The earlier studies of maps with memory were done on one-dimensional maps which were not derived from differential equations [55, 56, 57, 58, 59, 60]. Most results were obtained for the generalizations of the logistic map and the main general result is that the presence of memory makes systems more stable. The initial study of the FSM in [38, 39, 54] was concentrated on the investigation of the fixed/periodic points' stability of the FSM, new properties and new types of attractors.

The results obtained in [38, 39, 54] revealed new unusual properties of the fractional attractors of the FSM, which are different not only from the properties of the fixed/periodic points of the non-dissipative systems (the SM in our case), but also from the properties of attractors of the regular (not fractional) dissipative systems (e.g., the DSM). The most unusual observed property of the FSMs is the existence (and persistence) of the new type of attractors – cascade of bifurcations type trajectories (CBTT). A cascade of bifurcations, when with the change in the value of a parameter a system undergoes a sequence of period-doubling bifurcations, is a well-known pathway of a transition form order to chaos. In the case of the CBTT the period doubling occurs without a change in a system parameter and is an internal property of a system. In Section 3 we summarize results of the extensive numerical investigation of three fractional maps to demonstrate the properties of fractional attractors. An analysis of the conditions for the CBTTs' appearance will be presented in the following publications.

2. Basic equations

The standard map in the form

$$p_{n+1} = p_n - K \sin x_n, \quad x_{n+1} = x_n + p_{n+1} \pmod{2\pi} \tag{2.1}$$

can be derived from the differential equation

$$\ddot{x} + K \sin(x) \sum_{n=0}^{\infty} \delta\Big(\frac{t}{T} - (n+\varepsilon)\Big) = 0, \tag{2.2}$$

where $\varepsilon \to 0+$, following the steps proposed by Tarasov in [26].

The equations for the Riemann-Liouville FSM (FSMRL) were obtained in [37] and [26]. Following the steps proposed [26], the FSMRL in the form which converges to the SM as $\alpha \to 2$ can be derived from the differential equation with the Riemann-Liouville fractional derivative describing a kicked system

$${}_0D_t^{\alpha} x + K \sin(x) \sum_{n=0}^{\infty} \delta\Big(\frac{t}{T} - (n+\varepsilon)\Big) = 0, \quad (1 < \alpha \le 2), \tag{2.3}$$

where $\varepsilon \to 0+$, with the initial conditions

$$({}_0D_t^{\alpha-1} x)(0+) = p_1, \quad ({}_0D_t^{\alpha-2} x)(0+) = b, \tag{2.4}$$

where

$$\begin{aligned} {}_0D_t^{\alpha} x(t) &= D_t^n \; {}_0I_t^{n-\alpha} x(t) \\ &= \frac{1}{\Gamma(n-\alpha)} \frac{d^n}{dt^n} \int_0^t \frac{x(\tau) d\tau}{(t-\tau)^{\alpha-n+1}} \quad (n-1 < \alpha \le n), \end{aligned} \tag{2.5}$$

$D_t^n = d^n/dt^n$, and ${}_0I_t^{\alpha}$ is a fractional integral.

After integration of equation (2.3) the FSMRL can be written in the form

$$p_{n+1} = p_n - K \sin x_n, \tag{2.6}$$

$$x_{n+1} = \frac{1}{\Gamma(\alpha)} \sum_{i=0}^{n} p_{i+1} V_{\alpha}^1(n-i+1), \quad \pmod{2\pi}, \tag{2.7}$$

where

$$V_\alpha^k(m) = m^{\alpha-k} - (m-1)^{\alpha-k} \tag{2.8}$$

and momentum $p(t)$ is defined as

$$p(t) = {}_0D_t^{\alpha-1}x(t). \tag{2.9}$$

Here it is assumed that $T = 1$ and $1 < \alpha \leq 2$. The condition $b = 0$ is required in order to have solutions bounded at $t = 0$ for $\alpha < 2$ [38]. In this form the FSMRL equations in the limiting case $\alpha = 2$ coincide with the equations for the standard map under the condition $x_0 = 0$. For consistency and in order to compare corresponding results for all three maps (the SM, the FSMRL, and the Caputo FSM (FSMC)) all trajectories considered in this article have the initial condition $x_0 = 0$.

Following the steps proposed [26], the FSMC in the form which converges to the SM as $\alpha \to 2$ can be derived from the differential equation similar to (2.3) but with the Caputo fractional derivative

$${}_0^CD_t^\alpha x + K\sin(x)\sum_{n=0}^{\infty}\delta\Big(\frac{t}{T} - (n+\varepsilon)\Big) = 0, \quad (1 < \alpha \leq 2) \tag{2.10}$$

where $\varepsilon \to 0+$, with the initial conditions

$$p(0) = ({}_0^CD_t^1 x)(0) = (D_t^1 x)(0) = p_0, \quad x(0) = x_0, \tag{2.11}$$

where

$$\begin{aligned} {}_0^CD_t^\alpha x(t) &= {}_0I_t^{n-\alpha}\ D_t^n x(t) \\ &= \frac{1}{\Gamma(n-\alpha)}\int_0^t \frac{D_\tau^n x(\tau)d\tau}{(t-\tau)^{\alpha-n+1}} \quad (n-1 < \alpha \leq n). \end{aligned} \tag{2.12}$$

Integrating equation (2.10) with the momentum defined as $p = \dot{x}$ and assuming $T = 1$ and $1 < \alpha \leq 2$, one can derive the FSMC in the form

$$p_{n+1} = p_n - \frac{K}{\Gamma(\alpha-1)}\Big[\sum_{i=0}^{n-1} V_\alpha^2(n-i+1)\sin x_i + \sin x_n\Big], \quad (\text{mod } 2\pi), \tag{2.13}$$

$$x_{n+1} = x_n + p_0$$

$$-\frac{K}{\Gamma(\alpha)}\sum_{i=0}^{n} V_\alpha^1(n-i+1)\sin x_i, \quad (\text{mod } 2\pi). \tag{2.14}$$

It is important to note that the FSMC ((2.13), (2.14)) can be considered on a torus (x and p mod 2π), a cylinder (x mod 2π), or in an unbounded phase space, whereas the FSMRL ((2.6), (2.7)) can be considered only in a cylindrical or an unbounded phase space. The FSMRL has no periodicity in p and cannot be considered on a torus. This fact is related to the definition of momentum (2.9) and initial conditions (2.4). The comparison of the phase portraits of two FSMs is still possible if we compare the values of the x coordinates on the trajectories corresponding to the same values of the maps' parameters.

The DSM in the form

$$X_{n+1} = X_n + P_{n+1}, \tag{2.15}$$

$$P_{n+1} = -bP_n - Z\,\sin(X_n) \tag{2.16}$$

can be derived integrating the differential equation of the kicked damped rotator (see for example [39])

$$\ddot{X} + q\dot{X} = \varepsilon \sin(X) \sum_{n=0}^{\infty} \delta\Big(t - n\Big). \tag{2.17}$$

Two forms of the FDSM were derived in [7, 37, 39]. The form which has been investigated numerically in [39]

$$X_{n+1} = \frac{\mu^{-1}}{\Gamma(\alpha - 1)} \sum_{k=0}^{n} P_{k+1} W_\alpha(q, k - n - 1), \tag{2.18}$$

$$P_{n+1} = -bP_n - Z\sin(X_n), \tag{2.19}$$

where functions $W_\alpha(a, b)$ are defined by

$$W_\alpha(a, b) = a^{1-\alpha} e^{a(b+1)} \Big[\Gamma(\alpha - 1, ab) - \Gamma(\alpha - 1, a(b + 1))\Big] \tag{2.20}$$

and $\Gamma(a, b)$ is the incomplete Gamma function

$$\Gamma(a, b) = \int_b^\infty y^{a-1} e^{-y} dy \tag{2.21}$$

has been derived from the following fractional generalization of (2.17)

$${}_0D_t^\alpha X(t) - q\,{}_0D_t^\beta X(t) = \varepsilon \sin(X) \sum_{n=0}^{\infty} \delta(t - n), \tag{2.22}$$

where

$$q \in \mathbb{R}, \quad 1 < \alpha \leq 2, \quad \beta = \alpha - 1.$$

Here

$$\mu = (1 - e^{-q})/q, \quad Z = -\mu\varepsilon e^q. \tag{2.23}$$

Further in this paper we will also use $K = \varepsilon \exp(q)$ and $\Gamma = -q$.

3. Fractional attractors

3.1. Standard map: Fixed and periodic points

It appears that the dependence of the SM's fixed point (0,0) stability properties on the map parameter K plays an important role when transition to the FSMs is considered. (0,0) SM fixed point is stable for $0 < K < 4$. At $K = 4$ it becomes unstable but period two ($T = 2$) antisymmetric trajectory

$$p_{n+1} = -p_n, \quad x_{n+1} = -x_n \tag{3.1}$$

appears which is stable for $4 < K < 2\pi$. At $K = 2\pi$ this T=2 trajectory becomes unstable but it gives birth to two $T = 2$ trajectories [41, 44] with

$$p_{n+1} = -p_n, \quad x_{n+1} = x_n - \pi. \tag{3.2}$$

These $T = 2$ trajectories become unstable when $K \approx 6.59$ (see Figure 4a), at the point where $T = 4$ stable trajectories are born. This period doubling cascade of bifurcations sequence continues with $T = 8$ and $T = 16$ stable trajectories appearing at $K \approx 6.63$ and $K \approx 6.6344$ correspondingly. It continues until at $K \approx 6.6345$ all periodic points become unstable and corresponding islands disappear. This scenario of the elliptic-hyperbolic point transitions with the births of the double periodicity islands inside the original island has been investigated in [61] and applied to investigate the SM stochasticity at low values of the parameter $K < 4$.

3.2. Phase space at low K (stable (0,0) fixed point)

Stability of the fixed point (0,0) has been considered in [38, 54]. It has been shown that for fractional values of α (we consider only $1 < \alpha \leq 2$) this point turns from elliptic (for $\alpha = 2$) into a sink which is stable for

$$K < K_{c1}(\alpha) = \frac{2\Gamma(\alpha)}{V_{\alpha l}}, \tag{3.3}$$

where

$$V_{\alpha l} = \sum_{k=1}^{\infty}(-1)^{k+1}V_{\alpha}^{1}(k) \tag{3.4}$$

and can be calculated numerically (see Figure 4a). It can be shown that $K_{c1}(2) = 4$, which corresponds to the SM case. The structure of the FSM phase space for $K < K_{c1}$ preserves some features which exist in the $\alpha = 2$ case. Namely, stable higher period points, which exist in the SM case, still exist in the FSM, but they exist in the asymptotic sense and they transform into sinks or, in the FSMRL case, into attracting slow ($p_n \sim n^{2-\alpha}$) diverging trajectories (ASDT) (see Figure 1). Pure chaotic trajectories disappear. Each attractor has its own basin of attraction. The traces of the SM chaotic sea exist in the following sense: initially close trajectories, which do not start from any basin of attraction, may fall into absolutely different attractors. In most of the cases attracting points themselves do not belong to their own basins of attraction: trajectory which starts from an attracting point may fall into a different attractor. The FSMRL trajectories which converge to the fixed point follow two different routs. Trajectories which start from the basin of attraction converge according to the power law $x_n \sim n^{-1-\alpha}$ and $p_n \sim n^{-\alpha}$, while those starting from the outside of the basin of attraction are attracting slow converging trajectories (ASCT) with $x_n \sim n^{-\alpha}$ and $p_n \sim n^{1-\alpha}$. All FSMC trajectories converging to $(0,0)$ fixed point follow the slow power law $x_n \sim n^{1-\alpha}$ and $p_n \sim n^{1-\alpha}$ (see Figure 2). This convergence may correspond to the recently introduced for the fractional dynamic systems notion of the generalized Mittag-Leffler stability [62], for which the power-law stability is a special case.

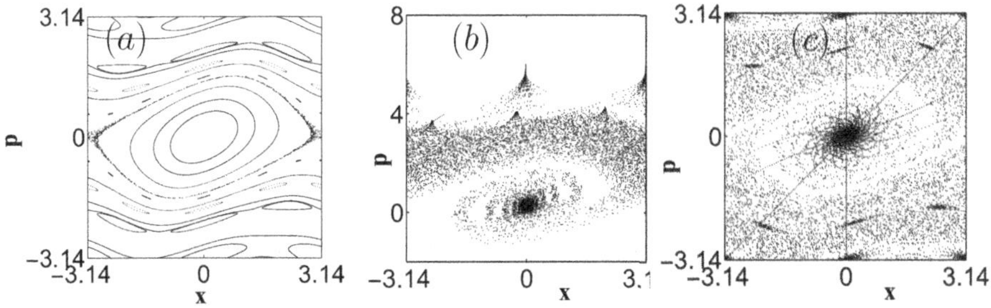

FIGURE 1. (a). 1000 iterations on each of 25 trajectories for the SM with $K = 0.6$. The main features are (0,0) fixed point and T=2 and T=3 trajectories; (b). 400 iterations on each trajectory with $p_0 = 2 + 0.04i$, $0 \leq i < 50$ for the FSMRL case $K = 0.6$, $\alpha = 1.9$. Trajectories converging to the fixed point and ASDTs of period 2 and 3 are present; (c). 100 iterations on each FSMC trajectory with $p_0 = -3.14 + 0.0314i$, $0 \leq i < 200$ for the same case as in Fig. 1b ($K = 0.6$, $\alpha = 1.9$) considered on a torus. In this case all trajectories converge to the fixed point, period two and period three stable attracting points.

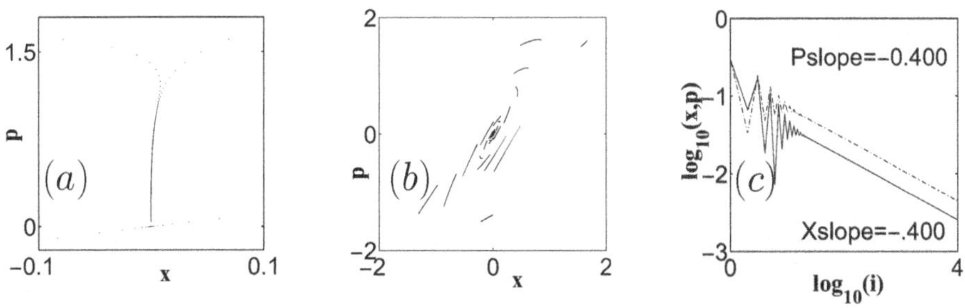

FIGURE 2. Convergence of trajectories to the fixed point $(0, 0)$ at $K < K_{c1}$: (a). Two trajectories for the FSMRL with $K = 2$, $\alpha = 1.4$, and 10^5 iterations on each trajectory. The bottom one with $p_0 = 0.3$ is a fast converging trajectory. The upper trajectory with $p_0 = 5.3$ is an example of the FSMRL's ASCT; b). Evolution of the FSMC trajectories with $p_0 = 1.6 + 0.002i$, $0 \leq i < 50$ for the case $K = 3$, $\alpha = 1.9$. The line segments correspond to the nth iteration on the set of trajectories with close initial conditions; c). x and p time dependence for the FSMC with $K = 2$, $\alpha = 1.4$, $x_0 = 0$, and $p_0 = 0.3$.

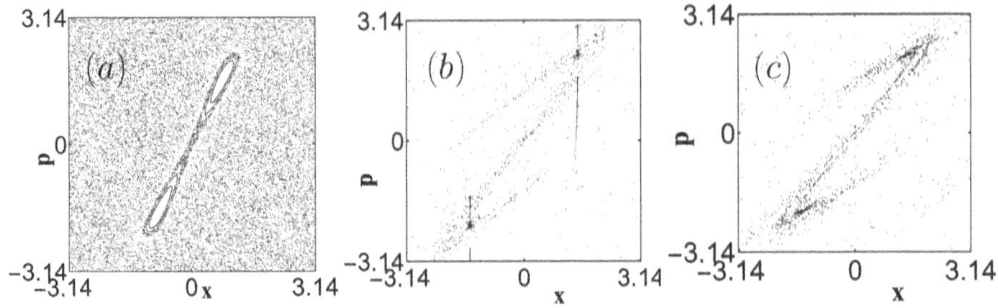

FIGURE 3. Stable antisymmetric $x_{n+1} = -x_n$, $p_{n+1} = -p_n$ period $T = 2$ trajectories for $K = 4.5$: (a). 1000 iterations on each of 25 trajectories for the SM with $K = 4.5$. The only feature is a system of two islands associated with the period two elliptic point; (b). The FSMRL stable $T = 2$ antisymmetric sink for $\alpha = 1.8$. 500 iterations on each of 25 trajectories: $p_0 = 0.0001 + 0.08i$, $0 \leq i < 25$. Slow and fast converging trajectories; (c). The FSMC stable $T = 2$ antisymmetric sink for $\alpha = 1.8$. 1000 iterations on each of 10 trajectories: $p_0 = -3.1415 + 0.628i$, $0 \leq i < 10$.

3.3. Phase space at $K_{c1} < K < K_{c2}$ (stable $T = 2$ antisymmetric trajectory)

Numerical simulations confirm that, as in the SM case, in the FSMs a period $T = 2$ antisymmetric trajectory (sink) exists (asymptotically) for $K_1(\alpha) < K < K_2(\alpha)$ ($4 < K < 2\pi$ in the SM case) with

$$p_{n+1} = -p_n, \quad x_{n+1} = -x_n. \tag{3.5}$$

Asymptotic existence and stability of this sink is a result of the gradual transformation of the SM's elliptic point with the decrease in the order of derivative from $\alpha = 2$ (see Figure 3). As in the fixed point case, there are two types of convergence of the trajectories to the FSMRL $T = 2$ sink: fast with

$$\delta x_n \sim n^{-1-\alpha}, \quad \delta p_n \sim n^{-\alpha} \tag{3.6}$$

and slow with

$$\delta x_n \sim n^{-\alpha}, \quad \delta p_n; \sim n^{1-\alpha} \tag{3.7}$$

and

$$\delta x_n \sim n^{1-\alpha}, \quad \delta p_n \sim n^{1-\alpha} \tag{3.8}$$

convergence in the FSMC case [54].

Let us consider the FSMRL and assume the asymptotic existence of the $T = 2$ sink (x_l, p_l) and $(-x_l, -p_l)$. In this case the limit $n \to \infty$ in (2.6) and (2.7) gives

$$p_l = \frac{K}{2}\sin(x_l), \tag{3.9}$$

$$x_l = \frac{K}{2\Gamma(\alpha)}\sin(x_l)\sum_{k=1}^{\infty}(-1)^{k+1}V_\alpha^1(k). \tag{3.10}$$

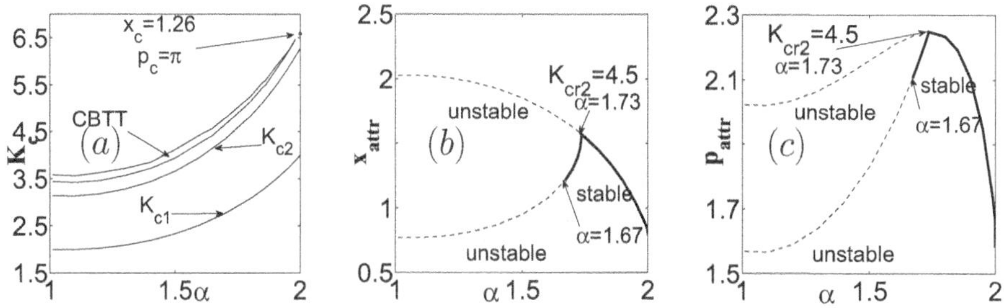

FIGURE 4. Critical values in the FSMs' (α, K)-space. The results obtained by the numerical simulations of equations (3.12), (3.16), (3.11), (3.15), (3.9) and confirmed by the direct simulations of the FSMs: (a). The $(0,0)$ fixed point is stable below $K = K_{c1}$ curve. It becomes unstable at $K = K_{c1}$ and gives birth to the antisymmetric $T = 2$ sink which is stable at $K_{c1} < K < K_{c2}$. A pair of $T = 2$ sinks with $x_{n+1} = x_n - \pi$, $p_{n+1} = -p_n$ is stable in the band above $K = K_{c2}$ curve. Cascade of bifurcations type trajectories (CBTTs) are the only phase space features that appear and exist in the narrow band which ends at the cusp on the figure. (x_c, p_c) is the point at which the SM's $T = 2$ elliptic points with $x_{n+1} = x_n - \pi$, $p_{n+1} = -p_n$ become unstable and bifurcate (see Sec. 3.1); (b). α-dependence of the $T = 2$ sinks' x-coordinate for $K = 4.5$. The transition from the antisymmetric sink (upper curve) to $x_{n+1} = x_n - \pi$, $p_{n+1} = -p_n$ point takes place at $\alpha = 1.73$. Solid lines represent areas of stability; (c). $K = 4.5$ α-dependence of the $T = 2$ sinks' p-coordinate.

The equation for x_l takes the form

$$x_l = \frac{K}{2\Gamma(\alpha)} V_{\alpha l} \sin(x_l), \tag{3.11}$$

from which the condition of the existence of the $T = 2$ sink is

$$K > K_{c1}(\alpha) = \frac{2\Gamma(\alpha)}{V_{\alpha l}}. \tag{3.12}$$

From (3.12) it follows that $K_1(\alpha) = K_{c1}(\alpha)$.

Remarks

1) The $T = 2$ sink exists only in the asymptotic sense. On a trajectory which starts from point (x_0, p_0)=(x_l, p_l) $x_2 \neq x_l$ and $p_2 \neq p_l$.
2) In spite of the fact that in the derivation of (3.9) and (3.10) we used the limit $n \to \infty$, computer simulations demonstrate convergence of all trajectories to the limiting values in the very good agreement with (3.6), (3.7), and (3.8).
3) Direct computations show that $K_1(\alpha) = K_{c1}(\alpha)$ is valid for the FSMC too.

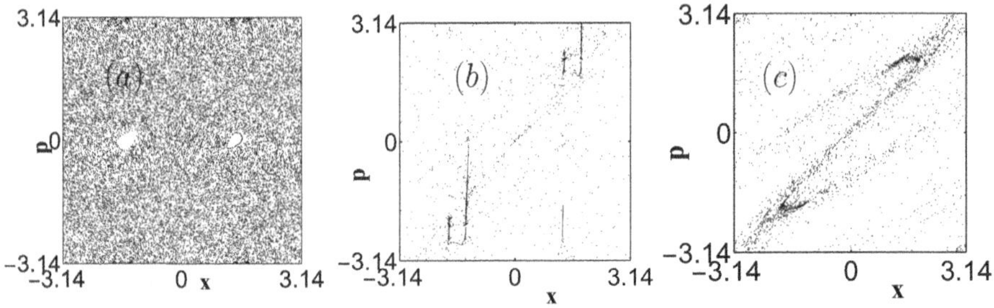

FIGURE 5. Stable $x_{n+1} = x_n - \pi$, $p_{n+1} = -p_n$ period $T = 2$ trajectories for $K > K_{c2}$: (a). 500 iterations on each of 50 trajectories for the SM with $K = 6.4$. The main features are two accelerator mode sticky islands around points $(-1.379, 0)$ and $(1.379, 0)$ which define the dynamics. Additional features – dark spots at the top and the bottom of the figure (which are clear on a zoom) – two systems of $T = 2$ tiny islands associated with two $T = 2$ elliptic points: $(1.379, \pi)$, $(1.379 - \pi, -\pi)$ and $(\pi - 1.379, \pi)$, $(-1.379, -\pi)$; (b). Two FSMRL's stable $T = 2$ sinks for $K = 4.5$, $\alpha = 1.71$. 500 iterations on each of 25 trajectories: $p_0 = 0.0001 + 0.08i$, $0 \le i < 25$; (c). Two FSMC's stable $T = 2$ sinks for $K = 4.5$, $\alpha = 1.71$. 1000 iterations on each of 10 trajectories: $p_0 = -3.1415 + 0.628i$, $0 \le i < 10$.

3.4. Phase space at $K > K_{c2}$

The SM's $T = 2$ antisymmetric trajectory becomes unstable when $K = 2\pi$, at the point in phase space where a pair of $T = 2$ trajectories with $x_{n+1} = x_n - \pi$, $p_{n+1} = -p_n$ appears. Numerical simulations show (see Figure 5) that the FSMs demonstrate similar behavior. Let us assume that the FSMRL equations (2.6) and (2.7) have an asymptotic solution

$$p_n = (-1)^n p_l, \quad x_n = x_l - \frac{\pi}{2}[1 - (-1)^n]. \tag{3.13}$$

Then it follows from (2.6) that relationship (3.9) $p_l = K/2\sin(x_l)$ holds in this case too. Simulations similar to those presented in Figure 2b [63] show that for $K > K_{c2}$ (see Figure 4a) the FSMRL has an asymptotic solution

$$p_n = (-1)^n p_l + An^{1-\alpha} \tag{3.14}$$

with the same A for both even and odd values of n. Substituting (3.14) in (2.7) and considering limit $n \to \infty$ one can derive (see [63])

$$\sin(x_l) = \frac{\pi\Gamma(\alpha)}{KV_{\alpha l}}, \tag{3.15}$$

which has solutions for

$$K > K_{c2} = \frac{\pi\Gamma(\alpha)}{V_{\alpha l}} \tag{3.16}$$

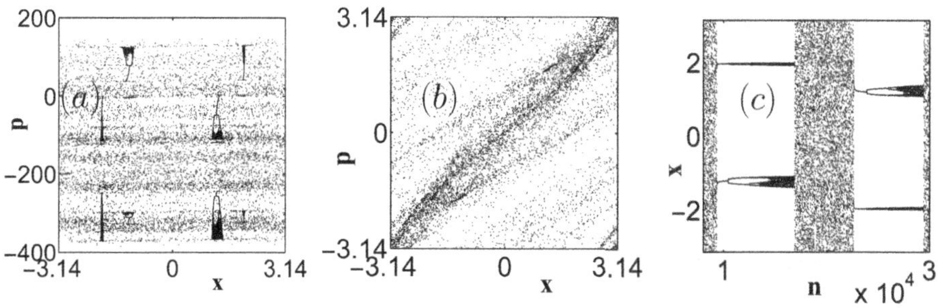

FIGURE 6. Cascade of bifurcations type trajectories (CBTT) from iterations on a single trajectory with $K = 4.5$, $\alpha = 1.65$, $x_0 = 0$, and $p_0 = 0.3$: (a). 25000 iterations for the FSMRL case. The trajectory occasionally sticks to one of the cascade of bifurcations type trajectories but later always returns to the chaotic sea; (b). 30000 iterations for the FSMC case. The CBTTs can hardly be recognized on the full phase portrait but can be seen on the x of n dependence in Figure 6c; (c). Time dependence of the coordinate x in Figure 6b.

(see Figure 4) and the value of A can also be calculated:

$$A = \frac{2x_l - \pi}{2\Gamma(2-\alpha)}. \tag{3.17}$$

These results are in a good agreement with the direct FSMRL numerical simulations and hold for the FSMC.

3.5. Cascade of bifurcations type trajectories

In the SM further increase in K at $K \approx 6.59$ causes an elliptic-hyperbolic point transition when $T = 2$ points become unstable and stable $T = 4$ elliptic points appear. A period doubling cascade of bifurcations leads to the disappearance of the corresponding islands of stability in the chaotic sea at $K \approx 6.6344$. The cusp in Figure 4a points to approximately this spot ($\alpha = 2$, $K \approx 6.63$). Inside of the band leading to the cusp a new type of the FSM's attractors appears: cascade of bifurcations type trajectories (CBTT) (see Figure 6). In CBTTs period doubling cascade of bifurcations occurs on a single trajectory without any change in the map parameter. In some cases CBTTs behave like sticky islands of the Hamiltonian dynamics: occasionally a trajectory enters a CBTT and then leaves it entering the chaotic sea. Near the cusp CBTTs are barely distinguishable – the relative time trajectories spend in CBTTs is small. With decrease in α the relative time trajectories spend in CBTTs increases and at α close to one a trajectory enters a CBTT after a few iterations and stays there over the longest computational time we were able to run our codes – 500000 iterations. A typical FSMC's CBTT of this kind is presented in the Figure 7c.

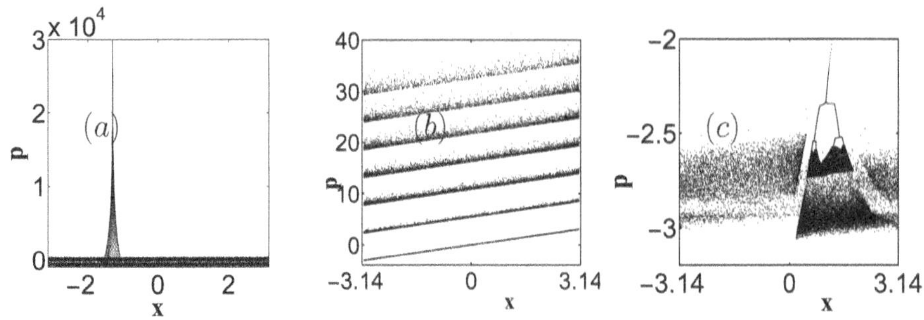

FIGURE 7. (a). A single FSMRL ballistic trajectory for $K = 1.999$, $\alpha = 6.59$; (b). Seven disjoint chaotic attractors for the case FSMRL with $K = 4.5$ and $\alpha = 1.02$. 1000 iterations on each of 20 trajectories: $x_0 = 0$ and $p_0 = 0.0001 + 1.65i$, $0 \le i < 20$; (c). 20000 iterations on each of two overlapping independent attractors for the FSMC case with $K = 4.5$ and $\alpha = 1.02$. The CBTT has $p_0 = -1.8855$ and the chaotic attractor $p_0 = -2.5135$ ($x_0 = 0$).

CBTTs, easily recognizable in the phase space of the FSMRL (Figure 6a), often almost impossible to distinguish in the phase space of the FSMC (Figure 6b). But they clearly reveal themselves on a coordinate versus time (step of iteration n) plot (Figure 6c).

3.6. More FSM's attractors

The main feature of the SM's phase space at $K > 2\pi$, which defines dynamics and transport and is well investigated (see for example [64]), is the presence of the sticky accelerator mode islands. Slight decrease in the order of derivative turns accelerator mode islands into attracting accelerator mode trajectories (AMT) (Figure 7a). Depending on the values of K and α trajectories either permanently sink into this attractor, or intermittently stick to it and return to the chaotic sea. AMTs do not exist (we were unable to find them) for $\alpha < 1.995$. In the (α, K)-plot (Figure 4a) the area of the AMT existence is a part ($K > 2\pi$) of a very narrow strip near $\alpha = 2$ line. In the area of the α, K-plot above the CBTT strip (we considered only $K < 7$) different kinds of chaotic attractors (or possibly pure chaotic behavior) can be found. Two examples are presented in Figure 7. Figure 7c contains two overlapping attractors one of which is a CBTT. Existence of the overlapping attractors and intersecting trajectories is a feature of maps with memory which is impossible in a regular dynamical systems. The simplest example of intersecting trajectories can be constructed by starting one trajectory at an arbitrary point (x_0, p_0) and then the second at $(x'_0, p'_0) = (x_1, p_1)$.

3.7. The fractional dissipative standard map (FDSM)

Consideration of the FDSM in [39] demonstrated that already a small deviation of α from 2 leads to significant changes in properties of the map. For example, a

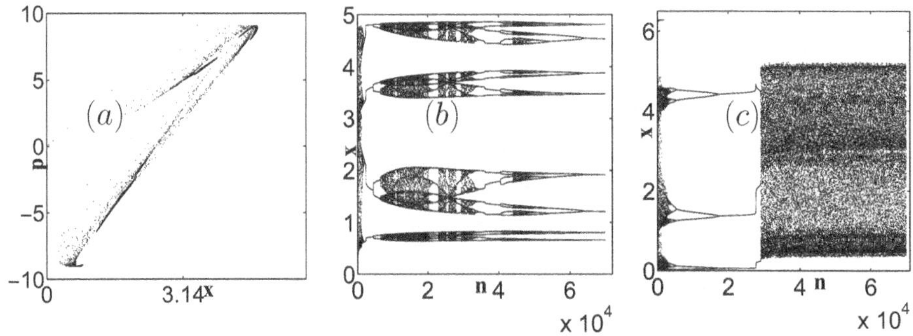

FIGURE 8. The FDSM (in all cases $\Gamma = 5$ and $\Omega = 0$): (a). Phase space for the case $K = 9.1$, $\alpha = 1.2$; (b). 70000 iterations on a single trajectory for the same case as in Figure 8a ($K = 9.1$, $\alpha = 1.2$). A cascade of bifurcations appears from chaos and then converges into a periodic trajectory; (c). 70000 iterations on a single trajectory for the case $K = 10$, $\alpha = 1.2$. A chaotic attractor appears in the FDSM's phase space after a chaotic trajectory through an inverse cascade of bifurcations forms a period $T = 3$ trajectory.

window of ballistic motion which exists in Zaslavsky Map near $K = 4\pi$ is closing when $\alpha < 1.996$. Further decrease in α at $K \approx 4\pi$ produces different kinds of chaotic attractors and sinks [39].

Some of the new features we observed, which are presented in Figure 8, include inverse cascade of bifurcations and trajectories which start as a CBTT and then converge as an inverse cascade of bifurcations. In phase space those trajectories may look like chaotic attractors.

4. Conclusion

We studied three different fractional maps which describe nonlinear systems with memory under periodic perturbations (kicks in our case). The types of solutions which we found in all maps include periodic sinks, attracting slow diverging trajectories, attracting accelerator mode trajectories, chaotic attractors, and cascade of bifurcations type trajectories. New features discovered include attractors which overlap, trajectories which intersect, and CBTTs.

The models with a similar behavior may include periodically kicked media which can be described by FDEs: viscoelastic materials [6] or dielectrics [65]. Experiments can be proposed to observe different kinds of solution, for example CBTTs, in those media.

Nonlinear models with memory which have chaotic and periodic solutions, where cascades of bifurcations considered as a result of a change in a system parameter, are used in population biology and epidemiology [66]. Our calculations show that such behavior can be a consequence of the essential internal properties of the systems with memory and suggest construction of new fractional models of biological systems.

Our computational results are based on the runs over more that 10^5 periods of perturbations. Due to the integro-differential nature of the fractional derivatives, at present time it is impossible to solve FDEs over more than 10^3 periods of oscillations. Our results suggest a new way of looking for a particular solutions of the FDEs. For example, we may suggest that CBTTs can be found at the values of system parameters, where in corresponding integer system a period doubling cascade of bifurcations leads to the disappearance of a system of islands.

Acknowledgment

The authors express their gratitude to the administration of the Stern College, Courant Institute, and DOE Grant DE-FG0286ER53223 for the financial support.

References

[1] V.E. Tarasov, *Fractional Dynamics: Application of Fractional Calculus to Dynamics of Particles, Fields and Media.* Springer, HEP, 2011.

[2] V.E. Tarasov, *Theoretical Physics Models with Integro-Differentiation of Fractional Order.* IKI, RCD, 2011 (in Russian).

[3] R. Herrmann, *Fractional Calculus: An Introduction for Physicists.* World Scientific, Singapore, 2011.

[4] R. Caponetto, G. Dongola, and L. Fortuna, *Fractional Order Systems: Modeling and Control Applications (World Scientific Series on Nonlinear Science Series a).* World Scientific, 2010.

[5] I. Petras, *Fractional-Order Nonlinear Systems.* Springer, 2011.

[6] F. Mainardi, *Fractional Calculus and Waves in Linear Viscoelasticity: An Introduction to Mathematical Models.* Imperial College Press, London, 2010.

[7] A.C.J. Luo and V. Afraimovich (eds.), *Long-range Interaction, Stochasticity and Fractional Dynamics.* Springer, 2010.

[8] G.M. Zaslavsky, *Hamiltonian Chaos and Fractional Dynamics.* Oxford University Press, Oxford, 2005.

[9] N. Laskin, G.M. Zaslavsky, *Nonlinear fractional dynamics on a lattice with long-range interactions.* Physica A **368** (2006), 38–4.

[10] V.E. Tarasov, G.M. Zaslavsky, *Fractional dynamics of coupled oscillators with long-range interaction.* Chaos **16** (2006), 023110.

[11] N. Korabel, G.M. Zaslavsky, *Transition to chaos in discrete nonlinear Schrödinger equation with long-range interaction.* Physica A **378** (2007), 223–237.

[12] G.M. Zaslavsky, M. Edelman, V.E. Tarasov, *Dynamics of the chain of forced oscillators with long-range interaction: From synchronization to chaos.* Chaos **17** (2007), 043124.

[13] V.E. Tarasov, G.M. Zaslavsky, *Fractional dynamics of systems with long-range space interaction and temporal memory.* Physica A **383** (2007), 291–308.

[14] I. Podlubny, *Fractional Differential Equations.* Academic Press, San Diego, 1999.

[15] R.R. Nigmatullin, *Fractional integral and its physical interpretation.* Theoretical and Mathematical Physics **90** (1992), 242–251.

[16] F.Y. Ren, Z.G. Yu, J. Zhou, A. Le Mehaute, R.R. Nigmatullin, *The relationship between the fractional integral and the fractal structure of a memory set.* Physica A **246** (1997), 419–429.

[17] W.Y. Qiu, J. Lu, *Fractional integrals and fractal structure of memory sets.* Phys. Lett. A **272** (2000), 353–358.

[18] R.R. Nigmatullin, *Fractional kinetic equations and 'universal' decoupling of a memory function in mesoscale region.* Physica A **363** (2006), 282–298.

[19] V.E. Tarasov, G.M. Zaslavsky *Fractional dynamics of systems with long-range space interaction and temporal memory.* Physica A **383** (2007), 291–308.

[20] A. Carpinteri, F. Mainardi, (eds.), *Fractals and Fractional Calculus in Continuum Mechanics.* Springer, Wien, 1997.

[21] B.N. Lundstrom, A.L. Fairhall, M. Maravall, *Multiple time scale encoding of slowly varying whisker stimulus envelope in cortical and thalamic neurons in vivo.* J. Neuroscience, **30** (2010), 5071–5077.

[22] B.N. Lundstrom, M.H. Higgs, W.J. Spain, A.L. Fairhall, *Fractional differentiation by neocortical pyramidal neurons.* Nature Neuroscience **11** (2008), 1335–1342.

[23] A.A. Kilbas, H.M. Srivastava, J.J. Trujillo, *Theory and Application of Fractional Differential Equations.* Elsevier, Amsterdam, 2006.

[24] A.A. Kilbas, B. Bonilla, J.J. Trujillo, *Nonlinear differential equations of fractional order is space of integrable functions.* Doklady Mathematics **62** (2000), 222–226, Translated from Doklady Akademii Nauk **374** (2000), 445–449. (in Russian).

[25] A.A. Kilbas, B. Bonilla, J.J. Trujillo, *Existence and uniqueness theorems for nonlinear fractional differential equations.* Demonstratio Mathematica **33** (2000), 583–602.

[26] V.E. Tarasov, *Differential equations with fractional derivative and universal map with memory.* Journal of Physics A **42** (2009), 465102.

[27] A. Wineman, *Nonlinear viscoelastic membranes.* Computers and Mathematics with Applications **53** (2007), 168–181.

[28] A. Wineman, *Nonlinear Viscoelastic Solids – A Review.* Mathematics and Mechanics of Solids **14** (2009), 300–366

[29] F. Hoppensteadt, *Mathematical Theories of Populations: Demographics, Genetics, and Epidemics.* SIAM, Philadelphia, 1975.

[30] F. Brauer, C. Castillo-Chavez, *Mathematical Models in Population Biology and Epidemiology.* Springer, New York, 2001.

[31] V. Gafiychuk, B. Datsko, *Mathematical modeling of different types of instabilities in time fractional reaction-diffusion systems.* Computers and Mathematics with Applications **59** (2010), 1001–1007.

[32] V. Gafiychuk, B. Datsko, V. Meleshko, D. Blackmore, *Analysis of the solutions of coupled nonlinear fractional reaction-diffusion equations.* Chaos, Solitons & Fractals **41** (2009), 1095–1104.

[33] V. Gafiychuk, B. Datsko, *Stability analysis and limit cycle in fractional system with Brusselator nonlinearities.* Phys. Let. A **372** (2008), 4902–4904.

[34] V. Gafiychuk, B. Datsko, V. Meleshko, *Analysis of fractional order Bonhoeffer-van der Pol oscillator.* Physica A **387** (2008), 418–424.

[35] G.M. Zaslavsky, A.A. Stanislavsky, M. Edelman, *Chaotic and pseudochaotic attractors of perturbed fractional oscillator.* Chaos **16** (2006), 013102.

[36] M.S. Tavazoei, M. Haeri, *Chaotic attractors in incommensurate fractional order systems.* Physica D **237** (2008), 2628–2637.

[37] V.E. Tarasov, G.M. Zaslavsky, *Fractional equations of kicked systems and discrete maps.* J. Phys. A **41** (2008), 435101.

[38] M. Edelman, V.E. Tarasov, *Fractional standard map.* Phys. Let. A **374** (2009), 279–285.

[39] V.E. Tarasov, M. Edelman, *Fractional dissipative standard map.* Chaos **20** (2010), 023127.

[40] V.E. Tarasov, *Discrete map with memory from fractional differential equation of arbitrary positive order.* Journal of Mathematical Physics. **50** (2009), 122703.

[41] B.V. Chirikov, *A universal instability of many dimensional oscillator systems.* Phys. Rep. **52** (1979), 263–379.

[42] G.M. Zaslavsky, *The simplest case of a strange attractor.* Phys. Lett. A **69** (1978), 145–147.

[43] G.M. Zaslavsky, Kh.-R. Ya. Rachko, *Singularities of transition to a turbulent motion.* Sov. Phys. JETP **49** (1979), 1039–1044.

[44] A.J. Lichtenberg, M.A. Lieberman, *Regular and Chaotic Dynamics.* Springer, Berlin, 1992.

[45] G.M. Zaslavsky, *Hamiltonian Chaos and Fractional Dynamics.* Oxford University Press, Oxford, 2005.

[46] E. Ott, *Strange Attractors and Chaotic Motions of Dynamical Systems.* Rev. Mod. Phys. **53** (1981), 655–671.

[47] V. Afraimovich, Sze-Bi Hsu, *Lectures on Chaotic Dynamical Systems.* Amer. Math. Society. International Press, Providence, 2002.

[48] P. Grassberger and I. Procaccia, *Measuring the strangeness of strange attractors.* Physica D **9** (1983), 189-208.

[49] D.A. Russel, J.D. Hanson, and E. Ott, *Dimension of strange attractors.* PRL **45** (1980), 1175–1178.

[50] F. Haake, *Quantum Signatures of Chaos.* Springer, Berlin, 2000.

[51] G.M. Zaslavsky, M. Edelman, *Superdiffusion in the Dissipative Standard Map.* Chaos **18** (2008), 033116.

[52] Q. Wang and L.-S. Young, *From invariant curves to strange attractors.* Commun. in Math. Phys. **225** (2002), 275–304.

[53] Q. Wang and L.-S. Young, *Strange attractors in periodically-kicked limit cycles and Hopf bifurcations.* Commun. in Math. Phys. **240** (2003), 509–529.

[54] M. Edelman, *Fractional Standard Map: Riemann-Liouville vs. Caputo.* Commun. Nonlin. Sci. Numer. Simul, **16** (2011), 4573-4580.

[55] A. Fulinski, A.S. Kleczkowski, *Nonlinear maps with memory.* Physica Scripta **35** (1987), 119–122.

[56] E. Fick, M. Fick, G. Hausmann, *Logistic equation with memory.* Phys. Rev. A **44** (1991), 2469–2473.

[57] K. Hartwich, E. Fick, *Hopf bifurcations in the logistic map with oscillating memory* Phys. Lett. A **177** (1993), 305–310.

[58] M. Giona, *Dynamics and relaxation properties of complex systems with memory.* Nonlinearity **4** (1991), 911–925.

[59] J.A.C. Gallas, *Simulating memory effects with discrete dynamical systems.* Physica A **195** (1993), 417–430; *Erratum.* Physica A **198** (1993), 339–339.

[60] A.A. Stanislavsky, *Long-term memory contribution as applied to the motion of discrete dynamical system.* Chaos **16** (2006), 043105.

[61] G. Schmidt, *Stochasticity and fixed-point transitions.* Phys. Rev. A **22** (1980), 2849–2854.

[62] Y. Li, Y.Q. Chen, I. Podlubny, *Stability of fractional-order nonlinear dynamic systems: Lyapunov direct method and generalized Mittag-Leffler stability.* Comput. Math. Appl. **59** (2010), 1810–1821.

[63] M. Edelman, *Cascade of bifurcation type trajectories in fractional dynamical systems* submitted to Chaos.

[64] G.M. Zaslavsky, M. Edelman, B.A. Niyazov, *Self-Similarity, renormalization, and phase space nonuniformity of Hamiltonian chaotic dynamics.* Chaos **7** (1997), 159–181.

[65] V.E. Tarasov, *Universal electromagnetic waves in dielectrics.* J Phys.: Condens. Matter **20** (2008), 175223.

[66] F. Hoppensteadt, *A nonlinear renewal equation with periodic and chaotic solutions.* SIAM-AMS Proc. **10** (1976), 51–60.

Mark Edelman and Laura Anna Taieb
Stern College for Women at Yeshiva University
245 Lexington Ave.
New York, NY 10016, USA
e-mail: `edelman@cims.nyu.edu`
`taieb@yu.edu`

Operator Theory:
Advances and Applications, Vol. 229, 157–173

Stability, Structural Stability and Numerical Methods for Fractional Boundary Value Problems

Neville J. Ford and M. Luísa Morgado

Abstract. In this work, we investigate the stability and the structural stability of a class of fractional boundary value problems. We approximate the solution by using a wide range of numerical methods illustrating our theoretical results.

Mathematics Subject Classification (2010). Primary 65L05; Secondary 34D10.

Keywords. Fractional differential equations, boundary value problems, numerical methods, stability.

1. Introduction

In this paper we consider nonlinear fractional boundary value problems (FBVPs) of the form

$$D_*^\alpha y(t) = f(t, y(t)), \quad t \in [0, T] \tag{1.1}$$

$$y(a) = y_a, \tag{1.2}$$

where we assume that $0 < \alpha < 1$, $0 < a < T$ and f is a continuous function. D_*^α is the Caputo derivative of y, of order $\alpha \notin \mathbb{N}$ ([1]), defined by

$$D_*^\alpha y(t) := D^\alpha(y - T[y])(t)$$

where $T[y]$ is the Taylor polynomial of degree $\lfloor \alpha \rfloor$ for y, centered at 0, and D^α is the Riemann-Liouville derivative of order α [12]. The latter is defined by $D^\alpha := D^{\lceil \alpha \rceil} J^{\lceil \alpha \rceil - \alpha}$, with J^β being the Riemann-Liouville integral operator,

$$J^\beta y(t) := \frac{1}{\Gamma(\beta)} \int_0^t (t-s)^{\beta-1} y(s) ds$$

and $D^{\lceil \alpha \rceil}$ is the classical integer order derivative. Here, as usual, $\lfloor \alpha \rfloor$ denotes the biggest integer smaller than α, and $\lceil \alpha \rceil$ represents the smallest integer greater than or equal to α.

In [4], Diethelm and Ford studied problem (1.1)–(1.2) in the case where $a = 0$, that is they considered an initial value problem and they have analysed not only the issues of existence and uniqueness of the solution, but also the dependence of the solution on the parameters in the differential equation.

Concerning the case where $a > 0$, Diethelm and Ford have recently investigated the uniqueness of the solution (see [5]). They have also proposed a prototype numerical method based on a shooting argument. But in that paper the authors have not considered the existence of the solution.

Recently, in [10], we have established sufficient conditions for the existence and uniqueness of the solution of problem (1.1)–(1.2). We have also proposed a range of numerical methods to approximate the solution of such problems.

The question of the structural stability of a problem is often neglected but is a key necessary feature for the effective application of numerical schemes. Numerical methods, by their very nature, can have the effect of perturbing the problem being solved. If the perturbed problem may have solutions that are quite different from the unperturbed equation, then the result of applying a numerical scheme may be disastrous. The analysis of the stability and structural stability was not provided in our previous work [10].

Hence, this work continues the investigation initiated in [5] and [10].

The paper is organized in the following way: We begin, in Section 2, with a summary of the main results obtained in [5] and [10]. To prepare for the construction of robust numerical methods, in Section 3 we investigate the sensitivity of the solution with respect to perturbations on the given parameters of the problem. To be specific, we analyse the dependence of the solution on the value of the boundary condition y_a, on the order of the derivative α and on the right-hand side function $f(t, y)$. Finally, in Section 4 we consider and compare a range of numerical methods to approximate the solution of the considered value problem and investigate if the proposed numerical schemes preserve the theoretical stability results obtained in Section 3.

2. Existence and uniqueness of the solution

In [5] the authors proved that under some simple natural conditions on f, there is at most one initial value y_0 for which the solution of the problem

$$D_*^\alpha y(t) = f(t, y(t)), \quad t \in [0, T] \tag{2.1}$$

$$y(0) = y_0, \tag{2.2}$$

satisfies $y(a) = y_a$. The fundamental theorem that shows that the problem under consideration is well posed is the following:

Theorem 2.1 ([5]). *Let $0 < \alpha < 1$ and assume $f : [0, b] \times [c, d] \to \mathbb{R}$ to be continuous and satisfy a Lipschitz condition with respect to the second variable. Consider two solutions y_1 and y_2 to the differential equation*

$$D_*^\alpha y_j(t) = f(t, y_j(t)) \qquad (j = 1, 2) \tag{2.3}$$

subject to the initial conditions $y_j(0) = y_{j0}$, respectively, where $y_{10} \neq y_{20}$. Then, for all t where both $y_1(t)$ and $y_2(t)$ exist, we have $y_1(t) \neq y_2(t)$.

In other words, if we know the value of a solution y to the equation (1.1) at $t = a$ then there is at most one corresponding value of $y(s)$ at any $s \in [0, a]$.

The following lemma, establishing an equivalence between a fractional initial value problem and a Volterra differential equation, is a widely known result in Fractional Calculus:

Lemma 2.2. *If the function f is continuous, the initial value problem (2.1)–(2.2) is equivalent to the following Volterra integral equation of the second kind:*

$$y(t) = y_0 + \frac{1}{\Gamma(\alpha)} \int_0^t (t-s)^{\alpha-1} f(s, y(s))ds. \tag{2.4}$$

In [10], we proved an analogous result for boundary value problems of the form (1.1)–(1.2):

Lemma 2.3. *If the function f is continuous and satisfies a Lipschitz condition with Lipschitz constant $L > 0$ with respect to its second argument:*

$$|f(t,y) - f(t,z)| \leq L\,|y - z|\,, \tag{2.5}$$

for some constant $L > 0$ independent of t, y and z, and if $\frac{2La^\alpha}{\Gamma(\alpha+1)} < 1$, then the boundary value problem (1.1)–(1.2) is equivalent to the integral equation

$$\begin{aligned} y(t) = y(a) &- \frac{1}{\Gamma(\alpha)} \int_0^a (a-s)^{\alpha-1} f(s, y(s))ds \\ &+ \frac{1}{\Gamma(\alpha)} \int_0^t (t-s)^{\alpha-1} f(s, y(s))ds \end{aligned} \tag{2.6}$$

We have also proved that if the conditions of Lemma 2.3 were satisfied, then equation (2.6) has a unique solution on $[0, a]$. As a consequence we find that every FBVP (1.1)–(1.2) coincides with a unique initial condition y_0:

$$y_0 = y(a) - \frac{1}{\Gamma(\alpha)} \int_0^a (a-s)^{\alpha-1} f(s, y(s))ds.$$

Then, as pointed out in that paper, having proved the existence of $y(0)$, existence and uniqueness results for $t > a$ could be inherited from the corresponding initial value problem theory (see [4]).

3. Dependence on the problem parameters

In this section we investigate the dependence of the solution on the given parameters of the problem, namely, on the boundary condition y_a, on the order of the derivative α and on the right-hand side function f. We only consider the nonlinear case. The simplest case of linear FBVPs was considered in [9].

We begin by analysing the change on the solution under a perturbation on the boundary condition imposed at $t = a > 0$. Given the two problems

$$\begin{aligned} D_*^\alpha y(t) &= f(t, y(t)), \quad t \in [0, T] \\ y(a) &= y_a \end{aligned} \tag{3.1}$$

and

$$\begin{aligned} D_*^\alpha z(t) &= f(t, z(t)), \quad t \in [0, T] \\ z(a) &= z_a \end{aligned} \tag{3.2}$$

consider, without loss of generality, the case where $y(a) \geq z(a)$. Remember, by [5], that we then have $y(t) \geq z(t)$ for any t where both solutions exist.

Taking Lemma 2.2 and the Lipschitz condition (2.5) into account, we have

$$\begin{aligned} y(t) - z(t) &= y(0) - z(0) + \frac{1}{\Gamma(\alpha)} \int_0^t (t-s)^{\alpha-1}(f(s, y(s)) - f(s, z(s)))ds \\ &\leq y(0) - z(0) + \frac{L}{\Gamma(\alpha)} \int_0^t (t-s)^{\alpha-1}|y(s) - z(s)|ds, \end{aligned}$$

and therefore, it follows by a Gronwall inequality (see [2]) that:

$$|y(t) - z(t)| \leq |y(0) - z(0)| e^{\left(\frac{L}{\alpha\Gamma(\alpha)} t^\alpha\right)}. \tag{3.3}$$

Analogously, we easily conclude that

$$|y(t) - z(t)| \geq |y(0) - z(0)| e^{\left(\frac{-L}{\alpha\Gamma(\alpha)} t^\alpha\right)}. \tag{3.4}$$

In particular, if we take $t = a$ in (3.4), we obtain

$$|y(0) - z(0)| \leq |y(a) - z(a)| e^{\left(\frac{L}{\alpha\Gamma(\alpha)} a^\alpha\right)}. \tag{3.5}$$

Substituting (3.5) in (3.3) we immediately obtain the following theorem.

Theorem 3.1. *Let $y(t)$ and $z(t)$ be the solutions of the two problems (3.1) and (3.2), respectively, where $0 < \alpha < 1$, f is a continuous function satisfying the Lipschitz condition (2.5) and $\frac{2La^\alpha}{\Gamma(\alpha+1)} < 1$. Then*

$$|y(t) - z(t)| \leq |y(a) - z(a)| e^{\left(\frac{L}{\alpha\Gamma(\alpha)}(t^\alpha + a^\alpha)\right)} \tag{3.6}$$

for any t where both solutions exist.

Taking into account this theorem we can conclude that

$$\|y - z\| = O\left(|y(a) - z(a)|\right),$$

on any compact interval where both solutions exist. Estimate (3.6) may be very rough. It is known from the theory of ODEs that if y_1 and y_2 are two solutions of the first-order differential equation $y'(t) = f(t, y(t))$, where f is a continuous function satisfying a Lipschitz condition with respect to its second argument, then the following inequality holds

$$|y_1(t) - y_2(t)| \leq |y_1(a) - y_2(a)| e^{L(t-a)}, \quad \forall t \geq a.$$

Next we investigate under what conditions a similar estimate holds in the fractional setting.

Theorem 3.2. *Let $y(t)$ and $z(t)$ be the solutions of the two problems* (3.1) *and* (3.2)*, respectively, where $0 < \alpha < 1$, f is a continuous function satisfying the Lipschitz condition* (2.5)*, $\frac{2La^\alpha}{\Gamma(\alpha+1)} < 1$ and the additional condition*

$$f(t,u) - f(t,v) \geq M(u-v), \quad \forall u \geq v, \tag{3.7}$$

for some $M > 0$. Then

$$|y(t) - z(t)| \leq |y(a) - z(a)| e^{\left(\frac{L}{\alpha\Gamma(\alpha)} h(t)\right)}, \quad \forall t \in [0,T], \tag{3.8}$$

where

$$h(t) = \begin{cases} (a-t)^\alpha + t^\alpha - a^\alpha, & t < a \\ (t-a)^\alpha, & t \geq a \end{cases}.$$

Proof. Let us assume without loss of generality that $y(a) \geq z(a)$. By [5] we know that we must have $y(t) \geq z(t)$, for every t where both solutions exist. Taking Lemma 2.3 into account, we may write

$$\begin{aligned} y(t) - z(t) = y(a) - z(a) &- \frac{1}{\Gamma(\alpha)} \int_0^a (a-s)^{\alpha-1} (f(s,y(s)) - f(s,z(s))) ds \\ &+ \frac{1}{\Gamma(\alpha)} \int_0^t (t-s)^{\alpha-1} (f(s,y(s)) - f(s,z(s))) ds. \end{aligned}$$

Let us first consider the case where $t \geq a$. Then

$$\begin{aligned} &y(t) - z(t) \\ &= y(a) - z(a) - \frac{1}{\Gamma(\alpha)} \int_0^a (a-s)^{\alpha-1} (f(s,y(s)) - f(s,z(s))) ds \\ &\qquad + \frac{1}{\Gamma(\alpha)} \int_0^a (t-s)^{\alpha-1} (f(s,y(s)) - f(s,z(s))) ds \\ &\qquad + \frac{1}{\Gamma(\alpha)} \int_a^t (t-s)^{\alpha-1} (f(s,y(s)) - f(s,z(s))) ds \\ &= y(a) - z(a) + \frac{1}{\Gamma(\alpha)} \int_0^a \left((t-s)^{\alpha-1} - (a-s)^{\alpha-1}\right) (f(s,y(s)) - f(s,z(s))) ds \\ &\qquad + \frac{1}{\Gamma(\alpha)} \int_a^t (t-s)^{\alpha-1} (f(s,y(s)) - f(s,z(s))) ds \end{aligned}$$

Since in this case $(t-s)^{\alpha-1} - (a-s)^{\alpha-1} \leq 0$ and $f(s,y(s)) - f(s,z(s)) \geq M > 0$ then

$$y(t) - z(t) \leq y(a) - z(a) + \frac{1}{\Gamma(\alpha)} \int_a^t (t-s)^{\alpha-1} (f(s,y(s)) - f(s,z(s))) ds.$$

Since we have assumed that $y(a) \geq z(a)$ we conclude

$$|y(t) - z(t)| \leq |y(a) - z(a)| + \frac{L}{\Gamma(\alpha)} \int_a^t (t-s)^{\alpha-1} |y(s) - z(s)| ds,$$

and the result follows by a Gronwall inequality.

On the other hand, if we now consider $t < a$ then using similar arguments we write

$$\begin{aligned} &y(t) - z(t) \\ &= y(a) - z(a) - \frac{1}{\Gamma(\alpha)} \int_0^t (a-s)^{\alpha-1} (f(s, y(s)) - f(s, z(s))) ds \\ &\qquad - \frac{1}{\Gamma(\alpha)} \int_t^a (a-s)^{\alpha-1} (f(s, y(s)) - f(s, z(s))) ds \\ &\qquad + \frac{1}{\Gamma(\alpha)} \int_0^t (t-s)^{\alpha-1} (f(s, y(s)) - f(s, z(s))) ds \\ &= y(a) - z(a) + \frac{1}{\Gamma(\alpha)} \int_0^t \left((t-s)^{\alpha-1} - (a-s)^{\alpha-1} \right) (f(s, y(s)) - f(s, z(s))) ds \\ &\qquad - \frac{1}{\Gamma(\alpha)} \int_t^a (a-s)^{\alpha-1} (f(s, y(s)) - f(s, z(s))) ds \\ &\leq y(a) - z(a) + \frac{1}{\Gamma(\alpha)} \int_0^t \left((t-s)^{\alpha-1} - (a-s)^{\alpha-1} \right) (f(s, y(s)) - f(s, z(s))) ds, \end{aligned}$$

and the desired result follows once again by a Gronwall inequality. □

Next, we investigate the dependence of the solution on the order of the derivative. As pointed out in [4], when modelling certain processes, the order of the derivative is only known up to a certain accuracy. Therefore it is important in the fractional case, to know how the solution depends on that parameter.

Let us then consider problem (1.1)–(1.2) and the following perturbed one

$$D_*^{\alpha-\delta} z(t) = f(t, z(t)), \quad t \in [0, T] \tag{3.9}$$

$$z(a) = y_a. \tag{3.10}$$

Theorem 3.3. *Let y and z be the uniquely determined solutions of problems* (1.1)–(1.2) *and* (3.9)–(3.10), *respectively, and assume that the continuous function f satisfies a Lipschitz condition of the form* (2.5). *Then*

$$\|y - z\| = O(\delta)$$

on any compact interval where both solutions exist.

Proof. We follow the steps of Theorem 3.1 in [4]. Since

$$D_*^{\alpha-\delta} \left(z(t) - D_*^{\delta} y(t) \right) = f(t, z(t)) - f(t, y(t))$$

and since $D_*^\delta y(t) = O(\delta)y(t) + y(t)$, we have

$$z(t) - y(t) = z(0) - y(0) \tag{3.11}$$
$$+ \frac{1}{\Gamma(\alpha-\delta)} \int_0^t (t-s)^{\alpha-\delta-1} \left(f(s,z(s)) - f(s,y(s))\right) ds + O(\delta)y(t).$$

The only difference between this equation and the one obtained in [4], is the first term on the right-hand side of (3.11), the quantity $z(0) - y(0)$. Since in [4], the authors have proved that $O(\delta)y(t) \le M\delta$, where M is a constant, in order to prove the theorem we just need to conclude that $|z(0) - y(0)| = O(\delta)$.

Since we have

$$\begin{aligned} z(0) - y(0) = &- \frac{1}{\Gamma(\alpha-\delta)} \int_0^a (a-s)^{\alpha-\delta-1} f(s,z(s)) ds \\ &+ \frac{1}{\Gamma(\alpha)} \int_0^a (a-s)^{\alpha-1} f(s,y(s)) ds, \end{aligned}$$

then

$$\begin{aligned} |z(0) - y(0)| &\le \frac{\|f\|}{\Gamma(\alpha-\delta)} \int_0^a (a-s)^{\alpha-\delta-1} ds + \frac{\|f\|}{\Gamma(\alpha)} \int_0^a (a-s)^{\alpha-1} \\ &= \frac{\|f\|}{\Gamma(\alpha-\delta+1)} a^{\alpha-\delta} + \frac{\|f\|}{\Gamma(\alpha)} a^\alpha \\ &\le C\delta, \end{aligned}$$

where C is a suitable constant depending on α, a and $\|f\|$. We can now write

$$|z(t) - y(t)| \le (M+C)\delta + \frac{L}{\Gamma(\alpha-\delta)} \int_0^t (t-s)^{\alpha-\delta-1} |z(t) - y(t)| ds.$$

It follows, by a Gronwall inequality, that

$$|z(t) - y(t)| \le (M+C)\delta e^{\left(\frac{L}{\Gamma(\alpha-\delta+1)} t^{\alpha-\delta}\right)}.$$

By taking the infinity-norm on a compact interval where both solutions exist, the theorem is proved. □

Finally, we consider a perturbation on the right-hand side function f.

Theorem 3.4. *Let y and z be the uniquely determined solutions of problems* (1.1)–(1.2) *and*

$$\begin{aligned} D_*^\alpha z(t) &= \tilde{f}(t, z(t)), \quad t \in [0,T] \\ z(a) &= y_a. \end{aligned}$$

respectively, and assume that the continuous function f satisfies a Lipschitz condition of the form (2.5). *Then*

$$\|y - z\| = O\left(\left\|f - \tilde{f}\right\|\right)$$

on any compact interval where both solutions exist.

Proof. Again, we follow the steps in [4]. Since

$$\begin{aligned} z(t) - y(t) &= y(0) - z(0) + \frac{1}{\Gamma(\alpha)} \int_0^t (t-s)^{\alpha-1} \left(f(s, y(s)) - \tilde{f}(s, z(s)) \right) ds \\ &= y(0) - z(0) + \frac{1}{\Gamma(\alpha)} \int_0^t (t-s)^{\alpha-1} \left(f(s, y(s)) - f(s, z(s)) \right) ds \\ &\quad + \frac{1}{\Gamma(\alpha)} \int_0^t (t-s)^{\alpha-1} \left(f(s, z(s)) - \tilde{f}(s, z(s)) \right) ds, \end{aligned}$$

we have

$$|y(t) - z(t)| \leq |y(0) - z(0)| + \frac{L}{\Gamma(\alpha)} \int_0^t (t-s)^{\alpha-1} |y(s) - z(s)| \, ds + D \left\| f - \tilde{f} \right\|,$$

where D is a constant. On the other hand, we have

$$y(0) - z(0) = -\frac{1}{\Gamma(\alpha)} \int_0^a (a-s)^{\alpha-1} \left(f(s, y(s)) - \tilde{f}(s, z(s)) \right) ds,$$

and therefore

$$|y(0) - z(0)| \leq \frac{1}{\Gamma(\alpha)} \left\| f - \tilde{f} \right\| \int_0^a (a-s)^{\alpha-1} ds = \frac{1}{\Gamma(\alpha+1)} \left\| f - \tilde{f} \right\| a^\alpha.$$

We can then write

$$|y(t) - z(t)| \leq O\left(\left\| f - \tilde{f} \right\| \right) + \frac{L}{\Gamma(\alpha)} \int_0^t (t-s)^{\alpha-1} |y(s) - z(s)| \, ds$$

and the desired result follows by a Gronwall inequality. □

4. Numerical examples

In this section we investigate whether natural numerical methods preserve the underlying stability of the problem.

From the results in [5] we know that for the solution of (1.1) that passes through the point (a, y_a), we are able to find at most one point $(0, y_0)$ that also lies on the same solution trajectory. In that paper that was carried out by using a shooting algorithm based on the bisection method. In what follows we will use a different approach where the bisection is replaced by the secant method. We shall also consider a range of simple algorithms for evaluating the solution.

The methods that we used to solve the initial value problems are listed below:

1) The first method we have considered was the fractional Adams scheme of [6], the one also used in [5];
2) This is a finite difference method based on the definition of the Grunwald-Letnikov operator (see, for example, [8]);
3) Fractional backward difference based on quadrature (see, for example, [3], [8]);
4) The fractional BDF method due to Lubich with $p = 3$ ([7], [11]).

First, we will be interested in investigating how small perturbations in the boundary condition will affect the values of the numerical solution in the whole interval where the differential equation is defined.

Let us consider the boundary value problem

$$D_*^{0.5}y(t) = \frac{1}{4}\left(-y(t) + t^2\right) + \frac{2}{\Gamma(2.5)}t^{1.5}, \quad t \in [0,1] \tag{4.1}$$

$$y(0.5) = 0.25, \tag{4.2}$$

whose analytical solution is $y(t) = t^2$. Let us also consider another problem where the boundary condition suffers a perturbation:

$$D_*^{0.5}z(t) = \frac{1}{4}\left(-z(t) + t^2\right) + \frac{2}{\Gamma(2.5)}t^{1.5}, \quad t \in [0,1] \tag{4.3}$$

$$z(0.5) = 0.25 + \epsilon, \tag{4.4}$$

The obtained approximated values for $y(0)$, $z(0)$, $y(1)$, $z(1)$ and $\|y - z\|_\infty$ are presented in Tables 1–8. Here, as usual, $\|y - z\|_\infty = \max_i |y_i - z_i|$, where y_i and z_i are the obtained numerical approximations of $y(t)$ and $z(t)$ at the discretization points t_i.

h	y_0	$z(0)$	$y(1)$	$z(1)$	$\|y-z\|_\infty$
$\frac{1}{10}$	-0.0054849330	0.1152990040	1.0030265377	1.0960650891	0.1207839
$\frac{1}{20}$	-0.0017210176	0.1190499124	1.0010149424	1.0940478654	0.1207709
$\frac{1}{40}$	-0.0005496786	0.1202165327	1.0003461746	1.0933770885	0.1207662
$\frac{1}{80}$	-0.0001789695	0.1205855492	1.0001194115	1.0931496117	0.1207645
$\frac{1}{160}$	-0.0000593227	0.1207045924	1.0000415048	1.0930714521	0.1207639
$\frac{1}{320}$	-0.0000199673	0.1207437331	1.0000145011	1.0930443589	0.1207637

TABLE 1. Numerical solution of problems (4.1)–(4.2) and (4.3)–(4.4) with $\epsilon = 0.1$ (method 1 to solve the IVP)

$z(0.5)$	$h = 0.1$	$h = 0.05$	$h = 0.01$
0.25	1.0030265377	1.0010149424	1.0000849174
0.255	1.0076784652	1.0056665885	1.0047364219
0.26	1.0123303928	1.0103182347	1.0093879263
0.265	1.0169823204	1.0149698808	1.0140394308
0.27	1.0216342480	1.0196215270	1.0186909352
0.275	1.0262861755	1.0242731731	1.0233424397

TABLE 2. Approximate values of $z(1)$ as $z(0.5)$ varies in problem (4.3)–(4.4) (method 1 to solve the IVP)

h	y_0	$z(0)$	$y(1)$	$z(1)$	$\|y-z\|_\infty$
$\frac{1}{10}$	-0.0253617838	0.0901790683	1.0216695974	1.1135543428	0.11554085
$\frac{1}{20}$	-0.0128768752	0.1041962790	1.0108130585	1.1029834366	0.11799207
$\frac{1}{40}$	-0.0065049880	0.1116788859	1.0053960318	1.0978061473	0.12098759
$\frac{1}{80}$	-0.0032750609	0.1156902566	1.0026937881	1.0952844389	0.12315925
$\frac{1}{160}$	-0.0016452209	0.1178637577	1.0013453037	1.0940659509	0.12471531
$\frac{1}{320}$	-0.0008252492	0.1190611268	1.0006720717	1.0934846232	0.12582340

TABLE 3. Numerical solution of problems (4.1)–(4.2) and (4.3)–(4.4) with $\epsilon = 0.1$ (method 2 to solve the IVP)

$z(0.5)$	$h = 0.1$	$h = 0.05$	$h = 0.01$
0.25	1.0216695974	1.0108130585	1.0021541258
0.255	1.0262638347	1.0154215774	1.0067859964
0.26	1.0308580719	1.0200300963	1.0114178670
0.265	1.0354523092	1.0246386152	1.0160497376
0.27	1.0400465465	1.0292471341	1.0206816082
0.275	1.0446407838	1.0338556530	1.0253134788

TABLE 4. Approximate values of $z(1)$ as $z(0.5)$ varies in problem (4.3)–(4.4) (method 2 to solve the IVP)

h	y_0	$z(0)$	$y(1)$	$z(1)$	$\|y-z\|_\infty$
$\frac{1}{10}$	-0.0103262980	0.1093043160	1.0042139824	1.0969612170	0.1196306
$\frac{1}{20}$	-0.0038807170	0.1163222634	1.0014768777	1.0943661559	0.1202030
$\frac{1}{40}$	-0.0014256161	0.1190605864	1.0005196320	1.0934799409	0.1204862
$\frac{1}{80}$	-0.0005166701	0.1201093552	1.0001832602	1.0931786389	0.1206260
$\frac{1}{160}$	-0.0001856936	0.1205095259	1.0000647110	1.0930774233	0.1206952
$\frac{1}{320}$	-0.0000663842	0.1206631666	1.0000228645	1.0930441677	0.1207296

TABLE 5. Numerical solution of problems (4.1)–(4.2) and (4.3)–(4.4) with $\epsilon = 0.1$ (method 3 to solve the IVP)

According to the numerical results in Tables 1, 3, 5 and 7, we see that $\|y - z\|_\infty \cong 0.12$ for all the considered methods, independently of the used step size h.

This is in fact in agreement with (3.6), whose right-hand side is, in this case, given approximately by 0.16.

$z(0.5)$	$h = 0.1$	$h = 0.05$	$h = 0.01$
0.25	1.0042139824	1.0014768777	1.0001310646
0.255	1.0088513442	1.0061213416	1.0047811812
0.26	1.0134887059	1.0107658055	1.0094312978
0.265	1.0181260676	1.0154102694	1.0140814145
0.27	1.0227634294	1.0200547333	1.0187315311
0.275	1.0274007911	1.0246991972	1.0233816477

TABLE 6. Approximate values of $z(1)$ as $z(0.5)$ varies in problem (4.3)–(4.4) (method 3 to solve the IVP)

h	y_0	$z(0)$	$y(1)$	$z(1)$	$\|y-z\|_\infty$
$\frac{1}{10}$	6.383×10^{-18}	0.1230002694	1.0000000000	1.0935742723	0.1230003
$\frac{1}{20}$	3.382×10^{-17}	0.1218749008	1.0000000000	1.0933073938	0.1218749
$\frac{1}{40}$	$< \times 10^{-25}$	0.1213130381	1.0000000000	1.0931669747	0.1213130
$\frac{1}{80}$	-1.17×10^{-17}	0.1210364412	1.0000000000	1.0930979133	0.1210364
$\frac{1}{160}$	5.551×10^{-16}	0.1208994229	1.0000000000	1.0930637117	0.1208994
$\frac{1}{320}$	6.707×10^{-17}	0.1208313115	1.0000000000	1.0930467121	0.1208313

TABLE 7. Numerical solution of problems (4.1)–(4.2) and (4.3)–(4.4) with $\epsilon = 0.1$ (method 4 to solve the IVP)

$z(0.5)$	$h = 0.1$	$h = 0.05$	$h = 0.01$
0.25	1.0000000000	1.0000000000	1.0000000000
0.255	1.0046787136	1.0046653697	1.0046542102
0.26	1.0093574272	1.0093307394	1.0093084203
0.265	1.0140361408	1.0139961091	1.0139626305
0.27	1.0187148545	1.0186614788	1.0186168407
0.275	1.0233935681	1.0233268485	1.0232710508

TABLE 8. Approximate values of $z(1)$ as $z(0.5)$ varies in problem (4.3)–(4.4) (method 4 to solve the IVP)

With $\epsilon = 0.001$ and for different step size h, the values of $\|y - z\|_\infty = \max_i |y_i - z_i|$, for problems (4.1)–(4.2) and (4.3)–(4.4) are plotted in Figure 1.

Note that in equation (4.1) although the right-hand side function $f(t, y) = \frac{1}{4}\left(-y + t^2\right) + \frac{2}{\Gamma(2.5)} x^{1.5}$ satisfies a Lipschitz condition with respect to its second argument, it does not satisfy a condition of form (3.7) and therefore we can not apply Theorem 3.2. So next, instead of problems (4.1)–(4.2) and (4.3)–(4.4), we

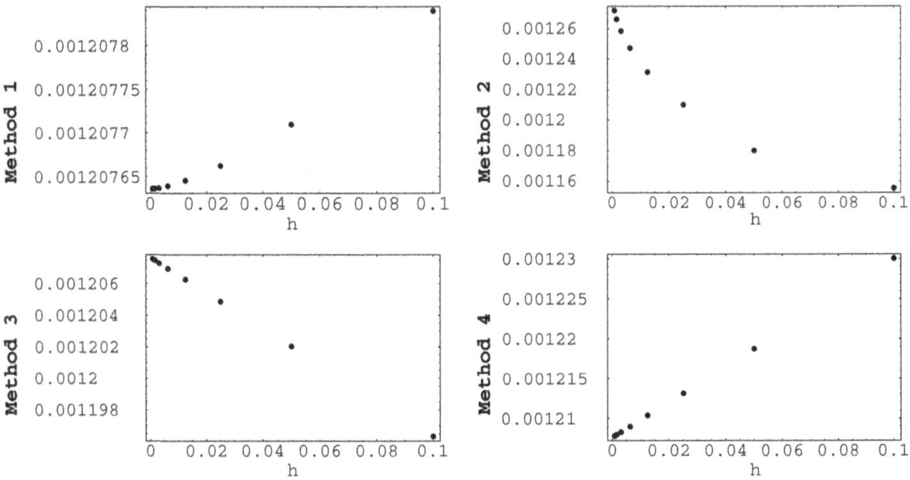

FIGURE 1. $\|y - z\|_\infty$ for problems (4.1)–(4.2) and (4.3)–(4.4) with $\epsilon = 0.001$.

consider the slightly different ones, for which all the conditions of theorem 3.2 hold:

$$D_*^{0.5} y(t) = \frac{1}{4}\left(y(t) - t^2\right) + \frac{2}{\Gamma(2.5)} t^{1.5}, \quad t \in [0,1] \tag{4.5}$$

$$y(0.5) = 0.25, \tag{4.6}$$

whose analytical solution is $y(t) = t^2$, and

$$D_*^{0.5} z(t) = \frac{1}{4}\left(z(t) - t^2\right) + \frac{2}{\Gamma(2.5)} t^{1.5}, \quad t \in [0,1] \tag{4.7}$$

$$z(0.5) = 0.25 + \epsilon. \tag{4.8}$$

The values of $\|y - z\|_\infty = \max_i |y_i - z_i|$, for problems (4.5)–(4.6) and (4.7)–(4.8) with $\epsilon = 0.001$, for different step-sizes h are plotted in Figure 2.

Now, let us assume that the order of the derivative is not a precise parameter in problem (4.1)–(4.2), that is let us consider the perturbed problem

$$D_*^{0.5+\delta} z(t) = \frac{1}{4}\left(-z(t) + t^2\right) + \frac{2}{\Gamma(2.5)} t^{1.5}, \quad t \in [0,1] \tag{4.9}$$

$$z(0.5) = 0.25, \tag{4.10}$$

and let us investigate how the solution is influenced as δ varies. In Tables 9–12 we display some approximations for the solution at $t = 1$, for each of the considered methods.

The numerical results in Tables 9–12, suggest a linear dependence of the solution on the change of the order of the derivative (which is in agreement with the obtained theoretical results) independently of the step size h.

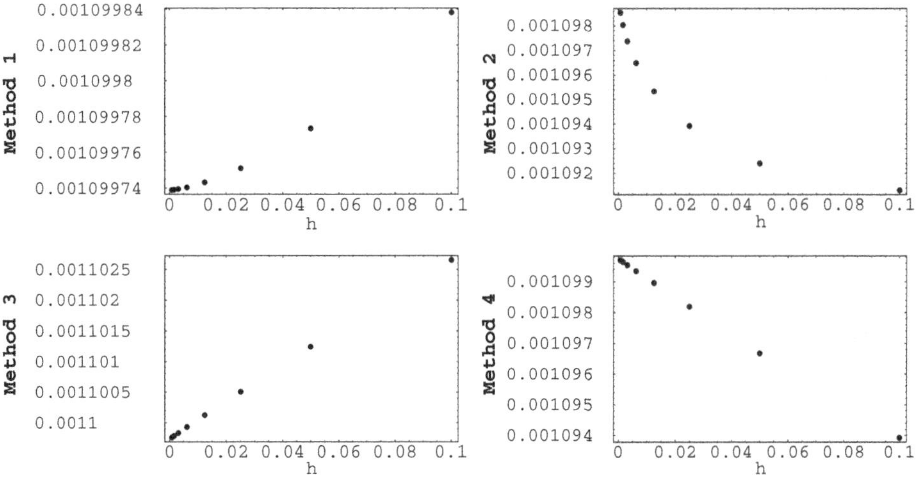

FIGURE 2. $\|y - z\|_\infty$ for problems (4.5)–(4.6) and (4.7)–(4.8) with $\epsilon = 0.001$.

This can, in fact, be observed in Figure 3 where, for problem (4.3)–(4.4), the values of $y(1)$ are presented, as α varies.

α	$h = 0.1$	$h = 0.05$	$h = 0.01$
0.5	1.0030265377	1.0010149424	1.0000849174
0.55	0.9801168033	0.9783362615	0.9775469250
0.6	0.9566946604	0.9551082706	0.9544343212
0.65	0.9328416892	0.9314185402	0.9308393324
0.7	0.9086470413	0.9073608852	0.9068593219
0.75	0.8842051162	0.8830334204	0.8825951869

TABLE 9. Approximate values of $z(1)$ as the order of the derivative (α) varies in problem (4.9)–(4.10) (method 1 to solve the IVP)

α	$h = 0.1$	$h = 0.05$	$h = 0.01$
0.5	1.0216695974	1.01081305854	1.0021541258
0.55	1.0016836686	0.98952041057	0.9798395008
0.6	0.9812243353	0.96770940423	0.9569580681
0.65	0.9603737723	0.94547410934	0.9336109471
0.7	0.9392085808	0.92290603211	0.9099026255
0.75	0.9177983522	0.90009116441	0.8859369581

TABLE 10. Approximate values of $z(1)$ as the order of the derivative (α) varies in problem (4.9)–(4.10) (method 2 to solve the IVP)

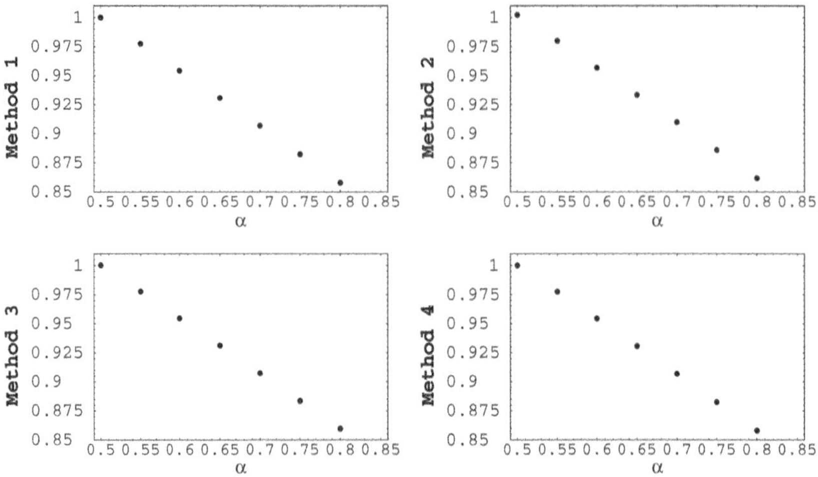

FIGURE 3. Approximate values of $y(1)$ for problem (4.3)–(4.4) as α varies, with $h = 0.01$.

α	$h = 0.1$	$h = 0.05$	$h = 0.01$
0.5	1.0042139824	1.0014768777	1.0001310646
0.55	0.9833621572	0.9796203509	0.9776850542
0.6	0.9623924898	0.9574112393	0.9546967148
0.65	0.9414623864	0.9349898685	0.9312723319
0.7	0.9207311610	0.9125041586	0.9075262987
0.75	0.9003575136	0.8901071952	0.8835800426

TABLE 11. Approximate values of $z(1)$ as the order of the derivative (α) varies in problem (4.9)–(4.10) (method 3 to solve the IVP)

α	$h = 0.1$	$h = 0.05$	$h = 0.01$
0.5	1.0000000000	1.0000000000	1.0000000000
0.55	0.9775753486	0.9775287122	0.9774900389
0.6	0.9545612500	0.9544726079	0.9544000388
0.65	0.9310453068	0.9309221716	0.9308226826
0.7	0.9071208720	0.9069732378	0.9068556280
0.75	0.8828856873	0.8827255289	0.8825999930

TABLE 12. Approximate values of $z(1)$ as the order of the derivative (α) varies in problem (4.9)–(4.10) (method 4 to solve the IVP)

Finally, let us assume that the right-hand side function in (4.1) may include small perturbations on the given data. Consider, for example the perturbed problem

$$D_*^{0.5}z(t) = \frac{1}{4}\left((-1+\eta)z(t) + t^2\right) + \frac{2}{\Gamma(2.5)}t^{1.5}, \quad t \in [0,1] \tag{4.11}$$

$$z(0.5) = 0.25, \tag{4.12}$$

and let us analyse how the solution varies as η varies. Some numerical examples for each of the considered methods are presented in Tables 13–16.

η	$h = 0.1$	$h = 0.05$	$h = 0.01$
0.00	1.0216695974	1.0108130585	1.0021541258
0.005	1.0222566738	1.0113807981	1.0027061428
0.01	1.0228444824	1.0119492258	1.0032588129
0.015	1.0234330243	1.0125183426	1.0038121372
0.02	1.0240223008	1.0130881498	1.0043661167
0.025	1.0246123132	1.0136586485	1.0049207525
0.03	1.0252030629	1.0142298399	1.0054760457

TABLE 13. Approximate values of $z(1)$ as the right-hand side function varies in problem (4.11)–(4.12) (method 1 to solve the IVP)

η	$h = 0.1$	$h = 0.05$	$h = 0.01$
0.00	1.0148798261	1.0074099033	1.0014735296
0.005	1.0165687087	1.0090604992	1.0030924942
0.01	1.0182639376	1.0107172033	1.0047173721
0.015	1.0199655467	1.0123800474	1.0063481936
0.02	1.0216735706	1.0140490639	1.0079849893
0.025	1.0233880437	1.0157242851	1.0096277899
0.03	1.0251090009	1.0174057439	1.0112766264

TABLE 14. Approximate values of $z(1)$ as the right-hand side function varies in problem (4.11)–(4.12) (method 2 to solve the IVP)

η	$h = 0.1$	$h = 0.05$	$h = 0.01$
0.00	1.0042139824	1.0014768777	1.0001310646
0.005	1.0047712945	1.0020281270	1.0006791657
0.01	1.0053292724	1.0025800279	1.0012279114
0.015	1.0058879170	1.0031325814	1.0017773026
0.02	1.0064472295	1.0036857885	1.0023273404
0.025	1.0070072110	1.0042396503	1.0028780258
0.03	1.0075678626	1.0047941679	1.0034293599

TABLE 15. Approximate values of $z(1)$ as the right-hand side function varies in problem (4.11)–(4.12) (method 3 to solve the IVP)

η	$h = 0.1$	$h = 0.05$	$h = 0.01$
0.00	1.0000000000	1.0000000000	1.0000000000
0.005	1.0005470969	1.0005474302	1.0005477003
0.01	1.0010948435	1.0010955073	1.0010960451
0.015	1.0016432410	1.0016442323	1.0016450354
0.02	1.0021922904	1.0021936062	1.0021946723
0.025	1.0027419926	1.0027436300	1.0027449568
0.03	1.0032923488	1.0032943049	1.0032958900

TABLE 16. Approximate values of $z(1)$ as the right-hand side function varies in problem (4.11)–(4.12) (method 4 to solve the IVP)

5. Conclusions

In this paper we have continued the investigation in [10], where existence and uniqueness results were stated for fractional boundary value problems of the form (1.1)–(1.2).

Here we proved that this kind of problem is stable in the sense that small perturbations of the given parameters, namely of the boundary condition, of the order of the derivative and of the right-hand side function, will produce small changes in the solution.

Some numerical results were carried out, by using a wide range of numerical methods, to illustrate and confirm the theoretical results.

Acknowledgment

M.L. Morgado acknowledges financial support from FCT, Fundação para a Ciência e Tecnologia, through grant SFRH/BPD/46530/2008.

References

[1] M. Caputo, *Elasticity e Dissipazione*, Zanichelli, Bologna, 1969.

[2] C. Corduneanu, *Principles of Differential and Integral Equations.* Allyn and Bacon, Boston, 1971.

[3] K. Diethelm, *An algorithm for the numerical solution of differential equations of fractional order* Electr. Trans. Numer. Anal. **5** (1997), 1–6.

[4] K. Diethelm and N.J. Ford, *Analysis of fractional differential equations* J. Math. Anal. Appl. **265** (2002), 229–248.

[5] K. Diethelm, and N.J. Ford, *Volterra integral equations and fractional calculus: Do neighbouring solutions intersect?* Journal of Integral Equations and Applications, to appear.

[6] K. Diethelm, N.J. Ford, and A.D. Freed, *A predictor-corrector approach for the numerical solution of fractional differential equations* Nonlinear Dynamics. **29** (2002), 3–22.

[7] K. Diethelm, J.M. Ford, N.J. Ford and M. Weilbeer, *Pitfalls in fast numerical solvers for fractional differential equations* J. Comp. Appl. Mathem. **186** 2 (2006), 482–503.

[8] N.J. Ford and J.A. Connolly, *Comparison of numerical methods for fractional differential equations* Commun. Pure Appl. Anal. **5**, no 2 (2006), 289–307.

[9] N.J. Ford and M.L. Morgado, *Structural Stability for Fractional Differential Equations as Boundary Value Problems* AIP Conference Proceedings **1281**, (2010), 1191–1194, 8th ICNAAM 2010, 19–25 September, Rhodes, Greece.

[10] N.J. Ford and M.L. Morgado, *Fract. Calc. Appl. Anal.*, Vol. 14, No 4 (2011), pp. 554–567; DOI: 10.2478/s13540-011-0034-4.

[11] E. Hairer, Ch. Lubich and M. Schlichte, *Fast numerical solution of weakly singular Volterra integral equations* J. Comp. Appl. Mathem. **23** (1988), 87–98.

[12] S.G. Samko, A.A. Kilbas and O.I. Marichev, *Fractional Integrals and Derivatives: Theory and Applications.* Gordon and Breach, Yverdon, 1993.

Neville J. Ford
Department of Mathematics
University of Chester
CH1 4BJ, UK
e-mail: `n.ford@chester.ac.uk`

M. Luísa Morgado
CEMAT, IST, Lisbon
and Department of Mathematics
University of Trás-os-Montes e Alto Douro
P-5001-801, Vila Real, Portugal
e-mail: `luisam@utad.pt`

Operator Theory:
Advances and Applications, Vol. 229, 175–199

On the Boundedness of the Fractional Maximal Operator, Riesz Potential and Their Commutators in Generalized Morrey Spaces

Vagif S. Guliyev and Parviz S. Shukurov

Dedicated to the 70th birthday of Prof. S. Samko

Abstract. In the paper the authors find conditions on the pair (φ_1, φ_2) which ensure the Spanne type boundedness of the fractional maximal operator M_α and the Riesz potential operator I_α from one generalized Morrey spaces M_{p,φ_1} to another M_{q,φ_2}, $1 < p < q < \infty$, $1/p - 1/q = \alpha/n$, and from M_{1,φ_1} to the weak space WM_{q,φ_2}, $1 < q < \infty$, $1 - 1/q = \alpha/n$. We also find conditions on φ which ensure the Adams type boundedness of the M_α and I_α from $M_{p,\varphi^{\frac{1}{p}}}$ to $M_{q,\varphi^{\frac{1}{q}}}$ for $1 < p < q < \infty$ and from $M_{1,\varphi}$ to $WM_{q,\varphi^{\frac{1}{q}}}$ for $1 < q < \infty$. As applications of those results, the boundeness of the commutators of operators M_α and I_α on generalized Morrey spaces is also obtained. In the case $b \in BMO(\mathbb{R}^n)$ and $1 < p < q < \infty$, we find the sufficient conditions on the pair (φ_1, φ_2) which ensures the boundedness of the operators $M_{b,\alpha}$ and $[b, I_\alpha]$ from M_{p,φ_1} to M_{q,φ_2} with $1/p - 1/q = \alpha/n$. We also find the sufficient conditions on φ which ensures the boundedness of the operators $M_{b,\alpha}$ and $[b, I_\alpha]$ from $M_{p,\varphi^{\frac{1}{p}}}$ to $M_{q,\varphi^{\frac{1}{q}}}$ for $1 < p < q < \infty$. In all cases conditions for the boundedness are given in terms of Zygmund-type integral inequalities on (φ_1, φ_2) and φ, which do not assume any assumption on monotonicity of φ_1, φ_2 and φ in r. As applications, we get some estimates for Marcinkiewicz operator and fractional powers of the some analytic semigroups on generalized Morrey spaces.

Mathematics Subject Classification (2010). Primary 42B20, 42B25, 42B35.

Keywords. Fractional maximal operator; Riesz potential operator; generalized Morrey space; commutator; BMO space.

The research of V. Guliyev was partially supported by the grant of Science Development Foundation under the President of the Republic of Azerbaijan project EIF-2010-1(1)-40/06-1 and by the Scientific and Technological Research Council of Turkey (TUBITAK Project No: 110T695).

1. Introduction

The theory of boundedness of classical operators of the real analysis, such as the maximal operator, fractional maximal operator, Riesz potential and the singular integral operators etc, from one weighted Lebesgue space to another one is well studied by now. These results have good applications in the theory of partial differential equations. However, in the theory of partial differential equations, along with weighted Lebesgue spaces, general Morrey-type spaces also play an important role.

For $x \in \mathbb{R}^n$ and $r > 0$, we denote by $B(x,r)$ the open ball centered at x of radius r, and by ${}^{\complement}B(x,r)$ denote its complement. Let $|B(x,r)|$ be the Lebesgue measure of the ball $B(x,r)$.

Let $f \in L_1^{\mathrm{loc}}(\mathbb{R}^n)$. The fractional maximal operator M_α and the Riesz potential I_α are defined by

$$M_\alpha f(x) = \sup_{t>0} |B(x,t)|^{-1+\frac{\alpha}{n}} \int_{B(x,t)} |f(y)|dy, \qquad 0 \le \alpha < n,$$

$$I_\alpha f(x) = \int_{\mathbb{R}^n} \frac{f(y)dy}{|x-y|^{n-\alpha}}, \qquad 0 < \alpha < n.$$

If $\alpha = 0$, then $M \equiv M_0$ is the Hardy-Littlewood maximal operator.

It is well known that fractional maximal operator, Riesz potential and Calderón-Zygmund operators play an important role in harmonic analysis (see [22, 29, 30]).

In [3] (see also [17]), we prove the boundedness of the maximal operator M and the Calderón-Zygmund operators T from one generalized Morrey space M_{p,φ_1} to another M_{p,φ_2}, $1 < p < \infty$, and from M_{1,φ_1} to the weak space WM_{1,φ_2}. In the case $b \in BMO(\mathbb{R}^n)$, we find the sufficient conditions on the pair (φ_1, φ_2) which ensure the boundedness of the operators M_b and $[b,T]$ from M_{p,φ_1} to M_{p,φ_2}, $1 < p < \infty$, where $M_b f(x) = M\big((b(\cdot) - b(x))f\big)(x)$ and $[b,T]f(x) = b(x)\,Tf(x) - T(bf)(x)$.

In this work, we prove the boundedness of the operators M_α and I_α, $\alpha \in (0,n)$ from one generalized Morrey space M_{p,φ_1} to another one M_{q,φ_2}, $1 < p < q < \infty$, $1/p - 1/q = \alpha/n$, and from M_{1,φ_1} to the weak space WM_{q,φ_2}, $1 < q < \infty$, $1 - 1/q = \alpha/n$. We also prove the Adams type boundedness of the operators M_α and I_α from $M_{p,\varphi^{\frac{1}{p}}}$ to $M_{q,\varphi^{\frac{1}{q}}}$ for $1 < p < q < \infty$ and from $M_{1,\varphi}$ to $WM_{q,\varphi^{\frac{1}{q}}}$ for $1 < q < \infty$. In the case $b \in BMO(\mathbb{R}^n)$, $1 < p < q < \infty$, we find the sufficient conditions on the pair (φ_1, φ_2) which ensures the boundedness of the commutator of operators $M_{b,\alpha}$ and $[b, I_\alpha]$ from M_{p,φ_1} to M_{q,φ_2}, $1 < p < q < \infty$, $1/p - 1/q = \alpha/n$, where $M_{b,\alpha} f(x) = M_\alpha\big((b(\cdot) - b(x))f\big)(x)$ and $[b, I_\alpha]f(x) = b(x)\,I_\alpha f(x) - I_\alpha(bf)(x)$. We also find the sufficient conditions on φ which ensures the boundedness of the operators $M_{b,\alpha}$ and $[b, I_\alpha]$ from $M_{p,\varphi^{\frac{1}{p}}}$ to $M_{q,\varphi^{\frac{1}{q}}}$ for $1 < p < q < \infty$. Finally, as applications we apply this result to several particular operators such as Marcinkiewicz operator and fractional powers of the some analytic semigroups.

By $A \lesssim B$ we mean that $A \leq CB$ with some positive constant C independent of appropriate quantities. If $A \lesssim B$ and $B \lesssim A$, we write $A \approx B$ and say that A and B are equivalent.

2. Morrey spaces

The classical Morrey spaces $M_{p,\lambda}$ were originally introduced by Morrey in [24] to study the local behavior of solutions to second-order elliptic partial differential equations. For the properties and applications of classical Morrey spaces, we refer the readers to [24, 26].

We denote by $M_{p,\lambda} \equiv M_{p,\lambda}(\mathbb{R}^n)$ the Morrey space, the space of all functions $f \in L_p^{\rm loc}(\mathbb{R}^n)$ with finite quasinorm

$$\|f\|_{M_{p,\lambda}} = \sup_{x\in\mathbb{R}^n,\, r>0} r^{-\frac{\lambda}{p}} \|f\|_{L_p(B(x,r))},$$

where $1 \leq p < \infty$ and $0 \leq \lambda \leq n$.

Note that $M_{p,0} = L_p(\mathbb{R}^n)$ and $M_{p,n} = L_\infty(\mathbb{R}^n)$. If $\lambda < 0$ or $\lambda > n$, then $M_{p,\lambda} = \Theta$, where Θ is the set of all functions equivalent to 0 on $\mathbb{R}^n$.

We also denote by $WM_{p,\lambda} \equiv WM_{p,\lambda}(\mathbb{R}^n)$ the weak Morrey space of all functions $f \in WL_p^{\rm loc}(\mathbb{R}^n)$ for which

$$\|f\|_{WM_{p,\lambda}} = \sup_{x\in\mathbb{R}^n,\, r>0} r^{-\frac{\lambda}{p}} \|f\|_{WL_p(B(x,r))} < \infty,$$

where $WL_p(B(x,r))$ denotes the weak L_p-space of measurable functions f for which

$$\begin{aligned}\|f\|_{WL_p(B(x,r))} &\equiv \|f\chi_{B(x,r)}\|_{WL_p(\mathbb{R}^n)} \\ &= \sup_{t>0} t \left|\{y \in B(x,r) : |f(y)| > t\}\right|^{1/p}.\end{aligned}$$

The classical result by Hardy-Littlewood-Sobolev states that if $1 < p < q < \infty$, then the operator I_α is bounded from $L_p(\mathbb{R}^n)$ to $L_q(\mathbb{R}^n)$ if and only if $\alpha = n\left(\frac{1}{p} - \frac{1}{q}\right)$ and for $p = 1 < q < \infty$, the operator I_α is bounded from $L_1(\mathbb{R}^n)$ to $WL_q(\mathbb{R}^n)$ if and only if $\alpha = n\left(1 - \frac{1}{q}\right)$. S. Spanne and D.R. Adams [1] studied boundedness of the Riesz potential in Morrey spaces. Their results can be summarized as follows.

Theorem 2.1 (Spanne, but published by Peetre [26]). *Let $0 < \alpha < n$, $1 < p < \frac{n}{\alpha}$, $0 < \lambda < n - \alpha p$. Moreover, let $\frac{1}{p} - \frac{1}{q} = \frac{\alpha}{n}$ and $\frac{\lambda}{p} = \frac{\mu}{q}$. Then for $p > 1$, the operator I_α is bounded from $M_{p,\lambda}$ to $M_{q,\lambda}$ and for $p = 1$, I_α is bounded from $M_{1,\lambda}$ to $WM_{q,\lambda}$.*

Theorem 2.2 (Adams [1]). *Let $0 < \alpha < n$, $1 < p < \frac{n}{\alpha}$, $0 < \lambda < n - \alpha p$ and $\frac{1}{p} - \frac{1}{q} = \frac{\alpha}{n-\lambda}$. Then for $p > 1$, the operator I_α is bounded from $M_{p,\lambda}$ to $M_{q,\lambda}$ and for $p = 1$, I_α is bounded from $M_{1,\lambda}$ to $WM_{q,\lambda}$.*

Recall that, for $0 < \alpha < n$,

$$M_\alpha f(x) \le v_n^{\frac{\alpha}{n}-1} I_\alpha(|f|)(x),$$

hence Theorems 2.1 and 2.2 also imply boundedness of the fractional maximal operator M_α, where v_n is the volume of the unit ball in $\mathbb{R}^n$.

3. Generalized Morrey spaces

We find it convenient to define the generalized Morrey spaces in the form as follows.

Definition 3.1. Let $\varphi(x,r)$ be a positive measurable function on $\mathbb{R}^n \times (0,\infty)$ and $1 \le p < \infty$. We denote by $M_{p,\varphi} \equiv M_{p,\varphi}(\mathbb{R}^n)$ the generalized Morrey space, the space of all functions $f \in L_p^{\rm loc}(\mathbb{R}^n)$ with finite quasinorm

$$\|f\|_{M_{p,\varphi}} = \sup_{x\in\mathbb{R}^n, r>0} \varphi(x,r)^{-1} \, |B(x,r)|^{-\frac{1}{p}} \, \|f\|_{L_p(B(x,r))}.$$

Also by $WM_{p,\varphi} \equiv WM_{p,\varphi}(\mathbb{R}^n)$ we denote the weak generalized Morrey space of all functions $f \in WL_p^{\rm loc}(\mathbb{R}^n)$ for which

$$\|f\|_{WM_{p,\varphi}} = \sup_{x\in\mathbb{R}^n, r>0} \varphi(x,r)^{-1} \, |B(x,r)|^{-\frac{1}{p}} \, \|f\|_{WL_p(B(x,r))} < \infty.$$

According to this definition, we recover the spaces $M_{p,\lambda}$ and $WM_{p,\lambda}$ under the choice $\varphi(x,r) = r^{\frac{\lambda-n}{p}}$:

$$M_{p,\lambda} = M_{p,\varphi}\Big|_{\varphi(x,r)=r^{\frac{\lambda-n}{p}}},$$

$$WM_{p,\lambda} = WM_{p,\varphi}\Big|_{\varphi(x,r)=r^{\frac{\lambda-n}{p}}}.$$

In [15]–[18], [20], [23] and [25] there were obtained sufficient conditions on φ_1 and φ_2 for the boundedness of the maximal operator M and Calderón-Zygmund operator from M_{p,φ_1} to M_{p,φ_2}, $1 < p < \infty$ and of the fractional maximal operator M_α and Riesz potential operator I_α from M_{p,φ_1} to M_{q,φ_2}, $1 < p < q < \infty$ (see also [5]–[9]). In [25] the following condition was imposed on $\varphi(x,r)$:

$$c^{-1}\varphi(x,r) \le \varphi(x,t) \le c\,\varphi(x,r) \tag{3.1}$$

whenever $r \le t \le 2r$, where $c(\ge 1)$ does not depend on t, r and $x \in \mathbb{R}^n$, jointly with the condition:

$$\int_r^\infty \varphi(x,t)^p \frac{dt}{t} \le C\,\varphi(x,r)^p, \tag{3.2}$$

for the singular integral operator T, and the condition

$$\int_r^\infty t^{\alpha p}\varphi(x,t)^p \frac{dt}{t} \le C\, r^{\alpha p}\varphi(x,r)^p \tag{3.3}$$

for the Riesz potential operator I_α, where $C(> 0)$ does not depend on r and $x \in \mathbb{R}^n$.

4. Boundedness of the fractional maximal operator in generalized Morrey spaces

4.1. Spanne type result

Sufficient conditions on φ for the boundedness of M and M_α in generalized Morrey spaces $\mathcal{M}_{p,\varphi}(\mathbb{R}^n)$ have been obtained in [2], [4], [5], [6], [8], [17], [18], [23], [25].

The following lemma is true.

Lemma 4.1. *Let $1 \le p < \infty$, $0 \le \alpha < \frac{n}{p}$, $\frac{1}{q} = \frac{1}{p} - \frac{\alpha}{n}$. Then for any ball $B = B(x,r)$ in $\mathbb{R}^n$ the inequality*

$$\|M_\alpha f\|_{L_q(B(x,r))} \lesssim \|f\|_{L_p(B(x,2r))} + r^{\frac{n}{q}} \sup_{t>2r} t^{-n+\alpha} \|f\|_{L_1(B(x,t))} \tag{4.1}$$

holds for all $f \in L_p^{\mathrm{loc}}(\mathbb{R}^n)$.

Moreover, the inequality

$$\|M_\alpha f\|_{WL_q(B(x,r))} \lesssim \|f\|_{L_1(B(x,2r))} + r^{\frac{n}{q}} \sup_{t>2r} t^{-n+\alpha} \|f\|_{L_1(B(x,t))} \tag{4.2}$$

holds for all $f \in L_1^{\mathrm{loc}}(\mathbb{R}^n)$.

Proof. Let $1 < p < q < \infty$ and $\frac{1}{p} - \frac{1}{q} = \frac{\alpha}{n}$. For arbitrary ball $B = B(x,r)$ let $f = f_1 + f_2$, where $f_1 = f\chi_{2B}$ and $f_2 = f\chi_{\complement(2B)}$.

$$\|M_\alpha f\|_{L_q(B)} \le \|M_\alpha f_1\|_{L_q(B)} + \|M_\alpha f_2\|_{L_q(B)}.$$

By the continuity of the operator $M_\alpha : L_p(\mathbb{R}^n) \to L_q(\mathbb{R}^n)$ we have

$$\|M_\alpha f_1\|_{L_q(B)} \lesssim \|f\|_{L_p(2B)}.$$

Let y be an arbitrary point from B. If $B(y,t) \cap {}^\complement(2B) \ne \emptyset$, then $t > r$. Indeed, if $z \in B(y,t) \cap {}^\complement(2B)$, then $t > |y-z| \ge |x-z| - |x-y| > 2r - r = r$.

On the other hand, $B(y,t) \cap {}^\complement(2B) \subset B(x,2t)$. Indeed, $z \in B(y,t) \cap {}^\complement(2B)$, then we get $|x-z| \le |y-z| + |x-y| < t + r < 2t$.

Hence

$$\begin{aligned} M_\alpha f_2(y) &= \sup_{t>0} \frac{1}{|B(y,t)|^{1-\alpha/n}} \int_{B(y,t)\cap {}^\complement(2B)} |f(z)|dz \\ &\le 2^{n-\alpha} \sup_{t>r} \frac{1}{|B(x,2t)|^{1-\alpha/n}} \int_{B(x,2t)} |f(z)|dz \\ &= 2^{n-\alpha} \sup_{t>2r} \frac{1}{|B(x,t)|^{1-\alpha/n}} \int_{B(x,t)} |f(z)|dz. \end{aligned}$$

Therefore, for all $y \in B$ we have

$$M_\alpha f_2(y) \le 2^{n-\alpha} \sup_{t>2r} \frac{1}{|B(x,t)|^{1-\alpha/n}} \int_{B(x,t)} |f(z)|dz. \tag{4.3}$$

Thus

$$\|M_\alpha f\|_{L_q(B)} \lesssim \|f\|_{L_p(2B)} + |B|^{\frac{1}{q}} \left(\sup_{t>2r} \frac{1}{|B(x,t)|^{1-\alpha/n}} \int_{B(x,t)} |f(z)|dz \right).$$

Let $p=1$. It is obvious that for any ball $B=B(x,r)$

$$\|M_\alpha f\|_{WL_q(B)} \leq \|M_\alpha f_1\|_{WL_q(B)} + \|M_\alpha f_2\|_{WL_q(B)}.$$

By the continuity of the operator $M_\alpha : L_1(\mathbb{R}^n) \to WL_q(\mathbb{R}^n)$ we have

$$\|M_\alpha f_1\|_{WL_q(B)} \lesssim \|f\|_{L_1(2B)}.$$

Then by (4.3) we get the inequality (4.2). □

Lemma 4.2. *Let $1 \leq p < \infty$, $0 \leq \alpha < \frac{n}{p}$, $\frac{1}{q} = \frac{1}{p} - \frac{\alpha}{n}$. Then for any ball $B = B(x,r)$ in $\mathbb{R}^n$, the inequality*

$$\|M_\alpha f\|_{L_q(B)} \lesssim r^{\frac{n}{q}} \sup_{t>2r} t^{-\frac{n}{q}} \|f\|_{L_p(B(x,t))} \tag{4.4}$$

holds for all $f \in L_p^{\rm loc}(\mathbb{R}^n)$.

Moreover, the inequality

$$\|M_\alpha f\|_{WL_q(B)} \lesssim r^{\frac{n}{q}} \sup_{t>2r} t^{-\frac{n}{q}} \|f\|_{L_1(B(x,t))} \tag{4.5}$$

holds for all $f \in L_1^{\rm loc}(\mathbb{R}^n)$.

Proof. Let $1 < p < \infty$, $0 < \alpha < \frac{n}{p}$, $\frac{1}{q} = \frac{1}{p} - \frac{\alpha}{n}$. Denote

$$\mathcal{M}_1 := |B|^{\frac{1}{q}} \left(\sup_{t>2r} \frac{1}{|B(x,t)|^{1-\alpha/n}} \int_{B(x,t)} |f(z)|dz \right),$$

$$\mathcal{M}_2 := \|f\|_{L_p(2B)}.$$

Applying Hölder's inequality, we get

$$\mathcal{M}_1 \lesssim |B|^{\frac{1}{q}} \left(\sup_{t>2r} \frac{1}{|B(x,t)|^{\frac{1}{q}}} \left(\int_{B(x,t)} |f(z)|^p dz \right)^{\frac{1}{p}} \right).$$

On the other hand,

$$\begin{aligned} |B|^{\frac{1}{q}} & \left(\sup_{t>2r} \frac{1}{|B(x,t)|^{\frac{1}{q}}} \left(\int_{B(x,t)} |f(z)|^p dz \right)^{\frac{1}{p}} \right) \\ & \gtrsim |B|^{\frac{1}{q}} \left(\sup_{t>2r} \frac{1}{|B(x,t)|^{\frac{1}{q}}} \right) \|f\|_{L_p(2B)} \approx \mathcal{M}_2. \end{aligned}$$

Since by Lemma 4.1

$$\|M_\alpha f\|_{L_q(B)} \leq \mathcal{M}_1 + \mathcal{M}_2,$$

we arrive at (4.4).

Let $p=1$. The inequality (4.5) directly follows from (4.2). □

Theorem 4.3. *Let $1 \le p < \infty$, $0 \le \alpha < \frac{n}{p}$, $\frac{1}{q} = \frac{1}{p} - \frac{\alpha}{n}$, and (φ_1, φ_2) satisfies the condition*

$$\sup_{r<t<\infty} t^{\alpha}\, \varphi_1(x,t) \le C\, \varphi_2(x,r), \tag{4.6}$$

where C does not depend on x and r. Then for $p > 1$, M_α is bounded from M_{p,φ_1} to M_{q,φ_2} and for $p = 1$, M_α is bounded from M_{1,φ_1} to WM_{q,φ_2}.

Proof. By Lemma 4.2 we get

$$\begin{aligned}
\|M_\alpha f\|_{M_{q,\varphi_2}(\mathbb{R}^n)} &\lesssim \sup_{x\in\mathbb{R}^n, r>0} \varphi_2(x,r)^{-1} \Big(\sup_{t>r} t^{-\frac{n}{q}} \|f\|_{L_p(B(x,t))}\Big) \\
&\lesssim \sup_{x\in\mathbb{R}^n, r>0} \varphi_2(x,r)^{-1} \sup_{t>r} t^{\frac{n}{p}-\frac{n}{q}}\, \varphi_1(x,t) \Big(\varphi_1(x,r)^{-1}\, t^{-\frac{n}{p}} \|f\|_{L_p(B(x,t))}\Big) \\
&\lesssim \|f\|_{M_{p,\varphi_1}(\mathbb{R}^n)} \sup_{x\in\mathbb{R}^n, r>0} \varphi_2(x,r)^{-1} \sup_{t>r} t^{\alpha}\, \varphi_1(x,t) \\
&\lesssim \|f\|_{M_{p,\varphi_1}(\mathbb{R}^n)}
\end{aligned}$$

if $p \in (1,\infty)$ and

$$\begin{aligned}
\|M_\alpha f\|_{WM_{q,\varphi_2}(\mathbb{R}^n)} &\lesssim \sup_{x\in\mathbb{R}^n, r>0} \varphi_2(x,r)^{-1} \Big(\sup_{t>r} t^{-\frac{n}{q}} \|f\|_{L_1(B(x,t))}\Big) \\
&\lesssim \|f\|_{M_{1,\varphi_1}(\mathbb{R}^n)} \sup_{x\in\mathbb{R}^n, r>0} \varphi_2(x,r)^{-1} \sup_{t>r} t^{\alpha}\, \varphi_1(x,t) \\
&\lesssim \|f\|_{M_{1,\varphi_1}(\mathbb{R}^n)}
\end{aligned}$$

if $p = 1$. □

In the case $\alpha = 0$ and $p = q$ from Theorem 4.3 we get the following corollary, which was proved in [3].

Corollary 4.1. *Let $1 \le p < \infty$ and let (φ_1, φ_2) satisfy the condition*

$$\sup_{r<t<\infty} \varphi_1(x,t) \le C\, \varphi_2(x,r), \tag{4.7}$$

where C does not depend on x and r. Then for $p > 1$, M is bounded from M_{p,φ_1} to M_{p,φ_2} and for $p = 1$, M is bounded from M_{1,φ_1} to WM_{1,φ_2}.

4.2. Adams type result

The following is a result of Adams type for the fractional maximal operator.

Theorem 4.4. *Let $1 \le p < q < \infty$, $0 < \alpha < \frac{n}{p}$ and let $\varphi(x,t)$ satisfy the condition*

$$\sup_{r<t<\infty} \varphi(x,t) \le C\, \varphi(x,r), \tag{4.8}$$

and

$$\sup_{r<t<\infty} t^{\alpha} \varphi(x,t)^{\frac{1}{p}} \le C r^{-\frac{\alpha p}{q-p}}, \tag{4.9}$$

where C does not depend on $x \in \mathbb{R}^n$ and $r > 0$.

Then the operator M_α is bounded from $M_{p,\varphi^{\frac{1}{p}}}$ to $M_{q,\varphi^{\frac{1}{q}}}$ for $p > 1$ and from $M_{1,\varphi}$ to $WM_{q,\varphi^{\frac{1}{q}}}$.

Proof. Let $1 < p < q < \infty$, $0 < \alpha < \frac{n}{p}$ and $f \in M_{p,\varphi^{\frac{1}{p}}}$. Write $f = f_1 + f_2$, where $B = B(x,r)$, $f_1 = f\chi_{2B}$ and $f_2 = f\chi_{\complement(2B)}$.

For $M_\alpha f_2(x)$ and for all $y \in B$ from (4.3) we have

$$M_\alpha(f_2)(y) \leq 2^{n-\alpha} \sup_{t>2r} \frac{1}{|B(x,t)|^{1-\alpha/n}} \int_{B(x,t)} |f(z)|dz$$
$$\lesssim \sup_{t>2r} t^{-\frac{n}{q}} \|f\|_{L_p(B(x,t))}. \tag{4.10}$$

Then from conditions (4.9) and (4.10) we get

$$\begin{aligned} M_\alpha f(x) &\lesssim r^\alpha Mf(x) + \sup_{t>2r} t^{\alpha-\frac{n}{p}} \|f\|_{L_p(B(x,t))} \\ &\leq r^\alpha Mf(x) + \|f\|_{M_{p,\varphi^{\frac{1}{p}}}} \sup_{t>2r} t^\alpha \varphi(x,t)^{\frac{1}{p}} \\ &\lesssim r^\alpha Mf(x) + r^{-\frac{\alpha p}{q-p}} \|f\|_{M_{p,\varphi^{\frac{1}{p}}}}. \end{aligned}$$

Consequently choosing $r = \left(\frac{\|f\|_{M_{p,\varphi^{1/p}}}}{Mf(x)}\right)^{\frac{q-p}{\alpha q}}$ for every $x \in \mathbb{R}^n$, we have

$$|M_\alpha f(x)| \lesssim (Mf(x))^{\frac{p}{q}} \|f\|_{M_{p,\varphi^{\frac{1}{p}}}}^{1-\frac{p}{q}}.$$

Hence the statement of the theorem follows in view of the boundedness of the maximal operator M in $M_{p,\varphi^{\frac{1}{p}}}$ provided by Corollary 4.1 in virtue of condition (4.8).

$$\begin{aligned} \|M_\alpha f\|_{M_{q,\varphi^{\frac{1}{q}}}} &= \sup_{x\in\mathbb{R}^n,\, t>0} \varphi(x,t)^{-\frac{1}{q}} t^{-\frac{n}{q}} \|M_\alpha f\|_{L_q(B(x,t))} \\ &\lesssim \|f\|_{M_{p,\varphi^{\frac{1}{p}}}}^{1-\frac{p}{q}} \sup_{x\in\mathbb{R}^n,\, t>0} \varphi(x,t)^{-\frac{1}{q}} t^{-\frac{n}{q}} \|Mf\|_{L_p(B(x,t))}^{\frac{p}{q}} \\ &= \|f\|_{M_{p,\varphi^{\frac{1}{p}}}}^{1-\frac{p}{q}} \left(\sup_{x\in\mathbb{R}^n,\, t>0} \varphi(x,t)^{-\frac{1}{p}} t^{-\frac{n}{p}} \|Mf\|_{L_p(B(x,t))}\right)^{\frac{p}{q}} \\ &= \|f\|_{M_{p,\varphi^{\frac{1}{p}}}}^{1-\frac{p}{q}} \|Mf\|_{M_{p,\varphi^{\frac{1}{p}}}}^{\frac{p}{q}} \lesssim \|f\|_{M_{p,\varphi^{\frac{1}{p}}}}, \end{aligned}$$

if $1 < p < q < \infty$ and

$$\begin{aligned} \|M_\alpha f\|_{WM_{q,\varphi^{\frac{1}{q}}}} &= \sup_{x\in\mathbb{R}^n,\, t>0} \varphi(x,t)^{-\frac{1}{q}} t^{-\frac{n}{q}} \|M_\alpha f\|_{WL_q(B(x,t))} \\ &\lesssim \|f\|_{M_{1,\varphi}}^{1-\frac{1}{q}} \sup_{x\in\mathbb{R}^n,\, t>0} \varphi(x,t)^{-\frac{1}{q}} t^{-\frac{n}{q}} \|Mf\|_{WL_1(B(x,t))}^{\frac{1}{q}} \\ &= \|f\|_{M_{1,\varphi}}^{1-\frac{1}{q}} \left(\sup_{x\in\mathbb{R}^n,\, t>0} \varphi(x,t)^{-1} t^{-n} \|Mf\|_{WL_1(B(x,t))}\right)^{\frac{1}{q}} \\ &= \|f\|_{M_{1,\varphi}}^{1-\frac{1}{q}} \|Mf\|_{WM_{1,\varphi}}^{\frac{1}{q}} \lesssim \|f\|_{M_{1,\varphi}}, \end{aligned}$$

if $1 < q < \infty$. □

In the case $\varphi(x,t) = t^{\lambda-n}$, $0 < \lambda < n$ from Theorem 4.4 we get the following Adams type result for the fractional maximal operator.

Corollary 4.2. *Let $0 < \alpha < n$, $1 < p < \frac{n}{\alpha}$, $0 < \lambda < n - \alpha p$ and $\frac{1}{p} - \frac{1}{q} = \frac{\alpha}{n-\lambda}$. Then for $p > 1$, the operator M_α is bounded from $M_{p,\lambda}$ to $M_{q,\lambda}$ and for $p = 1$, M_α is bounded from $M_{1,\lambda}$ to $WM_{q,\lambda}$.*

5. Riesz potential operator in the spaces $M_{p,\varphi}$

5.1. Spanne type result

In [25] the following statements was proved by Riesz potential operator I_α.

Theorem 5.5. *Let $1 < p < \infty$, $0 < \alpha < \frac{n}{p}$, $\frac{1}{q} = \frac{1}{p} - \frac{\alpha}{n}$ and $\varphi(x,r)$ satisfy the conditions (3.1) and (3.3). Then the operator I_α is bounded from $M_{p,\varphi}$ to $M_{q,\varphi}$.*

The following statements, containing results obtained in [23], [25] were proved in [15, 17] (see also [5]–[9], [16, 18]).

Theorem 5.6. *Let $1 \le p < \infty$, $0 < \alpha < \frac{n}{p}$, $\frac{1}{q} = \frac{1}{p} - \frac{\alpha}{n}$ and let (φ_1, φ_2) satisfy the condition*

$$\int_t^\infty r^\alpha \varphi_1(x,r) \frac{dr}{r} \le C\, \varphi_2(x,t), \tag{5.1}$$

where C does not depend on x and t. Then the operators M_α and I_α are bounded from M_{p,φ_1} to M_{q,φ_2} for $p > 1$ and from M_{1,φ_1} to WM_{q,φ_2}.

5.2. Adams type result

The following is a result of Adams type.

Theorem 5.7. *Let $1 \le p < \infty$, $0 < \alpha < \frac{n}{p}$, $q > p$ and let $\varphi(x,t)$ satisfy the conditions (4.8) and*

$$\int_r^\infty t^\alpha\, \varphi(x,t)^{\frac{1}{p}} \frac{dt}{t} \le C r^{-\frac{\alpha p}{q-p}}, \tag{5.2}$$

where C does not depend on $x \in \mathbb{R}^n$ and $r > 0$.

Then the operator I_α is bounded from $M_{p,\varphi^{\frac{1}{p}}}$ to $M_{q,\varphi^{\frac{1}{q}}}$ for $p > 1$ and from $M_{1,\varphi}$ to $WM_{q,\varphi^{\frac{1}{q}}}$.

Proof. Let $1 < p < \infty$, $0 < \alpha < \frac{n}{p}$, $q > p$ and $f \in M_{p,\varphi^{\frac{1}{p}}}$. Write $f = f_1 + f_2$, where $B = B(x,r)$, $f_1 = f\chi_{2B}$ and $f_2 = f\chi_{\complement(2B)}$.

For $I_\alpha f_2(x)$ we have

$$\begin{aligned}
|I_\alpha f_2(x)| &\le \int_{\complement B(x,2r)} |x-y|^{\alpha-n} |f(y)| dy \lesssim \int_{\complement B(x,2r)} |f(y)| dy \int_{|x-y|}^\infty t^{\alpha-n-1} dt \\
&\lesssim \int_{2r}^\infty \left(\int_{2r<|x-y|<t} |f(y)| dy \right) t^{\alpha-n-1} dt \\
&\lesssim \int_r^\infty t^{\alpha-\frac{n}{p}-1} \|f\|_{L_p(B(x,t))} dt.
\end{aligned} \tag{5.3}$$

Then from condition (5.2) and inequality (5.3) we get

$$
\begin{aligned}
|I_\alpha f(x)| &\lesssim r^\alpha\, Mf(x) + \int_r^\infty t^{\alpha-\frac{n}{p}-1}\|f\|_{L_p(B(x,t))}dt \\
&\le r^\alpha\, Mf(x) + \|f\|_{M_{p,\varphi}} \int_r^\infty t^\alpha \varphi(x,t)^{\frac{1}{p}}\frac{dt}{t} \\
&\lesssim r^\alpha\, Mf(x) + r^{-\frac{\alpha p}{q-p}}\,\|f\|_{M_{p,\varphi^{\frac{1}{p}}}}. \qquad (5.4)
\end{aligned}
$$

Hence choosing $r = \left(\frac{\|f\|_{M_{p,\varphi^{1/p}}}}{Mf(x)}\right)^{\frac{q-p}{\alpha q}}$ for every $x \in \mathbb{R}^n$, we have

$$
|I_\alpha f(x)| \lesssim (Mf(x))^{\frac{p}{q}}\,\|f\|_{M_{p,\varphi^{\frac{1}{p}}}}^{1-\frac{p}{q}}.
$$

Consequently the statement of the theorem follows in view of the boundedness of the maximal operator M in $M_{p,\varphi^{\frac{1}{p}}}$ provided by Corollary 4.1 in virtue of condition (4.8).

$$
\begin{aligned}
\|I_\alpha f\|_{M_{q,\varphi^{\frac{1}{q}}}} &= \sup_{x\in\mathbb{R}^n,\, t>0} \varphi(x,t)^{-\frac{1}{q}} t^{-\frac{n}{q}}\,\|I_\alpha f\|_{L_q(B(x,t))} \\
&\lesssim \|f\|_{M_{p,\varphi^{\frac{1}{p}}}}^{1-\frac{p}{q}} \sup_{x\in\mathbb{R}^n,\, t>0} \varphi(x,t)^{-\frac{1}{q}} t^{-\frac{n}{q}}\,\|Mf\|_{L_p(B(x,t))}^{\frac{p}{q}} \\
&= \|f\|_{M_{p,\varphi^{\frac{1}{p}}}}^{1-\frac{p}{q}} \left(\sup_{x\in\mathbb{R}^n,\, t>0} \varphi(x,t)^{-\frac{1}{p}} t^{-\frac{n}{p}}\,\|Mf\|_{L_p(B(x,t))}\right)^{\frac{p}{q}} \\
&= \|f\|_{M_{p,\varphi^{\frac{1}{p}}}}^{1-\frac{p}{q}}\,\|Mf\|_{M_{p,\varphi^{\frac{1}{p}}}}^{\frac{p}{q}} \\
&\lesssim \|f\|_{M_{p,\varphi^{\frac{1}{p}}}},
\end{aligned}
$$

if $1 < p < q < \infty$ and

$$
\begin{aligned}
\|I_\alpha f\|_{WM_{q,\varphi^{\frac{1}{q}}}} &= \sup_{x\in\mathbb{R}^n,\, t>0} \varphi(x,t)^{-\frac{1}{q}} t^{-\frac{n}{q}}\,\|I_\alpha f\|_{WL_q(B(x,t))} \\
&\lesssim \|f\|_{M_{1,\varphi}}^{1-\frac{1}{q}} \sup_{x\in\mathbb{R}^n,\, t>0} \varphi(x,t)^{-\frac{1}{q}} t^{-\frac{n}{q}}\,\|Mf\|_{WL_1(B(x,t))}^{\frac{1}{q}} \\
&= \|f\|_{M_{1,\varphi}}^{1-\frac{1}{q}} \left(\sup_{x\in\mathbb{R}^n,\, t>0} \varphi(x,t)^{-1} t^{-n}\|Mf\|_{WL_1(B(x,t))}\right)^{\frac{1}{q}} \\
&= \|f\|_{M_{1,\varphi}}^{1-\frac{1}{q}}\,\|Mf\|_{WM_{1,\varphi}}^{\frac{1}{q}} \\
&\lesssim \|f\|_{M_{1,\varphi}},
\end{aligned}
$$

if $1 < q < \infty$. □

In the case $\varphi(x,t) = t^{\lambda-n}$, $0 < \lambda < n$ from Theorem 5.7 we get Adams Theorem 2.2.

6. Commutators of fractional maximal operators in the spaces $M_{p,\varphi}$

6.1. Spanne type result

The theory of commutator was originally studied by Coifman, Rochberg and Weiss in [12]. Since then, many authors have been interested in studying this theory. When $1 < p < \infty$, $0 < \alpha < \frac{n}{p}$ and $\frac{1}{q} = \frac{1}{p} - \frac{\alpha}{n}$, Chanillo [10] proved that the commutator operator $[b, I_\alpha]f$ is bounded from $L_p(\mathbb{R}^n)$ to $L_q(\mathbb{R}^n)$ whenever $b \in BMO(\mathbb{R}^n)$.

First we introduce the definition of the space of $BMO(\mathbb{R}^n)$.

Definition 6.2. Suppose that $f \in L_1^{\text{loc}}(\mathbb{R}^n)$, and let

$$\|f\|_* = \sup_{x\in\mathbb{R}^n, r>0} \frac{1}{|B(x,r)|} \int_{B(x,r)} |f(y) - f_{B(x,r)}| dy < \infty,$$

where

$$f_{B(x,r)} = \frac{1}{|B(x,r)|} \int_{B(x,r)} f(y) dy.$$

Define

$$BMO(\mathbb{R}^n) = \{f \in L_1^{\text{loc}}(\mathbb{R}^n) \ : \ \|f\|_* < \infty\}.$$

If one regards two functions whose difference is a constant as one, then space $BMO(\mathbb{R}^n)$ is a Banach space with respect to norm $\|\cdot\|_*$.

Remark 6.1.

(1) The John-Nirenberg inequality: there are constants $C_1, C_2 > 0$, such that for all $f \in BMO(\mathbb{R}^n)$ and $\beta > 0$

$$|\{x \in B \ : \ |f(x) - f_B| > \beta\}| \le C_1 |B| e^{-C_2\beta/\|f\|_*}, \ \forall B \subset \mathbb{R}^n.$$

(2) The John-Nirenberg inequality implies that

$$\|f\|_* \approx \sup_{x\in\mathbb{R}^n, r>0} \left(\frac{1}{|B(x,r)|} \int_{B(x,r)} |f(y) - f_{B(x,r)}|^p dy \right)^{\frac{1}{p}} \tag{6.1}$$

for $1 < p < \infty$.

(3) Let $f \in BMO(\mathbb{R}^n)$. Then there is a constant $C > 0$ such that

$$\left| f_{B(x,r)} - f_{B(x,t)} \right| \le C \|f\|_* \ln \frac{t}{r} \text{ for } 0 < 2r < t, \tag{6.2}$$

where C is independent of f, x, r and t.

For the sublinear commutator of the fractional maximal operator

$$M_{b,\alpha}(f)(x) = \sup_{t>0} |B(x,t)|^{-1+\frac{\alpha}{n}} \int_{B(x,t)} |b(x) - b(y)||f(y)| dy$$

in [13] the following statement was proved, containing the result in [23, 25].

Theorem 6.8. *Let $1 < p < \infty$, $0 < \alpha < \frac{n}{p}$, $\frac{1}{q} = \frac{1}{p} - \frac{\alpha}{n}$, $b \in BMO(\mathbb{R}^n)$, $\varphi(x,r)$ satisfies the conditions (3.1) and (3.3). Then the operator $M_{b,\alpha}$ is bounded from $M_{p,\varphi}$ to $M_{q,\varphi}$.*

Lemma 6.3. *Let $1 < p < \infty$, $0 < \alpha < \frac{n}{p}$, $\frac{1}{q} = \frac{1}{p} - \frac{\alpha}{n}$, $b \in BMO(\mathbb{R}^n)$.*

Then the inequality

$$\|M_{b,\alpha}f\|_{L_q(B(x_0,r))} \lesssim \|b\|_* r^{\frac{n}{q}} \sup_{t>2r} \left(1 + \ln \frac{t}{r}\right) t^{-\frac{n}{q}} \|f\|_{L_p(B(x_0,t))}$$

holds for any ball $B(x_0,r)$ and for all $f \in L_p^{\rm loc}(\mathbb{R}^n)$.

Proof. Let $1 < p < \infty$, $0 < \alpha < \frac{n}{p}$, $\frac{1}{q} = \frac{1}{p} - \frac{\alpha}{n}$. Write $f = f_1 + f_2$, where $B = B(x_0,r)$, $f_1 = f\chi_{2B}$ and $f_2 = f\chi_{\complement(2B)}$. Hence,

$$\|M_{b,\alpha}f\|_{L_q(B)} \le \|M_{b,\alpha}f_1\|_{L_q(B)} + \|M_{b,\alpha}f_2\|_{L_q(B)}.$$

From the boundedness of $M_{b,\alpha}$ from $L_p(\mathbb{R}^n)$ to $L_q(\mathbb{R}^n)$ it follows that:

$$\begin{aligned}\|M_{b,\alpha}f_1\|_{L_q(B)} &\le \|M_{b,\alpha}f_1\|_{L_q(\mathbb{R}^n)} \\ &\lesssim \|b\|_* \|f_1\|_{L_p(\mathbb{R}^n)} = \|b\|_* \|f\|_{L_p(2B)}.\end{aligned}$$

For $x \in B$ we have

$$\begin{aligned}M_{b,\alpha}f_2(x) &\lesssim \sup_{t>0} \frac{1}{|B(x,t)|^{1-\alpha/n}} \int_{B(x,t)} |b(y) - b(x)|\, f_2(y)|dy \\ &= \sup_{t>0} \frac{1}{|B(x,t)|^{1-\alpha/n}} \int_{B(x,t)\cap \complement(2B)} |b(y) - b(x)|\, |f(y)|dy.\end{aligned}$$

Let x be an arbitrary point from B. If $B(x,t) \cap \{\complement(2B)\} \neq \emptyset$, then $t > r$. Indeed, if $y \in B(x,t) \cap \{\complement(2B)\}$, then $t > |x-y| \ge |x_0 - y| - |x_0 - x| > 2r - r = r$.

On the other hand, $B(x,t) \cap \{\complement(2B)\} \subset B(x_0, 2t)$. Indeed, $y \in B(x,t) \cap \{\complement(2B)\}$, then we get $|x_0 - y| \le |x-y| + |x_0 - x| < t + r < 2t$.

Hence

$$\begin{aligned}M_{b,\alpha}(f_2)(x) &= \sup_{t>0} \frac{1}{|B(x,t)|^{1-\alpha/n}} \int_{B(x,t)\cap \complement(2B)} |b(y) - b(x)|\, |f(y)|dy \\ &\le 2^{n-\alpha} \sup_{t>r} \frac{1}{|B(x_0,2t)|^{1-\alpha/n}} \int_{B(x_0,2t)} |b(y) - b(x)|\, |f(y)|dy \\ &= 2^{n-\alpha} \sup_{t>2r} \frac{1}{|B(x_0,t)|^{1-\alpha/n}} \int_{B(x_0,t)} |b(y) - b(x)|\, |f(y)|dy.\end{aligned}$$

Therefore, for all $x \in B$ we have

$$M_{b,\alpha}(f_2)(x) \le 2^{n-\alpha} \sup_{t>2r} \frac{1}{|B(x_0,t)|^{1-\alpha/n}} \int_{B(x_0,t)} |b(y) - b(x)|\, |f(y)|dy. \tag{6.3}$$

Then

$$\begin{aligned}\|M_{b,\alpha}f_2\|_{L_q(B)} &\lesssim \left(\int_B \left(\sup_{t>2r} \frac{1}{|B(x_0,t)|^{1-\alpha/n}} \int_{B(x_0,t)} |b(y)-b(x)|\,|f(y)|dy\right)^q dx\right)^{\frac{1}{q}} \\ &\le \left(\int_B \left(\sup_{t>2r} \frac{1}{|B(x_0,t)|^{1-\alpha/n}} \int_{B(x_0,t)} |b(y)-b_B|\,|f(y)|dy\right)^q dx\right)^{\frac{1}{q}} \\ &\quad + \left(\int_B \left(\sup_{t>2r} \frac{1}{|B(x_0,t)|^{1-\alpha/n}} \int_{B(x_0,t)} |b(x)-b_B|\,|f(y)|dy\right)^q dx\right)^{\frac{1}{q}} \\ &= J_1 + J_2.\end{aligned}$$

Let us estimate J_1.

$$\begin{aligned}J_1 &= r^{\frac{n}{q}} \sup_{t>2r} \frac{1}{|B(x_0,t)|^{1-\alpha/n}} \int_{B(x_0,t)} |b(y)-b_B|\,|f(y)|dy \\ &\approx r^{\frac{n}{q}} \sup_{t>2r} t^{\alpha-n} \int_{B(x_0,t)} |b(y)-b_B|\,|f(y)|dy.\end{aligned}$$

Applying Hölder's inequality and by (6.1), (6.2), we get

$$\begin{aligned}J_1 &\lesssim r^{\frac{n}{q}} \sup_{t>2r} t^{\alpha-n} \int_{B(x_0,t)} |b(y)-b_{B(x_0,t)}|\,|f(y)|dy \\ &\quad + r^{\frac{n}{q}} \sup_{t>2r} t^{\alpha-n} |b_{B(x_0,r)}-b_{B(x_0,t)}| \int_{B(x_0,t)} |f(y)|dy \\ &\lesssim r^{\frac{n}{q}} \sup_{t>2r} t^{\alpha-\frac{n}{p}} \left(\frac{1}{|B(x_0,t)|} \int_{B(x_0,t)} |b(y)-b_{B(x_0,t)}|^{p'} dy\right)^{\frac{1}{p'}} \|f\|_{L_p(B(x_0,t))} \\ &\quad + r^{\frac{n}{q}} \sup_{t>2r} t^{\alpha-\frac{n}{p}} |b_{B(x_0,r)}-b_{B(x_0,t)}| \|f\|_{L_p(B(x_0,t))} \\ &\lesssim \|b\|_* r^{\frac{n}{q}} \sup_{t>2r} t^{\frac{n}{q}} \left(1+\ln\frac{t}{r}\right) \|f\|_{L_p(B(x_0,t))}.\end{aligned}$$

In order to estimate J_2 note that

$$\begin{aligned}J_2 &= \left(\int_B |b(x)-b_B|^q dx\right)^{\frac{1}{q}} \sup_{t>2r} t^{\alpha-n} \int_{B(x_0,t)} |f(y)|dy \\ &\lesssim \|b\|_* r^{\frac{n}{q}} \sup_{t>2r} t^{\frac{n}{q}} \|f\|_{L_p(B(x_0,t))}.\end{aligned}$$

Summing up J_1 and J_2, for all $p \in (1,\infty)$ we get

$$\|M_{b,\alpha}f_2\|_{L_q(B)} \lesssim \|b\|_* r^{\frac{n}{q}} \sup_{t>2r} t^{\frac{n}{q}} \left(1+\ln\frac{t}{r}\right) \|f\|_{L_p(B(x_0,t))}. \tag{6.4}$$

Finally,

$$\|M_{b,\alpha}f\|_{L_q(B)} \lesssim \|b\|_* \|f\|_{L_p(2B)} + \|b\|_* r^{\frac{n}{q}} \sup_{t>2r} t^{\frac{n}{q}} \Big(1 + \ln \frac{t}{r}\Big) \|f\|_{L_p(B(x_0,t))}$$

$$\lesssim \|b\|_* r^{\frac{n}{q}} \sup_{t>2r} t^{\frac{n}{q}} \Big(1 + \ln \frac{t}{r}\Big) \|f\|_{L_p(B(x_0,t))}. \qquad \square$$

The following theorem is true.

Theorem 6.9. *Let* $1 < p < \infty$, $0 \le \alpha < \frac{n}{p}$, $\frac{1}{q} = \frac{1}{p} - \frac{\alpha}{n}$, $b \in BMO(\mathbb{R}^n)$ *and let* (φ_1, φ_2) *satisfy the condition*

$$\sup_{r<t<\infty} \Big(1 + \ln \frac{t}{r}\Big) t^{\alpha} \varphi_1(x,t) \le C \varphi_2(x,r), \tag{6.5}$$

where C *does not depend on* x *and* r.

Then the operator $M_{b,\alpha}$ *is bounded from* M_{p,φ_1} *to* M_{q,φ_2}. *Moreover*

$$\|M_{b,\alpha}f\|_{M_{q,\varphi_2}} \lesssim \|b\|_* \|f\|_{M_{p,\varphi_1}}.$$

Proof. The statement of Theorem 6.9 follows by Lemma 6.3 in the same manner as in the proof of Theorem 4.3. $\square$

In the case $\alpha = 0$ and $p = q$ from Theorem 6.9 we get the following corollary.

Corollary 6.3. *Let* $1 < p < \infty$, $b \in BMO(\mathbb{R}^n)$ *and let* (φ_1, φ_2) *satisfy the condition*

$$\sup_{r<t<\infty} \Big(1 + \ln \frac{t}{r}\Big) \varphi_1(x,t) \le C \varphi_2(x,r), \tag{6.6}$$

where C *does not depend on* x *and* r.

Then the operator M_b *is bounded from* M_{p,φ_1} *to* M_{p,φ_2}. *Moreover*

$$\|M_b f\|_{M_{p,\varphi_2}} \lesssim \|b\|_* \|f\|_{M_{p,\varphi_1}}.$$

6.2. Adams type result

The following is a result of Adams type.

Theorem 6.10. *Let* $1 < p < q < \infty$, $0 < \alpha < \frac{n}{p}$, $b \in BMO(\mathbb{R}^n)$ *and let* $\varphi(x,t)$ *satisfy the conditions*

$$\sup_{r<t<\infty} \Big(1 + \ln \frac{t}{r}\Big)^p \varphi(x,t) \le C \varphi(x,r) \tag{6.7}$$

and

$$\sup_{r<t<\infty} \Big(1 + \ln \frac{t}{r}\Big) t^{\alpha} \varphi(x,t)^{\frac{1}{p}} \le C r^{-\frac{\alpha p}{q-p}}, \tag{6.8}$$

where C *does not depend on* $x \in \mathbb{R}^n$ *and* $r > 0$.

Then $M_{b,\alpha}$ *is bounded from* $M_{p,\varphi^{\frac{1}{p}}}$ *to* $M_{q,\varphi^{\frac{1}{q}}}$.

Proof. Let $1 < p < q < \infty$, $0 < \alpha < \frac{n}{p}$ and $f \in M_{p,\varphi^{\frac{1}{p}}}$. For arbitrary $x_0 \in \mathbb{R}^n$, set $B = B(x_0, r)$ for the ball centered at x_0 and of radius r. Write $f = f_1 + f_2$ with $f_1 = f\chi_{2B}$ and $f_2 = f\chi_{\complement_{(2B)}}$.

For $x \in B$ we have

$$\begin{aligned}|M_{b,\alpha} f_2(x)| &\lesssim \sup_{t>0} t^{\alpha-n} \int_{B(x,t)} |b(y) - b(x)|\, |f_2(y)| dy \\ &\approx \sup_{t>2r} t^{\alpha-n} \int_{B(x,t)} |b(y) - b(x)|\, |f_2(y)| dy.\end{aligned}$$

Applying Hölder's inequality and by (6.1), (6.2), we get

$$\begin{aligned}|M_{b,\alpha} f_2(x)| \lesssim &\sup_{t>2r} t^{\alpha-n} \int_{B(x_0,t)} |b(y) - b_{B(x_0,t)}|\, |f(y)| dy \\ &+ \sup_{t>2r} t^{\alpha-n} |b_{B(x_0,r)} - b_{B(x_0,t)}| \int_{B(x_0,t)} |f(y)| dy \\ \lesssim &\sup_{t>2r} t^{\alpha-\frac{n}{p}} \left(\frac{1}{|B(x_0,t)|} \int_{B(x_0,t)} |b(y) - b_{B(x_0,t)}|^{p'} dy \right)^{\frac{1}{p'}} \|f\|_{L_p(B(x_0,t))} \\ &+ \sup_{t>2r} t^{\alpha-\frac{n}{p}} |b_{B(x_0,r)} - b_{B(x_0,t)}| \|f\|_{L_p(B(x_0,t))} \\ \lesssim &\|b\|_* \sup_{t>2r} t^{\alpha-\frac{n}{p}} \left(1 + \ln \frac{t}{r}\right) \|f\|_{L_p(B(x_0,t))}.\end{aligned}$$

Consequently, for all $p \in (1, \infty)$ and $x \in B$ we get

$$|M_{b,\alpha} f_2(x)| \lesssim \|b\|_* \sup_{t>2r} t^{\alpha-\frac{n}{p}} \left(1 + \ln \frac{t}{r}\right) \|f\|_{L_p(B(x_0,t))}. \tag{6.9}$$

Then from conditions (6.8) and (6.9) we get

$$\begin{aligned}M_{b,\alpha} f(x) &\lesssim \|b\|_* r^{\alpha} M_b f(x) + \|b\|_* \sup_{t>2r} t^{\alpha-\frac{n}{p}} \left(1 + \ln \frac{t}{r}\right) \|f\|_{L_p(B(x_0,t))} \\ &\leq \|b\|_* r^{\alpha} M_b f(x) + \|b\|_* \|f\|_{M_{p,\varphi^{\frac{1}{p}}}} \sup_{t>r} \left(1 + \ln \frac{t}{r}\right) t^{\alpha} \varphi(x,t)^{\frac{1}{p}} \\ &\lesssim \|b\|_* r^{\alpha} M_b f(x) + \|b\|_* r^{-\frac{\alpha p}{q-p}} \|f\|_{M_{p,\varphi^{\frac{1}{p}}}}.\end{aligned} \tag{6.10}$$

Hence choose $r = \left(\frac{\|f\|_{M_{p,\varphi^{\frac{1}{p}}}}}{M_b f(x)} \right)^{\frac{q-p}{\alpha q}}$ for every $x \in B$, we have

$$|M_{b,\alpha} f(x)| \lesssim \|b\|_* (M_b f(x))^{\frac{p}{q}} \|f\|_{M_{p,\varphi^{\frac{1}{p}}}}^{1-\frac{p}{q}}.$$

Hence the statement of the theorem follows in view of the boundedness of the commutator of maximal operator M_b in $M_{p,\varphi^{\frac{1}{p}}}$ provided by Corollary 6.3 in virtue of condition (6.7).

$$\begin{aligned}
\|M_{b,\alpha}f\|_{M_{q,\varphi^{\frac{1}{q}}}} &= \sup_{x\in\mathbb{R}^n,\ r>0} \varphi(x,r)^{-\frac{1}{q}} r^{-\frac{n}{q}} \|M_{b,\alpha}f\|_{L_q(B(x,r))} \\
&\lesssim \|b\|_* \|f\|_{M_{p,\varphi^{\frac{1}{p}}}}^{1-\frac{p}{q}} \sup_{x\in\mathbb{R}^n,\ r>0} \varphi(x,r)^{-\frac{1}{q}} r^{-\frac{n}{q}} \|M_b f\|_{L_p(B(x,r))}^{\frac{p}{q}} \\
&= \|b\|_* \|f\|_{M_{p,\varphi^{\frac{1}{p}}}}^{1-\frac{p}{q}} \left(\sup_{x\in\mathbb{R}^n,\ r>0} \varphi(x,r)^{-\frac{1}{p}} r^{-\frac{n}{p}} \|M_b f\|_{L_p(B(x,r))} \right)^{\frac{p}{q}} \\
&= \|b\|_* \|f\|_{M_{p,\varphi^{\frac{1}{p}}}}^{1-\frac{p}{q}} \|M_b f\|_{M_{p,\varphi^{\frac{1}{p}}}}^{\frac{p}{q}} \\
&\lesssim \|b\|_* \|f\|_{M_{p,\varphi^{\frac{1}{p}}}}.
\end{aligned}$$

□

In the case $\varphi(x,t) = t^{\lambda-n}$, $0 < \lambda < n$ from Theorem 6.10 we get the following Adams type result for the commutator of fractional maximal operator.

Corollary 6.4. *Let $0 < \alpha < n$, $1 < p < \frac{n}{\alpha}$, $0 < \lambda < n - \alpha p$, $\frac{1}{p} - \frac{1}{q} = \frac{\alpha}{n-\lambda}$ and $b \in BMO(\mathbb{R}^n)$. Then, the operator $M_{b,\alpha}$ is bounded from $M_{p,\lambda}$ to $M_{q,\lambda}$.*

7. Commutators of Riesz potential operators in the spaces $M_{p,\varphi}$

7.1. Spanne type result

In [13] the following statement was proved for the commutators of Riesz potential operators, containing the result in [23, 25].

Theorem 7.11. *Let $1 < p < \infty$, $0 < \alpha < \frac{n}{p}$, $\frac{1}{q} = \frac{1}{p} - \frac{\alpha}{n}$, $b \in BMO(\mathbb{R}^n)$, $\varphi(x,r)$ satisfies the conditions (3.1) and (3.3). Then the operator $[b, I_\alpha]$ is bounded from $M_{p,\varphi}$ to $M_{q,\varphi}$.*

Lemma 7.4. *Let $1 < p < \infty$, $0 < \alpha < \frac{n}{p}$, $\frac{1}{q} = \frac{1}{p} - \frac{\alpha}{n}$, $b \in BMO(\mathbb{R}^n)$.*

Then the inequality

$$\|[b, I_\alpha]f\|_{L_q(B(x_0,r))} \lesssim \|b\|_* r^{\frac{n}{q}} \int_{2r}^{\infty} \Big(1 + \ln\frac{t}{r}\Big) t^{-\frac{n}{q}-1} \|f\|_{L_p(B(x_0,t))} dt$$

holds for any ball $B(x_0,r)$ and for all $f \in L_p^{\mathrm{loc}}(\mathbb{R}^n)$.

Proof. Let $1 < p < \infty$, $0 < \alpha < \frac{n}{p}$, $\frac{1}{q} = \frac{1}{p} - \frac{\alpha}{n}$. Write $f = f_1 + f_2$, where $B = B(x_0,r)$, $f_1 = f\chi_{2B}$ and $f_2 = f\chi_{\complement(2B)}$. Hence,

$$\|[b, I_\alpha]f\|_{L_q(B)} \le \|[b, I_\alpha]f_1\|_{L_q(B)} + \|[b, I_\alpha]f_2\|_{L_q(B)}.$$

From the boundedness of $[b, I_\alpha]$ from $L_p(\mathbb{R}^n)$ to $L_q(\mathbb{R}^n)$ it follows that:

$$\begin{aligned}
\|[b, I_\alpha]f_1\|_{L_q(B)} &\le \|[b, I_\alpha]f_1\|_{L_q(\mathbb{R}^n)} \\
&\lesssim \|b\|_* \|f_1\|_{L_p(\mathbb{R}^n)} = \|b\|_* \|f\|_{L_p(2B)}.
\end{aligned}$$

For $x \in B$ we have

$$
\begin{aligned}
|[b, I_\alpha]f_2(x)| &\lesssim \int_{\mathbb{R}^n} \frac{|b(y)-b(x)|}{|x-y|^{n-\alpha}} |f(y)|dy \\
&\approx \int_{\complement_{(2B)}} \frac{|b(y)-b(x)|}{|x_0-y|^{n-\alpha}} |f(y)|dy.
\end{aligned}
$$

Then

$$
\begin{aligned}
\|[b, I_\alpha]f_2\|_{L_q(B)} &\lesssim \left(\int_B \left(\int_{\complement_{(2B)}} \frac{|b(y)-b(x)|}{|x_0-y|^{n-\alpha}} |f(y)|dy \right)^q dx \right)^{\frac{1}{q}} \\
&\le \left(\int_B \left(\int_{\complement_{(2B)}} \frac{|b(y)-b_B|}{|x_0-y|^{n-\alpha}} |f(y)|dy \right)^q dx \right)^{\frac{1}{q}} \\
&\quad + \left(\int_B \left(\int_{\complement_{(2B)}} \frac{|b(x)-b_B|}{|x_0-y|^{n-\alpha}} |f(y)|dy \right)^q dx \right)^{\frac{1}{q}} \\
&= J_1 + J_2.
\end{aligned}
$$

Let us estimate J_1.

$$
\begin{aligned}
J_1 &= r^{\frac{n}{q}} \int_{\complement_{(2B)}} \frac{|b(y)-b_B|}{|x_0-y|^{n-\alpha}} |f(y)|dy \\
&\approx r^{\frac{n}{q}} \int_{\complement_{(2B)}} |b(y)-b_B||f(y)| \int_{|x_0-y|}^{\infty} \frac{dt}{t^{n+1-\alpha}} dy \\
&\approx r^{\frac{n}{q}} \int_{2r}^{\infty} \int_{2r \le |x_0-y| \le t} |b(y)-b_B||f(y)|dy \frac{dt}{t^{n+1-\alpha}} \\
&\lesssim r^{\frac{n}{q}} \int_{2r}^{\infty} \int_{B(x_0,t)} |b(y)-b_B||f(y)|dy \frac{dt}{t^{n+1-\alpha}}.
\end{aligned}
$$

Applying Hölder's inequality and by (6.1), (6.2), we get

$$
\begin{aligned}
J_1 &\lesssim r^{\frac{n}{q}} \int_{2r}^{\infty} \int_{B(x_0,t)} |b(y)-b_{B(x_0,t)}||f(y)|dy \frac{dt}{t^{n+1-\alpha}} \\
&\quad + r^{\frac{n}{q}} \int_{2r}^{\infty} |b_{B(x_0,r)}-b_{B(x_0,t)}| \frac{dt}{t^{n+1-\alpha}} \int_{B(x_0,t)} |f(y)|dy \\
&\lesssim r^{\frac{n}{q}} \int_{2r}^{\infty} \left(\int_{B(x_0,t)} |b(y)-b_{B(x_0,t)}|^{p'} dy \right)^{\frac{1}{p'}} \|f\|_{L_p(B(x_0,t))} \frac{dt}{t^{n+1-\alpha}} \\
&\quad + r^{\frac{n}{q}} \int_{2r}^{\infty} |b_{B(x_0,r)}-b_{B(x_0,t)}| \|f\|_{L_p(B(x_0,t))} \frac{dt}{t^{\frac{n}{p}+1-\alpha}} \\
&\lesssim \|b\|_* r^{\frac{n}{q}} \int_{2r}^{\infty} \left(1+\ln\frac{t}{r}\right) \|f\|_{L_p(B(x_0,t))} \frac{dt}{t^{\frac{n}{q}+1}}.
\end{aligned}
$$

In order to estimate J_2 note that

$$J_2 = \left(\int_B |b(x) - b_B|^q dx\right)^{\frac{1}{q}} \int_{\complement(2B)} \frac{|f(y)|}{|x_0 - y|^{n-\alpha}} dy.$$

By (6.1), we get

$$J_2 \lesssim \|b\|_* r^{\frac{n}{q}} \int_{\complement(2B)} \frac{|f(y)|}{|x_0 - y|^{n-\alpha}} dy.$$

By Fubini's theorem we have

$$\begin{aligned}\int_{\complement(2B)} \frac{|f(y)|}{|x_0 - y|^{n-\alpha}} dy &\approx \int_{\complement(2B)} |f(y)| \int_{|x_0-y|}^{\infty} \frac{dt}{t^{n+1-\alpha}} dy \\ &\approx \int_{2r}^{\infty} \int_{2r \le |x_0-y| \le t} |f(y)| dy \frac{dt}{t^{n+1-\alpha}} \\ &\lesssim \int_{2r}^{\infty} \int_{B(x_0,t)} |f(y)| dy \frac{dt}{t^{n+1-\alpha}}.\end{aligned}$$

Applying Hölder's inequality, we get

$$\int_{\complement(2B)} \frac{|f(y)|}{|x_0 - y|^{n-\alpha}} dy \lesssim \int_{2r}^{\infty} \|f\|_{L_p(B(x_0,t))} \frac{dt}{t^{\frac{n}{q}+1}}. \tag{7.1}$$

Thus, by (7.1)

$$J_2 \lesssim \|b\|_* r^{\frac{n}{q}} \int_{2r}^{\infty} \|f\|_{L_p(B(x_0,t))} \frac{dt}{t^{\frac{n}{q}+1}}.$$

Summing up J_1 and J_2, for all $p \in (1, \infty)$ we get

$$\|[b, I_\alpha] f_2\|_{L_q(B)} \lesssim \|b\|_* r^{\frac{n}{q}} \int_{2r}^{\infty} \left(1 + \ln \frac{t}{r}\right) \|f\|_{L_p(B(x_0,t))} \frac{dt}{t^{\frac{n}{q}+1}}. \tag{7.2}$$

Finally,

$$\begin{aligned}\|[b, I_\alpha] f\|_{L_q(B)} &\lesssim \|b\|_* \, \|f\|_{L_p(2B)} + \|b\|_* r^{\frac{n}{q}} \int_{2r}^{\infty} \left(1 + \ln \frac{t}{r}\right) \|f\|_{L_p(B(x_0,t))} \frac{dt}{t^{\frac{n}{q}+1}} \\ &\lesssim \|b\|_* r^{\frac{n}{q}} \int_{2r}^{\infty} \left(1 + \ln \frac{t}{r}\right) \|f\|_{L_p(B(x_0,t))} \frac{dt}{t^{\frac{n}{q}+1}}.\end{aligned}$$ □

The following theorem is true.

Theorem 7.12. *Let $1 < p < \infty$, $0 < \alpha < \frac{n}{p}$, $\frac{1}{q} = \frac{1}{p} - \frac{\alpha}{n}$, $b \in BMO(\mathbb{R}^n)$ and let (φ_1, φ_2) satisfy the condition*

$$\int_r^{\infty} \left(1 + \ln \frac{t}{r}\right) t^{\alpha} \varphi_1(x,t) \frac{dt}{t} \le C \varphi_2(x,r), \tag{7.3}$$

where C does not depend on x and r.

Then the operator $[b, I_\alpha]$ is bounded from M_{p,φ_1} to M_{q,φ_2}. Moreover

$$\|[b, I_\alpha] f\|_{M_{q,\varphi_2}} \lesssim \|b\|_* \, \|f\|_{M_{p,\varphi_1}}.$$

Proof. The statement of Theorem 7.12 follows from Lemma 7.4.

$$\begin{aligned}\|[b,I_\alpha]f\|_{M_{q,\varphi_2}} &= \sup_{x\in\mathbb{R}^n,\, r>0} \varphi_2(x,r)^{-1} r^{-\frac{n}{q}} \|[b,I_\alpha]f\|_{L_q(B(x,r))} \\ &\lesssim \|b\|_* \sup_{x\in\mathbb{R}^n,\, r>0} \varphi_2(x,r)^{-1} \int_{2r}^\infty \Big(1+\ln\frac{t}{r}\Big) \|f\|_{L_p(B(x_0,t))} \frac{dt}{t^{\frac{n}{q}+1}} \\ &\lesssim \|b\|_* \|f\|_{M_{p,\varphi_1}} \sup_{x\in\mathbb{R}^n,\, r>0} \varphi_2(x,r)^{-1} \int_{2r}^\infty \Big(1+\ln\frac{t}{r}\Big) t^\alpha \varphi_2(x,t)\frac{dt}{t} \\ &\lesssim \|b\|_* \|f\|_{M_{p,\varphi_1}}. \end{aligned}$$

□

7.2. Adams type result

The following is a result of Adams type.

Theorem 7.13. *Let $1<p<q<\infty$, $0<\alpha<\frac{n}{p}$, $b\in BMO(\mathbb{R}^n)$ and let $\varphi(x,t)$ satisfy the conditions (6.7) and*

$$\int_r^\infty \Big(1+\ln\frac{t}{r}\Big) t^\alpha \varphi(x,t)^{\frac{1}{p}} \frac{dt}{t} \le C r^{-\frac{\alpha p}{q-p}}, \tag{7.4}$$

where C does not depend on $x\in\mathbb{R}^n$ and $r>0$. Then $[b,I_\alpha]$ is bounded from $M_{p,\varphi^{\frac{1}{p}}}$ to $M_{q,\varphi^{\frac{1}{q}}}$.

Proof. Let $1<p<q<\infty$, $0<\alpha<\frac{n}{p}$ and $f\in M_{p,\varphi^{\frac{1}{p}}}$. For arbitrary $x_0\in\mathbb{R}^n$, set $B=B(x_0,r)$ for the ball centered at x_0 and of radius r. Write $f=f_1+f_2$ with $f_1=f\chi_{2B}$ and $f_2=f\chi_{\complement(2B)}$.

For $x\in B$ we have

$$\begin{aligned}|[b,I_\alpha]f_2(x)| &\lesssim \int_{\mathbb{R}^n} \frac{|b(y)-b(x)|}{|x-y|^{n-\alpha}}|f_2(y)|dy \\ &\approx \int_{\complement(2B)} \frac{|b(y)-b(x)|}{|x_0-y|^{n-\alpha}}|f(y)|dy.\end{aligned}$$

Analogously to Section 7.1, for all $p\in(1,\infty)$ and $x\in B$ we get

$$|[b,I_\alpha]f_2(x)| \lesssim \|b\|_* \int_{2r}^\infty \Big(1+\ln\frac{t}{r}\Big) t^{\alpha-\frac{n}{p}-1} \|f\|_{L_p(B(x_0,t))} dt. \tag{7.5}$$

Then from conditions (7.4) and (7.5) we get

$$\begin{aligned}|[b,I_\alpha]f(x)| &\lesssim \|b\|_* r^\alpha M_b f(x) + \|b\|_* \int_r^\infty \Big(1+\ln\frac{t}{r}\Big) t^{\alpha-\frac{n}{p}-1}\|f\|_{L_p(B(x,t))}dt \\ &\le \|b\|_* r^\alpha M_b f(x) + \|b\|_* \|f\|_{M_{p,\varphi^{\frac{1}{p}}}} \int_r^\infty \Big(1+\ln\frac{t}{r}\Big) t^\alpha \varphi(x,t)\frac{dt}{t} \\ &\lesssim \|b\|_* r^\alpha M_b f(x) + \|b\|_* r^{-\frac{\alpha p}{q-p}} \|f\|_{M_{p,\varphi^{\frac{1}{p}}}}. \end{aligned} \tag{7.6}$$

Hence choose $r = \left(\frac{\|f\|_{M_{p,\varphi^{\frac{1}{p}}}}}{M_b f(x)} \right)^{\frac{q-p}{\alpha q}}$ for every $x \in B$, we have

$$|[b, I_\alpha]f(x)| \lesssim \|b\|_* \, (M_b f(x))^{\frac{p}{q}} \, \|f\|_{M_{p,\varphi^{\frac{1}{p}}}}^{1-\frac{p}{q}}.$$

Hence the statement of the theorem follows in view of the boundedness of the commutator of maximal operator M_b in $M_{p,\varphi^{\frac{1}{p}}}$ provided by Corollary 6.3 in virtue of condition (6.7)

$$\begin{aligned}
\|[b, I_\alpha]f\|_{M_{q,\varphi^{\frac{1}{q}}}} &= \sup_{x\in\mathbb{R}^n,\, r>0} \varphi(x,r)^{-\frac{1}{q}} r^{-\frac{n}{q}} \|[b, I_\alpha]f\|_{L_q(B(x,r))} \\
&\lesssim \|b\|_* \, \|f\|_{M_{p,\varphi^{\frac{1}{p}}}}^{1-\frac{p}{q}} \sup_{x\in\mathbb{R}^n,\, r>0} \varphi(x,r)^{-\frac{1}{q}} r^{-\frac{n}{q}} \|M_b f\|_{L_p(B(x,r))}^{\frac{p}{q}} \\
&= \|b\|_* \, \|f\|_{M_{p,\varphi^{\frac{1}{p}}}}^{1-\frac{p}{q}} \left(\sup_{x\in\mathbb{R}^n,\, r>0} \varphi(x,r)^{-\frac{1}{p}} r^{-\frac{n}{p}} \|M_b f\|_{L_p(B(x,r))} \right)^{\frac{p}{q}} \\
&= \|b\|_* \, \|f\|_{M_{p,\varphi^{\frac{1}{p}}}}^{1-\frac{p}{q}} \|M_b f\|_{M_{p,\varphi^{\frac{1}{p}}}}^{\frac{p}{q}} \lesssim \|b\|_* \, \|f\|_{M_{p,\varphi^{\frac{1}{p}}}}. \qquad \square
\end{aligned}$$

8. Some applications

In this section, we shall apply Theorems 4.4, 7.12 and 7.13 to several particular operators such as the Marcinkiewicz operator and fractional powers of the some analytic semigroups.

8.1. Marcinkiewicz operator

Let $S^{n-1} = \{x \in \mathbb{R}^n : |x| = 1\}$ be the unit sphere in $\mathbb{R}^n$ equipped with the Lebesgue measure $d\sigma$. Suppose that Ω satisfies the following conditions.

(a) Ω is the homogeneous function of degree zero on $\mathbb{R}^n \setminus \{0\}$, that is,

$$\Omega(\mu x) = \Omega(x), \text{ for any } \mu > 0, x \in \mathbb{R}^n \setminus \{0\}.$$

(b) Ω has mean zero on S^{n-1}, that is,

$$\int_{S^{n-1}} \Omega(x') d\sigma(x') = 0.$$

(c) $\Omega \in \mathrm{Lip}_\gamma(S^{n-1})$, $0 < \gamma \le 1$, that is there exists a constant $M > 0$ such that,

$$|\Omega(x') - \Omega(y')| \le M|x' - y'|^\gamma \text{ for any } x', y' \in S^{n-1}.$$

In 1958, Stein [28] defined the Marcinkiewicz integral of higher dimension μ_Ω as

$$\mu_\Omega(f)(x) = \left(\int_0^\infty |F_{\Omega,t}(f)(x)|^2 \frac{dt}{t^3} \right)^{1/2},$$

where

$$F_{\Omega,t}(f)(x) = \int_{|x-y|\le t} \frac{\Omega(x-y)}{|x-y|^{n-1}} f(y)dy.$$

Since Stein's work in 1958, the continuity of Marcinkiewicz integral has been extensively studied as a research topic and also provides useful tools in harmonic analysis [22, 29, 30].

The Marcinkiewicz operator is defined by (see [31])

$$\mu_{\Omega,\alpha}(f)(x) = \left(\int_0^\infty |F_{\Omega,\alpha,t}(f)(x)|^2 \frac{dt}{t^3} \right)^{1/2},$$

where

$$F_{\Omega,\alpha,t}(f)(x) = \int_{|x-y|\le t} \frac{\Omega(x-y)}{|x-y|^{n-1-\alpha}} f(y)dy.$$

Note that $\mu_\Omega f = \mu_{\Omega,0} f$.

The sublinear commutator of the operator $\mu_{\Omega,\alpha}$ is defined by

$$[b, \mu_{\Omega,\alpha}](f)(x) = \left(\int_0^\infty |F^b_{\Omega,t,\alpha}(f)(x)|^2 \frac{dt}{t^3} \right)^{1/2},$$

where

$$F^b_{\Omega,t,\alpha}(f)(x) = \int_{|x-y|\le t} \frac{\Omega(x-y)}{|x-y|^{n-1-\alpha}} [b(x) - b(y)] f(y)dy.$$

Let H be the space $H = \{h : \|h\| = (\int_0^\infty |h(t)|^2 dt/t^3)^{1/2} < \infty\}$. Then, it is clear that $\mu_{\Omega,\alpha}(f)(x) = \|F_{\Omega,\alpha,t}(x)\|$.

By Minkowski inequality and the conditions on Ω, we get

$$\begin{aligned} \mu_{\Omega,\alpha}(f)(x) &\le \int_{\mathbb{R}^n} \frac{|\Omega(x-y)|}{|x-y|^{n-1-\alpha}} |f(y)| \left(\int_{|x-y|}^\infty \frac{dt}{t^3} \right)^{1/2} dy \\ &\le C \int_{\mathbb{R}^n} \frac{|f(y)|}{|x-y|^{n-\alpha}} dy. \end{aligned}$$

It is known that for $b \in BMO(\mathbb{R}^n)$ the operators $\mu_{\Omega,\alpha}$ and $[b, \mu_{\Omega,\alpha}]$ are bounded from $L_p(\mathbb{R}^n)$ to $L_q(\mathbb{R}^n)$ for $p > 1$, and bounded from $L_1(\mathbb{R}^n)$ to $WL_q(\mathbb{R}^n)$ (see [31]), then from Theorems 4.4 and 7.12 we get

Corollary 8.5. *Let $1 \le p < \infty$, $0 < \alpha < \frac{n}{p}$, $\frac{1}{q} = \frac{1}{p} - \frac{\alpha}{n}$ and (φ_1, φ_2) satisfy condition (5.1) and Ω satisfies conditions (a)–(c). Then $\mu_{\Omega,\alpha}$ is bounded from M_{p,φ_1} to M_{q,φ_2} for $p > 1$ and from M_{1,φ_1} to WM_{q,φ_2}.*

Corollary 8.6. *Let $1 \le p < q < \infty$, $0 < \alpha < \frac{n}{p}$, φ satisfy conditions (4.7), (5.2) and Ω be satisfies the conditions (a)–(c). Then $\mu_{\Omega,\alpha}$ is bounded from $M_{p,\varphi^{\frac{1}{p}}}$ to $M_{q,\varphi^{\frac{1}{q}}}$ for $p > 1$ and from $M_{1,\varphi}$ to $M_{q,\varphi^{\frac{1}{q}}}$ for $p = 1$.*

Corollary 8.7. *Let $1 < p < \infty$, $0 \le \alpha < \frac{n}{p}$, $\frac{1}{q} = \frac{1}{p} - \frac{\alpha}{n}$, (φ_1, φ_2) satisfy condition* (7.3), *$b \in BMO(\mathbb{R}^n)$ and Ω be satisfies conditions* (a)–(c)*. Then $[b, \mu_{\Omega,\alpha}]$ is bounded from M_{p,φ_1} to M_{q,φ_2}.*

Corollary 8.8. *Let $1 < p < q < \infty$, $0 < \alpha < \frac{n}{p}$, φ satisfy conditions* (6.7), (7.4), *$b \in BMO(\mathbb{R}^n)$ and Ω be satisfies the conditions* (a)–(c)*. Then $[b, \mu_{\Omega,\alpha}]$ is bounded from $M_{p,\varphi^{\frac{1}{p}}}$ to $M_{q,\varphi^{\frac{1}{q}}}$.*

8.2. Fractional powers of the some analytic semigroups

The theorems of the previous sections can be applied to various operators which are estimated from above by Riesz potentials. We give some examples.

Suppose that L is a linear operator on L_2 which generates an analytic semigroup e^{-tL} with the kernel $p_t(x, y)$ satisfying a Gaussian upper bound, that is,

$$|p_t(x, y)| \le \frac{c_1}{t^{n/2}} e^{-c_2 \frac{|x-y|^2}{t}} \tag{8.1}$$

for $x, y \in \mathbb{R}^n$ and all $t > 0$, where c_1, $c_2 > 0$ are independent of x, y and t.

For $0 < \alpha < n$, the fractional powers $L^{-\alpha/2}$ of the operator L are defined by

$$L^{-\alpha/2} f(x) = \frac{1}{\Gamma(\alpha/2)} \int_0^\infty e^{-tL} f(x) \frac{dt}{t^{-\alpha/2+1}}.$$

Note that if $L = -\Delta$ is the Laplacian on $\mathbb{R}^n$, then $L^{-\alpha/2}$ is the Riesz potential I_α. See, for example, Chapter 5 in [29].

Property (8.1) is satisfied for large classes of differential operators (see, for example [7]). In [7] also other examples of operators which are estimates from above by Riesz potentials are given. In these cases Theorems 5.6, 5.7, 7.12 and 7.13 are also applicable for proving boundedness of those operators and commutators from M_{p,φ_1} to M_{q,φ_2} and from $M_{p,\varphi^{\frac{1}{p}}}$ to $M_{q,\varphi^{\frac{1}{q}}}$.

Corollary 8.9. *Let condition* (8.1) *be satisfied. Moreover, let $1 \le p < \infty$, $0 < \alpha < \frac{n}{p}$, $\frac{1}{q} = \frac{1}{p} - \frac{\alpha}{n}$, (φ_1, φ_2) satisfy condition* (5.1)*. Then $L^{-\alpha/2}$ is bounded from M_{p,φ_1} to M_{q,φ_2} for $p > 1$ and from M_{1,φ_1} to WM_{q,φ_2} for $p = 1$.*

Proof. Since the semigroup e^{-tL} has the kernel $p_t(x, y)$ which satisfies condition (8.1), it follows that

$$|L^{-\alpha/2} f(x)| \lesssim I_\alpha(|f|)(x)$$

(see [14]). Hence by the aforementioned theorems we have

$$\|L^{-\alpha/2} f\|_{M_{q,\varphi_2}} \lesssim \|I_\alpha(|f|)\|_{M_{q,\varphi_2}} \lesssim \|f\|_{M_{p,\varphi_1}}. \qquad \square$$

Corollary 8.10. *Let condition* (8.1) *be satisfied. Moreover, let $1 \le p < q < \infty$, $0 < \alpha < \frac{n}{p}$, φ satisfy conditions* (4.7) *and* (5.2)*. Then $L^{-\alpha/2}$ is bounded from $M_{p,\varphi^{\frac{1}{p}}}$ to $M_{q,\varphi^{\frac{1}{q}}}$ for $p > 1$ and from M_{1,φ_1} to $WM_{q,\varphi^{\frac{1}{q}}}$ for $p = 1$.*

Let b be a locally integrable function on $\mathbb{R}^n$, the commutator of b and $L^{-\alpha/2}$ is defined as follows

$$[b, L^{-\alpha/2}]f(x) = b(x)L^{-\alpha/2}f(x) - L^{-\alpha/2}(bf)(x).$$

In [14] extended the result of [10] from $(-\Delta)$ to the more general operator L defined above. More precisely, they showed that when $b \in BMO(\mathbb{R}^n)$, then the commutator operator $[b, L^{-\alpha/2}]$ is bounded from $L_p(\mathbb{R}^n)$ to $L_q(\mathbb{R}^n)$ for $1 < p < \infty$ and $\frac{1}{q} = \frac{1}{p} - \frac{\alpha}{n}$. Then from Theorems 7.12 and 7.13 we get

Corollary 8.11. *Let condition* (8.1) *be satisfied. Moreover, let* $1 < p < \infty$, $0 < \alpha < \frac{n}{p}$, $\frac{1}{q} = \frac{1}{p} - \frac{\alpha}{n}$, $b \in BMO(\mathbb{R}^n)$, *and* (φ_1, φ_2) *satisfy condition* (7.3). *Then* $[b, L^{-\alpha/2}]$ *is bounded from* M_{p,φ_1} *to* M_{q,φ_2}.

Corollary 8.12. *Let condition* (8.1) *be satisfied. Moreover, let* $1 < p < q < \infty$, $0 < \alpha < \frac{n}{p}$, $b \in BMO(\mathbb{R}^n)$, *and* φ *satisfy conditions* (6.7) *and* (7.4). *Then* $[b, L^{-\alpha/2}]$ *is bounded from* $M_{p,\varphi^{\frac{1}{p}}}$ *to* $M_{q,\varphi^{\frac{1}{q}}}$.

Acknowledgment

The authors would like to express their gratitude to the referees for his very valuable comments and suggestions.

References

[1] D.R. Adams, *A note on Riesz potentials*, Duke Math. **42** (1975) 765–778.

[2] Ali Akbulut, V.S. Guliyev and R. Mustafayev, *Boundedness of the maximal operator and singular integral operator in generalized Morrey spaces*, preprint, Institute of Mathematics, AS CR, Prague. 2010-1-26, 1–15.

[3] Ali Akbulut, V.S. Guliyev and R. Mustafayev, *On the Boundedness of the maximal operator and singular integral operators in generalized Morrey spaces*, Mathematica Bohemica, 2011, 1–17.

[4] V.I. Burenkov, H.V. Guliyev, *Necessary and sufficient conditions for boundedness of the maximal operator in the local Morrey-type spaces*, Studia Mathematica 163 (2) (2004), 157–176.

[5] V.I. Burenkov, H.V. Guliyev, V.S. Guliyev, *Necessary and sufficient conditions for the boundedness of the fractional maximal operator in the local Morrey-type spaces*, Dokl. Akad. Nauk **74** (1) (2006) 540–544.

[6] V.I. Burenkov, H.V. Guliyev, V.S. Guliyev, *Necessary and sufficient conditions for boundedness of the fractional maximal operators in the local Morrey-type spaces*, J. Comput. Appl. Math. **208** (1) (2007), 280–301.

[7] V.I. Burenkov, V.S. Guliyev, *Necessary and sufficient conditions for the boundedness of the Riesz potential in local Morrey-type spaces*, Potential Anal. **30** (3) (2009), 211–249.

[8] V. Burenkov, A. Gogatishvili, V.S. Guliyev, R. Mustafayev, *Boundedness of the fractional maximal operator in local Morrey-type spaces*, Complex Var. Elliptic Equ. **55** (8-10) (2010), 739–758.

[9] V. Burenkov, A. Gogatishvili, V.S. Guliyev, R. Mustafayev, *Boundedness of the fractional maximal operator in local Morrey-type spaces*, Potential Anal. 35 (2011), no. 1, 67–87.

[10] S. Chanillo, *A note on commutators*, Indiana Univ. Math. J., 23 (1982), 7–16.

[11] F. Chiarenza, M. Frasca, *Morrey spaces and Hardy-Littlewood maximal function*, Rend Mat. **7** (1987) 273–279.

[12] R. Coifman, R. Rochberg, G. Weiss, *Factorization theorems for Hardy spaces in several variables*, Ann. of Math. **103** (2) (1976) 611–635.

[13] Y. Ding, D. Yang, Z. Zhou, *Boundedness of sublinear operators and commutators on* $L^{p,\omega}(\mathbb{R}^n)$, Yokohama Math. J. **46** (1998) 15–27.

[14] X.T. Duong, L.X. Yan, *On commutators of fractional integrals*, Proc. Amer. Math. Soc. **132** (12) (2004), 3549–3557.

[15] V.S. Guliyev, *Integral operators on function spaces on the homogeneous groups and on domains in* $\mathbb{R}^n$. Doctor's degree dissertation, Mat. Inst. Steklov, Moscow, 1994, 329 pp. (in Russian).

[16] V.S. Guliyev, *Function spaces, Integral Operators and Two Weighted Inequalities on Homogeneous Groups. Some Applications*, Casioglu, Baku, 1999, 332 pp. (in Russian).

[17] V.S. Guliyev, *Boundedness of the maximal, potential and singular operators in the generalized Morrey spaces*, J. Inequal. Appl. 2009, Art. ID 503948, 20 pp.

[18] V.S. Guliyev, J. Hasanov, Stefan Samko, *Boundedness of the maximal, potential and singular operators in the generalized variable exponent Morrey spaces*, Math. Scand. **197** (2) (2010) 285–304.

[19] V.S. Guliyev, S.S. Aliyev, T. Karaman, *Boundedness of a class of sublinear operators and their commutators on generalized Morrey spaces*, Abstr. Appl. Anal. vol. 2011, Art. ID 356041, 18 pp. doi:10.1155/2011/356041

[20] Y. Lin, *Strongly singular Calderón-Zygmund operator and commutator on Morrey type spaces*, Acta Math. Sin. (Engl. Ser.) **23** (11) (2007) 2097–2110.

[21] G. Lu, S. Lu, D. Yang, *Singular integrals and commutators on homogeneous groups*, Analysis Mathematica, **28** (2002) 103–134.

[22] S. Lu, Y. Ding, D. Yan, *Singular integrals and related topics*, World Scientific Publishing, Singapore, 2006.

[23] T. Mizuhara, *Boundedness of some classical operators on generalized Morrey spaces*, Harmonic Analysis (S. Igari, editor), ICM 90 Satellite Proceedings, Springer-Verlag, Tokyo (1991) 183–189.

[24] C.B. Morrey, *On the solutions of quasi-linear elliptic partial differential equations*, Trans. Amer. Math. Soc. **43** (1938) 126–166.

[25] E. Nakai, *Hardy–Littlewood maximal operator, singular integral operators and Riesz potentials on generalized Morrey spaces*, Math. Nachr. **166** (1994) 95–103.

[26] J. Peetre, *On the theory of* $M_{p,\lambda}$, J. Funct. Anal. **4** (1969) 71–87.

[27] F. Soria, G. Weiss, *A remark on singular integrals and power weights*, Indiana Univ. Math. J. **43** (1994) 187–204.

[28] E.M. Stein, *On the functions of Littlewood-Paley, Lusin, and Marcinkiewicz*, Trans. Amer. Math. Soc. **88** (1958) 430–466.

[29] Stein, E.M., *Singular integrals and differentiability of functions*, Princeton University Press, Princeton, NJ, 1970.

[30] E.M. Stein, *Harmonic Analysis: Real Variable Methods, Orthogonality and Oscillatory Integrals*, Princeton Univ. Press, Princeton NJ, 1993.

[31] A. Torchinsky and S. Wang, *A note on the Marcinkiewicz integral*, Colloq. Math. **60/61** (1990) 235–243.

Vagif S. Guliyev
Ahi Evran University, Department of Mathematics
Kirsehir, Turkey

and

Institute of Mathematics and Mechanics
Academy of Sciences of Azerbaijan
F. Agayev St. 9
Baku, AZ 1141, Azerbaijan
e-mail: `vagif@guliyev.com`

Parviz S. Shukurov
Institute of Mathematics and Mechanics
Academy of Sciences of Azerbaijan
F. Agayev St. 9
Baku, AZ 1141, Azerbaijan
e-mail: `matematik80@mail.ru`

Operator Theory:
Advances and Applications, Vol. 229, 201–214

Existence of Solutions of a Class of Nonlinear Singular Equations in Lorentz Spaces

L. Huang, K. Murillo and E.M. Rocha

Abstract. We consider the following nonlinear elliptic Dirichlet problem involving a Leray-Lions type differential operator

$$-\operatorname{div}(\psi(x,u(x),\nabla u(x)))+a(x)u(x)=f(x),\quad \text{in}\quad \Omega,\quad u\in W_0^{1,p}(\Omega),$$

where $\Omega\subset\mathbb{R}^N$ is a bounded domain with smooth boundary, $2\leq p<N$, $a\in L^\infty_{\mathrm{loc}}(\Omega;\mathbb{R}_0^+)$ and $f\in L^{q,q_1}(\Omega)$ is a function in a Lorentz space. We show the existence of a solution $u\in W_0^{1,p}(\Omega)\cap L^{r,s}(\Omega)$ and an *a priori estimate* for the solution with respect to the Lorentz space norm of $f\in L^{q,q_1}(\Omega)$, for suitable values p,q,q_1,r and s.

Mathematics Subject Classification (2010). Primary 35J60; Secondary 46E30.

Keywords. Elliptic equations, Leray-Lions operator, Lorentz spaces, existence of solutions.

1. Introduction

Let $\Omega\subset\mathbb{R}^N$ be a bounded domain with smooth boundary. Here we study the existence of solutions $u\in W_0^{1,p}(\Omega)$ for the Dirichlet nonlinear problem

$$\begin{cases}-\operatorname{div}(\psi(x,u(x),\nabla u(x)))+a(x)u=f(x) & \text{in}\quad \Omega,\\ \qquad\qquad\qquad\qquad\qquad\qquad u=0 & \text{on}\quad \partial\Omega,\end{cases}\qquad (P_\Omega)$$

where $2\leq p<N$, $\Psi(u)=-\operatorname{div}(\psi(x,u(x),\nabla u(x)))$ is a Leray-Lions type operator, $a\in L^\infty_{\mathrm{loc}}(\Omega)$ with $a(x)\geq 0$ for all $x\in\Omega$, $f\in L^{q,q_1}(\Omega)$ is a function in a Lorentz space with $p'\leq q<\frac{N}{p}$ and $q_1=\frac{(N-p)q}{N-pq}$.

This work was supported by *FEDER* funds through *COMPETE* – Operational Programme Factors of Competitiveness ("Programa Operacional Factores de Competitividade") and by Portuguese funds through the *Center for Research and Development in Mathematics and Applications* (University of Aveiro) and the Portuguese Foundation for Science and Technology ("FCT – Fundação para a Ciência e a Tecnologia"), within project PEst-C/MAT/UI4106/2011 with COMPETE number FCOMP-01-0124-FEDER-022690; and under the PhD fellowships SFRH/BD/33401/2008 and SFRH/BD/51162/2010.

The operator $\Psi : X \to X^*$ defined by $\Psi(u) = -\operatorname{div}(\psi(x, u(x), \nabla u(x)))$ is called a *Leray-Lions operator* if satisfies the following conditions:

(i) *Carathéodory function*: the function $\psi : \Omega \times \mathbb{R} \times \mathbb{R}^N \to \mathbb{R}^N$ satisfies
   - the function $x \mapsto \psi(x, s, \xi)$ is measurable for all $(s, \xi) \in \mathbb{R} \times \mathbb{R}^N$;
   - the function $(s, \xi) \mapsto \psi(x, s, \xi)$ is continuous for almost all $x \in \Omega$;

(ii) *Elliptic condition*: there exists $\alpha > 0$ such that
$$\psi(x, s, \xi)\xi \geq \alpha|\xi|^p$$
for all $(x, s, \xi) \in \Omega \times \mathbb{R} \times \mathbb{R}^N$;

(iii) *Growth condition*: there exist $\beta > 0$ and $c \in L^{p'}(\Omega)$ such that
$$|\psi(x, s, \xi)| \leq c(x) + \beta(|s|^{p-1} + |\xi|^{p-1})$$
for all $(x, s, \xi) \in \Omega \times \mathbb{R} \times \mathbb{R}^N$;

(iv) *Monotonicity condition*: for $\xi, \eta \in \mathbb{R}^N$, $\xi \neq \eta$ and a.e. for $x \in \Omega$
$$[\psi(x, s, \xi) - \psi(x, s, \eta)](\xi - \eta) > 0.$$

Common examples of Leray-Lions operators are the generalized mean curvature operator (i.e., $\psi(x, s, \xi) = (1 + |\xi|^2)^{(p-2)/2}\xi)$ and the p-Laplacian (i.e., $\psi(x, s, \xi) = |\xi|^{p-2}\xi)$, but weighted versions of this operators can also be considered, besides others.

Lorentz spaces $L^{p,q}(\Omega)$, which are a generalization of Lebesgue spaces, were introduced by G.G. Lorentz [17]. For a measurable function $f : \Omega \to \mathbb{R}$, we define the *distribution function*
$$d^f_\Omega(t) = |\{x \in \Omega; |f(x)| > t\}|$$
with $t \geq 0$ and the *non-increasing rearrangement* of f to be
$$f^*_\Omega(s) = \sup\left\{t > 0 : d^f_\Omega(t) > s\right\}$$
with $0 \leq s \leq |\Omega|$, then the Lorentz space $L^{p,q}(\Omega)$ is the collection of measurable functions f on Ω such that $\|f\|_{L^{p,q}(\Omega)} < \infty$, where
$$\|f\|_{L^{p,q}(\Omega)} = \begin{cases} \left(\int_0^\infty \left(s^{1/p} f^{**}(s)\right)^q \frac{ds}{s}\right)^{\frac{1}{q}} & \text{if } 1 \leq q < \infty, \\ \sup_{s>0}\left\{s^{\frac{1}{p}} f^{**}(s)\right\} & \text{if } q = \infty; \end{cases}$$
with $1 \leq p < \infty$ and $f^{**}(s) = \frac{1}{s}\int_0^s f^*_\Omega(t)dt$.

Problem (P_Ω) is well defined since (by standard arguments) the left-hand side of (P_Ω) is in $W^{-1,p'}(\Omega)$ and, for $f \in L^{q,q_1}(\Omega)$, $f \in W^{-1,p'}(\Omega)$. In fact, since $p' \leq q \leq q_1$, from the Lorentz spaces scale (see below Lemma 2.6), we have $L^{q,q_1}(\Omega) \subset L^{\bar{q},p'}(\Omega)$, for any $\bar{q} < q$, so for $\bar{q} = p'$ we get
$$f \in L^{q,q_1}(\Omega) \subset L^{p',p'}(\Omega) \equiv L^{p'}(\Omega) \equiv (L^p(\Omega))^* \subset \left(W^{1,p}_0(\Omega)\right)^* \equiv W^{-1,p'}(\Omega).$$
Problem (P_Ω) has an additional difficulty since $a \in L^\infty_{\text{loc}}(\Omega)$, so the standard definition of weak solution may not make sense (i.e., with test functions in $W^{1,p}_0(\Omega)$).

Let $(\Omega_n)_{n\in\mathbb{N}}$ be a increasing sequence of open subsets of Ω, such that $\overline{\Omega}_n \subseteq \Omega_{n+1}$ and $\Omega = \bigcup_{n=1}^{\infty} \Omega_n$. We use the following notion of solution: $u \in W_0^{1,p}(\Omega)$ is a *weak solution* of problem (P_Ω) if satisfies

$$\int_\Omega \psi(x, u(x), \nabla u(x))\nabla\varphi + a(x)u\varphi\, dx = \int_\Omega f\varphi\, dx \tag{1.1}$$

for all $\varphi \in \bigcup_{n=1}^{\infty} W_0^{1,p}(\Omega_n)$. By problem (P_{Ω_n}) we mean the same problem as (P_Ω) but defined on Ω_n.

The main result of this work is the following.

Theorem 1.1. *Let $2 \leq p < N$, $p' \leq q < \frac{N}{p}$, $\sigma = (N - pq)^{-1}$, $q_1 = \sigma(N - p)q$, $r = \sigma N(q-1)q$, $s = \sigma(N-p)(p-1)q$, and $a \in L^\infty_{\mathrm{loc}}(\Omega; \mathbb{R}_0^+)$. If $f \in L^{q,q_1}(\Omega)$ then there exists (at least) one solution $u \in W_0^{1,p}(\Omega) \cap L^{r,s}(\Omega)$ of the problem (P_Ω) and the solution satisfies the a priori estimate*

$$\|u\|_{L^{r,s}(\Omega)} \leq C\|f\|^{p'/p}_{L^{q,q_1}(\Omega)} \quad (\textit{for some } C > 0). \tag{1.2}$$

If $a \in L^\infty(\Omega; \mathbb{R}_0^+)$ then the solution u is unique and $\int_\Omega fu\, dx \geq 0$.

Lemma 1.2 (Approximation of the solution). *Let $u \in W_0^{1,p}(\Omega)$ be the solution given in Theorem 1.1. For any $n \in \mathbb{N}$, the problem (P_{Ω_n}) has a unique solution $u_n \in W_0^{1,p}(\Omega_n)$ and the sequence $(u_n)_{n\in\mathbb{N}} \subset W_0^{1,p}(\Omega)$ converges strongly to u.*

Remark 1.3 (Linear case). Let $N > 4$, $2 \leq q < \frac{N}{2}$, $\sigma = (N-2q)^{-1}$, $\mu_1 = \sigma Nq$, $\mu_2 = \sigma(N-2)q$, $\Psi(u) = -\Delta u$, and $a \in L^\infty_{\mathrm{loc}}(\Omega; \mathbb{R}_0^+)$. If $f \in L^{q,\mu_2}(\Omega)$ then there exists (at least) one solution $u \in H_0^1(\Omega) \cap L^{\mu_1,\mu_2}(\Omega)$ which satisfies the *a priori estimate*

$$\|u\|_{L^{\mu_1,\mu_2}(\Omega)} \leq C\|f\|_{L^{q,\mu_2}(\Omega)} \quad (\text{for some } C > 0).$$

There are many results concerning the existence of solutions to boundary value problems for nonlinear second-order elliptic equations, here we are interested in the special situation where singular terms and terms in Lorentz spaces appear.

Singular elliptic problems depend deeply on the type of singularities involved. The best studied singular term in the literature is a negative power of the solution, i.e., considering the Dirichlet problem of the form $-\Delta_p u(x) = \beta(x)u(x)^{-\eta} + f(x, u(x))$, with $\eta \geq 0$, which was first studied in the context of semilinear equations ($p = 2$). Among the first works in this direction are the papers of Crandall, Rabinowitz, and Tártar [8] and Stuart [25]. Since then, there have been several other papers on the subject. We mention the relevant works of Coclite and Palmieri [7], Diaz, Morel, and Oswald [9], Lair and Shaker [15], Shaker [22], Shi and Yao [23], Sun, Wu, and Long [26], and Zhang [28]. In particular, Lair and Shaker [15] assumed that $f \equiv 0$ and $\beta \in L^2(\Omega)$ and established the existence of a unique positive weak solution. Their result was extended by Shi and Yao [23] to the case of a "sublinear" reaction, namely when $f(x, u) = \lambda u^{r-1}$ with $\lambda > 0$ and $1 < r \leq 2$. The case of a "superlinear-subcritical" reaction, i.e., when $2 < r < 2^*$, where 2^* is the critical Sobolev exponent, was investigated by Coclite and Palmieri [7] under the assumption that $\beta \equiv 1$. In both works (i.e., [7] and [23]), it is shown that there exists a critical value $\lambda^* > 0$ of the parameter, such that for every $\lambda \in (0, \lambda^*)$, the problem admits a nontrivial positive solution. Subsequently, Sun, Wu, and Long [26] using the Ekeland variational principle, obtained two nontrivial positive weak solutions for more general functions β. The work of Zhang [28] extended their result to more general nonnegative superlinear perturbations, using critical point theory on closed convex sets. For the same problem but driven by the p-Laplacian, we mention the works of Agarwal, Lü, and O'Regan [1], Agarwal and O'Regan [2], where $N = 1$ (ordinary differential equations), and Perera and Silva [19], Perera and Zhang [20], where $N \geq 2$ (partial differential equations) and the reaction term has the form $\beta(x)u(x)^{-\eta} + \lambda f(x, u(x))$ with $\lambda > 0$. For such a parametric nonlinearity, the authors prove existence and multiplicity results (two positive weak solutions), valid for all $\lambda \in (0, \lambda^*)$. Moreover, the perturbation term f exhibits a strict $(p-1)$-superlinear growth near $+\infty$ and, more precisely, it satisfies on $[0, +\infty)$, the well-known Ambrosetti-Rabinowitz condition. Chen, Papageorgiou and Rocha [4] considered the reaction term nonparametric and the perturbation as $(p-1)$-linear near $+\infty$ and proved the existence of an ordered pair of smooth positive strong solutions.

For other type of singularities, we mention the work of Giachetti and Segura de Leon [12], where they obtained for a problem involving a Leray-Lions operator plus the term $\frac{\operatorname{sig}(u-1)}{|u-1|^K}|\nabla u|^2 - f$ the existence of a weak solution $u \in H_0^1(\Omega)$ when $f \in L^m(\Omega)$, where $m \geq \frac{2N}{N+2}$, by using Stampacchia theorem to show that the gradient goes to zero faster than $|u-1|^K$ so, in fact, the term does not blow up. Problems involving a Hardy-type singular term $-\frac{\lambda}{|x|^2}u$, where $0 \in \Omega$, a term with critical Sobolev exponent (compactness loss) and other subcritical terms where study by Chen and Rocha [5], showing the existence of four solutions where at least one is sign changing. Moreover, if in the previous problem we add the singularity $-\mu u^{-q}$ with $0 < q < 1$, Chen and Rocha [6] proved the existence of two positive solutions under adequate hypotheses. In both works, Nehari optimization techniques and precise estimates of the energy of critical points are important tools.

Elliptic problems with terms in Lorentz spaces are considerably less common, mainly because of the use of non-increasing rearrangements in their definition which are relevant for radially symmetric problems but limits the application of other techniques. Consider the degenerate linear version of problem P_Ω without singularity, i.e., $a \equiv 0$ and the Leray-Lions operator has the form $\psi(x, \xi) = M(x)\xi$ with M is a symmetric matrix in $L^\infty(\Omega)^{N \times N}$ satisfying the ellipticity condition, there exists $\alpha > 0$ such that $M(x)\xi \cdot \xi \geq \alpha|\xi|^2$ for $x \in \Omega$ and $\xi \in \mathbb{R}^N$. Karch and Ricciardi [14], for the particular set of equations $\frac{\partial}{\partial x_i}\psi(x, \nabla(x)) = \frac{\partial}{\partial x_i} f$ showed that weak solutions are differentiable almost everywhere when $u \in H^1_{\mathrm{loc}}(\Omega)$ and $\frac{\partial}{\partial x_i} f \in L^{n,1}_{\mathrm{loc}}(\Omega)$. For $f = \operatorname{div} F$ with $F \in (L^2(\Omega))^N$ and $a \equiv 0$, Napoli and Mariani [18] using the Lax-Milgram Lemma obtained the existence of a unique solution in $H_0^1(\Omega)$ and for $F \in L^q(\Omega)$, with $q > 2$ used the Stampacchia result (Theorem 4.2 [24]) to improve summability. For the nonlinear setting, still with $a \equiv 0$, involving a Leray-Lions operator and $F \in (L^{p'}(\Omega))^N$, Napoli and Mariani [18] using the Leray-Lions theorem [16], showed also the existence of a unique solution in $W_0^{1,p}(\Omega)$.

This work generalizes some previous results, e.g., in Napoli and Mariani [18], besides others. As far as we are aware, there are no results in the literature simultaneously combining $a \in L^\infty_{\rm loc}(\Omega; \mathbb{R}_0^+)$ and f in a Lorentz space. The main point here is to take advantage of the best fitted embedding of the Sobolev space $W_0^{1,p}(\Omega)$ into a Lorentz space, compared with the standard Sobolev embedding into a Lebesgue space.

The paper is organized as follows. Section 2 introduces basic notation and results relevant in subsequence pages. In Section 3, we first prove the existence of a solution for the problem (P_Ω) and then obtain the estimative and respective inclusion of the solution in a Lorentz space.

From now on, we use $\doteq$ to emphasize a new definition. By $\to$ (respectively, $\rightharpoonup$) we denoted strongly (respectively, weakly) convergence on a Banach space.

2. Preliminaries results

In this section, we give some preliminaries which play an important role in the method used to study problem (P_Ω).

Let $\Omega \subset \mathbb{R}^N$ be a bounded domain with smooth enough boundary, X be a separable reflexive Banach space and X^* denote its dual space. The dual product in $X^* \times X$ is denoted by $\langle x^*, x\rangle_{X^* \times X}$ and, when $X^* \times X$ is clear from the context, we just write $\langle x^*, x\rangle$ for $x^* \in X^*$ and $x \in X$. For an exponent $p > 1$, $p' \doteq p(p-1)^{-1}$ denotes its conjugate exponent and $p^* \doteq Np(N-p)^{-1}$ its critical exponent in $\mathbb{R}^N$.

Definition 2.1. Let $B : X \to X^*$ be an operator, then B is said to be

- *Coercive*: if $\lim\limits_{\|u\|\to\infty} \frac{\langle Bu,u\rangle}{\|u\|} = \infty$;
- (*strictly*) *Monotone*: if $\langle Bu - Bv, u - v\rangle \geq 0$ (respectively $>$), for all $u, v \in X$;
- *Hemicontinuous*: if $\lambda \in \mathbb{R} \mapsto \langle B(u + \lambda v), w\rangle$ is continuous, for all $u, v, w \in X$;
- *Radially continuous*: if $\lambda \in \mathbb{R} \mapsto \langle B(u + \lambda v), v\rangle$ is continuous, for all $u, v \in X$;
- *Pseudo-monotone*: if $u_n \rightharpoonup u$ and $\limsup\limits_{n\to\infty} \langle B(u_n), u_n - u\rangle \leq 0$ imply
$$\liminf_{n\to\infty} \langle Bu_n, u_n - v\rangle \geq \langle Bu, u - v\rangle$$
for all $v \in X$;
- (S_+)-*type operator*: if $u_n \rightharpoonup u$ and
$$\limsup_{n\to\infty} \langle Bu_n, u_n - u\rangle \leq 0$$
imply $u_n \to u$.

Lemma 2.2 (see [29]). *Any hemicontinuous and monotone operator is a pseudo-monotone operator.*

Definition 2.3. Let V be a linear subspace of X and $A : X \times V \to \mathbb{R}$, then A is said to be of *type M with respect to V* if for any sequence $(v_\lambda)_{\lambda \in \Lambda} \subset V$, $w \in X$ and $v^* \in V^*$, we have

(a) $v_\lambda \rightharpoonup w$;
(b) $A(v_\lambda, v) \to \langle v^*, v\rangle$ for all $v \in V$;
(c) $A(v_\lambda, v_\lambda) \to \langle \bar{v}^*, w\rangle$, where $\bar{v}^*$ is the extension of v^* on the closure of V;

imply that $A(w, v) = \langle v^*, v\rangle$ for all $v \in V$.

Lemma 2.4 (see [29]). *The operator Ψ is pseudo-monotone and a (S_+)-type operator.*

The following is a version of the Browder-Minty theorem ([11], [21]).

Theorem 2.5. *Let V bet a reflexive Banach space, let $A : V \to V^*$ be a radially continuous, coercive and monotone operator, then A is surjective.*

Lemma 2.6 ([13], [27]). *For Lorentz spaces, we have the following results:*

(a) *Using the definition above, it is possible to prove that $L^{p,p}(\Omega)$ coincides with the Lebesgue space $L^p(\Omega)$ and $\|u\|_{L^{p,p}(\Omega)} = \|u\|_{L^p(\Omega)}$ for $u \in L^{p,p}(\Omega)$.*
(b) *(Duality) Let $1 < p < q < \infty$ then*
$$(L^{p,q}(\Omega))' = L^{p',q'}(\Omega).$$
(c) *Let $1 \le q_1 < p < q_2 < \infty$ and $p_1 < p$, then the following inclusions hold:*
$$L^{p,q_1}(\Omega) \subsetneq L^{p,p}(\Omega) \equiv L^p(\Omega) \subsetneq L^{p,q_2}(\Omega) \subsetneq L^{p,\infty}(\Omega) \subsetneq L^{p_1,q_1}(\Omega).$$
(d) *We have the following (Hölder type) inequality*
$$\|fg\|_{L^{p,q}(\Omega)} \le \|f\|_{L^{p_1,q_1}(\Omega)}\|g\|_{L^{p_2,q_2}(\Omega)},$$
where $\frac{1}{p} = \frac{1}{p_1} + \frac{1}{p_2}$ and $\frac{1}{q} = \frac{1}{q_1} + \frac{1}{q_2}$.
(e) *If $f \in L^{pm,qm}(\Omega)$ with $m > 0$, then $|f|^m \in L^{p,q}(\Omega)$ and*
$$\||f|^m\|_{L^{p,q}(\Omega)} = \|f\|^m_{L^{pm,qm}(\Omega)}.$$

Lemma 2.7. *Let $1 \le p < N$ then $W^{1,p}(R^N) \subset L^{p^*,p}(R^N)$ with continuous embedding, $W_0^{1,p}(\Omega) \subset L^{p^*,p}(\Omega)$ with continuous embedding, and when $\partial\Omega \in C^1$ the same result applies to $W^{1,p}(\Omega)$.*

To obtain the *a priori* estimate for the solution, truncation functions will be the main tool. For $k > 0$ and $x \in R$ define the *truncating function* $T_k : \mathbb{R} \to \mathbb{R}$ as
$$T_k(x) \doteq \begin{cases} -k & \text{if } x < -k, \\ x & \text{if } -k \le x \le k, \\ k & \text{if } x > k. \end{cases}$$
For any $u : \Omega \to \mathbb{R}$, by $T_k(u)$ we denote the map $T_k(u) : \Omega \to \mathbb{R}$ defined by $x \mapsto T_k(u(x))$.

Lemma 2.8. *The truncation function $\mathcal{T}_k$ satisfies*

(i) *For any $k > 0$, we have $x\mathcal{T}_k(x) \geq 0$;*
(ii) *If $\|\mathcal{T}_k(u)\|_{L^{p,q}(\Omega)} \leq C$ for any $k > 0$, then $u \in L^{p,q}(\Omega)$ and $\|u\|_{L^{p,q}(\Omega)} \leq C$.*

Remark 2.9. Note that if $u \in W^{j,p}(\Omega)$ by the Sobolev embedding we have $u \in L^{p^*}(\Omega) \equiv L^{p^*,p^*}(\Omega)$, but by the embedding to Lorentz spaces we have $u \in L^{p^*,q}(\Omega)$ with $p \leq q \leq p^*$. Hence there is an improvement in using the embedding to Lorentz spaces.

3. Proof of Theorem 1.1

Here we prove Theorem 1.1 which we divide in two parts: (1) the existence of the solution and (2) the *a priori* estimate (1.2). First part is mainly using ideas of An et al. [3] and Drivaliaris and Yannakakis [10]. The proof of the estimate is inspired by Napoli and Mariani [18].

3.1. Existence

Recall $\Omega \subset \mathbb{R}^N$ be a bounded domain with smooth boundary, $(\Omega_n)_{n\in\mathbb{N}}$ is a increasing sequence of open subsets of Ω, such that $\overline{\Omega}_n \subseteq \Omega_{n+1}$ and $\Omega = \bigcup_{n=1}^{\infty} \Omega_n$. For presentation clarity, from now on, we fix $X \doteq W_0^{1,p}(\Omega)$ and $X_n \doteq W_0^{1,p}(\Omega_n)$ for each $n \in \mathbb{N}$. Each X_n is a closed subspace of X by extending its elements by zero outside Ω_n and let $V \doteq \bigcup_{n=1}^{\infty} X_n$.

Define the map $T : X \to X^*$ by

$$T(u)(x) \doteq \Psi(u)(x) + a(x)u(x) = -\operatorname{div}(\psi(x, u(x), \nabla u(x)) + a(x)u(x),$$

for all $x \in \Omega$ and $u \in X$, and the operator $A : X \times V \to \mathbb{R}$ by

$$A(u, v) \doteq \langle T(u), v\rangle_{X^* \times X} = \int_\Omega \psi(x, u(x), \nabla u(x))\nabla v(x) + a(x)u(x)v(x)\, dx$$

for all $u \in X$ and $v \in V$. We also consider the operators $A_n : X_n \times X_n \to \mathbb{R}$, with $n \in \mathbb{N}$, defined by

$$A_n(u, v) \doteq \langle T(u), v\rangle_{X_n^* \times X_n} = \int_{\Omega_n} \psi(x, u(x), \nabla u(x))\nabla v(x) + a(x)u(x)v(x)\, dx$$

for all $u, v \in X_n$. Note that $A : X \times V \to \mathbb{R}$ is well defined but $A : X \times X \to \mathbb{R}$ is not, since $a \in L^\infty_{\text{loc}}(\Omega; \mathbb{R}_0^+)$. In fact, for any $v \in V$ there exists a $k \in \mathbb{N}$ such that $v \in X_k$ and

$$A(u, v) = \int_{\Omega_k} \psi(x, u(x), \nabla u(x))\nabla v(x) + a(x)u(x)v(x)\, dx < \infty.$$

Although $A_k(u, \varphi)$ is not well defined, for $u \in X$ and $v \in X_k$.

Claim 1. The operators A_n are coercive, for any $n \in \mathbb{N}$.

From the elliptic condition, we have for $u \in X_n$

$$A_n(u,u) \geq \alpha \int_\Omega |\nabla u|^p\, dx + \int_\Omega a\, u^2\, dx \geq \alpha \|u\|^p + \int_\Omega a\, u^2\, dx.$$

Thus, since $a \geq 0$, we get

$$\lim_{\|u\|\to\infty} \frac{A_n(u,u)}{\|u\|} \geq \lim_{\|u\|\to\infty} \alpha \|u\|^{p-1} = \infty.$$

Claim 2. We have $A_n(u,\cdot) \in {X_n}^*$ for all $n \in \mathbb{N}$ and all $u \in X_n$.

From $a \in L^\infty(\Omega_n)$, there exists a constant $c_a \geq 0$ such that

$$A_n(u,v) \leq \int_{\Omega_n} |\psi(x,u(x),\nabla u(x))\nabla v| dx + c_a \int_{\Omega_n} |uv| dx.$$

Since $\psi(x,u(x),\nabla u(x)) \in L^{p'}(\Omega_n)$ and $\nabla v \in L^p(\Omega_n)$, we can apply the Hölder inequality in the first term. For the second term, since $p' = \frac{p}{p-1}$, $p \geq 2$, and $p' < p$, we use the embedding $L^p(\Omega_{n_0}) \subset L^{p'}(\Omega_{n_0})$, i.e., if $u \in L^p(\Omega_{n_0})$, we can apply the Hölder inequality with $u \in L^{p'}(\Omega_n)$ and $v \in L^p(\Omega_n)$. Additionally from the Poincaré inequality, we obtain

$$\begin{aligned}
A_n(u,v) &\leq \|\psi(x,u(x),\nabla u(x))\|_{L^{p'}(\Omega_{n_0})}\|v\|_{L^p(\Omega_{n_0})} + c_a\|u\|_{L^{p'}(\Omega_{n_0})}\|v\|_{L^p(\Omega_{n_0})}\\
&\leq \left(\|\psi(x,u(x),\nabla u(x))\|_{L^{p'}(\Omega_{n_0})} + c_a\|u\|_{L^{p'}(\Omega_{n_0})}\right)\|v\|_{L^p(\Omega_{n_0})}\\
&\leq c_1\left(\|\psi(x,u(x),\nabla u(x))\|_{L^{p'}(\Omega_{n_0})} + c_a\|u\|_{L^{p'}(\Omega_{n_0})}\right)\|\nabla v\|_{L^p(\Omega_{n_0})}\\
&\leq c_2\|v\|_{W_0^{1,p}(\Omega_{n_0})}
\end{aligned}$$

for some $c_1, c_2 \geq 0$. Hence, we get that $A_n(x,\cdot)$ are bounded linear functionals on ${X_n}^*$. Since each $A_n(x,\cdot)$ is a bounded linear functional on ${X_n}^*$, the operators $T|_{X_n}$ are well defined for all $n \in \mathbb{N}$.

Claim 3. The operators $T|_{X_n}$ are monotone and hemicontinuous for all $n \in \mathbb{N}$.

This is clear from the fact that $T|_{X_n}$ are the sum of a Leray-Lions operator and a linear operator, i.e., each one is monotone and hemicontinuous.

Claim 4. There exists a solution $u_n \in X_n$ for each problem (P_{Ω_n}).

Since any hemicontinuous operator is radially continuous and $T|_{X_n}$ are monotone and coercive operators, $T|_{X_n}$ satisfy the conditions of the Browder-Minty theorem, so

$$\exists u_n \in X_n \quad \text{s.t.} \quad \langle T(u_n), v\rangle_{X_n^*\times X_n} = \langle f^*, v\rangle_{X_n^*\times X_n} \quad \text{for all } v \in X_n. \tag{3.1}$$

Recall that for any $v \in V \doteq \cup_{n=1}^\infty X_n$, there exists $\bar{n} \in \mathbb{N}$ such that $v \in X_{\bar{n}}$. By the definition of X_n, we know that $(X_n)_{n\in\mathbb{N}}$ is an upwards direct family of closed subspaces of X and hence for any $n \geq \bar{n}$, $v \in X_n$. Hence, by (3.1), we consecutively have

$$A(u_n, v) \to \langle f^*, v\rangle \quad \text{for all } n \geq \bar{n}, \tag{3.2}$$

$$A(u_n, v) \to \langle f^*, v\rangle \quad \text{for all } v \in V, \tag{3.3}$$

$$A(u_n, w) \to \langle f^*, w\rangle \quad \text{for all } w \in X, \tag{3.4}$$

since V is dense in X and $\int_\Omega avw\,dx = \int_{\Omega_n} avw\,dx$ for all $v \in X_n$ and $w \in X$.

Claim 5. The sequence of solutions u_n of (P_{Ω_n}) converges weakly in X, i.e., there exists $u \in X$ such $u_n \rightharpoonup u$.

From equation (3.4), setting $v = u_n$, we have $\langle T(u_n), u_n\rangle = \langle f^*, u_n\rangle$, which together with the coercivity of operator $T|_{X_n}$ gives that the sequence $(u_n)_{n\in\mathbb{N}}$ is bounded. If not, suppose that $\|u_n\| \to \infty$ then

$$\lim_{\|u_n\|\to\infty} \frac{\langle T(u_n), u_n\rangle}{\|u_n\|} \leq \lim_{\|u_n\|\to\infty} \frac{\|f^*\|\|u_n\|}{\|u_n\|} = \|f^*\| < \infty,$$

which is a contradiction with the operator to be coercive. Hence, since $X \equiv W_0^{1,p}(\Omega)$ is a separable reflexive Banach space, using Alaoglu's lemma we have that $(u_n)_{n\in\mathbb{N}}$ bounded implies $u_n \rightharpoonup u \in W_0^{1,p}(\Omega)$.

Claim 6. The sequence $(u_n)_{n\in\mathbb{N}}$ converges strongly in X.

By the weak convergence of the sequence $(u_n)_{n\in\mathbb{N}}$ to $u \in X$, we have $\langle f^*, u_n\rangle \to \langle f^*, u\rangle$. So, using (3.3) with $v = u_n$, we have

$$A(u_n, u_n) \to \langle f^*, u_n\rangle \to \langle f^*, u\rangle. \tag{3.5}$$

Using the compactness of the embedding $X \equiv W_0^{1,p}(\Omega)$ in $L^{p'}(\Omega)$, we conclude the strong convergence of $u_n \to u$ in X.

Alternatively the strong convergence of the sequence $(u_n)_{n\in\mathbb{N}} \subset X$ is ensured by the fact that the Leray-Lions operator Ψ is a (S_+)-type operator, $u_n \rightharpoonup u$ and, by (3.4), we have

$$\limsup_{n\to\infty} \langle T(u_n), u_n - u\rangle = \langle f^*, u_n - u\rangle = 0. \tag{3.6}$$

Note also that, since the Leray-Lions operator Ψ is a pseudo-monotone operator and (3.6), then $\Psi(u_n) \rightharpoonup \Psi(u)$ when $u_n \rightharpoonup u$.

Claim 7. The map $A : X \times V \to \mathbb{R}$ defined by $A(u, v) \doteq \langle T(u), v\rangle_{X^*\times X}$ is of type M in respect to V.

Let $(v_\lambda)_{\lambda\in\Lambda} \subset V$, $w \in X$ and $v^* \in V^*$. Assume the conditions (a)–(c) of Definition 2.3, then

$$A(v_\lambda, v) = \langle \Psi(v_\lambda), v\rangle + \langle a\, v, v_\lambda\rangle \to \langle \Psi(w), v\rangle + \langle a\, v, w\rangle = A(w, v)$$

since $v_\lambda \rightharpoonup w$ and $\Psi(v_\lambda) \rightharpoonup \Psi(w)$, which merging with (*b*) gives $A(w, v) = \langle v^*, v\rangle$ for all $v \in V$.

Claim 8. If $a \in L^\infty(\Omega; \mathbb{R}_0^+)$ then the solution u is unique and $\int_\Omega fu\,dx \geq 0$.

Let $J : W_0^{1,p}(\Omega) \to \mathbb{R}$ be defined by $J(u) \doteq \Psi(u) + a(x)u - f(x)$. Suppose $u_1, u_2 \in W_0^{1,p}(\Omega)$ are two solutions of problem P_Ω. Thus $\langle J(u_1), v\rangle = \langle J(u_2), v\rangle = 0$ for all $v \in W_0^{1,p}(\Omega)$. In particular, we have

$$\begin{aligned} &\langle J(u_2) - J(u_1), u_2 - u_1\rangle = 0 \\ \Leftrightarrow\quad &\langle \Psi(u_2) - \Psi(u_1), u_2 - u_1\rangle + \langle au_2 - au_1, u_2 - u_1\rangle = 0, \end{aligned}$$

and Ψ is a monotone operator, i.e., $\langle \Psi(u_2) - \Psi(u_1), u_2 - u_1\rangle \geq 0$, which implies $\langle a\,(u_2 - u_1), u_2 - u_1\rangle \leq 0$, but $a(x) \geq 0$ and $a \not\equiv 0$, hence $u_2 = u_1$. Moreover, by the ellipticity condition, if u is the solution of problem P_Ω we have

$$\begin{aligned} \langle J(u), u\rangle = 0 \quad \Leftrightarrow \quad &\int_\Omega fu - au^2\,dx = \int_\Omega \psi(x, u, \nabla u)\nabla u\,dx \\ \Rightarrow \quad &\int_\Omega fu - au^2\,dx \geq \alpha \int_\Omega |\nabla u|^p\,dx \geq 0. \end{aligned}$$

So $a \geq 0$ implies $\int_\Omega fu\,dx \geq 0$.

Proof of Theorem 1.1. The existence of a solution $u \in X$ is a direct consequence of Claim 5, (3.3), (3.5), and Claim 7. Claim 8 proves the last statement of the theorem and the *a priori* estimate will be shown in the next subsection.

Proof of Lemma 1.2. The statements are precisely Claims 4 and 6.

3.2. Estimate for the solution

In this subsection we will study an estimate for the solution of the problem (P_Ω). Let u be a solution of the problem (P_Ω) and set

$$\varphi_n \doteq \frac{1}{pm+1}\,|\mathcal{T}_k(u)|^{pm}\,\mathcal{T}_k(u)\chi_{\Omega_n}, \quad \text{for any } n \in \mathbb{N} \quad \text{and some} \quad m \in \mathbb{N},$$

where $\chi_\mathcal{S}$ is the characteristic function of the subset $\mathcal{S} \subset \mathbb{R}^N$, i.e.,

$$\chi_\mathcal{S} = \begin{cases} 1, & x \in \mathcal{S}. \\ 0, & x \in \mathbb{R}^N \backslash \mathcal{S}. \end{cases}$$

By the definition of $\mathcal{T}_k(u)$ and $u \in W_0^{1,p}(\Omega)$, we get that $\varphi_n \in V \doteq \bigcup_{n=1}^{\infty} X_n$ and

$$\nabla\varphi_n = |\mathcal{T}_k(u)|^{pm}\,\nabla\left(\mathcal{T}_k(u)\right)\chi_{\Omega_n} \quad \text{a.e. in } \Omega.$$

It follows from (1.1) that

$$\begin{aligned} &\int_{\Omega_n} \left(\psi(x, u(x), \nabla u(x)\right)|\mathcal{T}_k(u)|^{pm}\,\nabla\left(\mathcal{T}_k(u)\right)dx \\ &\quad + \frac{1}{pm+1}\int_{\Omega_n} a(x)u|\mathcal{T}_k(u)|^{pm}\mathcal{T}_k(u)\chi_{\Omega_n}\,dx \\ &= \frac{1}{pm+1}\int_{\Omega_n} f(x)\,|\mathcal{T}_k(u)|^{pm}\,\mathcal{T}_k(u)dx. \end{aligned} \tag{3.7}$$

We denote the first, second and last terms in (3.7), respectively, by I_1, I_2 and I_3. By the definition of $\mathcal{T}_k(u)$ and the ellipticity condition, we have

$$I_1 = \int_{\Omega_n} \psi(x,u(x),\nabla \mathcal{T}_k(u)) \left|\mathcal{T}_k(u)\right|^{pm} \nabla \mathcal{T}_k(u)dx \geq \alpha \int_\Omega \left|\nabla \mathcal{T}_k(u)\right|^p \left|\mathcal{T}_k(u)\right|^{pm} dx.$$

Note that

$$\left|\nabla\left(\mathcal{T}_k(u)\right)\right|^p \left|\mathcal{T}_k(u)\right|^{pm} = \left|\nabla\left(\frac{\left|\mathcal{T}_k(u)\right|^m \mathcal{T}_k(u)}{m+1}\right)\right|^p,$$

and then

$$I_1 \geq \alpha \int_{\Omega_n} \left|\nabla\left(\frac{\left|\mathcal{T}_k(u)\right|^m \mathcal{T}_k(u)}{m+1}\right)\right|^p dx = \alpha \left\|\frac{\left|\mathcal{T}_k(u)\right|^m \mathcal{T}_k(u)}{m+1}\right\|^p_{W_0^{1,p}(\Omega_n)}.$$

By the embedding of Sobolev spaces into Lorentz spaces (see Lemma 2.7), we obtain

$$\left\|\frac{\left|\mathcal{T}_k(u)\right|^m \mathcal{T}_k(u)}{m+1}\right\|^p_{L^{p^*,p}(\Omega_n)} \leq C \left\|\frac{\left|\mathcal{T}_k(u)\right|^m \mathcal{T}_k(u)}{m+1}\right\|^p_{W_0^{1,p}(\Omega_n)}.$$

Thus

$$I_1 \geq \widetilde{\alpha} \left\|\frac{\left|\mathcal{T}_k(u)\right|^m \mathcal{T}_k(u)}{m+1}\right\|^p_{L^{p^*,p}(\Omega_n)} = \frac{\widetilde{\alpha}}{(m+1)^p} \left\|\mathcal{T}_k(u)\right\|^{p(m+1)}_{L^{p^*(m+1),p(m+1)}(\Omega_n)}. \tag{3.8}$$

It follows from Remark 2.8 that

$$I_2 = \frac{1}{pm+1}\int_{\Omega_n} a(x)u\left|\mathcal{T}_k(u)\right|^{pm} \mathcal{T}_k(u)dx \geq 0.$$

Using (d) and (e) of Lemma 2.6, we have

$$\begin{aligned} I_3 &\leq \frac{1}{pm+1}\int_{\Omega_n} |f| \left|\mathcal{T}_k(u)\right|^{pm+1} dx \\ &\leq \frac{1}{pm+1}\|f\|_{L^{q,q_1}(\Omega_n)} \left\|\left|\mathcal{T}_k(u)\right|^{pm+1}\right\|_{L^{q',q_1'}(\Omega_n)} \\ &= \frac{1}{pm+1}\|f\|_{L^{q,q_1}(\Omega_n)} \left\|\mathcal{T}_k(u)\right\|_{L^{q'(pm+1),q_1'(pm+1)}(\Omega_n)}^{pm+1}. \end{aligned} \tag{3.9}$$

Combining (3.7)–(3.9) we arrive at

$$\frac{\widetilde{\alpha}}{(m+1)^p} \left\|\mathcal{T}_k(u)\right\|^{p(m+1)}_{L^{p^*(m+1),p(m+1)}(\Omega_n)} \leq \frac{1}{pm+1}\|f\|_{L^{q,q_1}(\Omega_n)} \left\|\mathcal{T}_k(u)\right\|^{pm+1}_{L^{r,s}(\Omega_n)}. \tag{3.10}$$

Now we choose the exponents such that

(i) $r \doteq p^*(m+1) = q'(pm+1)$,
(ii) $s \doteq p(m+1) = q_1'(pm+1)$.

Since that $p^* = \frac{pN}{N-p}$ and $q' = \frac{q}{q-1}$, from (i) we obtain

$$m = [Np(q-1) - q(N-p)]\sigma p^{-1}$$

where $\sigma = (N - pq)^{-1}$. Replacing the value of m, we obtain

$$r = \sigma N(q-1)q \quad \text{and} \quad s = \sigma(N-p)(p-1)q.$$

From (ii), we have

$$q_1 = \sigma(N-p)q.$$

Therefore, from (3.10), we get

$$\frac{\widetilde{\alpha}}{(m+1)^p} \|\mathcal{T}_k(u)\|^{p(m+1)}_{L^{r,s}(\Omega_n)} \le \frac{1}{pm+1} \|f\|_{L^{q,q_1}(\Omega_n)} \|\mathcal{T}_k(u)\|^{pm+1}_{L^{r,s}(\Omega_n)},$$

then

$$\|\mathcal{T}_k(u)\|^{p-1}_{L^{r,s}(\Omega_n)} \le C \|f\|_{L^{q,q_1}(\Omega_n)}.$$

By Remark 2.8, $u \in L^{r,s}(\Omega_n)$ and

$$\|u\|_{L^{r,s}(\Omega_n)} \le C \|f\|^{p'/p}_{L^{q,q_1}(\Omega_n)}. \tag{3.11}$$

For fixed $s \ge 0$, by monotone convergence properties of measures we obtain that

$$d^u_{\Omega_n}(s) \text{ increasingly converges to } d^u_{\Omega}(s) \text{ as } n \to \infty.$$

Therefore

$$u^{**}_{\Omega_n}(s) \text{ increasingly converges to } u^{**}_{\Omega}(s) \text{ as } n \to \infty.$$

Thus it follows from the Levi theorem that

$$\begin{aligned}
\lim_{n\to\infty} \|u\|_{L^{r,s}(\Omega_n)} &= \lim_{n\to\infty} \left(\int_0^\infty \left(s^{1/p} u^{**}_{\Omega_n}(s) \right)^q \frac{1}{s} ds \right)^{1/q} \\
&= \left(\int_0^\infty \left(s^{1/p} \lim_{n\to\infty} u^{**}_{\Omega_n}(s) \right)^q \frac{1}{s} ds \right)^{1/q} \\
&= \left(\int_0^\infty \left(s^{1/p} u^{**}_{\Omega}(s) \right)^q \frac{1}{s} ds \right)^{1/q} \\
&= \|u\|_{L^{r,s}(\Omega)}.
\end{aligned} \tag{3.12}$$

From (3.11) and (3.12) we get that

$$\|u\|_{L^{r,s}(\Omega)} \le C \|f\|^{p'/p}_{L^{q,q_1}(\Omega)}.$$

References

[1] R.P. Agarwal, H. Lü, and D. O'Regan, *Existence theorems for the one-dimensional singular p-Laplacian equation with sign changing nonlinearities.* Appl. Math. Comput. **143** (2003), 15–38.

[2] R.P. Agarwal and D. O'Regan, *Singular initial and boundary value problems with sign changing nonlinearities.* IMA J. Appl. Math. **65** (2000), 173–198.

[3] L.H. An, P.X. Du, D.M. Duc, and P.V. Tuoc, *Lagrange multipliers for functions derivable along directions in a linear subspace.* Proc. of AMS **133** (2005), 595–604.

[4] J. Chen, N. Papageorgiou, and E.M. Rocha, *Twin positive solutions for singular nonlinear elliptic equations.* Topol. Meth. in Nonl. Anal. **35**:1 (2010), 187–201.

[5] J. Chen and E.M. Rocha, *Four solutions of an inhomogeneous elliptic equation with critical exponent and singular term.* Nonl. Anal. TMA **71** (2009), 4739–4750.

[6] J. Chen and E.M. Rocha, *Positive solutions for elliptic problems with critical nonlinearity and combined singularity.* Mathematica Bohemica **135**:4 (2010), 413–422.

[7] M.M. Coclite and G. Palmieri, *On a singular nonlinear Dirichlet problem.* Comm. Partial Differential Equations **14** (1989), 1315–1327.

[8] M. Crandall, P. Rabinowitz, and L. Tártar, *On a Dirichlet problem with a singular nonlinearity.* Comm. Partial Differential Equations **2** (1977), 193–222.

[9] J.I. Diaz, J.-M. Morel, and L. Oswald, *An elliptic equation with singular nonlinearity.* Comm. Partial Differential Equations **12** (1987), 1333–1344.

[10] D. Drivaliaris and N. Yannakakis, *Generalizations of the Lax-Milgram Theorem.* Boundary Value Problems, (2007), 1–9 (doi:10.1155/2007/87104).

[11] H. Gajewski, K. Greger, K. Zacharias, *Nichtlineare Operator Gleichungen und Operator Differential Gleichungen.* Akademie-Verlag, Berlin, 1974.

[12] D. Giachetti and S. Segura de Leon, *Problemas Estacionarios con un Termino Dependiente Cuadraticamente del Gradiente y Conteniendo Singularidades.* XXI Congreso de Ecuaciones Diferenciales y Aplicaciones (2009), 1–6.

[13] R.A. Hunt, *On* $L(p,q)$ *spaces.* Enseign. Math. **2** (1966), 249–276.

[14] G. Karch and T. Ricciardi, *Note on Lorentz Spaces and Differentiability of Weak Solutions to the Elliptic Equations.* Bull. Polish Acad. Sci. **45** (1997), 1–5.

[15] A.V. Lair and A.W. Shaker, *Classical and weak solutions of a singular semilinear elliptic problem.* J. Math. Anal. Appl. **211** (1997), 371–385.

[16] J. Leray and J.L. Lions, *Quelques Résultats de Visik sur les Problèmes Elliptiques non Linéaires par les Méthodes de Minty-Browder.* Bull. Belg. Math. Soc. **93** (1965), 97–107.

[17] G.G. Lorentz, *Some new functional spaces.* Bull. Belg. Math. Soc. **2** (1950), 37–55.

[18] P. De Napoli and M.C. Mariani, *Solutions to equation of p-Laplacian type in Lorentz spaces.* Bull. Belg. Math. Soc. **8** (2001), 469–477.

[19] K. Perera and E.A.B. Silva, *Existence and multiplicity of positive solutions for singular quasilinear problems.* J. Math. Anal. Appl. **323** (2006), 1238–1252.

[20] K. Perera and Z. Zhang, *Multiple positive solutions of singular p-Laplacian problems by variational methods.* Boundary Value Problems **2005**:3 (2005), 377–382.

[21] T. Roubick, *Nonlinear Partial Differential Equations with Applications.* Birkhäuser, 2005.

[22] A.W. Shaker, *On singular semilinear elliptic equations.* J. Math. Anal. Appl. **173** (1993), 222–228.

[23] J. Shi and M. Yao, *On a singular nonlinear semilinear elliptic problem.* Proc. Royal Soc. Edinburgh Sect. A **128** (1998), 1389–1401.

[24] G. Stampacchia, *Le problème de Dirichlet pour les Equations Elliptiques du Second Ordre a Coefficients Discontinus – Ann. Inst. Fourier, Grenoble.* Bull. Belg. Math. Soc. **15** (1965), 189–258.

[25] C. Stuart, *Existence and approximation of solutions of nonlinear elliptic equations.* Math. Z. **147** (1976), 53–63.

[26] Y. Sun, S. Wu, and Y. Long, *Combined effects of singular and superlinear nonlinearities in some singular boundary value problems.* J. Differential Equations **176** (2001), 511–531.

[27] G. Talenti, *Inequalities in rearrangement-invariant function spaces.* G. Talenti: Inequalities in rearrangement invariant function spaces, in: Nonlinear Analysis, Function Spaces and Applications, Vol. 5, Proceedings of the Spring School held in Prague, May 1994, eds. M. Krbec et al., Prometheus Publishers, (1995), 177–230.

[28] Z. Zhang, *Critical points and positive solutions of singular elliptic boundary value problems.* J. Math. Anal. Appl. **302** (2005), 476–483.

[29] E. Zeidler, *Nonlinear functional analysis and its applications. II/B.* Nonlinear monotone operators, translated from the German by the author and Leo F. Boron. Springer-Verlag, New York, 1990.

L. Huang, K. Murillo and E.M. Rocha
Center for Research and Development
in Mathematics and Applications
Department of Mathematics
University of Aveiro
Campus de Santiago
Aveiro, 3810-193, Portugal
e-mail: `lirong@ua.pt`
`kellymurillo@ua.pt`
`eugenio@ua.pt`

Operator Theory:
Advances and Applications, Vol. 229, 215–231

Growth of Schrödingerian Subharmonic Functions Admitting Certain Lower Bounds

Alexander I. Kheyfits

Abstract. Matsaev's theorem on the growth of entire functions admitting some lower bounds is extended to subsolutions of the stationary Schrödinger equation

Mathematics Subject Classification (2010). Primary 31C05, 35J99; Secondary 30D15.

Keywords. Subharmonic functions; stationary Schrödinger operator; lower and upper bounds.

1. Introduction and statement of results

About a half-century ago V.I. Matsaev proved the following result.

Theorem M [14]. *If an entire function $f(z)$, $z = re^{\imath\theta}$, admits a lower bound*

$$\log |f(z)| \geq -C \frac{r^\rho}{|\sin \theta|^s} \tag{1.1}$$

with some constants $C > 0$, $\rho > 1$, and $s \geq 0$, then f is of order ρ and finite type.

Throughout, C stands for various insignificant constants, independent upon the arguments, whose values can change from occurrence to occurrence.

The proof of Theorem **M** is based on another Matsaev's result of independent interest, which asserts that a certain upper bound of a subharmonic function can be significantly improved.

The author is thankful to the referees for very careful reading of the manuscript and for pointing out Carleman's note [2]. The author is also indebted to Professors P. Agranovich and A. Rashkovskii for help and useful remarks and to The Center for Research and Development in Mathematics and Applications (University of Aveiro) for their friendliness and hospitality at IDOTA 2011. This work was partially supported by Bronx Community College/CUNY.

Theorem M1 [14], [11, p. 212, Theor. 3]. *Let $u(z)$ be a subharmonic function in the complex plane, which satisfies the estimate*

$$u(z) \leq C\frac{1+r^\rho}{|\sin\theta|^s}, \; z = re^{\imath\theta}, \; \rho > 1, \; s \geq 0.$$

Then $u(z)$ is of order ρ and finite type.

Remark 1.1. This result can be viewed as a far-reaching generalization of the famous Liouville theorem on the bounded entire functions, see, e.g., Carleman's paper [2].

The inequalities like (1.1) are crucial in many problems, since they are intrinsically connected with the estimates of the Cauchy type integrals. Specifically, Theorem **M** has found important applications in operator theory [4, 15], thus it is of interest to extend these results onto more general classes of functions. It can be easily verified though, that Theorem **M1** fails for functions analytic in a half-plane – see for instance examples in [9], therefore, some additional conditions are needed. Indeed, Govorov and Zhuravleva proved that Theorem **M1** can be extended to analytic functions in a half-plane, which are continuous up to, and satisfy a similar upper bound at the boundary.

Theorem GZ [6]. *Let a function $f(z)$, analytic in the open half-plane $\Im z > 0$ and continuous in $\Im z \geq 0$, satisfy*

$$\log|f(re^{\imath\theta})| \leq C\frac{r^\rho}{(\sin\theta)^s}, \quad 0 < \theta < \pi, \quad r > r_0 > 0,$$

and $\log|f(t)| \leq C|t|^\rho$ for real $t \in (-\infty, -r_0) \cup (r_0, \infty)$, where $s \geq 0$, $\rho > 1$, and $r_0 > 0$ are various constants. Then asymptotically

$$\max_{0\leq\theta\leq\pi} \log|f(re^{\imath\theta})| \leq Cr^\rho.$$

Sergienko extended Matsaev's theorem to meromorphic functions and the differences of subharmonic functions in the complex plane or half-plane and noticed, that in this case it was necessary to impose certain restrictions on the distribution of the Riesz masses of the functions involved – see [18, 19, 20] or the survey by I.V. Ostrovskii [16]. Let us remark that an entire function f in Theorem **M** can obviously have only real zeros, while a subharmonic function u in Theorem **M1** can have its Riesz associated masses distributed everywhere in the plane. Some interesting relevant results have also been proven by Yoshida [22].

Assumptions (1.1) and alike in the papers cited above are stated in the uniform norm. Rashkovskii has recently proved a version of the Matsaev theorem for subharmonic functions, where assumptions are imposed on the integral norm (1.2) of the negative part u^- of the function. Hereafter as usual, $a^+ = \max\{a; 0\}$ and $a = a^+ - a^-$.

Theorem R [17]. *Let a function u, subharmonic in the complex plane $\mathbb{C}$ and harmonic in $\mathbb{C} \setminus \mathbb{R}$, satisfy the inequality*

$$\int_{-\pi}^{\pi} u^-(re^{\imath\theta})\Phi(|\sin\theta|)d\theta \leq V(r), \ \forall r \geq r_0, \tag{1.2}$$

where the function $r^{-1-\delta}V(r)$ is increasing for some $\delta > 0$ and *Φ is a nonnegative nondecreasing function such that the integral $\int_0^1 \log^- \Phi(s)ds$ is convergent. Then there are constants $C > 0$ and $A \geq 1$, independent of u, such that*

$$u(re^{\imath\theta}) \leq CV(Ar), \ \forall r \geq r_1 = r_1(u).$$

In this paper, inspired by [17], we extend the theorems cited above onto functions subharmonic with respect to the stationary Schrödinger operator L_c, that is, onto weak solutions of the inequality

$$-L_c u \equiv \Delta u - c(x)u \geq 0,$$

Δ being the Laplacian, in the Euclidean space $\mathbb{R}^n$, $n \geq 2$, and in the half-space

$$\mathbb{R}^n_+ = \left\{x = (x_1, \ldots, x_{n-1}, x_n) \mid x_n > 0\right\}.$$

These subsolutions are called hereafter *c-subharmonic functions* or *subfunctions* for short. Solutions of the equation $\Delta u - c(x)u = 0$ are called *c-harmonic functions*. We denote the class of c-subfunctions in a domain Ω by $SbH_c(\Omega)$ and the class of c-harmonic functions by $H_c(\Omega)$. Differences of the subfunctions will be considered elsewhere.

Potential theory for the operator L_c was developed for the Kato class potentials by Cranston, Fabes, Zhao [3] and independently by Bramanti – see his survey [1]. Under our assumptions the potential theory for L_c was developed, also independently, by B.Ya. Levin and the present author, [12, 13]. We formulate here some necessary properties of the subfunctions; the reader can find a detailed exposition in [13].

In what follows we assume that the potential $c(x)$ is a nonnegative function such that $c \in L^p_{\rm loc}(\Omega)$ with $p > n$. Under our assumptions, the operator L_c can be extended in the standard way from the space $C_0^\infty(\Omega)$ to an essentially self-adjoint operator in $L^2(\Omega)$; we denote the extended operator L_c as well. The c-Green function $G(x, y)$ with the Dirichlet boundary conditions for the latter operator exists simultaneously with the harmonic Green function $g(x, y)$. The Green function vanishes almost everywhere at the boundary $\partial\Omega$, is positive in Ω, and possesses all the analytic properties, necessary in the subsequent proofs, see [13]. We normalize G as the harmonic Green function g:

$$G(x, y) \sim \gamma_n \psi_n(|x - y|) \quad \text{as} \quad |x - y| \to 0,$$

where $\gamma_2 = 1/(2\pi)$, $\psi_2(r) = \ln(1/r)$, $\gamma_n = 1/[(n-2)\sigma_n]$, $\psi_n = r^{2-n}$ for $n \geq 3$, and σ_n is the area of the unit sphere S in $\mathbb{R}^n$. The inner normal derivative

$$P(x, y) \equiv \frac{\partial G(x, y)}{\partial n(y)}, \ x \in \Omega, \ y \in \partial\Omega,$$

that is, the *c-Poisson kernel* for the domain Ω, is nonnegative. If the potential is radial, the c-Poisson kernel for a half-ball is explicitly computed in Section 2.

Let a c-subfunction u have a c-harmonic majorant in a domain Ω. Then similarly to the harmonic case $c = 0$, u has the nonnegative Riesz associated measure μ and the following Riesz decomposition is valid:

$$u(x) = h(x) - \gamma_n^{-1} \int_\Omega G_\Omega(x,y) d\mu(y),\ x \in \Omega,$$

where h is the least harmonic majorant of u and G_Ω is the Green function of L_c in Ω. The Riesz-Herglotz representation is also valid for subfunctions [10], and

$$h(x) = \int_{\partial\Omega} \left.\frac{\partial G_\Omega(x,y)}{\partial n(y)}\right|_{y\in\partial\Omega} d\nu(y), \tag{1.3}$$

where ν is a finite Borel measure on $\partial\Omega$. If $u \in SbH_c(\overline{\Omega})$ and the surface $\partial\Omega$ is sufficiently smooth (for example, if $\partial\Omega$ is a piecewise Lyapunov surface as in the case under consideration) then

$$\nu(e) = \int_e u(y) d\sigma(y),\ e \subset \partial\Omega,$$

here $d\sigma$ is the surface area measure on $\partial\Omega$.

To state our results, we need more notation. Let $B(x,r)$ be an open ball of radius r centered at a point $x \in \mathbb{R}^n$, $B(r) = B(0,r)$, $S(x,r) = \partial B(x,r)$, $S(r) = S(0,r)$, $S = S(1)$, $B_+(r) = B(r) \bigcap \mathbb{R}^n_+$, $S_+(r) = S(r) \bigcap \mathbb{R}^n_+$, $S_+ = S_+(1)$. Introduce in $\mathbb{R}^n$ spherical coordinates by

$$x = (x_1, \dots, x_n) = (r, \theta),\ \theta = (\theta_1, \theta_2, \dots, \theta_{n-1}),$$

such that $\cos\theta_1 = x_n/r$ for $0 \le \theta_1 \le \pi$. Let Δ^* be the Laplace-Beltrami operator on the unit sphere $S \subset \mathbb{R}^n$. The eigenvalue problem on S,

$$\Delta^* \varphi(\theta) + \lambda \varphi(\theta) = 0 \tag{1.4}$$

has the eigenvalues $\lambda_j = j(j+n-2)$, $j = 0, 1, 2, \dots$. Corresponding eigenfunctions are the classical spherical functions $\varphi_{j,\nu}$, $1 \le \nu \le \nu_j$, where $\nu_j = (n+2j-2)\frac{(n+j-3)!}{j!(n-2)!}$ are the multiplicities of λ_j, $\nu_0 = 1$.

Problem (1.4) on the half-sphere S_+ has the same strictly positive eigenvalues $\lambda_j = j(j+n-2)$, $j = 1, 2, \dots$, however with the multiplicities $\nu_j^+ = \frac{(n+j-3)!}{(j-1)!(n-2)!}$. On the half-sphere, the eigenfunctions of the latter problem are the spherical functions $\varphi_{j,\nu}$, which are even with respect to x_n, that is,

$$\varphi_{j,\nu}(\theta_1, \theta_2, \dots, \theta_{n-1}) = \varphi_{j,\nu}(\pi - \theta_1, \theta_2, \dots, \theta_{n-1}).$$

We usually suppress the subscript 0 or 1 and denote the principal eigenfunction by φ; it is rotationally symmetric, that is, $\varphi(\theta) = \varphi(\theta_1)$. For the entire unit sphere S, $\lambda_0 = 0$ and $\varphi_0(\theta) = \varphi(\theta) = const$; for the half-sphere S_+ the principal eigenvalue is $\lambda_1 = n-1$ and the principal eigenfunction is $\varphi(\theta_1) = \cos\theta_1$, $0 \le \theta_1 \le \pi/2$. Hence in our results the $\sin\theta$, which appears in (1.1)–(1.2), is replaced by $\cos\theta_1$.

Let $q(r)$ be any nonnegative continuous minorant of the potential c, $0 \leq q(r) \leq c(x)$, where $r = |x|$ is the Euclidean norm of x. Solutions of the ordinary differential equation

$$-y'' - (n-1)r^{-1}y' + \{\lambda r^{-2} + q(r)\}y(r) = 0,\ 0 < r < \infty, \tag{1.5}$$

play an essential role in our study. It is known (see, for example, [7]) that if $\lambda \geq 0$ and the potential $q(r)$ satisfies our assumptions, then equation (1.5) has a fundamental system of positive solutions $\{V,\ W\}$ such that $V(r)$ is non-decreasing,

$$0 \leq V(0^+) \leq V(r) \text{ as } 0 < r \to +\infty$$

and $W(r)$ is monotonically decreasing,

$$+\infty = W(0^+) > W(r) \searrow 0 \text{ as } 0 < r \to +\infty.$$

We always norm them by $V(1) = W(1) = 1$. To study the asymptotic behavior of the subfunctions, we suppose that the radial potential $q(r)$ has the following asymptotic properties.

There exists the finite limit

$$\lim_{r\to+\infty} r^2 q(r) = k < \infty$$

and moreover, the integral

$$\int^{\infty} |r^2 q(r) - k| \frac{dr}{r} < \infty$$

is convergent.

The class of potentials $c(x)$ in a domain Ω satisfying all the above conditions, is denoted hereafter by $\mathcal{C} = \mathcal{C}(\Omega)$ and the class of the radial potentials by $\mathcal{C}_r = \mathcal{C}_r(\Omega)$. For $q \in \mathcal{C}_r(\mathbb{R}^n_+)$, the asymptotic formulas in [21] imply that if in (1.5) $\lambda = n-1$, then

$$V(r) = V_1(r) = C\, r^{(2-n+\chi_k)\,/\,2}(1 + \bar{\bar{o}}(1)),\ r \to \infty \tag{1.6}$$

and

$$W(r) = W_1(r) = C\, r^{(2-n-\chi_k)\,/\,2}(1 + \bar{\bar{o}}(1)),\ r \to \infty, \tag{1.7}$$

where

$$\chi_k = \sqrt{(n-2)^2 + 4(n-1+k)} = \sqrt{n^2 + 4k}. \tag{1.8}$$

We also set

$$\rho_k \equiv \frac{2 - n + \chi_k}{2} \equiv \frac{2 - n + \sqrt{n^2 + 4k}}{2} \tag{1.9}$$

where χ_k is given by (1.8) and

$$\alpha_k^2 = (n-2)^2 + 4k. \tag{1.10}$$

Remark 1.2. As follows from (1.9), if $k = 0$, then the critical exponent $\rho_k = 1$ not only for $n = 2$ as in Theorem **M**, but for any dimension n of the space.

Remark 1.3. If $k = \infty$ and the potential satisfies mild regularity conditions, then the solutions V and W have the classical *JWKB*-asymptotics [7], and many conclusions of this paper remain valid as well, but we do not consider this case here.

If in (1.5) $\lambda = 0$, then the decreasing solution $W = W_0$ of (1.5) is a fundamental solution of the operator L_c and has the same singularity at $r = 0$ as the fundamental solution of the Laplacian.

Similarly to the case $c = 0$ [11], we say that a subfunction u is of a *growth order* ρ and *normal type* in an open cone, if in this cone asymptotically

$$u(x) \leq C\left(1 + |x|^{\rho}\right).$$

Now we can state our results – analogs of the theorems above for subfunctions in the half-space $\mathbb{R}^n_+$ and in the entire space $\mathbb{R}^n$. Denote by

$$K(r) = \partial B_+(r) \setminus S(r),\ 1 \leq r < \infty,$$

the "horizontal" part of the boundary of the half-ball $B_+(r)$ and set $K(1, r) = K(r) \setminus \overline{K(1)}$. The next result extends a special case of Theorem **R**.

Theorem 1.4. *Let a potential $c \in C(\mathbb{R}^n_+)$, and $u(x)$ be a c-harmonic function in $\mathbb{R}^n_+$, $n \geq 2$, which is continuous up to the boundary $\mathbb{R}^{n-1} = \partial\mathbb{R}^n_+$. Let also $|u(x)| \leq C$ in the unit half-ball $\overline{B_+}$.*

If the negative part of u has an integral estimate

$$\int_{S_+} u^-(r, \theta) \cos\theta_1 d\sigma(\theta) \leq C(1 + r^{\rho}),\ x \in \mathbb{R}^n_+$$

with $\rho > \rho_k$, and its boundary values satisfy

$$V_1(r) \int_{K(1,r)} u^-(y^{\circ}) \frac{W_1(t)}{t} dy^{\circ} \leq C\left(1 + r^{\rho}\right)$$

where $y^{\circ} = (y_1, \ldots, y_{n-1}, 0) \in \partial\mathbb{R}^n_+$, $t = |y^{\circ}|$, then everywhere in $\mathbb{R}^n_+$

$$\max_{\theta \in S} u(x) \leq C\left(1 + |x|^{\rho}\right),$$

that is, u has the growth of at most order ρ and normal type in $\mathbb{R}^n_+$.

Remark 1.5. If the function u is positive, then the integral conditions are clearly valid, thus $\max_{\theta \in S} u(x) \leq C\left(1 + |x|^{\rho}\right)$ for any $\rho \geq \rho_k$, and we see that a positive c-subfunction cannot grow faster than a certain specific rate (namely, r^{ρ_k} in our case) determined, through the principal eigenvalue, by the geometry of the domain, exactly as in the case of classical Riesz subharmonic functions. However, unlike the classical situation, in our case this limiting growth depends also on the potential $c(x)$ of the operator L_c.

Theorem 1.4 straightforwardly implies the next result.

Theorem 1.6. *Let a potential $c \in C(\mathbb{R}^n)$ and $u(x)$ be a c-harmonic function in $\mathbb{R}^n_+ \bigcup \mathbb{R}^n_-$ and continuous in the entire space $\mathbb{R}^n$. Assume that $|u(x)| \leq C$ in the unit ball B. If the negative part u^- satisfies*

$$\int_S u^-(y)|\cos\theta_1| d\sigma(\theta) \leq C(1 + r^{\rho}),\ 1 \leq r < \infty,$$

with $\rho > \rho_k$ and its values at $y^\circ \in K(1,r)$ (due to c-harmonicity, u cannot have the Riesz masses beyond $\partial\mathbb{R}^n_+$) satisfy

$$V_1(r)\int_{K(1,r)} u^-(y^\circ)\frac{W_1(t)}{t}dy^\circ \le C\,(1+r^\rho)\,,\ 1<r<\infty,$$

then everywhere in $\mathbb{R}^n$

$$\max_{\theta\in S} u(x) \le C\,(1+|x|^\rho)\,.$$

The next result is a generalization of Theorem 2 ([11, p. 209]). It is of independent interest and is the major ingredient of our proofs.

Theorem 1.7. *Let u be a q-harmonic function in the half-space $\mathbb{R}^n_+$, q-subharmonic in $\overline{\mathbb{R}^n_+}$ or continuous up to the boundary $\mathbb{R}^{n-1} = \partial\mathbb{R}^n_+$, where q is a radial potential, such that*

$$\int_{S_+} u^+(r,\theta)\cos\theta_1 d\sigma(\theta) \le C\,(1+r^\rho)\,, 1\le r<\infty, \tag{1.11}$$

$$V_1(r)\int_{K(1,r)} u^+(y^\circ)\frac{W_1(t)}{t}dy^\circ \le C(1+r^\rho),\ \ 1\le r<\infty, \tag{1.12}$$

and $|u(x)| \le C$ in $B_+(1)$. If $\rho > \rho_k$, where ρ_k is given by (1.9) and α_k by (1.10), then

$$u(x) \ge -C\frac{1+r^\rho}{(\cos\theta_1)^{(n+\alpha_k)/2}}$$

everywhere in $\mathbb{R}^n_+$.

Corollary 1.8. *Let u be a q-harmonic function of order $\rho > \rho_k$ and finite type in the half-space $\mathbb{R}^n_+$, c-subharmonic in $\overline{\mathbb{R}^n_+}$ or continuous up to the boundary, such that $|u(x)| \le C$ in $B_+(1)$. Then*

$$u(x) \ge -C\frac{1+r^\rho}{(\cos\theta_1)^{(n+\alpha_k)/2}}$$

everywhere in $\mathbb{R}^n_+$.

Carleman's formula (1.16) implies immediately (see Proposition 2.1 in Section 2) that u^+ in (1.11)–(1.12) can be replaced by u^-.

Corollary 1.9. *Let u be a q-harmonic function in the half-space $\mathbb{R}^n_+$ and q-subfunction in the closed half-space $\overline{\mathbb{R}^n_+}$, or continuous up to the boundary $\mathbb{R}^{n-1} = \partial\mathbb{R}^n_+$. If*

$$\int_{S_+} u^-(r,\theta)\cos\theta_1 d\sigma(\theta) \le C\,(1+r^\rho)\,, 1\le r<\infty,$$

where $\rho > \rho_k$,

$$V_1(r)\int_{K(1,r)} u^-(y^\circ)\frac{W_1(t)}{t}dy^\circ \le C(1+r^\rho),\ 1\le r<\infty,$$

and $|u(x)| \le C$ *in* $\overline{B_+(1)}$*, then*

$$u(x) \ge -C\frac{1+r^\rho}{(\cos\theta_1)^{(n+\alpha_k)/2}}$$

everywhere in $\mathbb{R}^n_+$.

Remark 1.10. It should be mentioned that if $\rho > \rho_k$, then $r^\rho/V_1(r) \nearrow \infty$ as $r \to \infty$; this follows from (1.6) and (1.8)–(1.9).

Now we state a generalization of Theorem **GZ**.

Theorem 1.11. *Let* u *be a c-harmonic function in* $\mathbb{R}^n_+$*, c-subharmonic in the closed half-space* $\overline{\mathbb{R}^n_+}$ *or continuous up to the boundary, such that*

$$u(x) \le C\frac{1+r^\rho}{(\cos\theta_1)^l},\ \forall x \in \mathbb{R}^n_+, \tag{1.13}$$

and

$$u(x^\circ) \le C\left(1+|x^\circ|^\rho\right),\ \forall x^\circ = (x_1,\dots,x_{n-1},0) \in \partial\mathbb{R}^n_+, \tag{1.14}$$

where $l \ge 0$ *and* $\rho > \rho_k$*. Then for all* $x \in \mathbb{R}^n_+$*,*

$$u(x) \le C\left(1+r^\rho\right).$$

The next result follows directly from Theorem 1.11.

Theorem 1.12. *Let* u *be c-harmonic in* $\mathbb{R}^n_+ \bigcup \mathbb{R}^n_-$*, c-subharmonic (or continuous) in* $\mathbb{R}^n$*, such that*

$$u(x) \le C\frac{1+r^\rho}{|\cos\theta_1|^l}$$

where $l \ge 0$ *and* $\rho > \rho_k$*. Then*

$$u(x) \le C\left(1+r^\rho\right)\ \textit{for all}\ x \in \mathbb{R}^n.$$

Remark 1.13. The results can be generalized by considering more flexible scales of growth; for example, the classical orders ρ can be replaced by the proximate orders $\rho(r)$. Moreover, instead of half-spaces we can consider any smooth cones.

Proofs of Theorems 1.4, 1.7 and 1.11 are given in Section 2.

In reasoning we mostly follow [11], however, we need some results about the subfunctions. First of all, we state a generalization of the Phragmén-Lindelöf principle for the subsolutions of the operator L_c [13]. The limit growth in this theorem is given by the growing solution, $V_1(r)$, of (1.5) with λ being the principal eigenvalue of problem (1.4). We state it here only for the half-space. In the case of a cone generated by any smooth spherical cap $D \subset S$, $\cos\theta_1$ in (1.15) has to be replaced by the principal eigenfunction φ of the Laplace-Beltrami operator in D, and λ in (1.5) by the corresponding eigenvalue [13].

Theorem LK1 [12], [13, Th. 11.3.1]. *Let* $u \in SbH_c(\mathbb{R}^n_+)$ *and* V_1 *be a growing solution of* (1.5) *with* $\lambda = n-1$. If

$$\limsup_{\mathbb{R}^n_+ \ni x \to x_0} u(x) \le A = \text{const}, \ \forall x_0 \in \partial\mathbb{R}^n_+$$

and

$$\liminf_{r\to\infty} \frac{1}{V_1(r)} \int_{S_+} u^+(r,\theta)\cos\theta_1 d\sigma(\theta) = 0 \tag{1.15}$$

then[1]

$$u(x) < A^+, \ \forall x \in \overline{\mathbb{R}^n_+}.$$

We will also use the following extensions [13] of two well-known integral formulas, namely the Carleman formula and Nevanlinna's representation. Here, we specialize these results for the case of half-balls and half-spaces.

Theorem LK2 [12], [13, Eq-n (11.54)]. *Denote*

$$C^R(r) = W_1(r) - \frac{W_1(R)}{V_1(R)} V_1(r)$$

and consider a function

$$C^R(r)\cos\theta_1, \ 1 \le r \le R, \ 0 \le \theta_1 \le \pi/2,$$

which is q-harmonic in $B_+(R)$. *Then the following generalization of the classical Carleman formula* [11, p. 187] *is valid for a q-subfunction* u *in* $\overline{B_+(R')} \setminus B(1)$, $1 < R < R'$,

$$\begin{aligned}\theta_n \textstyle\int_{B_+(R)} C^R(r)\cos\theta_1 d\mu(x) &= \tfrac{\kappa}{V_1(R)} \textstyle\int_{S_+} u(R,\theta)\cos\theta_1 d\sigma(\theta) \\ &+ \textstyle\int_{K(1,R)} u(r,\theta)\frac{1}{r} d\sigma(r,\theta) + A_u(R).\end{aligned} \tag{1.16}$$

Here μ *is the Riesz associated measure of* u *and*

$$\kappa = W_1(1)V_1'(1) - V_1(1)W_1'(1) = V_1'(1) - W_1'(1) > 0$$

the Wronskian of the basic solutions V_1 *and* W_1 *of* (1.5) *with* $\lambda = n-1$. *The quantity* $A_u(R) = \underline{O}(1)$ *as* $R \to \infty$.

It should be noted that if u is subharmonic or continuous in a closed domain, then Carleman's formula does not contain the terms with a boundary measure – cf. [5, Chap. 1].

[1]The integral in (1.15) is called the Nevanlinna norm of u.

2. Proofs

First we prove the following (almost obvious, in view of Carleman's formula (1.16)) proposition, which connects the growth of the positive and the negative parts of a subfunction in an integral norm.

Proposition 2.1. *Let u be a q-harmonic function in $\mathbb{R}^n_+$, q-subharmonic or continuous up to the boundary. Suppose that*

$$\int_{S_+} u^+(r,\theta)\cos\theta_1 d\sigma(\theta) \le C(1+r^\rho),\ 1\le r<\infty, \tag{2.1}$$

and

$$V_1(r)\int_{K(1,r)} u^+(y^\circ)\frac{W_1(t)}{t}dy^\circ \le C(1+r^\rho),\ 1\le r<\infty,\ t=|y^\circ|. \tag{2.2}$$

Then the negative part u^- of u satisfies similar bounds,

$$\int_{S_+} u^-(r,\theta)\cos\theta_1 d\sigma(\theta) \le C(1+r^\rho),\ 1\le r<\infty \tag{2.3}$$

and

$$V_1(r)\int_{K(1,r)} u^-(y^\circ)\frac{W_1(t)}{t}dy^\circ \le C(1+r^\rho),\ 1\le r<\infty,\ t=|y^\circ|. \tag{2.4}$$

Vice versa, under the same conditions, inequalities (2.3)–(2.4) *imply* (2.1)–(2.2). *Therefore, in either case, the function $u=u^+-u^-$ verifies*

$$\int_{S_+} |u(r,\theta)|\cos\theta_1 d\sigma(\theta) + V_1(r)\int_{K(1,r)} |u(y^\circ)|\frac{W_1(t)}{t}dy^\circ \le C\left(1+r^\rho\right).$$

Proof. Fix an $R>1$ and apply Carleman's formula (1.16) to the function u in the half-ball $\mathbf{B}_+(R)$; we notice that as in the harmonic case $c(r)=0$ (see above) the Riesz associated measure of u is zero and the singular part of the boundary measure (1.3) vanishes as well (cf. [5, Chap. 1]). Separating the positive and negative parts of $u=u^+-u^-$, we have

$$\begin{aligned}\frac{1}{V_1(R)}\int_{S_+} u^-(R,\theta)\cos\theta_1 d\sigma(\theta) &+ \frac{1}{\kappa}\int_{K(1,R)} u^-(y^\circ)\left(W_1(t)-\frac{W_1(R)}{V_1(R)}V_1(t)\right)\frac{dy^\circ}{t}\\ =\frac{1}{V_1(R)}\int_{S_+} u^+(R,\theta)\cos\theta_1 d\sigma(\theta)&\\ +\frac{1}{\kappa}\int_{K(1,R)} u^+(y^\circ)&\left(W_1(t)-\frac{W_1(R)}{V_1(R)}V_1(t)\right)\frac{dy^\circ}{t}+A(R),\end{aligned} \tag{2.5}$$

where $A(R)=O(1)$ as $R\to\infty$ and all integrals are positive.

Now by the assumptions,

$$\int_{S_+} u^+(R,\theta)\cos\theta_1 d\sigma(\theta) \le C\left(1+R^\rho\right)$$

and

$$\int_{K(1,R)} u^+(y^\circ)\left(W_1(t) - \frac{W_1(R)}{V_1(R)}V_1(t)\right)\frac{dy^\circ}{t} \le \int_{K(1,R)} u^+(y^\circ)\frac{W_1(t)}{t}dy^\circ \le C\frac{1+R^\rho}{V_1(R)}.$$

Hence, (2.5) implies the inequalities

$$\int_{S_+} u^-(R,\theta)\cos\theta_1 d\sigma(\theta) \le C(1+R^\rho)$$

and

$$\int_{K(1,R)} u^-(y^\circ)\left(W_1(t) - \frac{W_1(R)}{V_1(R)}V_1(t)\right)\frac{dy^\circ}{t} \le C\frac{1+R^\rho}{V_1(R)}.$$

Next we notice that if R_0 is big enough, then for $R \ge R_0 > 1$ and for all t, $1 \le t \le R$, due to the monotonicity of the solutions V_1 and W_1 the sequel inequality is valid,

$$\frac{1}{2} \le 1 - \frac{V_1(t)}{V_1(2R)}\frac{W_1(2R)}{W_1(t)} \le 1.$$

Therefore,

$$\begin{aligned}\int_{K(1,R)} u^-(y^\circ)\frac{W_1(t)}{t}dy^\circ &\le 2\int_{K(1,R)} u^-(y^\circ)\left(W_1(t) - \frac{W_1(2R)}{V_1(2R)}V_1(t)\right)\frac{\partial y^\circ}{t}\\ &\le 2\int_{K(1,2R)} u^-(y^\circ)\left(1 - \frac{W_1(2R)}{V_1(2R)}\frac{V_1(t)}{W_1(t)}\right)W_1(t)\frac{y^\circ}{t}\\ &\le C\frac{(2R)^\rho}{V_1(2R)} \le C\frac{1+R^\rho}{V_1(R)}.\end{aligned}$$

The conclusion follows since the term $A_u(R)$ is bounded as $R \to \infty$. □

Now we derive explicit formulas for the q-Poisson kernel $P_q(x,y) \equiv P(x,y) = \frac{\partial G_+(x,y)}{\partial n(y)}$ in an n-dimensional half-ball. If $q = 0$, these expressions can be found in [11, Sect. 24.3] for $n = 2$, and in [8, Eq-n (3.1.23)] for any $n \ge 2$.

Proposition 2.2. *Let $q(r) \in C_r(B_+(R))$. Then*

$$P_q(x,y) = \frac{R^2 - |x|^2}{R}\left\{\frac{|W_0'(|x-y|)|}{|x-y|} - \frac{|W_0'(|\overline{x}-y|)|}{|\overline{x}-y|}\right\}$$

for $x \in B_+(R)$ and $y \in S_+(R)$, and

$$P_q(x,y) = 2x_n\left\{\frac{|W_0'(|x-y^\circ|)|}{|x-y^\circ|} - \frac{|W_0'(|R\widetilde{x} - R^{-1}|x|y^\circ|)|}{|R\widetilde{x} - R^{-1}|x|y^\circ|}\right\}$$

for $x = |x|\widetilde{x} \in B_+(R)$, $|\widetilde{x}| = 1$, and $y = y^\circ \in K(R)$.

Proof. The q-Green function in the ball $B(R)$ with zero boundary conditions and the singularity at a point $y \in B(R)$ is readily verified to be

$$G_R(x,y) = W_0(|x-y|) - W_0\left(\frac{|x|}{R}|x' - y|\right)$$

where $x' = R^2|x|^{-2}x$ is the inversion of a point x in the sphere $S(R)$; here W_0, the decreasing solution of equation (1.5) with $\lambda = 0$, is a singular solution of the operator $-\Delta_q$. Now we can straightforwardly construct the Green function $G_+(x,y)$ for L_q in the half-ball $B_+(R)$ by the reflection method as the difference

$$G_+(x,y) = G_R(x,y) - G_R(\overline{x},y),$$

where $\overline{x} = (x_1,\ldots,x_{n-1},-x_n)$. Denote $W_0'(s) = dW_0(s)/ds$. Now differentiation with respect to the inward normal gives the q-Poisson kernel

$$P_q(x,y) = \frac{\partial G_+(x,y)}{\partial n(y)},\ x \in B_+(R),\ y \in \partial B_+(R),$$

for the Dirichlet problem in the half-ball $B_+(R)$ for the operator L_q, namely (we remind that $W'(s) < 0$)

$$P_q(x,y) = \frac{R^2 - |x|^2}{R}\left\{\frac{|W_0'(|x-y|)|}{|x-y|} - \frac{|W_0'(|\overline{x}-y|)|}{|\overline{x}-y|}\right\} \tag{2.6}$$

for $x \in B_+(R)$ and $y = y \in S_+(R)$, and

$$P_q(x,y) = 2x_n\left\{\frac{|W_0'(|x-y^\circ|)|}{|x-y^\circ|} - \frac{|W_0'(|R\widetilde{x} - R^{-1}|x|y^\circ|)|}{|R\widetilde{x} - R^{-1}|x|y^\circ|}\right\} \tag{2.7}$$

for $x \in B_+(R)$ and $y^\circ \in K(R)$. □

Now we prove our main results, starting with Theorem 1.7. In the proofs we use the following generalization of the classical Riesz decomposition theorem.

Theorem K [10]. *Let u be a q-subfunction in $B_+(R')$, $R' > R > 0$, having there a positive q-harmonic majorant. Then for $|x| = r < R$, $x_n > 0$,*

$$\begin{aligned} u(x) = \frac{1}{\sigma_n}\int_{B_+(R)} G(x,y)d\mu(y) &+ \int_{S_+(R)} \frac{\partial G(x,y)}{\partial n(y)} u(y)d\sigma(y) \\ &+ \int_{K(R)} \left.\frac{\partial G(x,y)}{\partial n(y)}\right|_{|y|=t} [u(t)dt + d\nu(t)], \end{aligned}$$

where ν is the measure in (1.3).

Proof of Theorem 1.7. Apply Theorem **K** to the function $-u(x)$, keeping in mind that u is q-harmonic in $\mathbb{R}^n_+$ and continuous up to the boundary, thus both measures μ and ν vanish:

$$\begin{aligned} -u(x) = &\int_{S_+(R)} (-u(y))\,P(x,y)\Big|_{y\in S_+(R)} d\sigma_R(y) \\ &+ \int_{K(1,R)} (-u(y^\circ))\,P(x,y^\circ)dy^\circ + A_u(R), \end{aligned}$$

where $P_q(x,y)$ is given by (2.6) or (2.7), respectively.

From here, since $P_q(x,y) \geq 0$ and $-u = u^- - u^+ \leq u^-$, we have

$$-u(x) \leq \int_{S_+(R)} u^-(y)P(x,y)\Big|_{|y|=R} d\sigma(y) + \int_{K(1,R)} u^-(y^\circ)P(x,y^\circ)dy^\circ + O(1)$$
$$\equiv u_1(x) + u_2(x) + O(1), \ |x| \to \infty. \tag{2.8}$$

We have to estimate the q-Poisson kernel P_q. If $|y| = R$, we apply Lagrange's mean value theorem to the function (Cf. (2.6))

$$l(x_n) = \frac{W_0'(|x-y|)}{|x-y|},$$

and get

$$\big|l(x_n) - l(\overline{x_n})\big| = 2x_n|x_n^* - y_n|\frac{\big||x^*-y|W_0''(|x^*-y|) - W_0'(|x^*-y|)\big|}{|x^*-y|^3},$$

where an intermediate value $x_n^* \in [-x_n, x_n]$ and $x^* = (x_1, \dots, x_{n-1}, x_n^*)$. For the singular solution W_0 of (1.5) we have (here α_k is given by (1.10))

$$W_0(t) \approx Ct^{(2-n-\alpha_k)/2}. \tag{2.9}$$

Asymptotic formula (2.9) can be twice differentiated, so that

$$tW_0''(t) - W_0'(t) \approx Ct^{-(n+\alpha_k)/2},$$

and the latter holds true also in the case $n = 2$ and $k = 0$, when $\alpha_k = 0$ and $W_0'(t) \approx Ct^{-1}$. Therefore,

$$\big|l(x_n) - l(\overline{x_n})\big| \leq Cx_n|x_n^* - y_n||x^* - y|^{-(n+6+\alpha_k)/2}$$

and

$$\begin{aligned} u_1(x) &= \int_{S_+(R)} u^-(R,\theta)P_q(x,y)\Big|_{y\in S_+(R)} d\sigma(\theta) \\ &\leq Cx_n\frac{R^2-r^2}{R}\int_{S_+(R)} u^-(R,\theta)|x_n^*-y_n|\frac{d\sigma(\theta)}{|x^*-y|^{(n+6+\alpha_k)/2}} \\ &\leq Cx_n\frac{R^2-r^2}{R}\int_{S_+(R)} u^-(R,\theta)R\cos\theta_1\frac{d\sigma(\theta)}{|x^*-y|^{(n+6+\alpha_k)/2}}. \end{aligned}$$

Given $x = (r,\phi)$, we fix $R = 2r$, thus $|x^* - y| \geq CR$, and deduce

$$\begin{aligned} u_1(x) &\leq C\frac{\cos\phi_1}{R^{(n+\alpha_k)/2}}\int_{S_+(R)} u^-(R,\theta)\cos\theta_1 d\sigma(\theta) \\ &= C\frac{\cos\phi_1}{R^{(n+\alpha_k)/2-n+1}}\int_{S_+} u^-(R,\theta)\cos\theta_1 d\sigma(\theta). \end{aligned}$$

Since $\frac{n+\alpha_k}{2} - n + 1 \geq 0$, we get the required bound of u_1,

$$u_1(x) \leq C\int_{S_+} u^-(R,\theta)\cos\theta_1 d\sigma(\theta) \leq Cr^\rho. \tag{2.10}$$

Now we take up $u_2(x)$. In this case

$$P_q(x,y)\big|_{y=y^\circ\in K(1,R)} = 2x_n\left\{\frac{\left|W_0'(|x-y^\circ|)\right|}{|x-y^\circ|} - \frac{\left|W_0'(|R\widetilde{x}-\frac{|x|y^\circ}{R}|)\right|}{|R\widetilde{x}-\frac{|x|y^\circ}{R}|}\right\}$$

where $\widetilde{x}=x/|x|$, $|\widetilde{x}|=1$. From here,

$$P_q(x,y^\circ)\le 2x_n\frac{\left|W_0'(|x-y^\circ|)\right|}{|x-y^\circ|}\le Cx_n\frac{\left|W_0'(|y^\circ|\cos\theta_1)\right|}{|y^\circ|\cos\theta_1},$$

since $|x-y^\circ|\ge|y^\circ|\cos\theta_1$. Recalling that $x_n=r\cos\theta_1$, $r=|x|$, $t=|y^\circ|$, and applying the asymptotic formula (1.7), we have

$$P_q(x,y^\circ)\big|_{y^\circ\in K(1,R)}\le C\frac{x_n}{\left(|y^\circ|\cos\theta_1\right)^{(n+\alpha_k)/2+1}} = C\frac{r}{t^{(n+2+\alpha_k)/2}\left(\cos\theta_1\right)^{(n+\alpha_k)/2}}. \tag{2.11}$$

It should be noticed that if $n=2$ and $k=0$, then in (1.5) $\lambda=0$, thus $\alpha_k=0$. Therefore, the right-hand side of (2.11) becomes $Cr/(t^2\cos\theta_1)$ and coincides with the estimate [11, p. 211, Eq-n (11)]. The same is valid for all other estimates above.

Finally we have

$$\begin{aligned}
u_2(x) &\le C\frac{r}{(\cos\theta_1)^{(n+\alpha_k)/2}}\int_{K(1,R)} u^-(y^\circ)\frac{dy^\circ}{t^{(n+2+\alpha_k)/2}}\\
&\le C\frac{r}{(\cos\theta_1)^{(n+\alpha_k)/2}}\int_{K(1,R)} u^-(y^\circ)\frac{W_1(t)}{t^2}dy^\circ\\
&\le C\frac{r}{(\cos\theta_1)^{(n+\alpha_k)/2}}\int_{K(1,R)} u^-(y^\circ)\frac{W_1(t)}{t^2}\left(1-\frac{V_1(t)}{V_1(2R)}\frac{W_1(2R)}{W_1(t)}\right)dy^\circ\\
&\le C\frac{r}{(\cos\theta_1)^{(n+\alpha_k)/2}}\int_{K(1,R)} u^-(y^\circ)\left(W_1(t)-\frac{V_1(t)}{V_1(2R)}W_1(2R)\right)\frac{dy^\circ}{t^2}\\
&\le C\frac{r}{(\cos\theta_1)^{(n+\alpha_k)/2}}\int_{K(1,2R)} u^-(y^\circ)\left(W_1(t)-\frac{W_1(2R)}{V_1(2R)}V_1(t)\right)\frac{dy^\circ}{t^2}\\
&=\int_1^{2R}\frac{1}{t}\left\{\left(\int_{S_t}u^-(t,\theta)d\sigma_t(\theta)\right)\left(W_1(t)-\frac{W_1(2R)}{V_1(2R)}V_1(t)\right)\right\}\frac{dt}{t}.
\end{aligned}$$

Integrating the latter by parts and employing asymptotic formulas (1.6)–(1.7), we derive the estimate

$$u_2(x)\le C\frac{R^\rho}{(\cos\theta_1)^{(n+\alpha_k)/2}}. \tag{2.12}$$

Inserting (2.10) and (2.12) into (2.8), we complete the proof of Theorem 1.7. □

To prove our next result, we need the following lemma, whose conclusion in the classical case $c(r)=0$ is clear.

Lemma 2.3. *For any $n \geq 2$ and $l \geq 0$ there exists a negative subharmonic function $h(x)$ in the unit ball $B(1)$ such that*

$$h(x) \leq -\frac{C}{(\cos\theta_1)^{2l}}.$$

Proof. If $n = 2$, such a function (even harmonic one) is well known, see for example, [11, sect. 24.6], therefore we consider only the case $n \geq 3$. For $x \in \mathbb{R}^n$ and a vector $\bar{a} = a(1, 0, \ldots, 0) \in \mathbb{R}^n$, where a scalar $a > 0$ will be specified later, we set

$$h_{\pm}(x) = |x \pm \bar{a}|^{-2l}, \quad \text{and} \quad h(x) = -h_+(x)h_-(x).$$

We have to verify that $\Delta h \geq 0$ in $B(1)$. Indeed, a direct computation shows

$$\Delta h(x) = 4l\left[(n-2)a^2 - (4l - n + 2)|x|^2\right]|x + \bar{a}|^{-2(l+1)}|x - \bar{a}|^{-2(l+1)}.$$

From the latter we see that if $l \leq (n-2)/4$, then h is subharmonic everywhere in $\mathbb{R}^n$. If $l > (n-2)/4$, then h is subharmonic in a ball

$$\left\{x \in \mathbb{R}^n : |x|^2 \leq \frac{n-2}{2-n-4l}a^2\right\}.$$

Thus, without loss of generality we assume $l > \frac{n-2}{2}$ and if we fix $a^2 > \frac{4l-n+2}{n-2} > 1$, then h is subharmonic in $\overline{B(1)}$.

To find the estimate of h, we assume that x is at the lateral surface of the cone of opening θ_1, $0 \leq \theta_1 \leq \pi/2$, with the vertex at the origin and the vertical axis $\theta_1 = 0$. Then $|x - \bar{a}|^2$ is the largest whenever x is opposite to $\bar{a}$. Hence

$$|x - \bar{a}|^2|x + \bar{a}|^2 \leq (2a)^2(x_1 \sin\alpha)^2,$$

where α is the angle between the vectors x and $-\bar{a}$. Therefore,

$$|x - \bar{a}|^2|x + \bar{a}|^2 \leq (2a)^2(\cos\theta_1 \ \sin\alpha)^2 \leq (2a)^2\cos^2\theta_1$$

and

$$h_+(x)h_-(x) \geq \frac{C}{|\cos\theta_1|^{2l}}.$$

□

Proof of Theorem 1.11. First we prove the statement for q-subfunctions with respect to a radial potential $q(r)$. Consider a function

$$u(rx) = -h(x) + [u(rx) + h(x)].$$

For $|x| \leq 1$, $r > 1$ and $x \in B(1)$, we have by virtue of Lemma 2.3,

$$u(rx) \leq C + C\max_{0 \leq \theta_1 \leq \pi}\left\{\frac{1 + r^\rho}{(\cos\theta_1)^l} - \frac{1}{(\cos\theta_1)^{2l}}\right\}.$$

Since $h(x) < 0$, $\Delta h - q_1(r)h(x) \geq 0$, hence $h \in SbH_{q_1}$, where the potential $q_1(r) = r^{-2}q(r|x|)$ with a fixed $r > 1$, thus $u(rx) \in SbH_{q_1}$. The right-hand side of the latter inequality has a maximum if

$$|\cos\theta_1|^l = \frac{2}{1 + r^\rho},$$

therefore by the maximum principle [13] applied to the q_1-subfunction $u(rx)$ in the unit ball $B(1)$ we conclude that $u(rx) < C\left(1 + r^{2\rho}\right)$.

Next, given a $\rho > 0$, we choose θ_1 close enough to $\pi/2$, so that the smallest eigenvalue λ_1 of the spherical domain $\pi/2 - \varepsilon < \theta_1 \le \pi/2$ is as large as we wish; we make $\lambda_1 > \chi_k$, thus $V_1(r) > r^{2\rho+\delta}$, $\delta > 0$ for all $r > 1$, and by the Phragmén-Lindelöf principle (Theorem **LK1**) we get

$$u(x) \le C\,(1 + r^{\rho})$$

for all $x \in \mathbb{R}^n_+$ with $\pi/2 - \varepsilon < \theta_1 \le \pi/2$; the same estimate for x with $0 \le \theta_1 \le \pi/2 - \varepsilon$ is now obvious.

Finally, let u be a c-subfunction with respect to any, not necessarily radial, potential $c(x)$. Then it is easy to check (see [13]) that $u^+ \in SbH_c$, therefore (ibid) $u^+ \in SbH_q$, where $q(r)$ is any nonnegative continuous minorant of $c(x)$. It is now clear that u^+ satisfies (1.13)–(1.14), whence

$$u(x) \le u^+(x) \le C(1 + r^{\rho}). \qquad \square$$

Proof of Theorem 1.4. Again, we first consider the case of radial potentials q. Now the function $-u(x)$ is q-harmonic in $\mathbb{R}^n_+$ and continuous up to the boundary $\mathbb{R}^{n-1} = \partial\mathbb{R}^n_+$. By Theorem 1.7 and Corollary 1.8 applied to $-u$, we get

$$u(x) \le C \frac{1 + r^{\rho}}{(\cos\theta_1)^{(n+\alpha_k)/2}}.$$

Now Theorem 1.4 in the case of radial potentials follows from Theorem 1.11. In the general case we proceed exactly as in the previous proof, that is, choose a radial minorant q and apply the already proven "radial" case of Theorem 1.4 to the q-harmonic function u^+. $\square$

References

[1] M. Bramanti, *Potential theory for stationary Schrödinger operators: a survey of results.* Le Matematiche **47** (1992), 25–61.

[2] T. Carleman, *Extension d'un théorème de Liouville.* Acta Math. **48** (1926), 363–366.

[3] M. Cranston, E. Fabes, Z. Zhao, *Conditional gauge and potential theory for the Schrödinger operator.* Trans. Amer. Math. Soc. **307** (1988), 415–425.

[4] I.Tz. Gohberg, M.G. Krein, *Introduction to the Theory of Linear Nonselfadjoint Operators in Hilbert Space.* Transl. Math. Monographs, Vol. 18, Amer. Math. Soc., Providence, RI, 1969.

[5] N.V. Govorov, *Riemann's Boundary Problem With Infinite Index.* Birkhäuser Verlag, Basel, 1994.

[6] N.V. Govorov, M.I. Zhuravleva, *On an upper bound of the module of a function analytic in a half-plane and in a plane with a cut.* Izv. Severo-Kavkaz. Nauchn. Tsentra Vyssh. Shkoly Estestv. Nauk. **4** (1973), 102–103.

[7] P. Hartman, *Ordinary Differential Equations.* John Wiley & Sons, New York – London – Sydney, 1964.

[8] L. Hörmander, *Notions of Convexity.* Birkhäuser, Boston – Basel – Berlin, 1994.

[9] A. Kheyfits, *Indicators of functions of order less than one that are analytic in an open half-plane and have completely regular growth in interior angles.* Math. USSR, Izv. **9** (1975), 850–860.

[10] A. Kheyfits, *The Riesz-Herglotz formula for generalized harmonic functions and their boundary behavior.* Soviet Math. Dokl. **44** (1992), 688–691.

[11] B.Ya. Levin, *Lectures on Entire Functions.* AMS, Providence, RI 1996.

[12] B.Ya. Levin, A.I. Kheyfits, *Asymptotic behavior of subfunctions of the Schrödinger operator in an n-dimensional cone.* Soviet Math. Dokl. **38** (1989), 109–112.

[13] B.Ya. Levin, A. Kheyfits, *Asymptotic behavior of subfunctions of time-Independent Schrödinger operator.* Chap. 11 in *Some Topics on Value Distribution and Differentiability in Complex and P-adic Analysis.* Science Press, Beijing, 2008, 323–397.

[14] V.I. Macaev, *On the growth of entire functions that admit a certain estimate from below.* Soviet Math. Dokl. **1** (1960), 548–552.

[15] V.I. Macaev, E.Z. Mogul'skii, *A division theorem for analytic functions with a given majorant, and some of its applications.* Investigations on linear operators and theory of functions, VI. Zap. Naucn. Sem. Leningrad. Otdel. Mat. Inst. Steklov. (LOMI). **56** (1976), 73–89.

[16] I.V. Ostrovskij, *The investigations of M.G. Krein in the theory of entire and meromorphic functions and their further development.* Ukrainian Math. J. **46** (1994), 87–100.

[17] A. Rashkovskii, *Classical and new loglog-theorems.* Expo. Math. **27** (2009), 271–287.

[18] E.N. Sergienko, *The growth of entire functions that admit a special estimation from below.* Teor. Funkcii Funkcional. Anal. i Prilozhen. **15** (1972), 78–96.

[19] E.N. Sergienko, *The growth of meromorphic functions that admit a special lower bound.* Teor. Funkcii Funkcional. Anal. i Prilozhen. **21** (1974), 83–104.

[20] E.N. Sergienko, *Growth of functions representable as the difference of subharmonic functions and admitting a special lower bound.* Teor. Funktsii Funktsional. Anal. i Prilozhen. **37** (1982), 116–122.

[21] G.M. Verzhbinskii, V.G. Maz'ya, *Asymptotic behavior of solutions of elliptic equations of the second order close to a boundary. I.* Sibirskii Math. J. **12** (1971), 874–899.

[22] H. Yoshida, *A boundedness criterion for subharmonic functions.* J. London Math. Soc. (2) **24** (1981), 148–160.

Alexander I. Kheyfits
The Graduate Center and Bronx Community College
of the City University of New York, USA
e-mail: akheyfits@gc.cuny.edu

Operator Theory:
Advances and Applications, Vol. 229, 233–251

The Riemann and Dirichlet Problems with Data from the Grand Lebesgue Spaces

V. Kokilashvili and V. Paatashvili

Dedicated to Professor Stefan Samko to mark his 70th birthday

Abstract. In Section 1, we present a solution of the following boundary value problem: find an analytic function Φ on the plane cut along a closed piecewise-smooth curve Γ which is represented by a Cauchy type integral with a density from the Grand Lebesgue Space $L^{p),\theta}(\Gamma)$ $(1 < p < \infty, 0 < \theta < \infty)$ and whose boundary values satisfy the conjugacy condition

$$\Phi^+(t) = G(t)\Phi^-(t) + g(t), \quad t \in \Gamma.$$

Here G and g are functions defined on Γ such that G is a piecewise continuous function, $G(t) \neq 0$ and $g \in L^{p),\theta}(\Gamma)$. The conditions for the problem to be solvable are established and the solutions are constructed in explicit form.

In Section 2, the Dirichlet problem for harmonic functions, real parts of Cauchy type integrals with densities from weighted generalized Grand Lebesgue Spaces is studied when boundary data belong to the same space.

Mathematics Subject Classification (2010). 47B38, 30E20, 42B20, 42B25, 46E30, 26D15.

Keywords. Boundary value problem, analytic function, Cauchy type integral, the Grand Lebesgue Space, piecewise smooth boundary, piecewise continuous coefficient, Dirichlet problem, Lyapunov contour.

1. The Riemann boundary value problem for analytic functions

Let Γ be a simple closed rectifiable curve dividing the plane into two domains of which D^+ is bounded and D^- unbounded, and $A(\Gamma)$ be some set of complex-valued analytic functions Φ on the plane cut along Γ and having boundary (in a certain sense) functions $\Phi^+(t)$ and $\Phi^-(t)$, $t \in \Gamma$. In the present paper in the capacity of $A(\Gamma)$ is regarded the class of analytic functions represented by the Cauchy type integral with density from the Grand Lebesgue Space. The functions Φ^+ and Φ^-

are assumed to be non-tangential (angular) boundary values of function $\Phi(z)$. It is well known, that for arbitrary rectifiable curve Γ the Cauchy type integral

$$\Phi(z) = \frac{1}{2\pi i} \int_{\Gamma} \frac{\varphi(\tau)}{\tau - t} d\tau$$

with $\varphi \in L^1(\Gamma)$ has almost everywhere the boundary values (see, e.g., [5], Section 6.2, Theorem 2.22).

According to the terminology used in [24], we call the following problem the boundary value problem of linear conjugation: Define a function $\Phi \in A(\Gamma)$ for which the condition

$$\Phi^+(t) = G(t)\Phi^-(t) + g(t), \quad t \in \Gamma, \tag{1}$$

is fulfilled, where $G(t)$ and $g(t)$ are given functions on Γ.

The problem of linear conjugation for analytic functions was for the first time encountered by B. Riemann [25]. Important results on which the posterior solution of problem (1) was based, were obtained by Yu. Sokhotskiĭ, D. Hilbert, I. Plemelj and T. Carleman. The solution of the Riemann problem in a continuous setting was given by F.D. Gakhov [8]. In the case when the continuity is violated only at a finite number of boundary points a complete solution of the problem was given by N. Muskhelishvili [24].

In [15], it is proposed to divide problems like (1) (and also other boundary value problems of function theory) into three groups depending on a character of the existence of boundary conditions: continuous when it is required that $\Phi(z)$ be continuously extendable on Γ both from D^+ and from D^-; piecewise-continuous if the function to be defined is continuously extendable everywhere on Γ except for a finite number of points on Γ near which it can be unbounded but of order less than one, and discontinuous problems – these are all other problems.

Continuous and piecewise-continuous problems of type (1) have been studied pretty well: the conditions of their solvability have been established and solutions constructed when $G(t)$, $g(t)$ are Hölder-continuous functions and $G(t) \neq 0$ or these conditions are not fulfilled at a finite number of boundary points. These and other results associated with boundary value problems of function theory and the related integral equations are presented in [8], [24].

Discontinuous problem of linear conjugation has been studied thoroughly when of the sought for function it is required to be representable by a Cauchy type integral with a density from the Lebesgue space $L^p(\Gamma;\rho)$ with power weight ρ. In this context, there arise quite interesting problems on admissible boundary curves and weights, oscillating coefficients and also some other problems. As to the solutions of these problems we refer to [12], [14], [15].

In recent years, many problems of the theory of equations of mathematical physics and boundary value problems of function theory have come to be investigated in the framework of nonstandard function spaces, see, e.g., [3], [7], [16], [18], [28], [29]. We have managed to investigate problem (1) and the Riemann–Hilbert problem

$$\operatorname{Re}\left[a(t)\Phi^{+}(t)\right] = b(t)$$

under the assumption that the sought function belongs to the class $K^{p(\cdot)}(\Gamma,\rho)$, i.e., it is representable by a Cauchy type integral with a density from the Lebesgue class $L^{p(t)}(\Gamma;\rho)$ with a variable exponent $p(t)$. Moreover, we have introduced new classes of analytic functions $E^{p(t)}(D)$ that are generalizations of the Smirnov classes $E^{p}(D)$ for variable exponents and obtained solutions of some boundary value problems in these classes, too [13], [17], [22].

For some aspects of the Riemann and Riemann-Hilbert problems we refer the reader also to [1] and [27].

In our present paper, the boundary value problem is investigated in the framework of other nonstandard Banach functional spaces, namely Grand Lebesgue Spaces. These spaces were introduced by T. Iwaniec and C. Sbordone in [11] and were treated in more general terms in [10]. The study of these space is nowadays one of the intensively developing directions of modern analysis. The necessity of the study of Grand Lebesgue Spaces is based on the fact that they do play an essential role in various areas of mathematics, in particular, in the PDE theory they are right spaces in which some nonlinear equations have been considered.

1.1. The class $L^{p),\theta}(\Gamma;\omega)$

Let Γ be a simple closed rectifiable curve of finite length, ω be a measurable function different from zero almost everywhere on Γ. We will say that a measurable function f on Γ belongs to the class $L^{p),\theta}(\Gamma;\omega)$ $(1<p<\infty,\ \theta>0)$ if

$$\|f\|_{L^{p),\theta}(\Gamma;\omega)} = \|f\omega\|_{L^{p),\theta}(\Gamma)} = \sup_{0<\varepsilon<p-1}\left(\frac{\varepsilon^{\theta}}{|\Gamma|}\int_{\Gamma}|f(t)\omega(t)|^{p-\varepsilon}ds\right)^{\frac{1}{p-\varepsilon}} < \infty, \tag{2}$$

where $|\Gamma|$ is the length of the curve Γ.

The space $L^{p),\theta}(\Gamma;\omega)$ is a Banach functional space.

The following continuous embeddings are fulfilled:

$$L^{p}(\Gamma;\omega) \subset L^{p),\theta}(\Gamma;\omega) \subset L^{p-\varepsilon}(\Gamma;\omega)\,.$$

It is well known [21] that, for instance, the function $g_{\lambda}(t) = t^{-\frac{1}{p}}\ln^{\lambda}\frac{2}{t} \in L^{p),1}(0,1)$ if and only if $\lambda \le 0$. Along with this, $g_{\lambda} \notin L^{p}(0,1)$.

In the previous studies, the discontinuous boundary value problems were studied under the assumption that the boundary function belongs to the class $L^{p}(\Gamma)$ $(1<p<\infty)$. Therefore functions of the form $g_{\lambda}(t)$ were inadmissible for the use as boundary functions. One of the novelties of the present paper is the solution of the boundary value problem of linear conjugation when boundary functions are taken from more wide functional classes, namely, from $L^{p),\theta}(\Gamma)$, $1<p<\infty$, $\theta>0$.

In this paper the Riemann boundary value problem is solved in the frame of more subtler scale of functional classes that it was known before. By our assumption the boundary function g is taken from the class $L^{p),\theta}(\Gamma)$ $(1<p<\infty,\ \theta>0)$. It is evident that, this condition implies belonging of g to the class $L^{p-\epsilon}(\Gamma)$ and, therefore, basing on our previous research (see [13], Chapter 2, Theorem 1.1) we are able to conclude the solvability in the class of analytic functions represented by the Cauchy type integral with density from $L^{p-\epsilon}(\Gamma)$. While in the present paper for $g \in L^{p),\theta}(\Gamma)$ we solve the problem in the class $K^{p),\theta}$, more narrower than $K^{p-\epsilon}(\Gamma)$ (Definitions of these classes will be given in what follows).

1.2. The classes $W^{p),\theta}(\Gamma)$ and $K^{p),\theta}(\Gamma)$

Let ω be a measurable function different from zero almost everywhere on Γ. We will write $\omega \in W^{p),\theta}(\Gamma)$, $1 < p < \infty$, $\theta > 0$, if the Cauchy singular operator

$$S_\Gamma : f \to S_\Gamma f, \quad (S_\Gamma f)(t) = \frac{1}{\pi i} \int_\Gamma \frac{f(\tau)\, d\tau}{\tau - t}, \qquad t \in \Gamma,$$

is continuous in $L^{p),\theta}(\Gamma, \omega)$. This definition is equivalent to the following: $\omega \in W^{p),\theta}(\Gamma)$ if and only if the operator

$$\frac{\omega(t)}{\pi i} \int_\Gamma \frac{f(\tau)}{\omega(\tau)} \frac{d\tau}{\tau - t}, \qquad t \in \Gamma,$$

is continuous in $L^{p),\theta}(\Gamma)$. By $W^p(\Gamma)$ we denote a set of weights ω for which the operator $\omega S_\Gamma\left(\frac{f}{\omega}\right)$ is bounded in $L^p(\Gamma)$.

Let

$$D(t,r) := \Gamma \cap B(t,r), \quad r > 0$$

where $B(t,r) = \{z \in \mathbf{C} : \ |z - t| < r\}$.

Let ν be the arc-length measure defined on Γ. We remind that a rectifiable curve Γ is called Carleson curve (regular curve) if there exists a constant $c_0 > 0$ such that

$$\nu D(t,r) \leq c_0 r$$

for arbitrary $t \in \Gamma$ and $r > 0$.

Theorem A ([20], [21]). *If Γ is a simple closed rectifiable Carleson curve, $t_k \in \Gamma$, $k = \overline{1,m}$, $\alpha_k \in \mathbb{R}$, $p > 1$, then the function*

$$\omega(t) = \prod_{k=1}^{m} |t - t_k|^{\alpha_k} \tag{3}$$

belongs to $W^{p),\theta}(\Gamma)$, $\theta > 0$, provided that the conditions

$$-\frac{1}{p} < \alpha_k < \frac{1}{p'}, \quad p' = \frac{p}{p-1}, \tag{4}$$

are fulfilled.

To investigate the Riemann problem, we need to have a more general function belonging to $W^{p),\theta}$, namely,

$$\rho(t) = \prod_{k=1}^{n} |t - t_k|^{\alpha_k} \exp \int_\Gamma \frac{\psi(t)}{\tau - t}\, d\tau,$$

where $-\frac{1}{p} < \alpha_k < \frac{1}{p'}$ and $\psi(t)$ is a real continuous function given on Γ.

In Section 1.5 below we give two proofs of this fact, of which one rests on the analysis of the boundary value problem and the other on the theory of weighted singular integrals in the Lebesgue spaces with weights from Muckenhoupt type classes. The investigation of each of the proofs of the above-mentioned facts is, in our opinion, of independent interest.

Define

$$K^{p),\theta}(\Gamma;\omega) = \left\{\Phi : \Phi(z) = \frac{1}{2\pi i}\int_\Gamma \frac{\varphi(\tau)}{\tau - z}\,d\tau,\ \ z \notin \Gamma,\ \varphi \in L^{p),\theta}(\Gamma;\omega)\right\},$$

$$p > 1,\ \theta > 0,$$

$$K^{p),\theta}(\Gamma) \equiv K^{p),\theta}(\Gamma;1), \quad K^p(\Gamma) = \left\{\Phi : \Phi(z) = \frac{1}{2\pi i}\int_\Gamma \frac{\varphi(\tau)}{\tau - z}\,d\tau,\ \varphi \in L^p(\Gamma)\right\}.$$

We need also the definition of Smirnov classes analytic functions.

Let D^+ be a simply connected internal domain bounded by a rectifiable curve Γ. Then $\Phi \in E^p(D^+)$, $p > 0$, if

$$\sup_r \int_{\Gamma_r} |\phi(z)|^p |dz| < \infty,$$

where Γ_r is the image of the circumference $|z| = r$ under a conformal mapping of $U = \{w : |w| = 1\}$ onto D^+.

If D^- is an external domain bounded by Γ, then $E^p(D^-)$ is a set of analytic in D^- functions ϕ for which $\Psi(w) = \Phi\left(\frac{1}{w} + z_0\right) \in E^p(D_1)$, $\Phi(\infty) = 0$, where D_1 is the finite domain into which the function $w = \frac{1}{z - z_0}$, $z_0 \overline{\in} D^-$, maps D^- (this class obviously does not depend on the choice of z_0). (See [9, Ch. X].)

It is clear, that the class $E^p(U)$ coincides with the well-known Hardy class H^p. For these classes we refer the reader to the book [4].

As is known (see [12, p. 29]), if Γ is a Carleson curve and $\varphi \in L^p(\Gamma)$, $p > 1$, then the function $(K_\Gamma\varphi)(z) = \frac{1}{2\pi i}\int_\Gamma \frac{\varphi(\tau)\,d\tau}{\tau - z}$ belongs to the Smirnov class $E^p(D^\pm)$. Hence it follows that if $\varphi \in L^{p),\theta}(\Gamma)$, then $(K_\Gamma\varphi) \in \bigcap_{p>\varepsilon>0} E^{p-\varepsilon}(D^\pm)$. Throughout this paper the following well-known fact will be used: the class E^1 coincides with the class of analytic functions represented by the Cauchy integral (see [9, Chapter X]).

1.3. One property of the operator S_Γ

Theorem 1. *Let Γ be a simple rectifiable curve, $p > 1$, $1/\omega \in L^{p'+\varepsilon}(\Gamma)$, $\varepsilon > 0$, and the operator S_Γ map the space $L^{p),\theta}(\Gamma;\omega)$ into $L^{p),\theta}(\Gamma;\omega)$, then the operator S_Γ is continuous in $L^{p),\theta}(\Gamma;\omega)$.*

Proof. First we will show that the operator S_Γ is measure-continuous, i.e., if f_n tends to f_0 in $L^{p),\theta}(\Gamma;\omega)$, then $S_\Gamma f_n$ tends to $S_\Gamma f_0$ with respect to measure. For this, we observe that $\forall f \in L^{p),\theta}(\Gamma;\omega)$ we have $f = \varphi/\omega$ where

$$\varphi \in L^{p),\theta}(\Gamma) \subset \bigcap_{0<\varepsilon<p-1} L^{p-\varepsilon}(\Gamma).$$

Since $\frac{1}{\omega} \in L^{p'+\varepsilon}(\Gamma)$, it follows that $f \in L^1(\Gamma)$, i.e., $L^{p),\theta}(\Gamma;\omega) \subset L^1(\Gamma)$. As the sequence $\{f_n\}_{n=1}^{\infty}$ converges to f_0 in $L^{p),\theta}(\Gamma;\omega)$, it converges to f_0 in $L^1(\Gamma)$ as well. Indeed, for η, $0<\eta<p-1$, we have

$$\int_\Gamma |f_n(t)-f_0(t)||dt| \le \|f_n - f_0\|_{L^{p-\eta}_\omega(\Gamma)} \cdot \left\|\frac{1}{\omega}\right\|_{L^{(p-\eta)'}(\Gamma)}$$

$$\le c\cdot \|f_n - f_0\|_{L^{p),\theta}_\omega(\Gamma)} \cdot \left\|\frac{1}{\omega}\right\|_{L^{p'+\epsilon}(\Gamma)} \to 0$$

when $n\to\infty$. Here we used the fact that $(p-\eta)' > p'$ and thus $(p-\eta)' = p'+\epsilon$ for some $\epsilon > 0$.

But the operator S_Γ is continuous from $L^1(\Gamma)$ into the space of convergence in measure ([12, Theorem 2.1, p. 21]). This implies the continuity of the operator S_Γ from $L^{p),\theta}(\Gamma;\omega)$ into the space of convergence in measure.

Let now $S_\Gamma(L^{p),\theta}(\Gamma;\omega)) \subset L^{p),\theta}(\Gamma;\omega)$. We will show that S_Γ is a closed operator in $L^{p),\theta}(\Gamma;\omega)$. Indeed, if $f_n \overset{L^{p),\theta}(\Gamma;\omega)}{\longrightarrow} f_0$ and $S_\Gamma f_n \overset{L^{p),\theta}(\Gamma;\omega)}{\longrightarrow} \psi$ and we know that $S_\Gamma f_n$ converges in measure to $S_\Gamma f_0$, then we conclude that $\psi = S_\Gamma f_0$. This means the closure of the operator S_Γ in $L^{p),\theta}(\Gamma;\omega)$. Since $L^{p),\theta}(\Gamma;\omega)$ is a Banach space, by virtue of the well-known theorem on a closed graph we conclude that S_Γ is a continuous operator in $L^{p),\theta}(\Gamma;\omega)$. □

1.4. Problem (1) in the class $K^{p),\theta}(\Gamma)$ for continuous G

The following theorem is proved in [18].

Theorem B. *Let Γ be a simple closed rectifiable Carleson curve, $G(t)\in C(\Gamma)$, $G(t)\neq 0$, $t\in\Gamma$, $p>1$, and $g(t)\in L^{p),\theta}(\Gamma)$. If $\varkappa = \operatorname{ind} G(t) = \frac{1}{2\pi}[\arg G(t)]_\Gamma$, where by the symbol $[\arg G(t)]_\Gamma$ is denoted the increment of argument along Γ.*

Then

i) for $\varkappa \ge 0$, problem (1) is unconditionally solved in $K^{p),\theta}(\Gamma)$ and its general solution is given by the equality

$$\Phi(z) = \frac{X(z)}{2\pi i}\int_\Gamma \frac{g(t)}{X^+(t)(t-z)} + X(z)Q_{\varkappa-1}(z), \tag{5}$$

where

$$X(z) = \begin{cases} \exp h(z), & z\in D^+, \\ (t-z_0)^{-\varkappa}\exp h(z), & z\in D^-,\quad z_0\in D^+, \end{cases} \tag{6}$$

$$h(z) = K_\Gamma\left(\ln G(t)(t-z_0)^{-\varkappa}\right)(z), \tag{7}$$

and $Q_{\varkappa-1}(z)$ is an arbitrary polynomial of order $\varkappa-1$ ($Q_{-1}(z)\equiv 0$).

ii) for $\varkappa<0$, problem (1) is solvable in $K^{p),\theta}(\Gamma)$ if and only if

$$\int_\Gamma \frac{g(t)}{X^+(t)}t^k dt = 0, \quad k=\overline{0,|\varkappa|-1}. \tag{8}$$

When these conditions are fulfilled, problem (1) *has a unique solution given by equality* (5) *for* $Q_{\varkappa-1}(z) \equiv 0$.

1.5. On a weight function from $W^{p),\theta}(\Gamma)$

Theorem 2. *If* Γ *is a simple closed rectifiable Carleson curve,* $\omega \in W^{p),\theta}(\Gamma)$, $p > 1$, $\frac{1}{\omega(t)} \in L^{p'+\varepsilon}(\Gamma)$, $\varepsilon > 0$, *and* ψ *is a real continuous function on* Γ, *then*

$$\rho(t) = \omega(t) \exp \int_\Gamma \frac{\psi(\tau)\, d\tau}{\tau - t} \in W^{p),\theta}(\Gamma). \tag{9}$$

Proof. Consider problem (1) in the class $K^{p),\theta}(\Gamma;\omega)$ when $G(t) = \exp 2\pi i \psi(t)$. The function G is continuous on Γ and $\varkappa(G) = 0$. Choose a rational function $\widetilde{G}(z)$ such that

$$\sup_{t\in\Gamma} \left| \frac{G(t)}{\widetilde{G}(t)} - 1 \right| < \frac{1}{2} \left(1 + \|S_\Gamma\|_{L^{p),\theta}(\Gamma;\omega)}\right)^{-1}. \tag{10}$$

It is obvious that $\operatorname{ind} \widetilde{G}(t) = 0$. Consider the function

$$\widetilde{X}(z) = \exp\left\{ \left(K_\Gamma(\ln \widetilde{G}(t)) \right)(z) \right\}, \quad z \notin \Gamma.$$

This function is continuous in the domains $D^\pm$-bounded by Γ and $\widetilde{X}(z) \neq 0$, $\widetilde{X}(\infty) = 1$. Since $\widetilde{X}^+(t)[\widetilde{X}^-(t)]^{-1} = \widetilde{G}(t)$, condition (1) can be written in the form

$$\left(\frac{\Phi}{\widetilde{X}}\right)^+ = \frac{G}{\widetilde{G}} \left(\frac{\Phi}{\widetilde{X}}\right)^- + \frac{g}{\widetilde{X}^+}, \tag{11}$$

i.e.,

$$F^+ = \frac{G}{\widetilde{G}} F^- + \frac{g}{\widetilde{X}^+},$$

where

$$F(z) = \frac{\Phi(z)}{\widetilde{X}(z)}.$$

Since $\Phi \in K^{p),\theta}(\Gamma;\omega)$, we have

$$\Phi(z) = \frac{1}{2\pi i} \int_\Gamma \frac{\varphi(t)\, dt}{t - z}, \quad \varphi \in L^{p),\theta}(\Gamma;\omega),$$

i.e., $\varphi = \mu/\omega$, where $\mu \in L^{p),\theta}(\Gamma)$ and, by assumption, $\frac{1}{\omega} \in L^{p'+\varepsilon}(\Gamma)$.

Since $L^{p),\theta}(\Gamma) \subset \bigcap_{0<\delta<p-1} L^{p-\delta}(\Gamma)$, $p > 1$, it is easy to establish that for some $\alpha > 1$, $\varphi \in L^\alpha(\Gamma)$. Since Γ is a Carleson curve, we have $\Phi \in E^\alpha(D^\pm)$ and, consequently, $F \in E^\alpha(D^\pm)$ (see Section 1.2). From the condition $\alpha > 1$ it follows that $F(z) = K_\Gamma(F^+ - F^-)(z) = \frac{1}{2\pi i} \int_\Gamma \frac{f(t)\, dt}{t-z}$, $f = F^+ - F^-$. The functions $F^\pm$ belong to $L^{p),\theta}(\Gamma)$. Indeed, by the Sokhotski–Plemelj formulas (see, e.g., [2], Theorem 6.7) we obtain

$$\Phi^\pm = \pm\frac{1}{2}\varphi + \frac{1}{2} S_\Gamma \varphi, \quad \varphi \in L^{p),\theta}(\Gamma;\omega)$$

and since $\omega \in W^{p),\theta}(\Gamma)$, we conclude that $\Phi^{\pm} \in L^{p),\theta}(\Gamma;\omega)$, whereas by virtue of the boundedness of the function $\frac{1}{\widetilde{X}(z)}$ the more so $F^{\pm} \in L^{p),\theta}(\Gamma;\omega)$. Thus F is a solution of problem (1) from the class $K^{p),\theta}(\Gamma;\omega)$. Let us write (11) in the form

$$F^+ - F^- = \left(\frac{G}{\widetilde{G}} - 1\right) F^- + g_1, \quad g_1 = \frac{g}{\widetilde{X}^+},$$

i.e.,

$$f = \left(\frac{G}{\widetilde{G}} - 1\right)\left(-\frac{1}{2} f + \frac{1}{2} S_\Gamma f\right) + g_1. \tag{12}$$

By virtue of condition (10), the right-hand part of (12) is the contracting operator in $L^{p),\theta}(\Gamma;\omega)$ and therefore equation (12) is uniquely solvable in $L^{p),\theta}(\Gamma;\omega)$. Therefore problem (11) and thereby problem (1) have a unique solution in $K^{p),\theta}(\Gamma;\omega)$. According to Theorem B, it is given by equality (5) where $Q_{\varkappa-1} = 0$. This means that the function

$$\Phi(z) = \frac{X(z)}{2\pi i} \int_\Gamma \frac{g(\tau)}{X^+(\tau)} \frac{d\tau}{\tau - z},$$

where $X(z)$ is given by equalities (6)–(7), belongs to $K^{p),\theta}(\Gamma;\omega)$ for any $g \in L^{p),\theta}(\Gamma;\omega)$. Hence it follows that the operator $X^+ S_\Gamma \frac{g}{X^+}$ acts from $L^{p),\theta}(\Gamma;\omega)$ into $L^{p),\theta}(\Gamma;\omega)$. Therefore

$$(Tf)(t) = \frac{\omega(t)X^+(t)}{\pi i} \int_\Gamma \frac{f(\tau)}{\omega(\tau)X^+(\tau)} \frac{d\tau}{\tau - t} \in L^{p),\theta}(\Gamma) \tag{13}$$

for any $f \in L^{p),\theta}(\Gamma)$.

By Theorem 1 we conclude that T is continuous in $L^{p),\theta}(\Gamma)$. Since $G(t) = \exp 2\pi i\psi(t)$, we have $X^+(t) \sim \exp \int_\Gamma \frac{\psi(\tau)\,d\tau}{\tau - t}$, $t \in \Gamma$ ($f \sim g$ means that $0 < \inf \left|\frac{f}{g}\right| \le \sup \left|\frac{f}{g}\right| < \infty$). Now the validity of Theorem 2 follows from (13) and Theorem 1. $\square$

As has been mentioned earlier, we are going to give another proof of the fact that a certain function belongs to the class $W^{p),\theta}(\Gamma)$. We begin with

Definition 1. Let Γ be a rectifiable curve. We say that $\omega \in A_p(\Gamma)$ if the condition

$$\sup_{\substack{z\in\Gamma\\ r>0}} \frac{1}{r} \int_{D(z,r)} \omega(t)\,ds \left(\frac{1}{r} \int_{D(z,r)} \omega^{1-p'}(t)ds\right)^{p-1} < \infty \quad (A_p(\Gamma) \text{ condition})$$

is fulfilled, where $D(z,r) = B(z,r) \cap \Gamma$ and $B(z,r)$ is a circle with center at the point z and radius r.

The following theorem is the known one.

Theorem C (**[2], [12]**). *Let $1 < p < \infty$. For a function ρ to belong to the class $W^p(\Gamma)$ it is necessary and sufficient that Γ be Carleson curve and the condition $\rho^p \in A_p(\Gamma)$ be fulfilled.*

In particular the function

$$\omega(t)=\prod_{k=1}^{n}|t-t_k|^{\alpha_k}$$

belongs to the class $W^p(\Gamma)$ if and only if Γ is Carleson curve and the condition

$$-\frac{1}{p}<\alpha_k<\frac{1}{p'}$$

is fulfilled.

Remark. Theorem C can be reformulated as follows: a Cauchy singular operator is bounded in the space $L^p_\omega(\Gamma)$ for $1<p<\infty$ if and only if Γ is a Carleson curve and $\omega\in A_p(\Gamma)$. Here the norm in the space $L^p_\omega(\Gamma)$ is understood in the sense

$$\|f\|_{L^p_\omega(\Gamma)}=\left(\int_\Gamma|f(t)|^p\omega(t)\,ds\right)^{\frac{1}{p}}.$$

A generalization of Theorem C in the latter formulation for the spaces $L^{p),\theta}_\omega(\Gamma)$, where the norm is understood in the sense

$$\|f\|_{L^{p),\theta}_\omega(\Gamma)}=\sup_{0<\varepsilon<p-1}\left(\frac{\varepsilon^\theta}{|\Gamma|}\int_\Gamma|f(t)|^{p-\varepsilon}\omega(t)\,ds\right)^{\frac{1}{p-\varepsilon}},$$

is obtained in [19].

Let now return to the classes $W^{p),\theta}$. Note that the equivalency

$$g\in L^{p),\theta}_\omega\Longleftrightarrow g\omega^{\frac{1}{p}}\in L^{p),\theta}$$

does not hold for the Grand Lebesgue Spaces (see, e.g., [6]).

Theorem 3. *Let for some sublinear operator T the weighted operator*

$$T_\rho: f\to\rho T\left(\frac{f}{\rho}\right)$$

be bounded in $L^p(\Gamma)$ $(1<p<\infty)$ for some ρ, $\rho^p\in A_p(\Gamma)$ and at the same time T_ρ be bounded in $L^{p-\sigma}(\Gamma)$ for $\rho^{p-\sigma}\in A_{p-\sigma}(\Gamma)$ for some small σ. Then T_ρ is bounded in $L^{p),\theta}(\Gamma)$, i.e., $\rho\in W^{p),\theta}(\Gamma)$.

Proof. We will use an idea developed in [6]. First we will prove that there exist numbers $\sigma\in(0,p-1)$ and $M>0$ such that

$$\|T_\rho\|_{L^{p-\varepsilon}(\Gamma)\to L^{p-\varepsilon}(\Gamma)}\le M \tag{14}$$

for any $\varepsilon\in(0,\sigma)$.

As is known, if $\rho^p\in A_p(\Gamma)$, then there exists a number $\sigma\in(0,p-1)$ such that $\rho^p\in A_{p-\sigma}(\Gamma)$, and also $\rho^{p\alpha}\in A_{p-\sigma}(\Gamma)$ for arbitrary α, $0<\alpha<1$. Assume now $\alpha=\frac{p-\sigma}{p}$. Then $0<\alpha<1$ and $\rho^{p-\sigma}\in A_{p-\sigma}(\Gamma)$.

By virtue of the condition of the theorem we have

$$\|T_\rho f\|_{L^p(\Gamma)}\le M_1\|f\|_{L^p(\Gamma)}$$

and

$$\|T_\rho f\|_{L^{p-\sigma}(\Gamma)} \le M_2 \|f\|_{L^{p-\sigma}(\Gamma)}\,.$$

For $p-\varepsilon$ with the condition $\varepsilon \in (0,\sigma)$ there exists $t_\varepsilon \in (0,1)$ such that

$$\frac{1}{p-\varepsilon} = \frac{t_\varepsilon}{p} + \frac{1-t_\varepsilon}{p-\sigma}\,.$$

By the Riesz–Thorin interpolation theorem, the inequality

$$\|T_\rho f\|_{L^{p-\varepsilon}(\Gamma)} \le M_1^{t_\varepsilon} M_2^{1-t_\varepsilon} \|f\|_{L^{p-\varepsilon}(\Gamma)}$$

is fulfilled for arbitrary $\varepsilon \in (0,\sigma)$. From the latter inequality we get (14).

We have

$$\|T_\rho f\|_{L^{p),\theta}(\Gamma)} \le \max\{A, B\}, \tag{15}$$

where

$$A = \sup_{0<\varepsilon\le\sigma} \varepsilon^{\frac{\theta}{p-\varepsilon}} \|T_\rho f\|_{L^{p-\varepsilon}(\Gamma)}$$

and

$$B = \sup_{\sigma<\varepsilon\le p-1} \varepsilon^{\frac{\theta}{p-\varepsilon}} \|T_\rho f\|_{L^{p-\varepsilon}(\Gamma)}.$$

Fix $\varepsilon \in (\sigma, p-1)$. Then $\frac{p-\sigma}{p-\varepsilon} > 1$.

Using the Hölder inequality with the exponent $\frac{p-\sigma}{p-\varepsilon}$ and observing that

$$\left(\frac{p-\sigma}{p-\varepsilon}\right)' = \frac{p-\sigma}{\varepsilon-\sigma}\,,$$

we find that

$$\|T_\rho f\|_{L^{p-\varepsilon}(\Gamma)} \le \|T_\rho f\|_{L^{p-\sigma}(\Gamma)} |\Gamma|^{\frac{\varepsilon-\sigma}{(p-\varepsilon)(p-\sigma)}} \le c\|T_\rho f\|_{L^{p-\sigma}(\Gamma)}, \tag{16}$$

where c is a constant not depending on f and ε. Thus

$$B \le c \sup_{\sigma<\varepsilon<p-1} \varepsilon^{\frac{\theta}{p-\varepsilon}} \sigma^{-\frac{\theta}{p-\sigma}} \sigma^{\frac{\theta}{p-\sigma}} \|T_\rho f\|_{L^{p-\sigma}(\Gamma)}.$$

Further, since $\varepsilon \in (\sigma, p-1)$, we have

$$0 < \frac{\varepsilon-\sigma}{(p-\sigma)(p-\varepsilon)} < \frac{p-1-\sigma}{p-\sigma}\,, \quad (p-1)\sigma^{-\frac{1}{p-\sigma}} > 1.$$

Then

$$\begin{aligned}\|T_\rho f\|_{L^{p),\theta}(\Gamma)} &\le cA \max\left\{1, \varepsilon^{\frac{\theta}{p-\varepsilon}} \sigma^{-\frac{\theta}{p-\sigma}}\right\} \\ &= \sup_{0<\varepsilon\le\sigma} \left(\varepsilon^{\frac{\theta}{p-\varepsilon}} \|T_\rho f\|_{L^{p-\varepsilon}(\Gamma)} \max\left\{1, \varepsilon^{\frac{\theta}{p-\varepsilon}} \sigma^{-\frac{\theta}{p-\sigma}}\right\}\right).\end{aligned}$$

Taking advantage of (14), from (15) and (16) we finally get

$$\|T_\rho f\|_{L^{p),\theta}(\Gamma)} \le c \max\left\{1, (p-1)\sigma^{-\frac{\theta}{p-\sigma}}\right\} \|f\|_{L^{p),\theta}(\Gamma)}. \qquad \square$$

On the other hand, applying the well-known extrapolation theorem due to Rubio de Francia [26] based on the proof of Theorem 3 we conclude that the following assertion is true.

Theorem 3′. *Let for some sublinear operator T the weighted operator*

$$T_\rho : f \to \rho T\left(\frac{f}{\rho}\right)$$

be bounded in $L^p(\Gamma)$ $(1 < p < \infty)$ for arbitrary ρ, $\rho^p \in A_p(\Gamma)$. Then T_ρ is bounded in $L^{p),\theta}(\Gamma)$ for all ρ, $\rho^p \in A_p(\Gamma)$.

Corollary 1. *Let $1 < p < \infty$, $\theta > 0$. Let, further, ψ be a real continuous function given on Γ.*

If Γ is a Carleson curve, then

$$\rho(t) = \prod_{k=1}^{n} |t - t_k|^{\alpha_k} \exp \int_\Gamma \frac{\psi(\tau)\, d\tau}{\tau - t},$$

where $-\frac{1}{p} < \alpha_k < \frac{1}{p'}$, belongs to the class $W^{p),\theta}(\Gamma)$.

Proof. We know [16] that $\rho \in W^p(\Gamma)$, i.e., $\rho^p \in A_p(\Gamma)$. Then Theorem 3 implies that $\rho^p \in W^{p),\theta}(\Gamma)$. □

1.6. Problem (1) with a piecewise-continuous coefficient $G(t)$ and $g \in L^{p),\theta}(\Gamma)$

First we give the condition with respect to the curve Γ which is assumed to be fulfilled when considering problem (1) in $K^{p),\theta}(\Gamma)$. Let Γ be a simple Carleson curve bounding the domain D^+. We assume that if $\arg(z - t_0)$, $z \in D^+$, changes continuously for any point $t_0 \in \Gamma$, then there exists a positive number M such that

$$|\arg(z - t_0)| < M \tag{17}$$

for arbitrary $z \in D^+$ and $t_0 \in \Gamma$.

Below we give one more requirement for Γ.

Now we introduce conditions for G. It is assumed that $G \in C(\Gamma; t_1, \dots, t_m)$, i.e., G is continuous on the closed arcs $[t_k, t_{k+1}]$, $t_{m+1} = t_1$. It is assumed that the points t_k are numbered according to the growth of the arc abscissa. Suppose that

$$\inf_{t\in\Gamma} |G(t)| > 0. \tag{18}$$

We will show that the method adopted in [8], [24], [15] for factorization of piecewise-continuous functions can be successfully applied in the case under consideration if the theorems proved above are used.

Let

$$\frac{G(t_k-)}{G(t_k+)} = e^{2\pi i \lambda_k}, \quad k = \overline{1,m}, \tag{19}$$

where $\lambda_k = \alpha_k + i\beta_k$ is a complex number whose real part is defined to an accuracy of an integer summand.

Assume that

$$\alpha_k \neq \frac{1}{p'} \pmod 1. \tag{20}$$

Then α_k can be chosen so that

$$-\frac{1}{p} < \alpha_k < \frac{1}{p'}. \tag{21}$$

Choose in D^+ a point z_0 and draw from it a simple smooth curve γ_k connecting it with $z = \infty$ and having only one common point t_k with Γ. The function $(z - z_0)^{\lambda_k}$ is analytic in the plane cut along γ_k and

$$(t_k - z_0)_+^{\lambda_k} = (t_k - z_0)_-^{\lambda_k} \exp(2\pi i \lambda_k),$$

where $(t_k - z_0)_\pm^{\lambda_k} = \lim\limits_{t \to t_k \pm,\, t\in\Gamma} (t - z_0)^{\lambda_k}$. This and (19) imply that the function

$$G_1(t) = \frac{G(t)}{\prod\limits_{k=1}^{m} (t - z_0)^{\lambda_k}} \tag{22}$$

is continuous on Γ ([24, p. 432]) and $G_1(t) \neq 0$, $t \in \Gamma$.

Let

$$\rho_k(z) = \begin{cases} (z - z_0)^{\lambda_k}, & z \in D^+, \\ \left(\frac{z-t_k}{z-z_0}\right)^{\lambda_k} = \frac{(z-t_k)^{\lambda_k}}{(z-z_0)^{\lambda_k}}, & z \in D^-, \end{cases} \qquad k = \overline{1, m}, \tag{23}$$

where the branch of functions $\left(\frac{z-t_k}{z-z_0}\right)^{\lambda_k}$ is chosen so that it tends to 1 as $z \to \infty$. These functions are analytic in the domains $D^\pm$.

We have

$$\rho_k(z) = e^{\alpha_k \ln|z-t_k| - \beta_k \arg(z-t_k)} e^{i(\beta_k \ln|z-t_k| + \alpha_k \arg(z-t_k))}.$$

By virtue of condition (17) we conclude that

$$|\rho_k(z)| \sim |z - t_k|^{\alpha_k}.$$

Let us now require that

$$(\rho_k(z) - 1) \in \bigcup_{\delta>0} E^\delta(D^\pm). \tag{24}$$

Since Γ has the Carleson property and using (21) the function $r(t) = |t - t_k|^{\alpha_k}$ belongs to $W^p(\Gamma)$ (i.e., the operator $Tf = rS_\Gamma \frac{1}{r} f$ is continuous in $L^p(\Gamma)$). Hence it follows that from some $\eta > 0$

$$\rho_k^\pm \in L^{p+\eta}(\Gamma), \quad \frac{1}{\rho_k^\pm} \in L^{p'+\eta}(\Gamma). \tag{25}$$

Since Γ is a Carleson curve, the domains $D^\pm$ are Smirnov domains (see, e.g., [12, p. 22]) and therefore to the functions $\rho_k(z)$ we can apply the well-known Smirnov's theorem which states that if $\Phi \in E^{p_1}(D^+)$ and $\Phi^+ \in L^{p_2}(\Gamma)$, $p_2 > p_1$, then $\Phi \in E^{p_2}(D^+)$ ([9], Ch. X). An analogous statement is true for D^-. By virtue of (24) and (25) we obtain

$$(\rho_k(z) - 1) \in E^{p+\eta}(D^\pm), \quad \left(\frac{1}{\rho_k(z)} - 1\right) \in E^{p'+\eta}(D^\pm). \tag{26}$$

Now assuming

$$\rho(z) = \prod_{k=1}^{m} \rho_k(z), \tag{27}$$

by virtue of (24) we obtain

$$(\rho(z) - 1) \in E^{p+\eta}(D^+), \quad \left(\frac{1}{\rho(z)} - 1\right) \in E^{p'+\eta}(D^-). \tag{28}$$

Consider the function

$$X_1(z) = \exp\{(K_\Gamma \ln G_1(t))(z)\} \tag{29}$$

and let

$$X(z) = \rho(z) X_1(z). \tag{30}$$

Since G_1 is continuous, we have $(X_1 - 1) \in \bigcap\limits_{\delta>1} E^\delta(D^\pm)$ (see [12, p. 96]) and therefore from (24) and (30) it follows that for some $\delta_0 > 0$,

$$(X - 1) \in E^{p+\delta_0}(D^\pm), \quad \left(\frac{1}{X} - 1\right) \in E^{p'+\delta_0}(D^\pm). \tag{31}$$

Let us introduce a new unknown function

$$\Phi_1(z) = \frac{\Phi(z)}{\rho(z)}.$$

Since $\Phi = K_\Gamma\varphi$, $\varphi \in \bigcap\limits_{0<\delta<p-1} L^{p-\delta}$, $p > 1$, and Γ is a Carleson curve, we have $\Phi \in \bigcap\limits_{0<\delta<p-1} E^{p-\delta}(D^\pm)$. This and (28) imply that $\Phi_1 \in E^{\delta_1}(D^\pm)$, for some $\delta_1 > 0$. Moreover, $\Phi_1^\pm(t) \in L^{1+\eta_1}(\Gamma)$, $\eta_1 > 0$. Thus, by virtue of Smirnov's theorem mentioned above, we obtain $\Phi_1 \in E^{1+\eta_1}(D^\pm)$ for some $\eta_1 > 0$ and in that case the representation $\Phi_1(z) = K_\Gamma(\Phi_1^+ - \Phi_1^-)$ implies $\Phi_1 \in K^{1+\eta_1}(\Gamma)$. Therefore Φ_1 is a solution of the class $K^{1+\eta_1}(\Gamma)$ for the following boundary value problem

$$\Phi_1^+ = G_1\Phi_1^- + g/\rho^+.$$

If $\varkappa = \operatorname{ind} G_1 \geq 0$, then all solutions of this problem are given by the equality

$$\Phi_1(z) = X_1(z) K_\Gamma\left(\frac{g}{\rho^+ X_1^+}\right)(z) + X_1(z) Q_{\varkappa-1}(z)$$

(see, e.g., [15, p. 113]).

Hence, taking the equality $\Phi(z) = \Phi_1(z)\rho(z)$ into account, we obtain that all solutions of problem (1) of the class $K^{p),\theta}(\Gamma)$ lie in the set of functions

$$\Phi(z) = \rho(z) X_1(z) K_\Gamma\left(\frac{g}{\rho^+ X_1^+}\right)(z) + \rho(z) X_1(z) Q_{\varkappa-1}(z). \tag{32}$$

Let us show that all functions given by equality (32) belong to the class $K^{p),\theta}(\Gamma)$.

In the first place we note that $\rho(z)X_1(z) = X(z)$ and from (31) it follows that $X(z)Q_{\varkappa-1}(z) \in E^p(D^{\pm})$ and $[X^+(t) - X^-(t)]Q_{\varkappa-1}(t) \in L^p(\Gamma)$. So, the second summands from (32) $\rho(z)X_1(z)Q_{\varkappa-1}(z)$ are contained in $K^p(\Gamma)$ and thereby in $K^{p),\theta}(\Gamma)$, too.

For the first summand $F(z) = \rho(z)X_1(z)K_\Gamma\left(\frac{g}{\rho^+X_1^+}\right)(z)$ from (32) we have

$$F^{\pm}(t) = \pm\frac{1}{2}\,g(t) + \frac{1}{2}\,\rho^+(t)X_1^+(t)S_\Gamma\left(\frac{g}{\rho^+X_1^+}\right)(t), \quad g \in L^{p)}(\Gamma). \tag{33}$$

Here $\rho^+ \in W^{p),\theta}(\Gamma)$ since $\rho(t) \sim \prod_{k=1}^{m} |t-t_k|^{\alpha_k}$ and, for the numbers α_k, conditions (21) are fulfilled (this is one of the claims of Theorem A). Hence, by assumption (22), it follows that $\frac{1}{\rho^+} \in L^{p'+\varepsilon}(\Gamma)$, $\varepsilon > 0$. Moreover, using the Sokhotski-Plemelj formulas (see, e.g., [2], Theorem 6.7), we get

$$X_1^+(t) = \sqrt{|G(t)|}\,\exp\frac{1}{2\pi}\int_\Gamma \frac{\arg G_1(\tau)(\tau-z_0)^{-\varkappa}}{\tau - r}\,d\tau = \sqrt{|G(t)|}\,\exp\int_\Gamma \frac{\psi(\tau)\,d\tau}{\tau - t},$$

where $\psi(\tau) = 2\pi\arg G_1(\tau)(\tau-z_0)^{-\varkappa}$ is a real continuous function. Thus $X_1^+ \sim \exp\int_\Gamma \frac{\psi(\tau)\,d\tau}{\tau-t}$. Therefore by Theorem 2 we conclude that $\rho^+X_1^+ \in W^{p),\theta}(\Gamma)$. By virtue of this fact, (33) implies that $F(z) \in K^{p),\theta}(\Gamma)$.

As a result we obtain that the functions Φ given by equality (32) belong to $K^{p),\theta}(\Gamma)$ for $\varkappa \ge 0$.

For $\varkappa < 0$, a function Φ given by equality (32) belongs to $K^{1+\eta_1}(\Gamma)$ if and only if $Q_{\varkappa-1}(z) \equiv 0$ and the conditions

$$\int_\Gamma \frac{g(\tau)}{\rho^+(\tau)X_1^+(\tau)}\,\tau^k d\tau = 0, \quad k = \overline{0, |\varkappa|-1}, \tag{34}$$

are fulfilled.

Assuming that these conditions are fulfilled, in the same manner as above we will show that $\Phi \in K^{p),\theta}(\Gamma)$.

So, if conditions (17) and (24) (for Γ) and conditions (18) and (20) (for G) are fulfilled, then we have the situation in which problem (1) is solvable.

We observe that any piecewise-smooth curve is a Carleson one and for such curves condition (17) is fulfilled. As to condition (24), it is fulfilled at least for those curves which do not have internal cusps.

Lemma 1. *If a domain D is bounded by a simple piecewise smooth closed curve Γ with one angular point t_0 where an angle value with respect to D is equal to $\nu\pi$, $0 < \nu \le 2$, then the function $\frac{1}{(z-t_0)^\lambda}$, $\lambda = \alpha + i\beta$, $\alpha \in \left(-\frac{1}{p}, \frac{1}{p'}\right)$, belongs to some class $E^\delta(D)$.*

Proof. Let Γ_r be the images of circumferences $|w| = r < 1$ for the conformal mapping $z = z(w)$ of the unit circle $\mathrm{U} = \{\mathrm{w} : |\mathrm{w}| < 1\}$ onto D. To prove the lemma, it suffices to show that

$$I = \sup_{0<r<1} \int_{\Gamma_r} \frac{|dz|}{|z-t_0|^{\alpha\delta}} < \infty, \quad \alpha \in \left(-\frac{1}{p}, \frac{1}{p'}\right), \quad \delta > 0.$$

Assuming $z(c) = t_0$, we obtain

$$I = \sup_{0<r<1} \int_0^{2\pi} \frac{|z'(w)|\,|dw|}{|z(w)-z(c)|^{\alpha\delta}}\,.$$

If $\alpha \le 0$, then $I < \infty$ since $z' \in H^1$ (see, e.g., [9, Ch. X]). So we should consider the case $\alpha \in \left(0, \frac{1}{p'}\right)$. Using the relations ([12, p. 153])

$$z'(w) = (w-c)^{\nu-1} z_0(w), \quad z(w)-z(c) = (w-c)^{\nu} z_1(w), \quad z_k^{\pm 1} \in \bigcap_{\delta>0} H^{\delta}, \ \ k = 0, 1,$$

we obtain

$$I = \sup_r \int_0^{2\pi} |w-c|^{\nu-1-\nu\alpha\delta} |z_0(w)|\,|z_1(w)|^{-\alpha\delta} |dw|\,.$$

Hence we see that $I < \infty$ if $\nu(1-\alpha\delta) - 1 > -1$, i.e., if $1 - \alpha\delta > 0$. Now, for δ we obtain the inequality $\delta < \frac{1}{\alpha}$ and since $\frac{1}{\alpha} > p'$, we finally have that $I < \infty$ for $\delta < p'$. □

Thus the following theorem is valid.

Theorem 4. *Let Γ be a piecewise-smooth simple closed curve without a zero internal angle and $G(t)$ be a function of the class $C(\Gamma; t_1, \dots, t_m)$, $t_k \in \Gamma$, given on Γ and being such that conditions* (18)–(20) *are fulfilled for it. Assume that $g \in L^{p),\theta}(\Gamma)$, $p > 1$, $\theta > 0$. Let further the function $\rho(z)$ be given by equalities* (23) *and* (27), *the function G_1 by equality* (22), *$X_1(z)$ by formula* (29) *and $\varkappa = \operatorname{ind} G_1$.*

Then for $\varkappa \ge 0$, problem (1) *is solvable in the class $K^{p),\theta}(\Gamma)$ and its general solution is given by equality* (32) *where $Q_{\varkappa-1}(z)$ is an arbitrary polynomial of order $\varkappa - 1$ ($Q_{-1}(z) \equiv 0$). If however $\varkappa < 0$, then the problem is solvable if and only if conditions* (34) *are fulfilled and in that case it is solvable uniquely. Then a solution is given by equality* (32) *where $Q_{\varkappa-1}(z) \equiv 0$.*

2. The Dirichlet problem

In this section it is assumed that D is the internal domain bounded by a simple closed Lyapunov curve Γ. Put

$$K^{p),\theta}(D;\omega) = \left\{\Phi : \Phi(z) = \frac{1}{2\pi i}\int_\Gamma \frac{\varphi(t)\,dt}{t-z},\ z \in D,\ \varphi \in L^{p),\theta}(\Gamma;\omega)\right\}, p>1,\ \theta>0,$$

$$e^{p),\theta}(D;\omega) = \left\{u : u = \operatorname{Re}\Phi,\ \ \Phi \in K^{p),\theta}(\Gamma;\omega)\right\}$$

and

$$h^{\delta}(U) = \left\{u : u = \operatorname{Re}\Phi,\ \Phi \in H^{\delta}\right\}, \quad \delta > 0.$$

Assume that the weight function

$$\omega(t) = \prod_{k=1}^{n} (t - t_k)^{\alpha_k}, \quad t_k \in \Gamma, \quad \alpha_k \in \left(\frac{1}{p}, \frac{1}{p'}\right).$$

Here we will solve the Dirichlet problem formulated as follows

$$\begin{aligned} \Delta \mathcal{U} = 0, \qquad & \mathcal{U} \in e^{p),\theta}(D;\omega), \\ \mathcal{U}|_\Gamma = f(t), \quad & f \in L^{p),\theta}(\Gamma;\omega). \end{aligned} \tag{35}$$

As in the preceding section, $z = z(w)$ denotes the conformal mapping of the unit circle U onto the domain D, and $w = w(z)$ is its inverse mapping. Consider the function

$$\rho(\tau) = \prod_{k=1}^{n} |\tau - \tau_k|^{\alpha_k}, \quad \tau_k = w(t_k). \tag{36}$$

Since Γ is a Lyapunov curve, by virtue of the well-known theorem due to Kellog (see, for example, [9], p.411) we have

$$|z(\tau) - z(\tau_k)| \approx |\tau - \tau_k| \quad \text{and} \quad \inf_{\tau\in\gamma} |z'(\tau)| > 0, \quad \sup_{\tau\in\gamma} |z'(\tau)| < \infty. \tag{37}$$

Hence it readily follows that

$$|\omega(z(\tau))| \sim |\rho(\tau)| \quad \text{and} \quad |\rho(w(t))| \sim |\omega(t)|. \tag{38}$$

Along with the function $\mathcal{U}(z)$, we will consider the harmonic function $u(w) = \mathcal{U}(z(w))$.

Lemma 2. *If $\mathcal{U} \in e^{p),\theta}(D;\omega)$ and $u(w) = \mathcal{U}(z(w))$, then $u \in e^{p),\theta}(\mathrm{U};\rho)$. Conversely, if $u \in e^{p),\theta}(\mathrm{U},\rho)$, then $\mathcal{U}(z) = u(w(z)) \in e^{p),\theta}(D;\omega)$.*

Proof. It suffices to establish that if $\mathcal{U}(z) = \operatorname{Re}\Phi(z)$, $\Phi \in K^{p),\theta}(D;\omega)$, then the function $\Psi(w) = \Phi(z(w))$ belongs to the class $K^{p),\theta}(\mathrm{U};\rho)$. Since $\Phi \in K^{p),\theta}(D;\omega)$, we have $\Phi(z) = (K_\Gamma\varphi)(z)$, $\varphi \in L^{p),\theta}(\Gamma;\omega)$. But then $\varphi\omega \in \bigcap_{0<\varepsilon<p} L^{p-\varepsilon}(\Gamma)$ and since $\frac{1}{\omega} \in L^{p'+\delta}(\Gamma)$, $\delta > 0$, there exists $\eta > 0$ such that $\varphi \in L^{1+\eta}(\Gamma)$. Therefore $\Phi \in E^{1+\eta}(D)$ (see [12, p. 29]). Thus $\Phi(z(w)) \sqrt[1+\eta]{z'(w)}$ belongs to the class $H^{1+\eta}$. Furthermore, by (37) we conclude that $\Psi(w) = \Phi(z(w)) \in H^1$ and therefore $\Psi(w) = (K_\gamma\Psi^+)(w)$.

For the function Ψ^+ we have

$$\begin{aligned} \int_\gamma |\Psi^+(\tau)\rho(\tau)|^{p-\varepsilon}|d\tau| &= \int_\Gamma \left|\Phi^+(z(\tau))\omega(z(\tau))\,\frac{\rho(\tau)}{\omega(z(\tau))}\right|^{p-\varepsilon} |d\tau| \\ &\le M \int_\gamma |\Phi^+(z(\tau))\omega(z(\tau))|^{p-\varepsilon}|d\tau| \\ &= M \int_\Gamma |\Phi^+(t)\omega(t)|^{p-\varepsilon} \frac{|dt|}{|z'(t)|} \le M_1 \int_\Gamma |\Phi^+(t)\omega(t)|^{p-\varepsilon} ds. \end{aligned}$$

Since a Cauchy singular integral operator is continuous in the space $L^{p),\theta}(\Gamma;\omega)$, the inclusion $K_\Gamma\varphi \in K^{p),\theta}(D;\omega)$ implies that $\Phi^+ \in L^{p),\theta}(D;\omega)$. Now from the above estimates we establish that $\Psi^+ \in L^{p),\theta}(\gamma;\rho)$, whereas from the equality $\Psi(w) = (K_\gamma\Psi^+)(w)$ we obtain $\Psi \in K^{p),\theta}(\mathrm{U};\rho)$.

The converse statement is proved analogously. □

By virtue of this lemma we have that problem (35) is equivalent to the problem

$$\begin{aligned} &\Delta u = 0, \quad u \in e^{p),\theta}(\mathrm{U};\rho), \\ &u|_\Gamma = f(z(\tau)), \quad f(z(\tau)) = g(\tau) \in L^{p),\theta}(\gamma;\rho). \end{aligned} \tag{39}$$

As easily seen, $g \in L^{1+\eta}(\gamma)$, for some $\eta > 0$, whereas from the proof of the lemma it follows that $e^{p),\theta}(\mathrm{U};\rho) \subset \mathrm{h}^{1+\eta}(\mathrm{U})$. Therefore if the solution $u(w)$ of problem (39) exists, it belongs to $h^{1+\eta}(\mathrm{U})$, too. But the function

$$u(w) = \mathrm{Re}\,\frac{1}{2\pi}\int_\gamma \frac{f(z(\tau))}{\tau}\,\frac{\tau + w}{\tau - w}\,d\tau \tag{40}$$

is the only solution of such kind (see [23]).

Furthermore,

$$\begin{aligned} \Psi(w) &= \frac{1}{2\pi}\int_\gamma \frac{f(z(\tau))}{\tau}\,\frac{\tau + w}{\tau - w}\,d\tau \\ &= \frac{1}{2\pi}\int_\gamma \frac{f(z(\tau))}{\tau - w}\,d\tau + \frac{w}{2\pi}\int_\gamma \frac{f(z(\tau))}{\tau}\,\frac{d\tau}{\tau - w}\,d\tau \\ &= \Psi_1(w) + \Psi_2(w). \end{aligned}$$

The function $\Psi(w)$ belongs to $K^{p),\theta}(\gamma;\rho)$ since both summands belong to the Hardy class H^1, and, by the continuity of the operator S_Γ in the space $L^{p),\theta}(\gamma;\rho)$, their boundary values belong to the class $L^{p),\theta}(\Gamma;\rho)$. Hence it follows that the function defined by equality (40) belongs to $e^{p),\theta}(\mathrm{U};\rho)$.

Thus we have proved

Theorem 5. *If D is the internal domain bounded by a closed Lyapunov curve, then the Dirichlet problem* (35) *is uniquely solvable and its solution is given by the equality*

$$\mathcal{U}(z) = \mathrm{Re}\,\frac{1}{2\pi}\int_\gamma \frac{f(z(\tau))}{\tau}\,\frac{\tau + w(z)}{\tau - w(z)}\,d\tau,$$

where $w = w(z)$ is the conformal mapping of D onto U.

Acknowledgement

The authors were partially supported by the Shota Rustaveli National Science Foundation (Project GNSF/ST09_23_3-100).

The authors express their gratitude to the referees for valuable remarks and suggestions.

References

[1] H.G.W. Begehr, *Complex analytic methods for partial differential equations.* An introductory text. World Scientific Publishing Co., Inc. River Edge, NJ, 1994.

[2] A. Böttcher and Y.I. Karlovich, *Carleson curves, Muckenhoupt weights, and Toeplitz operators.* Progress in Mathematics, 154. Birkhäuser Verlag, Basel, 1997.

[3] L. Diening, P. Harjulehto, P. Hästo, and M. Růžička, *Lebesgue and Sobolev spaces with variable exponent.* Lecture Notes in Math., vol. 2017, Springer-Verlag, 2011.

[4] P.L. Duren, *Theory of H^p spaces.* Pure and Applied Mathematics, Vol. 38. Academic Press, New York–London, 1970.

[5] E.M. Dynkin, Methods of the theory of singular integrals. (Hilbert transforms and Calderón–Zygmund theory.) (Russian) *Itogi Nauki i Tekhniki, Sovrem. Probl. Mat. Fund. Naprav.* 15, *Moscow,* 1987, 197–292.

[6] A. Fiorenza, B. Gupta and P. Jain, The maximal theorem in weighted grand Lebesgue spaces, *Studia Math.* **188**(2008), No. 2, 123–133.

[7] A. Fiorenza and C. Sbordone, *Existence and uniqueness results for solutions of nonlinear equations with right-hand side in L^1.* Studia Math. **127** (1998), no. 3, 223–231.

[8] F.D. Gakhov, *Boundary value problems.* (Russian) Third edition, revised and augmented Izdat. "Nauka", Moscow, 1977. English Translation of the second edition edited by I.N. Sneddon, Pergamon Press International Series of Monographs in Pure and Applied Mathematics, Volume **85** Addison Wesley 1965.

[9] G.M. Goluzin, *Geometrical theory of functions of a complex variable.* (Russian) Second edition. "Nauka", Moscow, 1966.

[10] L. Greco, T. Iwaniec, and C. Sbordone, *Inverting the p-harmonic operator.* Manuscripta Math. **92** (1997), no. 2, 249–258.

[11] T. Iwaniec and C. Sbordone, *On the integrability of the Jacobian under minimal hypotheses.* Arch. Rational Mech. Anal. **119** (1992), no. 2, 129–143.

[12] G. Khuskivadze, V. Kokilashvili, and V. Paatashvili, *Boundary value problems for analytic and harmonic functions in domains with nonsmooth boundaries. Applications to conformal mappings.* Mem. Differential Equations Math. Phys. **14** (1998), 195 pp.

[13] G. Khuskivadze, V. Kokilashvili, and V. Paatashvili, *The Dirichlet problem for variable exponent Smirnov class harmonic functions in doubly-connected domains.* Mem. Differential Equations Math. Phys. **52** (2011), 131–156.

[14] B.V. Khvedelidze, *Linear discontinuous boundary problems in the theory of functions, singular integral equations and some of their applications.* (Russian) Trudy Tbiliss. Mat. Inst. Razmadze **23** (1956), 3–158.

[15] B.V. Khvedelidze, *The method of Cauchy type integrals in discontinuous boundary value problems of the theory of holomorphic functions of a complex variable.* Itogi Nauki i Tekhniki. Ser. Sovrem. Probl. Mat., 7, VINITI, Moscow, 1975, 5–162; translation in J. Soviet Math. **7** (1977), no. 3, 309–414.

[16] V. Kokilashvili, V. Paatashvili, and S. Samko, *Boundary value problems for analytic functions in the class of Cauchy-type integrals with density in $L^{p(\cdot)}(\Gamma)$.* Bound. Value Probl. **2005**, no. 1, 43–71.

[17] V. Kokilashvili and V. Paatashvili, *The Riemann–Hilbert problem in weighted classes of Cauchy type integrals with density from* $L^{p(\cdot)}(\Gamma)$. Complex Anal. Oper. Theory **2** (2008), no. 4, 569–591.

[18] V. Kokilashvili, *The Riemann boundary value problem for analytic functions in the frame of grand* L^p *spaces.* Bull. Georgian Natl. Acad. Sci. (N.S.) **4** (2010), no. 1, 5–7.

[19] V. Kokilashvili, *Boundedness criteria for singular integrals in weighted grand Lebesgue spaces.* J. Math. Sci. **170** (2010), no. 1, 20–23.

[20] V. Kokilashvili and S. Samko, *Boundedness of weighted singular integral operators on Carleson curves in Grand Lebesgue spaces.* In: ICNAAM 2010: Intern. Conf. Nam. Anal. and Appl. Math. 1281, pp. 490–493. ATP Conf. Proc., 2010.

[21] V. Kokilashvili and S. Samko, *Boundedness of weighted singular integral operators in Grand Lebesgue spaces.* Georgian Math. J. **18** (2011), 259–269.

[22] V. Kokilashvili and V. Paatashvili, *The Dirichlet problem for harmonic functions from variable exponent Smirnov classes in domains with piecewise smooth boundary.* J. Math. Sci **172** (2011), no. 3, 1–21.

[23] Z. Meshveliani, V. Paatashvili, *On Smirnov classes of harmonic functions, and the Dirichlet problem.* Proc. A. Razmadze Math. Inst. **123** (2000), 61–91.

[24] N.I. Muskhelishvili, *Singular integral equations.* (Russian) Third, corrected and augmented edition. Izdat. "Nauka", Moscow, 1968. English Translation edited by J.R. Radok, Noordhoff International Publishing Leyden, 1977.

[25] B. Riemann, *Grundlagen Für eine allgemeine Theorie der Funktionen einer veränderlichen komplexen Grösse.* Worke, Leipzig, 1867.

[26] J.L. Rubio de Francia, *Factorization and extrapolation of weights.* Bull. Amer. Math. Soc. (N.S.), **7**(1982), 393–395.

[27] R. Wegmann, A.H.M. Murid, M.M.S. Nasser, *The Riemann-Hilbert problem and the generalized Neumann kernel.* Journal of Computational and Applied Mathematics. **182** (2005), 388–415.

[28] V.V. Zhikov, *Averaging of functionals of the calculus of variations and elasticity theory.* (Russian) Izv. Akad. Nauk SSSR Ser. Mat. **50** (1986), no. 4, 675–710.

[29] V.V. Zhikov, S.E. Pastukhova, *On the improved integrability of the gradient of solutions of elliptic equations with a variable nonlinearity exponent.* (Russian) Mat. Sb. **199** (2008), no. 12, 19–52; translation in Sb. Math. **199** (2008), no. 11-12, 1751–1782.

V. Kokilashvili and V. Paatashvili
A. Razmadze Mathematical Institute
I. Javakhishvili Tbilisi State University
2, University Str.
Tbilisi 0186, Georgia
e-mail: `kokil@rmi.ge`

Operator Theory:
Advances and Applications, Vol. 229, 253–268

Overview of Fractional h-difference Operators

Dorota Mozyrska and Ewa Girejko

Abstract. Fractional difference operators and their properties are discussed. We give a characterization of three operators that we call Grünwald-Letnikov, Riemann-Liouville and Caputo like difference operators. We show relations among them. In the paper, linear fractional h-difference equations are described. We give formulas of solutions to initial value problems. Crucial formulas are gathered in the tables presented in the last section of the paper.

Mathematics Subject Classification (2010). Primary 39A70; Secondary 26A33.

Keywords. Fractional difference operator, fractional difference equation, fractional derivative.

1. Introduction

The fractional calculus is a generalization of differential/integral calculus of integer order [2, 8, 17, 21]. The calculus is a very important mathematical research field that has its applications in such areas as engineering, in particular electronic devices, economics and physics [3, 7, 16, 18, 23]. One can consider different types of fractional operators that are associated with different fractional derivatives, namely Grünwald-Letnikov, Riemann-Liouville and Caputo ones. Similarly as in the continuous case there are various ways to introduce the fractional operators as a generalization of the corresponding derivatives/differences or integrals/sums. The most known fractional operators are: sum operator introduced by Miller and Ross in [16], Riemann-Liouville like difference operator explored by Atici and Eloe in [5, 6], Caputo like difference operator introduced in [1, 4, 11] and the last one, Grünwald-Letnikov like difference operator examined by Podlubny, Kaczorek, Ostalczyk (see [14, 19, 20]) and by many others. However, there are only few papers, like the ones by Bastos, Ferreira and Torres (see [9, 12]) that concern the development of h-sum and h-difference operators (namely Riemann-Liouville like difference operators). The variety of operators was the motivation for us to describe and to characterize three of them: Grünwald-Letnikov, Riemann-Liouville and Caputo like difference operators.

Recently one can find in the literature one more approach in this scientific area, namely in [10] authors present advanced results for linear fractional difference initial value problems on general time scale. They use a different definition of h-factorial functions from the one we use in this paper. Our approach arises from [1, 6, 9, 12]. Since these two approaches lead to different definitions, we decided not to compare them. We follow the notations introduced in [1, 6, 9, 12].

Researchers from the area of fractional differences can find the presentation of relations among the listed above operators useful. Moreover, the results presented in the paper can be applied to the theory of discrete fractional systems as well as to discrete control fractional systems. In the present paper we focus on the overview of the mentioned operators: we summarize but also develop what was done in this area so far. To our knowledge, there was a gap in the literature concerning the description of these three operators. To make it more general and flexible in the sense to get freedom in choosing a step, we decided to present all results on $h\mathbb{Z}$. Moreover, the case with different step h that is taken under our consideration gives us a task for the future: to examine the case when h tends to zero.

The manuscript consists of six sections. In the second one we gather all preliminary definitions and notations that are needed in the sequel. In this section we introduce the definition of Caputo like difference operator for any $h > 0$. We also develop some properties of this operator in Proposition 2.9. Section 3 concerns power rule formulas and all results presented in this section, besides Lemma 3.1 are due to the authors. In the fourth section we prove results on properties of the three mentioned operators. In Section 5, which is devoted to initial value problems and exponentials, almost all stated results are original excluding Proposition 5.4. The last section summarizes the paper gathering some crucial facts in tables.

2. Preliminaries

Let us denote by $\mathcal{F}_D$ the set of real-valued functions defined on D. Let $h > 0$ and put $(h\mathbb{N})_a := \{a, a+h, \dots\}$, with $a \in \mathbb{R}$. Let $\sigma(t) = t + h$ for $t \in (h\mathbb{N})_a$.

Definition 2.1. For a function $\varphi \in (h\mathbb{N})_a$ the forward h-difference operator is defined as

$$(\Delta_h \varphi)(t) = \frac{\varphi(\sigma(t)) - \varphi(t)}{h}, \quad t \in (h\mathbb{N})_a = \{a, a+h, a+2h, \dots\}$$

while the h-difference sum is given by

$$\left({}_a\Delta_h^{-1}\varphi\right)(t) = \sum_{k=\frac{a}{h}}^{\frac{t}{h}-1} \varphi(kh)h, \quad t \in (h\mathbb{N})_a = \{a, a+h, a+2h, \dots\}.$$

Definition 2.2. [9] For arbitrary $t, \alpha \in \mathbb{R}$ the h-factorial function is defined by

$$t_h^{(\alpha)} := h^\alpha \frac{\Gamma(\frac{t}{h}+1)}{\Gamma(\frac{t}{h}+1-\alpha)},$$

where Γ is the Euler gamma function with $\frac{t}{h}+1 \notin \mathbb{Z}_- \cup \{0\}$, and we use the convention that division at a pole yields zero.

Remark 2.3. [12] For $t \geq 0$ and $\alpha \in \mathbb{R}$, $\lim_{h\to 0} t_h^{(\alpha)} = t^\alpha$.

Definition 2.4. [9] For a function $\varphi \in \mathcal{F}_{(h\mathbb{N})_a}$ the *fractional h-sum of order* $\alpha > 0$ is given by

$$\left({}_a\Delta_h^{-\alpha}\varphi\right)(t) = \frac{h}{\Gamma(\alpha)} \sum_{k=\frac{a}{h}}^{\frac{t}{h}-\alpha} (t-\sigma(kh))_h^{(\alpha-1)}\varphi(kh),$$

with $\left({}_a\Delta_h^{0}\varphi\right)(t) = \varphi(t)$ and $\sigma(kh) = (k+1)h$.

Remark 2.5. Note that ${}_a\Delta_h^{-\alpha} : \mathcal{F}_{(h\mathbb{N})_a} \to \mathcal{F}_{(h\mathbb{N})_{a+\alpha h}}$.

As the first we present fractional h-difference Riemann-Liouville like operator. The definition of the operator can be found, for example, in [6] (for $h = 1$) or in [9] (for any $h > 0$).

Definition 2.6. Let $\alpha \in (n-1, n]$ and set $\mu = n - \alpha$, where $n \in \mathbb{N}$. The *Riemann-Liouville like fractional h-difference operator* ${}_a\Delta_h^\alpha\varphi$ of order α for a function $\varphi \in \mathcal{F}_{(h\mathbb{N})_a}$ is defined by

$$\left({}_a\Delta_h^{\alpha}\varphi\right)(t) = \left(\Delta_h^n\left({}_a\Delta_h^{-\mu}\varphi\right)\right)(t) = \frac{h}{\Gamma(\mu)}\Delta_h^n \sum_{k=\frac{a}{h}}^{\frac{t}{h}-\mu} (t-\sigma(kh))_h^{(\mu-1)}\varphi(kh). \tag{2.1}$$

Remark 2.7. Note that ${}_a\Delta_h^{\alpha} : \mathcal{F}_{(h\mathbb{N})_a} \to \mathcal{F}_{(h\mathbb{N})_{a+\nu}}$, where $\alpha \in (n-1, n]$ and $\nu = \mu h$.

Theorem 2.8. *Let* $\varphi : (h\mathbb{N})_a \to \mathbb{R}$ *and* $\alpha > 0$ *be given, with* $\alpha \in (n-1, n]$, $\mu = n - \alpha$ *and* $\nu = \mu h$, $\mu, \nu > 0$. *The following formula for the fractional difference* ${}_a\Delta_h^\alpha\varphi : (h\mathbb{N})_{a+\nu} \to \mathbb{R}$ *is equivalent to* (2.1):

$$\left({}_a\Delta_h^{\alpha}\varphi\right)(t) = \begin{cases} \frac{h}{\Gamma(-\alpha)} \sum_{s=\frac{a}{h}}^{\frac{t}{h}+\alpha} (t-\sigma(kh))_h^{(-\alpha-1)}\varphi(kh), & \alpha \in (n-1, n) \\ (\Delta_h^n\varphi)(t), & \alpha = n. \end{cases} \tag{2.2}$$

Proof. Let φ and α be as in the statement of the theorem. We show that (2.1) is equivalent to (2.2) on $(h\mathbb{N})_{a+\nu}$ for $\mu = n - \alpha$ and $\nu = \mu h$, $\mu, \nu > 0$.
The case $\alpha = n$ is obvious, since

$$\left({}_a\Delta_h^{\alpha}\varphi\right)(t) = (\Delta_h^n({}_a\Delta_h^{-\mu}\varphi))(t) = \left(\Delta_h^n\left({}_a\Delta_h^{-0}\varphi\right)\right)(t) = (\Delta_h^n\varphi)(t).$$

If $\alpha \in (n-1, n)$, then the direct application of (2.1) yields:

$$\left({}_a\Delta_h^{\alpha}\varphi\right)(t) = \left(\Delta_h^n\left({}_a\Delta_h^{-\mu}\varphi\right)\right)(t) = \left(\frac{\Delta_h^{n-1}}{\Gamma(\mu)}\Delta_h\left(\sum_{k=\frac{a}{h}}^{\frac{t}{h}-\mu}(t-\sigma(kh))_h^{(\mu-1)}\varphi(kh)h\right)\right).$$

Let us denote by $g(t,k)=(t-\sigma(kh))_h^{(\mu-1)}\varphi(kh)h$. Then taking the difference with respect to t of the function $f(t,k)=\sum_{k=\frac{a}{h}}^{\frac{t}{h}-\mu} g(t,k)$ we get

$$\Delta_h(f(\cdot,k))(t)=\sum_{k=\frac{a}{h}}^{\frac{t}{h}+1-\mu} g(t+h,k)-\sum_{k=\frac{a}{h}}^{\frac{t}{h}-\mu} g(t,k)=h\sum_{k=\frac{a}{h}}^{\frac{t}{h}-\mu}\Delta_h g(t,k)+g(t+h,t/h+1-\mu).$$

The h-difference Δ_h of g with respect to t has the following form

$$(\Delta_h g(\cdot,k))(t)=(\mu-1)(t-\sigma(kh))_h^{\mu-2}\varphi(kh)h$$

and

$$g(t+h,t/h+1-\mu)=(t+h-\sigma(t+h-\mu h))_h^{(\mu-1)}\varphi(t+h-\mu h)h\,.$$

In particular, we receive

$$\Delta_h\left(f(\cdot,k)\right)(t)=\sum_{k=\frac{a}{h}}^{\frac{t}{h}-\mu}(\mu-1)(t-\sigma(kh))_h^{(\mu-2)}\varphi(kh)h+h^{\mu-1}\Gamma(\mu)\varphi(t+h-\mu h)h\,.$$

Now we easily see that

$$\begin{aligned}\left({}_a\Delta_h^\alpha\varphi\right)(t)&=\Delta_h^{n-1}\left(\frac{\Delta_h f(\cdot,k)}{\Gamma(\mu)}\right)(t)\\&=\Delta_h^{n-1}\left(\sum_{k=\frac{a}{h}}^{\frac{t}{h}+1-\mu}\frac{h}{\Gamma(\mu-1)}(t-\sigma(kh))_h^{(\mu-2)}\varphi(kh)\right).\end{aligned}$$

Repeating the similar procedure $n-2$ times we get

$$\begin{aligned}\left({}_a\Delta_h^\alpha\varphi\right)(t)&=\Delta_h^{n-n}\left(\sum_{k=\frac{a}{h}}^{\frac{t}{h}+n-\mu}\frac{h}{\Gamma(\mu-n)}(t-\sigma(kh))_h^{(\mu-(n+1))}\varphi(kh)\right)\\&=\frac{h}{\Gamma(-\alpha)}\sum_{k=\frac{a}{h}}^{\frac{t}{h}+\alpha}(t-\sigma(kh))_h^{(-\alpha-1)}\varphi(kh)\,.\end{aligned}$$ □

The next definition concerns the fractional h-difference Caputo like operator and for the case $h=1$ one can find it, for example, in [11].

Definition 2.9. Let $\alpha\in(n-1,n]$ and set $\mu=n-\alpha$, where $n\in\mathbb{N}$. The *Caputo like h-difference operator* ${}_a\Delta_{h,*}^\alpha\varphi$ of order α for a function $\varphi\in\mathcal{F}_{(h\mathbb{N})_a}$ is defined by

$$\left({}_a\Delta_{h,*}^\alpha\varphi\right)(t)=\left({}_a\Delta_h^{-\mu}\left(\Delta_h^n\varphi\right)\right)(t)=\frac{h}{\Gamma(\mu)}\sum_{k=\frac{a}{h}}^{\frac{t}{h}-\mu}(t-\sigma(kh))_h^{(\mu-1)}(\Delta_h^n\varphi)(kh),\quad(2.3)$$

for $t\in(h\mathbb{N})_{a+\nu}$.

Remark 2.10. Note that ${}_a\Delta_{h,*}^\alpha:\mathcal{F}_{(h\mathbb{N})_a}\to\mathcal{F}_{(h\mathbb{N})_{a+\nu}}$, where $\alpha\in(0,1]$ and $\nu=\mu h$.

Since

$$(\Delta_h^n \varphi)(kh) = \frac{1}{h^n} \sum_{r=0}^{n} (-1)^{r+1} \binom{n}{r} \varphi(r+k)h$$

we get the following proposition:

Proposition 2.11. *Let $\alpha \in (n-1, n]$ and set $\mu = n - \alpha$, where $n \in \mathbb{N}$. The following formula is equivalent to* (2.3)*:*

$$\left({}_a\Delta_{h,*}^{\alpha} \varphi\right)(t) = \begin{cases} \frac{h^{1-n}}{\Gamma(n-\alpha)} \sum_{k=\frac{a}{h}}^{\frac{t}{h}-(n-\alpha)} (t - \sigma(kh))_h^{(n-\alpha-1)} & \\ \qquad \times \sum_{r=0}^{n} (-1)^{r+1} \binom{n}{r} \varphi(r+k)h, & \alpha \in (n-1, n) \\ (\Delta_h^n \varphi)(t), & \alpha = n. \end{cases}$$

The last operator that we take under our consideration is the fractional h-difference Grünwald-Letnikov like operator and the definition of the operator can be found, for example, in [14].

Definition 2.12. Let $\alpha \in \mathbb{R}$. The *Grünwald-Letnikov like h-difference operator* ${}_a\widetilde{\Delta}_h^{\alpha}$ of order α for a function $\varphi \in \mathcal{F}_{(h\mathbb{N})_a}$ is defined by

$$\left({}_a\widetilde{\Delta}_h^{\alpha} \varphi\right)(t) = \sum_{s=0}^{\frac{t-a}{h}} a_s^{(\alpha)} \varphi(t - sh) \tag{2.4}$$

where

$$a_s^{(\alpha)} = (-1)^s \binom{\alpha}{s} \frac{1}{h^{\alpha}}$$

with

$$\binom{\alpha}{s} = \begin{cases} 1 & \text{for } s = 0 \\ \frac{\alpha(\alpha-1)\ldots(\alpha-s+1)}{s!} & \text{for } s \in \mathbb{N}. \end{cases}$$

Remark 2.13. Note that: ${}_a\widetilde{\Delta}_h^{\alpha} : \mathcal{F}_{(h\mathbb{N})_a} \to \mathcal{F}_{(h\mathbb{N})_a}$, whatever α is used.

In [1] we find the following definition of the discrete Mittag-Leffler functions for scalar case.

Definition 2.14. For $\lambda \in \mathbb{R}$ and $\alpha, \beta, z \in \mathbb{C}$ with $\text{Re}(\alpha) > 0$, the *discrete Mittag-Leffler functions* are defined by

$$E_{(\alpha,\beta)}(\lambda, z) = \sum_{k=0}^{\infty} \lambda^k \frac{(z + (k-1)(\alpha-1))^{(k\alpha)} (z + k(\alpha-1))^{(\beta-1)}}{\Gamma(\alpha k + \beta)}.$$

For $\beta = 1$ we write $E_{(\alpha)} = E_{(\alpha,1)}$.

We need in the sequel the following form for the Mittag-Leffler two-parameters function:

$$E_{(\alpha,\alpha)}(\lambda, z) = \sum_{k=0}^{\infty} \lambda^k \frac{(z + k(\alpha-1))^{(k\alpha+\alpha-1)}}{\Gamma((k+1)\alpha)}.$$

3. Power rule formulas

Now our goal is to prove formulas for fractional operators acting on power functions. Firstly we easily notice that for $p \neq 0$ it holds

$$\Delta_h (t-a)_h^{(p)} = p(t-a)_h^{(p-1)} .$$

If $\alpha \neq 1$ and p be such that $p+\alpha+1 \notin \mathbb{Z}_-$ (is not nonpositive integer), then from [5, 11, 13], for $h=1$ and the function $\varphi(s) = (s-a+p)_{h=1}^{(p)}$ it holds

$$\left({}_a\Delta_{h=1}^{-\alpha}\varphi\right)(t) = \frac{\Gamma(p+1)}{\Gamma(p+\alpha+1)}(t-a)^{(p+\alpha)}, \tag{3.1}$$

where $t \in \mathbb{N}_{a+p+\alpha}$.

In particular, if $\varphi(s) \equiv \text{constant} = \lambda$, then (3.1) implies that

$$\left({}_a\Delta_{h=1}^{-\alpha}\lambda\right)(t) = \frac{\lambda}{\Gamma(\alpha+1)}(t-a)^{(\alpha)}$$

for $t \in \mathbb{N}_{a+\alpha}$. The formula (3.1) can be also presented, see [13], as

$${}_{a+p}\Delta_{h=1}^{-\alpha}(t-a)^{(p)} = p^{(-\alpha)}(t-a)^{(p+\alpha)}$$

for $t \in \mathbb{N}_{a+p+\alpha}$. In [12] the authors prove the following lemma that gives transition between fractional summation operators for any $h>0$ and $h=1$.

Lemma 3.1. *Let $\varphi \in \mathcal{F}_{(h\mathbb{N})_a}$ and $\alpha > 0$. Then,*

$$\left({}_a\Delta_h^{-\alpha}\varphi\right)(t) = h^\alpha \left({}_{\frac{a}{h}}\Delta_1^{-\alpha}\tilde{\varphi}\right)\left(\frac{t}{h}\right),$$

where $t \in (h\mathbb{N})_{a+\alpha h}$ and $\tilde{\varphi}(s) = \varphi(sh)$.

Proposition 3.2. *Let $a \in \mathbb{R}, \alpha > 0$ be given. Then,*

$${}_{a+ph}\Delta_h^{-\alpha}(t-a)_h^{(p)} = p^{(-\alpha)}(t-a)_h^{(p+\alpha)} \tag{3.2}$$

for $t \in (h\mathbb{N})_{a+ph+\alpha h}$.

Proof. Let $\varphi(s) = (sh-a)_h^{(p)} = h^p(s-\frac{a}{h})^{(p)}$. We use Lemma 3.1 and write

$$\begin{aligned}
{}_{a+ph}\Delta_h^{-\alpha}(t-a)_h^{(p)} &= h^\alpha \left({}_{\frac{a}{h}+p}\Delta_1^{-\alpha}\varphi\right)\left(\frac{t}{h}\right) \\
&= h^{\alpha+p}\left({}_{\frac{a}{h}+p}\Delta_1^{-\alpha}\left(s-\frac{a}{h}\right)^{(p)}\right)\left(\frac{t}{h}\right) \\
&= h^\alpha h^p p^{(-\alpha)}\left(\frac{t-a}{h}\right)^{(p+\alpha)} = p^{(-\alpha)}(t-a)_h^{(p+\alpha)}.
\end{aligned}$$

□

The formula (3.2) can be also presented, according to the case $h=1$ in [13], as

$$\left({}_a\Delta_h^{-\alpha}\psi\right)(t) = p^{(-\alpha)}(t-a)_h^{(p+\alpha)}, \tag{3.3}$$

for $\psi(s) = (s-a+ph)_h^{(p)}$ and $t \in \mathbb{N}_{a+ph+\alpha h}$.

From the application of the power rule (3.2) follows the rule for composing two fractional h-sums. The proof for the case $h=1$ can be found in [13].

Proposition 3.3. *Let φ be a real-valued function defined on $(h\mathbb{N})_a$, where $a,h\in\mathbb{R}, h>0$ and $\alpha,\beta>0$. Then the following equalities hold*

$$\left({}_{a+\beta h}\Delta_h^{-\alpha}\left({}_a\Delta_h^{-\beta}\varphi\right)\right)(t)=\left({}_a\Delta_h^{-(\alpha+\beta)}\varphi\right)(t)=\left({}_{a+\alpha h}\Delta_h^{-\beta}\left({}_a\Delta_h^{-\alpha}\varphi\right)\right)(t), \quad (3.4)$$

where $t\in(h\mathbb{N})_{a+(\alpha+\beta)h}$.

Proof. Let $\varphi:(h\mathbb{N})_a\to\mathbb{R}$ and $\alpha,\beta>0$. Then for $t\in(h\mathbb{N})_{a+(\alpha+\beta)h}$ holds

$$\begin{aligned}\psi(t)&:=\left({}_{a+\beta h}\Delta_h^{-\alpha}\left({}_a\Delta_h^{-\beta}\varphi\right)\right)(t)\\&=\frac{h^2}{\Gamma(\alpha)\Gamma(\beta)}\sum_{s=\frac{a}{h}+\beta}^{t/h-\alpha}(t-\sigma_h(sh))_h^{(\alpha-1)}\left(\sum_{r=\frac{a}{h}}^{s-\beta}(sh-\sigma_h(rh))_h^{(\beta-1)}\varphi(rh)\right)\\&=\frac{h^2}{\Gamma(\alpha)\Gamma(\beta)}\sum_{r=\frac{a}{h}}^{t/h-\alpha-\beta}\sum_{s=r+\beta}^{t/h-\alpha}(t-\sigma_h(sh))_h^{(\alpha-1)}(sh-\sigma_h(rh))_h^{(\beta-1)}\varphi(rh).\end{aligned}$$

Let $xh=(sh-\sigma_h(rh))=(s-r-1)h$, then

$$\begin{aligned}\psi(t)&:=\frac{h}{\Gamma(\beta)}\sum_{r=\frac{a}{h}}^{t/h-\alpha-\beta}\left(\frac{h}{\Gamma(\alpha)}\sum_{x=\beta-1}^{t/h-\alpha-r-1}(t-xh-rh-2h)_h^{(\alpha-1)}(xh)_h^{(\beta-1)}\right)\varphi(rh)\\&=\frac{h}{\Gamma(\beta)}\sum_{r=\frac{a}{h}}^{t/h-\alpha-\beta}\left(\frac{h}{\Gamma(\alpha)}\sum_{x=\beta-1}^{(t-rh-h)/h-\alpha}(t-rh-h-\sigma_h(xh))_h^{(\alpha-1)}(xh)_h^{(\beta-1)}\right)\varphi(rh)\\&=\frac{h}{\Gamma(\beta)}\sum_{r=\frac{a}{h}}^{t/h-\alpha-\beta}\left(\left({}_{(\beta-1)h}\Delta_h^{-\alpha}\left(t_h^{(\beta-1)}\right)(t-rh-h)\right)\varphi(rh)\right)\end{aligned}$$

and using formula (3.2)

$$\begin{aligned}\psi(t)&=\frac{h}{\Gamma(\beta)}\sum_{r=\frac{a}{h}}^{t/h-\alpha-\beta}\frac{\Gamma(\beta)}{\Gamma(\alpha+\beta)}(t-rh-h)_h^{(\alpha+\beta-1)}\varphi(rh)\\&=\frac{h}{\Gamma(\alpha+\beta)}\sum_{r=\frac{a}{h}}^{t/h-\alpha-\beta}(t-\sigma_h(rh))_h^{(\alpha+\beta-1)}\varphi(rh)\\&=\left({}_a\Delta_h^{-(\alpha+\beta)}\varphi\right)(t).\end{aligned}$$

Since α and β were chosen arbitrary, the last equality in the thesis also holds. □

In the sequel we use equalities (3.4) for $a=-(1-\alpha)h$ and $\beta=1-\alpha$. Then we have $a+\beta h=0$ and

$$\left({}_0\Delta_h^{-\alpha}\left({}_{(\alpha-1)h}\Delta_h^{-(1-\alpha)}\varphi\right)\right)(t)=\left({}_{(\alpha-1)h}\Delta_h^{-1}\varphi\right)(t)\,, \tag{3.5}$$

where $t\in(h\mathbb{N})_{\alpha h}=\{\alpha h,\alpha h+h,\dots\}$.

The important part of studying properties of operators is to describe their kernels. It is obvious that each of operators ${}_a\Delta_h^{-\alpha}$, ${}_a\Delta_{h,*}^{\alpha}$, ${}_a\widetilde{\Delta}_h^{\alpha}$ is linear. Hence kernels of the operators are nonempty.

Proposition 3.4. *Let φ be a real-valued function defined on $(h\mathbb{N})_a$, where $a,h\in\mathbb{R}, h>0$ and $\alpha\in(n-1,n]$ where $n\in\mathbb{N}$. Then the following relations hold.*

(i) *For $t\in(h\mathbb{N})_{a+\nu}$ and $\nu=(1-\alpha)h$, $({}_a\Delta_h^{\alpha}\varphi)(t)=0$ if and only if $\varphi(t)=\frac{\lambda}{\Gamma(\alpha)}(t-(1-\alpha)h-a)_h^{(\alpha-1)}$, $t\in\{a,a+h,\dots\}$ where λ is an arbitrary constant.*

(ii) $\left({}_a\Delta_{h,*}^{\alpha}\varphi\right)(t)=0\;\Leftrightarrow\;\varphi(t)=0.$

(iii) $\left({}_a\widetilde{\Delta}_h^{\alpha}\varphi\right)(t)=0\;\Leftrightarrow\;\varphi(t)=0.$

Proof. (i) The relation was proved in [9].
(ii) The relation is obvious since the operator is Caputo-like one.
(iii) Let φ and α be defined as in the statement of the theorem. A direct application of (2.4) yields:

$$\left(a\widetilde{\Delta}_h^{\alpha}\varphi\right)(a)=\frac{1}{h^\alpha}\varphi(a)=0\Leftrightarrow\varphi(a)=0$$

$$\left(a\widetilde{\Delta}_h^{\alpha}\varphi\right)(a+h)=\frac{1}{h^\alpha}(\varphi(a+h)-\alpha\varphi(a))=0\Leftrightarrow\varphi(a+h)=0.$$

And finally we get $\varphi(t)=\varphi(a+kh)=0$. □

Proposition 3.5. *Let $\alpha\in(n-1,n]$, where $n\in\mathbb{N}$ and $\nu=(n-\alpha)h$. Set $p\in\mathbb{Z}\backslash\{0,1,\dots,n-1\}$ and $p-\alpha+1\notin\mathbb{Z}$. Then for the function $\varphi(s)=(s-a)_h^{(p)}$, $s\in\mathbb{N}_a$ holds*

$$\left({}_{a+ph}\Delta_h^{\alpha}\varphi\right)(t)=p^{(\alpha)}(t-a)_h^{(p-\alpha)}\,, \tag{3.6}$$

where $t\in\mathbb{N}_{a+ph+\nu}$.

Proof. We can easily show that

$$\Delta_h(t-a)_h^{(p)}=p(t-a)_h^{(p-1)}$$

and further that

$$\Delta_h^n(t-a)_h^{(p)}=p^{(n)}(t-a)_h^{(p-n)}\,. \tag{3.7}$$

Then from Definition 2.6 and formulas (3.3), (3.7) it follows

$$\begin{aligned}\left({}_{a+ph}\Delta_h^{\alpha}\varphi\right)(t)&=\Delta_h^n\left({}_{a+ph}\Delta_h^{-\nu}\varphi\right)(t)\\&=p^{(-\nu)}\Delta_h^n(t-a)_h^{(p+\nu)}=p^{(-\nu)}(p+\nu)^{(n)}(t-a)_h^{(p+\nu-n)}\\&=p^{(\alpha)}(t-a)_h^{(p+\alpha)}\,.\end{aligned}$$

□

Remark 3.6. The power rule formula (3.6) we can also rewrite in the following way

$$\left({}_a\Delta_h^\alpha \psi\right)(t) = p^{(\alpha)}(t-a)_h^{(p-\alpha)},$$

where $\psi(s) = (s-a+ph)_h^{(p)}$ and $t \in \mathbb{N}_{a+ph+\nu}$.

With formulas (3.3) and (3.7) one can easily prove the following.

Proposition 3.7. *Let $\alpha \in (n-1, n]$, where $n \in \mathbb{N}$ and $\nu = (n-\alpha)h$. Set $p \in \mathbb{Z}\backslash\{0, 1, \ldots, n-1\}$ and $p-\alpha+1 \notin \mathbb{Z}$. Then for the function $\varphi(s) = (s-a)_h^{(p)}$, $s \in \mathbb{N}_a$ holds*

$$\left({}_{a+(p-n)h}\Delta_{h,*}^\alpha \varphi\right)(t) = p^{(\alpha)}(t-a)_h^{(p-\alpha)}, \tag{3.8}$$

where $t \in (h\mathbb{N})_{a+ph-\alpha h}$.

Proof. From Definition 2.9 and formulas (3.3), (3.7) follows

$$\begin{aligned}\left({}_{a+(p-n)h}\Delta_{h,*}^\alpha \varphi\right)(t) &= {}_{a+(p-n)h}\,\Delta_h^{-(n-\alpha)} p^{(n)}(t-a)_h^{(p-n)} \\ &= p^{(n)}(p-n)^{-(n-\alpha)}(t-a)_h^{(p-n+n-\alpha)} \\ &= p^{(\alpha)}(t-a)_h^{(p-\alpha)}.\end{aligned}$$

□

Remark 3.8. The power rule formula (3.8) we can also rewrite in the following way

$$\left({}_a\Delta_h^\alpha \psi\right)(t) = p^{(\alpha)}(t-a)_h^{(p-\alpha)},$$

where

$$\psi(s) = (s-a+(p-n)h)_h^{(p)} \qquad \text{and} \qquad t \in \mathbb{N}_{a+ph-\alpha h}.$$

4. Properties of the operators

Theorem 4.1. *For any $\alpha > 0$ and any positive integer p, the following equality holds:*

$$\left({}_a\Delta_h^{-\alpha}\Delta_h^p \varphi\right)(t) = \Delta_h^p\left({}_a\Delta_h^{-\alpha}\varphi\right)(t) - \sum_{k=0}^{p-1}\frac{(t-a)_h^{(\alpha-p+k)}}{\Gamma(\alpha+k-p+1)}(\Delta_h^k\varphi)(a) \tag{4.1}$$

where φ is defined on $(h\mathbb{N})_a$.

Proof. We prove the thesis by induction. Firstly, we should check if it works for $p=1$ but it was proved already by Bastos et al. in [9] that the following formula holds

$$\left({}_a\Delta_h^{-\alpha}\Delta_h \varphi\right)(t) = \Delta_h\left({}_a\Delta_h^{-\alpha}\varphi\right)(t) - \frac{\alpha}{\Gamma(\alpha+1)}(t-a)_h^{(\alpha-1)}\varphi(a). \tag{4.2}$$

Next, we assume that for $p \in \{1,2,\dots,n\}$ thesis holds. Then we prove that it is also true for $p = n+1$. We replace φ by $\Delta_h^n \varphi$ in (4.2):

$$\begin{aligned}
\left({}_a\Delta_h^{-\alpha}\Delta_h^{n+1}\varphi\right)(t) &= \left({}_a\Delta_h^{-\alpha}(\Delta_h(\Delta_h^n\varphi))\right)(t)\\
&= \left(\Delta_h\left({}_a\Delta_h^{-\alpha}(\Delta_h^n\varphi)\right)\right)(t) - \frac{1}{\Gamma(\alpha)}(t-a)_h^{(\alpha-1)}(\Delta_h^n\varphi)(a)\\
&= \Delta_h\left\{\Delta_h^n({}_a\Delta_h^{-\alpha}\varphi)(t) - \sum_{k=0}^{n-1}\frac{(t-a)_h^{(\alpha-n+k)}}{\Gamma(\alpha+k-n+1)}(\Delta_h^k\varphi)(a)\right\}\\
&\quad - \frac{1}{\Gamma(\alpha)}(t-a)_h^{(\alpha-1)}(\Delta_h^n\varphi)(a)\\
&= \Delta_h^{n+1}({}_a\Delta_h^{-\alpha}\varphi)(t) - \sum_{k=0}^{n-1}\frac{(t-a)_h^{(\alpha-n+k-1)}}{\Gamma(\alpha+k-n)}(\Delta_h^k\varphi)(a)\\
&\quad - \frac{1}{\Gamma(\alpha)}(t-a)_h^{(\alpha-1)}(\Delta_h^n\varphi)(a)\\
&= \Delta_h^{n+1}({}_a\Delta_h^{-\alpha}\varphi)(t) - \sum_{k=0}^{n}\frac{(t-a)_h^{(\alpha-n+k-1)}}{\Gamma(\alpha+k-n)}(\Delta_h^k\varphi)(a). \qquad \square
\end{aligned}$$

Corollary 4.2. *It is quite easy to notice that for $p = n$ and $\alpha \in (n-1, n]$ with $n \in \mathbb{N}$ the formula (4.1) can be rewritten as:*

$$\left({}_a\Delta_{h,*}^{\alpha}\varphi\right)(t) = \left({}_a\Delta_h^{\alpha}\varphi\right)(t) - \sum_{k=0}^{p-1}\frac{(t-a)_h^{(\alpha-p+k)}}{\Gamma(\alpha+k-p+1)}(\Delta_h^k\varphi)(a)$$

for φ defined on $(h\mathbb{N})_a$. In particular, for $p = 1$ and $\alpha \in (0,1]$ one can get

$$\left({}_a\Delta_{h,*}^{\alpha}\varphi\right)(t) = \left({}_a\Delta_h^{\alpha}\varphi\right)(t) - \frac{1-\alpha}{\Gamma(2-\alpha)}(t-a)_h^{(-\alpha)}\varphi(a),$$

where $t \in (h\mathbb{N})_a$ and $(t-a)_h^{(-\alpha)} = h^{-\alpha}\frac{\Gamma(\frac{t-a}{h}+1)}{\Gamma(\frac{t-a}{h}+1+\alpha)}$.

Remark 4.3. We can simplify the formula $\frac{1-\alpha}{\Gamma(2-\alpha)}$ and get $\frac{1}{\Gamma(1-\alpha)}$, but the current form gives immediately the equality $\left({}_a\Delta_{h,*}^{1}\varphi\right)(t) = \left({}_a\Delta_h^{1}\varphi\right)(t)$.

The next propositions give useful identities of transforming fractional difference equations into fractional summations.

Proposition 4.4. *Let $\alpha \in (0,1]$, $h > 0$, $\nu = (1-\alpha)h$ and x be a real-valued function defined on $(h\mathbb{N})_{-\nu}$. The following formula holds*

$$\left({}_0\Delta_h^{-\alpha}\left({}_{-\nu}\Delta_h^{\alpha}x\right)\right)(t) = x(t) - \frac{h^{1-\alpha}}{\Gamma(\alpha)}t_h^{(\alpha-1)}x(-\nu), \quad t \in (h\mathbb{N})_{\alpha h}.$$

Proof. Directly from definition of the operator ${}_{-\nu}\Delta_h^{\alpha}$ we have

$$\psi(t) := \left({}_0\Delta_h^{-\alpha}\left({}_{-\nu}\Delta_h^{\alpha}x\right)\right)(t) = \left({}_0\Delta_h^{-\alpha}\left(\Delta_h\left({}_{-\nu}\Delta_h^{-(1-\alpha)}x\right)\right)\right)(t).$$

Next using equation (4.2) we can state

$$\psi(t) = \Delta_h \left({}_0\Delta_h^{-\alpha}\varphi\right)(t) - \frac{t_h^{(\alpha-1)}}{\Gamma(\alpha)}\varphi(0),$$

where $\varphi(t) = \left({}_{-\nu}\Delta_h^{-(1-\alpha)}x\right)(t)$. The value $\varphi(0)$ we calculate in the following way. Firstly observe that $\varphi : (h\mathbb{N})_0 \to \mathbb{R}$. Then from the definition of the h-sum operator we have

$$\varphi(0) = \left({}_{-\nu}\Delta_h^{-(1-\alpha)}x\right)(0) = \frac{h}{\Gamma(1-\alpha)} \sum_{k=-\frac{\nu}{h}}^{-(1-\alpha)} \left(-\sigma_h(kh)\right)_h^{(-\alpha)} x(kh) = h^{1-\alpha}x(-\nu).$$

Further, from Proposition 3.3 and equation (3.5) we get

$$\begin{aligned}\psi(t) &= \left(\Delta_h \left({}_0\Delta_h^{-\alpha}\left({}_{-\nu}\Delta_h^{-(1-\alpha)}x\right)\right)\right)(t) - \frac{h^{1-\alpha}}{\Gamma(\alpha)} t_h^{(\alpha-1)} x(-\nu) \\ &= \left(\Delta_h\left({}_{-\nu}\Delta_h^{-1}x\right)\right)(t) - \frac{h^{1-\alpha}}{\Gamma(\alpha)} t_h^{(\alpha-1)} x(-\nu) \\ &= x(t) - \frac{h^{1-\alpha}}{\Gamma(\alpha)} t_h^{(\alpha-1)} x(-\nu) = x(t) - \frac{1}{\Gamma(\alpha)}\left(\frac{t}{h}\right)^{(\alpha-1)} x(-\nu).\end{aligned}$$

□

Proposition 4.5. *Let $\alpha \in (0,1]$, $h > 0$, $\nu = (1-\alpha)h$ and x be a real-valued function defined on $(h\mathbb{N})_{-\nu}$. The following formula holds*

$$\left({}_0\Delta_h^{-\alpha}\left({}_{-\nu}\Delta_{h,*}^{\alpha}x\right)\right)(t) = x(t) - x(-\nu), \quad t \in (h\mathbb{N})_{\alpha h}.$$

Proof. From definition of the operator ${}_{-\nu}\Delta_{h,*}^{\alpha}$ we have

$$\psi(t) := \left({}_0\Delta_h^{-\alpha}\left({}_{-\nu}\Delta_{h,*}^{\alpha}x\right)\right)(t) = \left({}_0\Delta_h^{-\alpha}\left({}_{-\nu}\Delta_h^{-(1-\alpha)}(\Delta_h x)\right)\right)(t).$$

From Proposition 3.3 and equation (3.5) we write

$$\begin{aligned}\psi(t) &= \left({}_{-\nu}\Delta_h^{-1}(\Delta_h x)\right)(t) = h\sum_{k=\alpha-1}^{t/h-1} (\Delta_h x)(kh) \\ &= \sum_{k=\alpha-1}^{t/h-1} \left(x(kh+h) - x(kh)\right) = x(t) - x(-\nu).\end{aligned}$$

□

5. Initial value problems and exponentials

In this Section we discuss initial value problem for orders $\alpha \in (0,1]$ that define fractional type of exponentials in the meaning of the solution of particular initial value problem of fractional order.

For the Riemann-Liouville type fractional difference, similarly as it is stated in case $h = 1$ in [13], we can formulate the following. Let $\alpha \in (0,1]$ and $\nu = (1-\alpha)h$, then using Lemma 4.4 we can write that the initial value problem

$$\left({}_{-\nu}\Delta_h^\alpha x\right)(t) = \lambda x(t-\nu)\,, \quad t \in (h\mathbb{N})_0 \tag{5.1}$$

$$x(-\nu) = a_0\,, \quad a_0 \in \mathbb{R} \tag{5.2}$$

has the unique solution given by the recurrence formula

$$x(t) = \frac{h^{1-\alpha} t_h^{(\alpha-1)}}{\Gamma(\alpha)} a_0 + \frac{h\lambda}{\Gamma(\alpha)} \sum_{s=0}^{t/h-\alpha} (t-\sigma(sh))_h^{(\alpha-1)} x(sh-\nu)\,, \tag{5.3}$$

where $t \in (h\mathbb{N})_{-\nu}$.

Proposition 5.1. *The linear initial value problem* (5.1) *has the unique solution given by the formula*

$$x(t) = E_{(\alpha,\alpha)}\left(\lambda h^\alpha, t/h\right) a_0\,,$$

where $t \in (h\mathbb{N})_{-\nu}$.

Proof. We adopt the proof for scalar case that one can find in [5]. Starting with the recurrence formula (5.3) we define the following sequence. Let $x_0(t) = \frac{h^{1-\alpha} t_h^{(\alpha-1)}}{\Gamma(\alpha)} a_0$, and for $m \in \mathbb{N}$ we set

$$x_m(t) = x_0(t) + \lambda \left({}_0\Delta_h^{-\alpha} x_{m-1}\right)(t-\nu)\,.$$

Then using the power rule (3.3) we see that

$$x_1(t) = h^{1-\alpha} a_0 \left(\frac{t_h^{(\alpha-1)}}{\Gamma(\alpha)} + \frac{\lambda}{\Gamma(2\alpha)} (t-\nu)_h^{(2\alpha-1)} \right)\,.$$

Applying the same power rule m-times it follows inductively that for $m \in \mathbb{N}_0$ holds

$$x_m(t) = h^{1-\alpha} a_0 \sum_{s=0}^{m} \frac{\lambda^s}{\Gamma((s+1)\alpha)} (t-s\nu)_h^{((s+1)\alpha-1)}\,.$$

Moreover, we can write that

$$x_m(t) = h^{1-\alpha} h^{\alpha-1} a_0 \sum_{s=0}^{m} \frac{(\lambda h^\alpha)^s}{\Gamma((s+1)\alpha)} (t/h + s(\alpha-1))^{((s+1)\alpha-1)}\,.$$

Now taking the limit with $m \to \infty$ we obtain

$$x(t) = a_0 \sum_{s=0}^{\infty} \frac{(\lambda h^\alpha)^s}{\Gamma((s+1)\alpha)} (t/h + s(\alpha-1))^{((s+1)\alpha-1)}\,.$$

Finally, using Definition 2.14 we have the thesis. □

For the Caputo type fractional difference we use the following form of systems, similarly as it is stated in case $h = 1$ in [1, 11].

Proposition 5.2. *Let* $\alpha \in (0,1]$ *and* $\nu = (1-\alpha)h$. *The initial value problem*

$$\left({}_{-\nu}\Delta_{h,*}^\alpha x\right)(t) = \lambda x(t-\nu)\,, \quad t \in \mathbb{N}_0 \tag{5.4}$$

$$x(-\nu) = a_0\,, \quad a_0 \in \mathbb{R} \tag{5.5}$$

has the unique solution given by the recurrence formula

$$x(t) = a_0 + \frac{h\lambda}{\Gamma(\alpha)} \sum_{s=0}^{t/h-\alpha} (t - \sigma(sh))_h^{(\alpha-1)} x(sh - \nu),$$

where $t \in (h\mathbb{N})_{-\nu}$.

Proposition 5.3. *The linear initial value problem* (5.4) *has the unique solution given by the formula*

$$x(t) = E_{(\alpha)}(\lambda h^\alpha, t/h)\, a_0,$$

where $t \in (h\mathbb{N})_{-\nu}$.

Proof. To get the thesis one can apply the method of successive approximations as it was done in [1, Example 20] using the crucial power rule formula. □

For the Grünwald-Letnikov type fractional difference we use the following form, see [14, 19, 22].

Proposition 5.4. *Let* $\alpha \in (0, 1]$. *The initial value problem*

$$\left({}_0\widetilde{\Delta}_h^\alpha x\right)(t + h) = \lambda x(t), \quad t \in (h\mathbb{N})_0 \tag{5.6}$$

$$x(0) = a_0, \quad a_0 \in \mathbb{R} \tag{5.7}$$

has the unique solution given by the recurrence formula

$$x(t + h) = \lambda x(t) - \sum_{s=1}^{t/h+1} (-1)^s \binom{\alpha}{s} x(t - sh + h),$$

where $t \in (h\mathbb{N})_0$.

Similarly as it was proved for the case $h = 1$ in, for example, [14] we can get the following result.

Proposition 5.5. *The linear initial value problem* (5.6) *has the unique solution given by the formula*

$$x(t) = \Phi(t) a_0,$$

where $t \in (h\mathbb{N})_0$ *and the transition function* Φ *is determined by the recurrence formula*

$$\Phi(t + h) = \lambda\Phi(t) + \sum_{s=1}^{t/h+1} (-1)^{s+1} \binom{\alpha}{s} \Phi(t - sh + h)$$

with $\Phi(0) = 1$.

Remark 5.6. To the authors knowledge there is unknown the exact formula for the transition function Φ described for example by a power series. The authors do not know how to derive it. The transition function Φ is widely used in the control theory by many authors, see for example [14, 15, 22].

6. Summary

Operator	From	To	the Kernel	$\alpha = 1$
${}_a\Delta_h^{-\alpha}$	$\mathcal{F}_{(h\mathbb{N})_a}$	$\mathcal{F}_{(h\mathbb{N})_{a+\alpha h}}$	It is not finitely generated	$\sum_{s=a}^{t-h} \varphi(s)h$
${}_a\Delta_h^{\alpha}$	$\mathcal{F}_{(h\mathbb{N})_a}$	$\mathcal{F}_{(h\mathbb{N})_{a+\nu}}$	$\frac{c}{\Gamma(\alpha)}(t-(1-\alpha)h-a)_h^{(\alpha-1)}$	$\Delta\varphi$
${}_a\Delta_{h,*}^{\alpha}$	$\mathcal{F}_{(h\mathbb{N})_a}$	$\mathcal{F}_{(h\mathbb{N})_{a+\nu}}$	0	$\Delta\varphi$
${}_a\widetilde{\Delta}_h^{\alpha}$	$\mathcal{F}_{(h\mathbb{N})_a}$	$\mathcal{F}_{(h\mathbb{N})_a}$	0	$\nabla\varphi$

TABLE 1. Domains and kernels of the operators for $\alpha \in (0,1]$ and $\nu = (1-\alpha)h$.

The operator	the IVP
${}_{-\nu}\Delta_h^{\alpha}$	$\begin{cases} ({}_{-\nu}\Delta_h^{\alpha}x)(t) = \lambda x(t-\nu), \\ x(-\nu) = a_0 \end{cases}$
${}_{-\nu}\Delta_{h,*}^{\alpha}$	$\begin{cases} ({}_{-\nu}\Delta_{h,*}^{\alpha}x)(t) = \lambda x(t-\nu), \\ x(-\nu) = a_0 \end{cases}$
${}_0\widetilde{\Delta}_h^{\alpha}$	$\begin{cases} \left({}_0\widetilde{\Delta}_h^{\alpha}x\right)(t+h) = \lambda x(t), \\ x(0) = a_0 \end{cases}$

TABLE 2. h-difference IVP problems for $\alpha \in (0,1]$, $\nu = (1-\alpha)h$ and $t \in (h\mathbb{N})_0$.

The operator	the solution to IVP
${}_{-\nu}\Delta_h^{\alpha}$	$x(t) = \frac{h^{1-\alpha}t_h^{(\alpha-1)}}{\Gamma(\alpha)}a_0 + \frac{h\lambda}{\Gamma(\alpha)}\sum_{s=0}^{t/h-\alpha}(t-\sigma(sh))_h^{(\alpha-1)}x(sh-\nu)$
${}_{-\nu}\Delta_{h,*}^{\alpha}$	$x(t) = a_0 + \frac{h\lambda}{\Gamma(\alpha)}\sum_{s=0}^{t/h-\alpha}(t-\sigma(sh))_h^{(\alpha-1)}x(sh-\nu)$
${}_0\widetilde{\Delta}_h^{\alpha}$	$x(t) = \Phi(t)a_0$, where $\Phi(t+h) = \lambda\Phi(t) + \sum_{s=1}^{t/h+1}(-1)^{s+1}\binom{\alpha}{s}\Phi(t-sh+h)$, $\Phi(0) = 1$

TABLE 3. Recurrence solution to h-difference IVP problems for $\alpha \in (0,1]$ and $\nu = (1-\alpha)h$.

The operator	The exponential	$\alpha = 1$
$_{-\nu}\Delta_h^\alpha$	$E_{(\alpha,\alpha)}(\lambda h^\alpha, t/h)$	$(1+\lambda h)^{\frac{t}{h}}$
$_{-\nu}\Delta_{h,*}^\alpha$	$E_{(\alpha)}(\lambda h^\alpha, t/h)$	
$_0\widetilde{\Delta}_h^\alpha$	The exact formula is unknown.	

TABLE 4. Fractional exponentials for $\alpha \in (0,1]$ and $\nu = (1-\alpha)h$.

Acknowledgment

The work was supported by Białystok University of Technology grant S/WI/2/2011. The second author (Ewa Girejko) was also supported by the post-doc fellowship SFRH/BPD/48439/2008 of the Portuguese Foundation for Science and Technology (FCT) and by *FEDER* funds through *COMPETE*-Operational Programme Factors of Competitiveness ("Programa Operacional Factores de Competitividade") and by Portuguese founds through the *Center for Research and Development in Mathematics and Applications* (University of Aveiro) and the Portuguese Foundation for Science and Technology ("FCT–Fundaçao para a Ciencia e a Tecnologia"), within project PEst-C/MAT/UI4106/2011 with COMPETE number FCOMP-01-0124-FEDER-022690.

References

[1] T. Abdeljawad, *On Riemann and Caputo fractional differences.* Comput. Math. Appl. **62** (3) (2011), 1602–1611.

[2] T. Abdeljawad, D. Baleanu, *Fractional differences and integration by parts.* J. Comput. Anal. Appl. **13** (3) (2011), 574–582.

[3] O.P. Agrawal, D. Baleanu. *A Hamiltonian formulation and a direct numerical scheme for fractional optimal control problems.* J. Vib. Control **13** (9-10) (2007), 1269–1281.

[4] G.A. Anastassiou, *Fractional Differentiation Inequalities*, Research Monograph, Springer, New York 2009.

[5] F.M. Atici, P.W. Eloe, *Initial value problems in discrete fractional calculus.* Proc. Amer. Math. Soc. **137** (3) (2008), 981–989.

[6] F.M. Atici, P.W. Eloe, *A Transform Method in Discrete Fractional Calculus.* Int. J. Difference Equ. **2** (2007), 165–176.

[7] D. Baleanu, *New applications of fractional variational principles.* Rep. Math. Phys. **61** (2) (2008), 199–206.

[8] D. Baleanu, K. Diethelm, E. Scalas, J.J. Trujillo, *Fractional Calculus Models and Numerical Method.* World Scientific Publishing, New York 2012.

[9] N.R.O. Bastos, R.A.C. Ferreira, D.F.M. Torres, *Necessary optimality conditions for fractional difference problems of the calculus of variations.* Discrete Contin. Dyn. Syst. **29** (2) (2011), 417–437.

[10] J. Čermák, T. Kisela, L. Nechvátal, *Discrete Mittag-Leffler functions in linear fractional difference equations.* Abst. Appl. Anal. (2011), 1–21.

[11] F. Chen, X. Luo, Y. Zhou, *Existence results for nonlinear fractional difference equation.* Advances in Difference Eq. article ID 713201, 12 p., doi: 10.1155/2011/713201 2011.

[12] R.A.C. Ferreira, D.F.M. Torres, *Fractional h-difference equations arising from the calculus of variations.* Appl. Anal. Discrete Math. **5** (1) (2011), 110–121.

[13] M.T. Holm, *The theory of discrete fractional calculus: Development and application.* University of Nebraska – Lincoln 2011.

[14] T. Kaczorek, *Positive 1D and 2D Systems* Springer-Verlag, London 2002.

[15] J. Klamka, *Controllability of nonlinear discrete systems.* Int. J. Appl. Math. Comput. Sci. **12** (2) (2002), 173–180.

[16] K.S. Miller, B. Ross, *Fractional difference calculus.* Proceedings of the International Symposium on Univalent Functions, Fractional Calculus and their Applications, Nihon University, Kōriyama, Japan (1989), 139–152.

[17] K.S. Miller, B. Ross, *An introduction to the fractional calculus and fractional differential equations* Wiley, New York 1993.

[18] M.D. Ortigueira, *Fractional central differences and derivatives.* J. Vib. Control **14** (9-10) (2008), 1255–1266.

[19] P. Ostalczyk, *The non-integer difference of the discrete time function and its application to the control system synthesis.* Int. J. Syst. Sci. **31** (12) (2000), 1551–1561.

[20] I. Podlubny, *Fractional Differential Equations.* AP, New York, NY 1999.

[21] S.G. Samko, A.A. Kilbas, O.I. Marichev, *Fractional integrals and derivatives.* Translated from the 1987 Russian original, Gordon and Breach, Yverdon 1993.

[22] D. Sierociuk, D. Dzieliński, *Fractional Kalman filter algorithm for the states parameters and order of fractional system estimation.* Int. J. Appl. Math. Comp. Sci. **16** (1) (2006), 129–140.

[23] M.F. Silva, J.A. Tenreiro Machado, R.S. Barbosa, *Using fractional derivatives in joint control of hexapod robots.* J. Vib. Control **14** (9-10) (2008), 1473–1485.

Dorota Mozyrska and Ewa Girejko
Faculty of Computer Science
Białystok University of Technology
Wiejska 45a
PL-15-351 Białystok, Poland
e-mail: `d.mozyrska@pb.edu.pl`
`e.girejko@pb.edu.pl`

Operator Theory:
Advances and Applications, Vol. 229, 269–289

A Singularly Perturbed Dirichlet Problem for the Poisson Equation in a Periodically Perforated Domain. A Functional Analytic Approach

Paolo Musolino

Abstract. Let Ω be a sufficiently regular bounded open connected subset of $\mathbb{R}^n$ such that $0 \in \Omega$ and that $\mathbb{R}^n \setminus \mathrm{cl}\Omega$ is connected. Then we take $(q_{11}, \ldots, q_{nn}) \in]0, +\infty[^n$ and $p \in Q \equiv \prod_{j=1}^{n}]0, q_{jj}[$. If ϵ is a small positive number, then we define the periodically perforated domain $\mathbb{S}[\Omega_{p,\epsilon}]^- \equiv \mathbb{R}^n \setminus \cup_{z \in \mathbb{Z}^n} \mathrm{cl}\bigl(p + \epsilon\Omega + \sum_{j=1}^{n}(q_{jj} z_j) e_j\bigr)$, where $\{e_1, \ldots, e_n\}$ is the canonical basis of $\mathbb{R}^n$. For ϵ small and positive, we introduce a particular Dirichlet problem for the Poisson equation in the set $\mathbb{S}[\Omega_{p,\epsilon}]^-$. Namely, we consider a Dirichlet condition on the boundary of the set $p + \epsilon\Omega$, together with a periodicity condition. Then we show real analytic continuation properties of the solution as a function of ϵ, of the Dirichlet datum on $p + \epsilon\partial\Omega$, and of the Poisson datum, around a degenerate triple with $\epsilon = 0$.

Mathematics Subject Classification (2010). 35J25; 31B10; 45A05; 47H30.

Keywords. Dirichlet problem; singularly perturbed domain; Poisson equation; periodically perforated domain; real analytic continuation in Banach space.

1. Introduction

In this article, we consider a Dirichlet problem for the Poisson equation in a periodically perforated domain with small holes. We fix once for all a natural number

$$n \in \mathbb{N} \setminus \{0, 1\} \qquad \text{and} \qquad (q_{11}, \ldots, q_{nn}) \in]0, +\infty[^n$$

and a periodicity cell

$$Q \equiv \Pi_{j=1}^{n}]0, q_{jj}[\, .$$

Then we denote by $\mathrm{meas}(Q)$ the measure of the fundamental cell Q, and by ν_Q the outward unit normal to ∂Q, where it exists. We denote by q the $n \times n$ diagonal matrix defined by $q \equiv (\delta_{i,j}q_{jj})_{i,j=1,\dots,n}$, where $\delta_{i,j} = 1$ if $i = j$ and $\delta_{i,j} = 0$ if $i \neq j$, for all $i,j \in \{1,\dots,n\}$. Clearly, $q\mathbb{Z}^n \equiv \{qz : z \in \mathbb{Z}^n\}$ is the set of vertices of a periodic subdivision of $\mathbb{R}^n$ corresponding to the fundamental cell Q. Let $m \in \mathbb{N} \setminus \{0\}$, $\alpha \in]0,1[$. Then we take a point $p \in Q$ and a bounded open connected subset Ω of $\mathbb{R}^n$ of class $C^{m,\alpha}$ such that $\Omega^- \equiv \mathbb{R}^n \setminus \mathrm{cl}\Omega$ is connected and that $0 \in \Omega$. Here 'cl' denotes the closure. If $\epsilon \in \mathbb{R}$, then we set

$$\Omega_{p,\epsilon} \equiv p + \epsilon\Omega .$$

Then we take $\epsilon_0 > 0$ such that $\mathrm{cl}\Omega_{p,\epsilon} \subseteq Q$ for $|\epsilon| < \epsilon_0$, and we introduce the periodically perforated domain

$$\mathbb{S}[\Omega_{p,\epsilon}]^- \equiv \mathbb{R}^n \setminus \bigcup_{z\in\mathbb{Z}^n} \mathrm{cl}(\Omega_{p,\epsilon} + qz) ,$$

for $\epsilon \in]-\epsilon_0, \epsilon_0[$. Let $\rho > 0$. Next we fix a function g_0 in the Schauder space $C^{m,\alpha}(\partial\Omega)$ and a function f_0 in the Roumieu class $C^0_{q,\omega,\rho}(\mathbb{R}^n)_0$ of real analytic periodic functions with vanishing integral on Q (see (2.1) and (2.3) in Section 2.) For each triple $(\epsilon, g, f) \in]0,\epsilon_0[\times C^{m,\alpha}(\partial\Omega) \times C^0_{q,\omega,\rho}(\mathbb{R}^n)_0$ we consider the Dirichlet problem

$$\begin{cases} \Delta u(x) = f(x) & \forall x \in \mathbb{S}[\Omega_{p,\epsilon}]^- , \\ u(x + qe_j) = u(x) & \forall x \in \mathrm{cl}\mathbb{S}[\Omega_{p,\epsilon}]^- , \quad \forall j \in \{1,\dots,n\} , \\ u(x) = g\big((x-p)/\epsilon\big) & \forall x \in \partial\Omega_{p,\epsilon} , \end{cases} \tag{1.1}$$

where $\{e_1,\dots,e_n\}$ is the canonical basis of $\mathbb{R}^n$. If $(\epsilon, g, f) \in]0,\epsilon_0[\times C^{m,\alpha}(\partial\Omega) \times C^0_{q,\omega,\rho}(\mathbb{R}^n)_0$, then problem (1.1) has a unique solution in $C^{m,\alpha}(\mathrm{cl}\mathbb{S}[\Omega_{p,\epsilon}]^-)$, and we denote it by $u[\epsilon, g, f](\cdot)$ (cf. Proposition 2.2.)

Then we pose the following questions.

(i) Let x be fixed in $\mathbb{R}^n \setminus (p + q\mathbb{Z}^n)$. What can be said on the map $(\epsilon, g, f) \mapsto u[\epsilon, g, f](x)$ when (ϵ, g, f) approaches $(0, g_0, f_0)$?

(ii) What can be said on the map $(\epsilon, g, f) \mapsto \int_{Q\setminus\mathrm{cl}\Omega_{p,\epsilon}} |D_x u[\epsilon, g, f](x)|^2 \, dx$ when (ϵ, g, f) approaches $(0, g_0, f_0)$?

(iii) What can be said on the map $(\epsilon, g, f) \mapsto \int_{Q\setminus\mathrm{cl}\Omega_{p,\epsilon}} u[\epsilon, g, f](x) \, dx$ when (ϵ, g, f) approaches $(0, g_0, f_0)$?

Questions of this type have long been investigated with the methods of Asymptotic Analysis, which aim at computing, for example, an asymptotic expansion of the value of the solution at a fixed point in terms of the parameter ϵ. Here, we mention, *e.g.*, Ammari and Kang [1, Ch. 5], Kozlov, Maz'ya, and Movchan [13], Maz'ya, Nazarov, and Plamenewskij [26], Ozawa [30], Ward and Keller [36]. We also mention the vast literature of Calculus of Variations and of Homogenization Theory, where the interest is focused on the limiting behaviour as the singular perturbation parameter degenerates (cf., *e.g.*, Cioranescu and Murat [4, 5].)

Here instead we wish to characterize the behaviour of $u[\epsilon, g, f](\cdot)$ at $(\epsilon, g, f) = (0, g_0, f_0)$ by a different approach. Thus for example, if we consider a certain function, say $F(\epsilon, g, f)$, relative to the solution such as, for example, one of those considered in questions (i)–(iii) above, we would try to prove that $F(\cdot,\cdot,\cdot)$ can be continued real analytically around $(\epsilon, g, f) = (0, g_0, f_0)$. We observe that our approach does have its advantages. Indeed, if we know that $F(\epsilon, g, f)$ equals, for $\epsilon > 0$, a real analytic operator of the variable (ϵ, g, f) defined on a whole neighborhood of $(0, g_0, f_0)$, then, for example, we could infer the existence of a sequence of real numbers $\{a_j\}_{j=0}^{\infty}$ and of a positive real number $\epsilon' \in]0, \epsilon_0[$ such that

$$F(\epsilon, g_0, f_0) = \sum_{j=0}^{\infty} a_j \epsilon^j \qquad \forall \epsilon \in]0, \epsilon'[\,,$$

where the power series in the right-hand side converges absolutely on the whole of $]-\epsilon', \epsilon'[$. Such a project has been carried out by Lanza de Cristoforis in several papers for problems in a bounded domain with a small hole (cf., *e.g.*, Lanza [16, 17, 20, 21].) In the frame of linearized elastostatics, we mention, *e.g.*, Dalla Riva and Lanza [7, 8], and for a boundary value problem for the Stokes system in a singularly perturbed exterior domain we refer to Dalla Riva [6]. Instead, for periodic problems, we mention [24, 29], where Dirichlet and nonlinear Robin problems for the Laplace equation have been considered.

As far as problems in periodically perforated domains are concerned, we mention, for instance, Ammari and Kang [1, Ch. 8]. We also observe that boundary value problems in domains with periodic inclusions can be analysed, at least for the two-dimensional case, with the method of functional equations. Here we mention, *e.g.*, Castro and Pesetskaya [2], Castro, Pesetskaya, and Rogosin [3], Drygas and Mityushev [10], Mityushev and Adler [28], Rogosin, Dubatovskaya, and Pesetskaya [34]. In connection with doubly periodic problems for composite materials, we mention the monograph of Grigolyuk and Fil'shtinskij [12].

We now briefly outline our strategy. By means of a periodic analog of the Newtonian potential, we can convert problem (1.1) into an auxiliary Dirichlet problem for the Laplace equation. Next we analyze the dependence of the solution of the auxiliary problem upon the triple (ϵ, g, f) by exploiting the results of [29] on the homogeneous Dirichlet problem and of Lanza [18] on the Newtonian potential in the Roumieu classes, and then we deduce the representation of $u[\epsilon, g, f](\cdot)$ in terms of ϵ, g, and f, and we prove our main results.

This article is organized as follows. Section 2 is a section of preliminaries. In Section 3, we formulate the auxiliary Dirichlet problem for the Laplace equation and we show that the solutions of the auxiliary problem depend real analytically on ϵ, g, and f. In Section 4, we show that the results of Section 3 can be exploited to prove Theorem 4.1 on the representation of $u[\epsilon, g, f](\cdot)$, Theorems 4.2 and 4.4 on the representation of the energy integral and of the integral of the solution, respectively. At the end of the paper, we have included an Appendix on a slight variant of a result on composition operators of Preciso (cf. Preciso [31, Prop. 4.2.16, p. 51], Preciso [32, Prop. 1.1, p. 101]), which belongs to a different flow of ideas and which accordingly we prefer to discuss separately.

2. Notation and preliminaries

We denote the norm on a normed space $\mathcal{X}$ by $\|\cdot\|_{\mathcal{X}}$. Let $\mathcal{X}$ and $\mathcal{Y}$ be normed spaces. We endow the space $\mathcal{X} \times \mathcal{Y}$ with the norm defined by $\|(x,y)\|_{\mathcal{X}\times\mathcal{Y}} \equiv \|x\|_{\mathcal{X}} + \|y\|_{\mathcal{Y}}$ for all $(x,y) \in \mathcal{X} \times \mathcal{Y}$, while we use the Euclidean norm for $\mathbb{R}^n$. For standard definitions of Calculus in normed spaces, we refer to Prodi and Ambrosetti [33]. The symbol $\mathbb{N}$ denotes the set of natural numbers including 0. A dot "$\cdot$" denotes the inner product in $\mathbb{R}^n$. Let A be a matrix. Then A_{ij} denotes the (i,j)-entry of A, and the inverse of the matrix A is denoted by A^{-1}. Let $\mathbb{D} \subseteq \mathbb{R}^n$. Then $\mathrm{cl}\mathbb{D}$ denotes the closure of $\mathbb{D}$ and $\partial\mathbb{D}$ denotes the boundary of $\mathbb{D}$. For all $R > 0$, $x \in \mathbb{R}^n$, x_j denotes the jth coordinate of x, $|x|$ denotes the Euclidean modulus of x in $\mathbb{R}^n$, and $\mathbb{B}_n(x,R)$ denotes the ball $\{y \in \mathbb{R}^n : |x-y| < R\}$. Let Ω be an open subset of $\mathbb{R}^n$. The space of m times continuously differentiable real-valued functions on Ω is denoted by $C^m(\Omega,\mathbb{R})$, or more simply by $C^m(\Omega)$. $\mathcal{D}(\Omega)$ denotes the space of functions of $C^\infty(\Omega)$ with compact support. The dual $\mathcal{D}'(\Omega)$ denotes the space of distributions in Ω. Let $r \in \mathbb{N} \setminus \{0\}$. Let $f \in (C^m(\Omega))^r$. The sth component of f is denoted f_s, and Df denotes the Jacobian matrix $\left(\frac{\partial f_s}{\partial x_l}\right)_{\substack{s=1,\dots,r,\\ l=1,\dots,n}}$. Let $\eta \equiv (\eta_1,\dots,\eta_n) \in \mathbb{N}^n$, $|\eta| \equiv \eta_1 + \cdots + \eta_n$. Then $D^\eta f$ denotes $\frac{\partial^{|\eta|} f}{\partial x_1^{\eta_1}\dots\partial x_n^{\eta_n}}$. The subspace of $C^m(\Omega)$ of those functions f whose derivatives $D^\eta f$ of order $|\eta| \leq m$ can be extended with continuity to $\mathrm{cl}\Omega$ is denoted $C^m(\mathrm{cl}\Omega)$. The subspace of $C^m(\mathrm{cl}\Omega)$ whose functions have mth order derivatives that are Hölder continuous with exponent $\alpha \in]0,1]$ is denoted $C^{m,\alpha}(\mathrm{cl}\Omega)$ (cf., *e.g.*, Gilbarg and Trudinger [11].) The subspace of $C^m(\mathrm{cl}\Omega)$ of those functions f such that $f_{|\mathrm{cl}(\Omega\cap\mathbb{B}_n(0,R))} \in C^{m,\alpha}(\mathrm{cl}(\Omega \cap \mathbb{B}_n(0,R)))$ for all $R \in]0,+\infty[$ is denoted $C^{m,\alpha}_{\mathrm{loc}}(\mathrm{cl}\Omega)$. Let $\mathbb{D} \subseteq \mathbb{R}^r$. Then $C^{m,\alpha}(\mathrm{cl}\Omega,\mathbb{D})$ denotes $\{f \in (C^{m,\alpha}(\mathrm{cl}\Omega))^r : f(\mathrm{cl}\Omega) \subseteq \mathbb{D}\}$.

Now let Ω be a bounded open subset of $\mathbb{R}^n$. Then $C^m(\mathrm{cl}\Omega)$ and $C^{m,\alpha}(\mathrm{cl}\Omega)$ are endowed with their usual norm and are well known to be Banach spaces (cf., *e.g.*, Troianiello [35, §1.2.1].) We say that a bounded open subset Ω of $\mathbb{R}^n$ is of class C^m or of class $C^{m,\alpha}$, if its closure is a manifold with boundary imbedded in $\mathbb{R}^n$ of class C^m or $C^{m,\alpha}$, respectively (cf., *e.g.*, Gilbarg and Trudinger [11, §6.2].) We denote by ν_Ω the outward unit normal to $\partial\Omega$. For standard properties of functions in Schauder spaces, we refer the reader to Gilbarg and Trudinger [11] and to Troianiello [35] (see also Lanza [14, §2, Lem. 3.1, 4.26, Thm. 4.28], Lanza and Rossi [25, §2].)

If $\mathbb{M}$ is a manifold imbedded in $\mathbb{R}^n$ of class $C^{m,\alpha}$, with $m \geq 1$, $\alpha \in]0,1]$, one can define the Schauder spaces also on $\mathbb{M}$ by exploiting the local parametrizations. In particular, one can consider the spaces $C^{k,\alpha}(\partial\Omega)$ on $\partial\Omega$ for $0 \leq k \leq m$ with Ω a bounded open set of class $C^{m,\alpha}$, and the trace operator of $C^{k,\alpha}(\mathrm{cl}\Omega)$ to $C^{k,\alpha}(\partial\Omega)$ is linear and continuous. We denote by $d\sigma$ the area element of a manifold imbedded in $\mathbb{R}^n$.

We note that throughout the paper "analytic" means "real analytic". For the definition and properties of real analytic operators, we refer to Prodi and Ambrosetti [33, p. 89]. In particular, we mention that the pointwise product in

Schauder spaces is bilinear and continuous, and thus analytic (cf., *e.g.*, Lanza and Rossi [25, p. 141].)

For all bounded open subsets Ω of $\mathbb{R}^n$ and $\rho > 0$, we set

$$C^0_{\omega,\rho}(\mathrm{cl}\Omega) \equiv \left\{ u \in C^\infty(\mathrm{cl}\Omega) \colon \sup_{\beta \in \mathbb{N}^n} \frac{\rho^{|\beta|}}{|\beta|!} \|D^\beta u\|_{C^0(\mathrm{cl}\Omega)} < +\infty \right\},$$

and

$$\|u\|_{C^0_{\omega,\rho}(\mathrm{cl}\Omega)} \equiv \sup_{\beta \in \mathbb{N}^n} \frac{\rho^{|\beta|}}{|\beta|!} \|D^\beta u\|_{C^0(\mathrm{cl}\Omega)} \qquad \forall u \in C^0_{\omega,\rho}(\mathrm{cl}\Omega),$$

where $|\beta| \equiv \beta_1 + \cdots + \beta_n$, for all $\beta \equiv (\beta_1, \dots, \beta_n) \in \mathbb{N}^n$. As is well known, the Roumieu class $\left(C^0_{\omega,\rho}(\mathrm{cl}\Omega), \|\cdot\|_{C^0_{\omega,\rho}(\mathrm{cl}\Omega)}\right)$ is a Banach space.

A straightforward computation based on the inequality $\binom{\beta}{\alpha} \leq \binom{|\beta|}{|\alpha|}$, which holds for α, $\beta \in \mathbb{N}^n$, $0 \leq \alpha \leq \beta$, shows that the pointwise product is bilinear and continuous from $\left(C^0_{\omega,\rho}(\mathrm{cl}\Omega)\right)^2$ to $C^0_{\omega,\rho'}(\mathrm{cl}\Omega)$ for all $\rho' \in]0, \rho[$.

Let $\mathbb{D} \subseteq \mathbb{R}^n$ be such that $qz + \mathbb{D} \subseteq \mathbb{D}$ for all $z \in \mathbb{Z}^n$. We say that a function f from $\mathbb{D}$ to $\mathbb{R}$ is q-periodic if $f(x+qe_j) = f(x)$ for all $x \in \mathbb{D}$ and for all $j \in \{1, \dots, n\}$. If $k \in \mathbb{N}$, then we set

$$C^k_q(\mathbb{R}^n) \equiv \{u \in C^k(\mathbb{R}^n) : u \text{ is } q - \text{periodic}\}.$$

We also set

$$C^\infty_q(\mathbb{R}^n) \equiv \{u \in C^\infty(\mathbb{R}^n) : u \text{ is } q - \text{periodic}\}.$$

Similarly, if $\rho > 0$, we set

$$C^0_{q,\omega,\rho}(\mathbb{R}^n) \equiv \left\{ u \in C^\infty_q(\mathbb{R}^n) \colon \sup_{\beta \in \mathbb{N}^n} \frac{\rho^{|\beta|}}{|\beta|!} \|D^\beta u\|_{C^0(\mathrm{cl}Q)} < +\infty \right\}, \tag{2.1}$$

and

$$\|u\|_{C^0_{q,\omega,\rho}(\mathbb{R}^n)} \equiv \sup_{\beta \in \mathbb{N}^n} \frac{\rho^{|\beta|}}{|\beta|!} \|D^\beta u\|_{C^0(\mathrm{cl}Q)} \qquad \forall u \in C^0_{q,\omega,\rho}(\mathbb{R}^n).$$

As can be easily seen, the periodic Roumieu class $\left(C^0_{q,\omega,\rho}(\mathbb{R}^n), \|\cdot\|_{C^0_{q,\omega,\rho}(\mathbb{R}^n)}\right)$ is a Banach space. Obviously, the restriction operator from $C^0_{q,\omega,\rho}(\mathbb{R}^n)$ to $C^0_{\omega,\rho}(\mathrm{cl}\mathbb{D})$ is linear and continuous for all bounded open subsets $\mathbb{D}$ of $\mathbb{R}^n$ and $\rho > 0$. Similarly, if $\rho > 0$, if $\mathbb{D}$ is a bounded open subset of $\mathbb{R}^n$ such that $\mathrm{cl}Q \subseteq \mathbb{D}$, and if $f \in C^0_q(\mathbb{R}^n)$ is such that $f_{|\mathrm{cl}\mathbb{D}} \in C^0_{\omega,\rho}(\mathrm{cl}\mathbb{D})$, then clearly $f \in C^0_{q,\omega,\rho}(\mathbb{R}^n)$.

If Ω is an open subset of $\mathbb{R}^n$, $k \in \mathbb{N}$, $\beta \in]0, 1]$, we set

$$C^{k,\beta}_b(\mathrm{cl}\Omega) \equiv \{u \in C^{k,\beta}(\mathrm{cl}\Omega) : D^\gamma u \text{ is bounded } \forall \gamma \in \mathbb{N}^n \text{ such that } |\gamma| \leq k\},$$

and we endow $C^{k,\beta}_b(\mathrm{cl}\Omega)$ with its usual norm

$$\|u\|_{C^{k,\beta}_b(\mathrm{cl}\Omega)} \equiv \sum_{|\gamma| \leq k} \sup_{x \in \mathrm{cl}\Omega} |D^\gamma u(x)| + \sum_{|\gamma| = k} |D^\gamma u : \mathrm{cl}\Omega|_\beta \qquad \forall u \in C^{k,\beta}_b(\mathrm{cl}\Omega),$$

where $|D^\gamma u : \mathrm{cl}\Omega|_\beta$ denotes the β-Hölder constant of $D^\gamma u$.

Next we turn to periodic domains. If $\mathbb{I}$ is an arbitrary subset of $\mathbb{R}^n$ such that $\mathrm{cl}\mathbb{I} \subseteq Q$, then we set

$$\mathbb{S}[\mathbb{I}] \equiv \bigcup_{z\in\mathbb{Z}^n} (qz + \mathbb{I}) = q\mathbb{Z}^n + \mathbb{I}\,, \qquad \mathbb{S}[\mathbb{I}]^- \equiv \mathbb{R}^n \setminus \mathrm{cl}\mathbb{S}[\mathbb{I}]\,.$$

We note that if $\mathbb{R}^n \setminus \mathrm{cl}\mathbb{I}$ is connected, then $\mathbb{S}[\mathbb{I}]^-$ is connected.

If $\mathbb{I}$ is an open subset of $\mathbb{R}^n$ such that $\mathrm{cl}\mathbb{I} \subseteq Q$ and if $k \in \mathbb{N}$, $\beta \in]0,1]$, then we set

$$C_q^{k,\beta}(\mathrm{cl}\mathbb{S}[\mathbb{I}]) \equiv \left\{u \in C_b^{k,\beta}(\mathrm{cl}\mathbb{S}[\mathbb{I}]) : u \text{ is } q-\text{periodic}\right\},$$

which we regard as a Banach subspace of $C_b^{k,\beta}(\mathrm{cl}\mathbb{S}[\mathbb{I}])$, and

$$C_q^{k,\beta}(\mathrm{cl}\mathbb{S}[\mathbb{I}]^-) \equiv \left\{u \in C_b^{k,\beta}(\mathrm{cl}\mathbb{S}[\mathbb{I}]^-) : u \text{ is } q-\text{periodic}\right\},$$

which we regard as a Banach subspace of $C_b^{k,\beta}(\mathrm{cl}\mathbb{S}[\mathbb{I}]^-)$.

The Laplace operator is well known to have a $\{0\}$-analog of a q-periodic fundamental solution, *i.e.*, a q-periodic tempered distribution $S_{q,n}$ such that

$$\Delta S_{q,n} = \sum_{z\in\mathbb{Z}^n} \delta_{qz} - \frac{1}{\mathrm{meas}(Q)}\,,$$

where δ_{qz} denotes the Dirac measure with mass in qz (cf., *e.g.*, [22, Thm. 3.1].) As is well known, $S_{q,n}$ is determined up to an additive constant, and we can take

$$S_{q,n}(x) = -\sum_{z\in\mathbb{Z}^n\setminus\{0\}} \frac{1}{\mathrm{meas}(Q)4\pi^2|q^{-1}z|^2} e^{2\pi i(q^{-1}z)\cdot x} \qquad \text{in } \mathcal{D}'(\mathbb{R}^n)$$

(cf., *e.g.*, Ammari and Kang [1, p. 53], [22, Thm. 3.1].) Clearly $S_{q,n}$ is even. Moreover, $S_{q,n}$ is real analytic in $\mathbb{R}^n \setminus q\mathbb{Z}^n$ and is locally integrable in $\mathbb{R}^n$.

Let S_n be the function from $\mathbb{R}^n \setminus \{0\}$ to $\mathbb{R}$ defined by

$$S_n(x) \equiv \begin{cases} \frac{1}{s_n}\log|x| & \forall x \in \mathbb{R}^n\setminus\{0\}, \quad \text{if } n=2\,, \\ \frac{1}{(2-n)s_n}|x|^{2-n} & \forall x \in \mathbb{R}^n\setminus\{0\}, \quad \text{if } n>2\,, \end{cases}$$

where s_n denotes the $(n-1)$-dimensional measure of $\partial\mathbb{B}_n$. S_n is well known to be the fundamental solution of the Laplace operator.

Then the function $S_{q,n} - S_n$ is analytic in $(\mathbb{R}^n \setminus q\mathbb{Z}^n) \cup \{0\}$ (cf., *e.g.*, Ammari and Kang [1, Lemma 2.39, p. 54], [22, Thm. 3.5].) Then we find convenient to set

$$R_n \equiv S_{q,n} - S_n \qquad \text{in } (\mathbb{R}^n \setminus q\mathbb{Z}^n) \cup \{0\}\,.$$

We now introduce the periodic simple layer potential. Let $\alpha \in]0,1[$, $m \in \mathbb{N}\setminus\{0\}$. Let $\mathbb{I}$ be a bounded open subset of $\mathbb{R}^n$ of class $C^{m,\alpha}$ such that $\mathrm{cl}\mathbb{I} \subseteq Q$. If $\mu \in C^{0,\alpha}(\partial\mathbb{I})$, we set

$$v_q[\partial\mathbb{I},\mu](x) \equiv \int_{\partial\mathbb{I}} S_{q,n}(x-y)\mu(y)\,d\sigma_y \qquad \forall x \in \mathbb{R}^n\,.$$

As is well known, if $\mu \in C^{m-1,\alpha}(\partial\mathbb{I})$, then the function $v_q^+[\partial\mathbb{I},\mu] \equiv v_q[\partial\mathbb{I},\mu]_{|\mathrm{cl}\mathbb{S}[\mathbb{I}]}$ belongs to $C_q^{m,\alpha}(\mathrm{cl}\mathbb{S}[\mathbb{I}])$, and the function $v_q^-[\partial\mathbb{I},\mu] \equiv v_q[\partial\mathbb{I},\mu]_{|\mathrm{cl}\mathbb{S}[\mathbb{I}]^-}$ belongs to

$C^{m,\alpha}_q(\mathrm{cl}\mathbb{S}[\mathbb{I}]^-)$. Similarly, we introduce the periodic double layer potential. If $\mu \in C^{0,\alpha}(\partial\mathbb{I})$, we set

$$w_q[\partial\mathbb{I},\mu](x) \equiv -\int_{\partial\mathbb{I}} (DS_{q,n}(x-y))\nu_{\mathbb{I}}(y)\mu(y)\,d\sigma_y \qquad \forall x \in \mathbb{R}^n .$$

If $\mu \in C^{m,\alpha}(\partial\mathbb{I})$, then the restriction $w_q[\partial\mathbb{I},\mu]_{|\mathbb{S}[\mathbb{I}]}$ can be extended uniquely to an element $w_q^+[\partial\mathbb{I},\mu]$ of $C^{m,\alpha}_q(\mathrm{cl}\mathbb{S}[\mathbb{I}])$, and the restriction $w_q[\partial\mathbb{I},\mu]_{|\mathbb{S}[\mathbb{I}]^-}$ can be extended uniquely to an element $w_q^-[\partial\mathbb{I},\mu]$ of $C^{m,\alpha}_q(\mathrm{cl}\mathbb{S}[\mathbb{I}]^-)$, and we have $w_q^\pm[\partial\mathbb{I},\mu] = \pm\frac{1}{2}\mu + w_q[\partial\mathbb{I},\mu]$ on $\partial\mathbb{I}$. Moreover, if $\mu \in C^{0,\alpha}(\partial\mathbb{I})$, then we have

$$w_q[\partial\mathbb{I},\mu](x) = -\sum_{j=1}^{n} \frac{\partial}{\partial x_j} \int_{\partial\mathbb{I}} S_{q,n}(x-y)(\nu_{\mathbb{I}}(y))_j\mu(y)\,d\sigma_y\,, \tag{2.2}$$

for all $x \in \mathbb{R}^n \setminus \partial\mathbb{S}[\mathbb{I}]$ (cf., *e.g.*, [22, Thm. 3.18].)

As we have done for the periodic layer potentials, we now introduce a periodic analog of the Newtonian potential. If $f \in C^0_q(\mathbb{R}^n)$, then we set

$$P_q[f](x) \equiv \int_Q S_{q,n}(x-y)f(y)\,dy \qquad \forall x \in \mathbb{R}^n .$$

Clearly, $P_q[f]$ is a q-periodic function on $\mathbb{R}^n$.

In the following theorem, we collect some elementary properties of the periodic Newtonian potential. A proof can be effected by splitting $S_{q,n}$ into the sum of S_n and R_n, and by exploiting the results of Lanza [18] on the (classical) Newtonian potential in the Roumieu classes and standard properties of integral operators with real analytic kernels and with no singularity (cf., *e.g.*, [23, §3] and [9].)

Theorem 2.1. *The following statements hold.*

(i) *Let $\beta \in]0,1]$. Let $f \in C^{0,\beta}_q(\mathbb{R}^n)$. Then $P_q[f] \in C^2_q(\mathbb{R}^n)$ and*

$$\Delta P_q[f](x) = f(x) - \frac{1}{\mathrm{meas}(Q)}\int_Q f(y)\,dy \qquad \forall x \in \mathbb{R}^n .$$

(ii) *Let $\rho > 0$. Then there exists $\rho' \in]0,\rho]$ such that $P_q[f] \in C^0_{q,\omega,\rho'}(\mathbb{R}^n)$ for all $f \in C^0_{q,\omega,\rho}(\mathbb{R}^n)$ and such that $P_q[\cdot]$ is linear and continuous from $C^0_{q,\omega,\rho}(\mathbb{R}^n)$ to $C^0_{q,\omega,\rho'}(\mathbb{R}^n)$.*

Let $m \in \mathbb{N}\setminus\{0\}$, $\alpha \in]0,1[$. If Ω is a bounded open subset of $\mathbb{R}^n$ of class $C^{m,\alpha}$, we find convenient to set

$$C^{m,\alpha}(\partial\Omega)_0 \equiv \left\{\phi \in C^{m,\alpha}(\partial\Omega) \colon \int_{\partial\Omega} \phi\,d\sigma = 0\right\} .$$

If $\rho > 0$, we also set

$$C^0_{q,\omega,\rho}(\mathbb{R}^n)_0 \equiv \left\{f \in C^0_{q,\omega,\rho}(\mathbb{R}^n) \colon \int_Q f\,dx = 0\right\} . \tag{2.3}$$

As the following proposition shows, a periodic Dirichlet boundary value problem for the Poisson equation in the perforated domain $\mathbb{S}[\mathbb{I}]^-$ has a unique solution in

$C^{m,\alpha}_q(\mathrm{cl}\mathbb{S}[\mathbb{I}]^-)$, which can be represented as the sum of a periodic double layer potential, of a constant, and of a periodic Newtonian potential.

Proposition 2.2. *Let $m \in \mathbb{N}\setminus\{0\}$, $\alpha \in]0,1[$. Let $\rho > 0$. Let $\mathbb{I}$ be a bounded connected open subset of $\mathbb{R}^n$ of class $C^{m,\alpha}$ such that $\mathbb{R}^n\setminus\mathrm{cl}\mathbb{I}$ is connected and that $\mathrm{cl}\mathbb{I} \subseteq Q$. Let $\Gamma \in C^{m,\alpha}(\partial\mathbb{I})$. Let $f \in C^0_{q,\omega,\rho}(\mathbb{R}^n)_0$. Then the following boundary value problem*

$$\begin{cases} \Delta u(x) = f(x) & \forall x \in \mathbb{S}[\mathbb{I}]^-\,, \\ u \text{ is } q\text{–periodic in } \mathrm{cl}\mathbb{S}[\mathbb{I}]^-\,, & \\ u(x) = \Gamma(x) & \forall x \in \partial\mathbb{I}\,, \end{cases}$$

has a unique solution $u \in C^{m,\alpha}_q(\mathrm{cl}\mathbb{S}[\mathbb{I}]^-)$. Moreover,

$$u(x) = w_q^-[\partial\mathbb{I},\mu](x) + \xi + P_q[f](x) \qquad \forall x \in \mathrm{cl}\mathbb{S}[\mathbb{I}]^-\,,$$

where (μ,ξ) is the unique solution in $C^{m,\alpha}(\partial\mathbb{I})_0 \times \mathbb{R}$ of the following integral equation

$$-\frac{1}{2}\mu(x) + w_q[\partial\mathbb{I},\mu](x) + \xi = \Gamma(x) - P_q[f](x) \qquad \forall x \in \partial\mathbb{I}\,.$$

Proof. A proof can be effected by considering the difference $u - P_q[f]$ and by solving the corresponding homogeneous problem (cf. [29, Prop. 2.10] and Theorem 2.1.) □

3. Formulation and analysis of an auxiliary problem

We shall consider the following assumptions for some $\alpha \in]0,1[$ and for some natural $m \geq 1$.

$$\begin{aligned} &\text{Let } \Omega \text{ be a bounded connected open subset of } \mathbb{R}^n \text{ of class } C^{m,\alpha} \\ &\text{such that } \mathbb{R}^n \setminus \mathrm{cl}\Omega \text{ is connected and that } 0 \in \Omega. \\ &\text{Let } p \in Q. \end{aligned} \tag{3.1}$$

If $\epsilon \in \mathbb{R}$ and (3.1) holds, we set

$$\Omega_{p,\epsilon} \equiv p + \epsilon\Omega\,.$$

Now let ϵ_0 be such that

$$\epsilon_0 > 0 \quad \text{and} \quad \mathrm{cl}\Omega_{p,\epsilon} \subseteq Q \quad \forall \epsilon \in]-\epsilon_0,\epsilon_0[\,. \tag{3.2}$$

A simple topological argument shows that if (3.1) holds, then $\mathbb{S}[\Omega_{p,\epsilon}]^-$ is connected, for all $\epsilon \in]-\epsilon_0,\epsilon_0[$. We also note that

$$\nu_{\Omega_{p,\epsilon}}(p + \epsilon t) = \mathrm{sgn}(\epsilon)\nu_\Omega(t) \qquad \forall t \in \partial\Omega\,,$$

for all $\epsilon \in]-\epsilon_0,\epsilon_0[\setminus\{0\}$, where $\mathrm{sgn}(\epsilon) = 1$ if $\epsilon > 0$, $\mathrm{sgn}(\epsilon) = -1$ if $\epsilon < 0$. Let $\rho > 0$. Then we shall consider also the following assumptions.

$$\text{Let } g_0 \in C^{m,\alpha}(\partial\Omega). \tag{3.3}$$

$$\text{Let } f_0 \in C^0_{q,\omega,\rho}(\mathbb{R}^n)_0. \tag{3.4}$$

If $(\epsilon, g, f) \in]0, \epsilon_0[\times C^{m,\alpha}(\partial\Omega) \times C^0_{q,\omega,\rho}(\mathbb{R}^n)_0$, then we denote by $u[\epsilon, g, f]$ the unique solution in $C^{m,\alpha}_q(\mathrm{cl}\mathbb{S}[\Omega_{p,\epsilon}]^-)$ of problem (1.1), and by $u_\#[\epsilon, g, f]$ the unique solution in $C^{m,\alpha}_q(\mathrm{cl}\mathbb{S}[\Omega_{p,\epsilon}]^-)$ of the following auxiliary boundary value problem

$$\begin{cases} \Delta u(x) = 0 & \forall x \in \mathbb{S}[\Omega_{p,\epsilon}]^- , \\ u \text{ is q-periodic in } \mathrm{cl}\mathbb{S}[\Omega_{p,\epsilon}]^- , \\ u(x) = g((x-p)/\epsilon) - P_q[f](x) & \forall x \in \partial\Omega_{p,\epsilon} . \end{cases} \tag{3.5}$$

Clearly,

$$u[\epsilon, g, f] = u_\#[\epsilon, g, f] + P_q[f] \qquad \text{on } \mathrm{cl}\mathbb{S}[\Omega_{p,\epsilon}]^- . \tag{3.6}$$

Let $f \in C^0_{q,\omega,\rho}(\mathbb{R}^n)$, $\epsilon \in]0, \epsilon_0[$. We note that

$$P_q[f](x) = P_q[f]\Big(p + \epsilon((x-p)/\epsilon)\Big) \qquad \forall x \in \partial\Omega_{p,\epsilon} .$$

Accordingly, the Dirichlet condition in problem (3.5) can be rewritten as

$$u(x) = \Big(g - P_q[f] \circ (p + \epsilon \mathrm{id}_{\partial\Omega})\Big)((x-p)/\epsilon) \qquad \forall x \in \partial\Omega_{p,\epsilon} ,$$

where $\mathrm{id}_{\partial\Omega}$ denotes the identity map in $\partial\Omega$. As a consequence, in order to study the dependence of $u_\#[\epsilon, g, f]$ upon (ϵ, g, f) around $(0, g_0, f_0)$, we can exploit the results of [29], concerning the dependence of the solution of the Dirichlet problem for the Laplace equation upon ϵ and the Dirichlet datum. In order to do so, we need to study the regularity of the map from $]-\epsilon_0, \epsilon_0[\times C^0_{q,\omega,\rho}(\mathbb{R}^n)$ to $C^{m,\alpha}(\partial\Omega)$ which takes (ϵ, f) to the function $P_q[f] \circ (p + \epsilon \mathrm{id}_{\partial\Omega})$.

Lemma 3.1. *Let $m \in \mathbb{N} \setminus \{0\}$, $\alpha \in]0,1[$. Let $\rho > 0$. Let (3.1)–(3.4) hold. Let $\mathrm{id}_{\partial\Omega}$, $\mathrm{id}_{\mathrm{cl}\Omega}$ denote the identity map in $\partial\Omega$ and in $\mathrm{cl}\Omega$, respectively. Then the following statements hold.*

(i) *The map from $]-\epsilon_0, \epsilon_0[\times C^0_{q,\omega,\rho}(\mathbb{R}^n)$ to $C^{m,\alpha}(\mathrm{cl}\Omega)$ which takes (ϵ, f) to $P_q[f] \circ (p + \epsilon \mathrm{id}_{\mathrm{cl}\Omega})$ is real analytic.*

(ii) *The map from $]-\epsilon_0, \epsilon_0[\times C^0_{q,\omega,\rho}(\mathbb{R}^n)$ to $C^{m,\alpha}(\partial\Omega)$ which takes (ϵ, f) to $P_q[f] \circ (p + \epsilon \mathrm{id}_{\partial\Omega})$ is real analytic.*

Proof. We first prove statement (i). By Theorem 2.1 (ii), there exists $\rho' \in]0, \rho]$ such that the linear map from $C^0_{q,\omega,\rho}(\mathbb{R}^n)$ to $C^0_{\omega,\rho'}(\mathrm{cl}Q)$ which takes f to $P_q[f]_{|\mathrm{cl}Q}$ is continuous. As a consequence, by Proposition A.1 of the Appendix, we immediately deduce that the map from $]-\epsilon_0, \epsilon_0[\times C^0_{q,\omega,\rho}(\mathbb{R}^n)$ to $C^{m,\alpha}(\mathrm{cl}\Omega)$ which takes (ϵ, f) to $P_q[f] \circ (p + \epsilon \mathrm{id}_{\mathrm{cl}\Omega})$ is real analytic. By the continuity of the trace operator from $C^{m,\alpha}(\mathrm{cl}\Omega)$ to $C^{m,\alpha}(\partial\Omega)$ and statement (i), we deduce the validity of (ii). □

Then we have the following lemma (cf. [29, Lem. 3.8].)

Lemma 3.2. *Let $m \in \mathbb{N} \setminus \{0\}$, $\alpha \in]0,1[$. Let $\rho > 0$. Let (3.1)–(3.4) hold. Let τ_0 be the unique solution in $C^{m-1,\alpha}(\partial\Omega)$ of the following problem*

$$\begin{cases} -\frac{1}{2}\tau(t) + \int_{\partial\Omega} (DS_n(t-s))\nu_\Omega(t)\tau(s)\, d\sigma_s = 0 & \forall t \in \partial\Omega , \\ \int_{\partial\Omega} \tau\, d\sigma = 1 . \end{cases}$$

Then equation

$$-\frac{1}{2}\theta(t) - \int_{\partial\Omega}(DS_n(t-s))\nu_\Omega(s)\theta(s)\,d\sigma_s + \xi = g_0(t) - P_q[f_0](p) \qquad \forall t \in \partial\Omega\,,$$

which we call the limiting equation, *has a unique solution in* $C^{m,\alpha}(\partial\Omega)_0 \times \mathbb{R}$, *which we denote by* $(\tilde{\theta}, \tilde{\xi})$. *Moreover,*

$$\tilde{\xi} = \int_{\partial\Omega} g_0\tau_0\,d\sigma - P_q[f_0](p)\,,$$

and the function $\tilde{u} \in C^{m,\alpha}_{\mathrm{loc}}(\mathbb{R}^n \setminus \mathrm{cl}\Omega)$, *defined by*

$$\tilde{u}(t) \equiv -\int_{\partial\Omega}(DS_n(t-s))\nu_\Omega(s)\tilde{\theta}(s)d\sigma_s \qquad \forall t \in \mathbb{R}^n \setminus \mathrm{cl}\Omega\,,$$

has a unique continuous extension to $\mathbb{R}^n \setminus \Omega$, *which we still denote by* $\tilde{u}$, *and such an extension is the unique solution in* $C^{m,\alpha}_{\mathrm{loc}}(\mathbb{R}^n \setminus \Omega)$ *of the following problem*

$$\begin{cases} \Delta u(t) = 0 & \forall t \in \mathbb{R}^n \setminus \mathrm{cl}\Omega\,,\\ u(t) = g_0(t) - \int_{\partial\Omega} g_0\tau_0\,d\sigma & \forall t \in \partial\Omega\,,\\ \lim_{t\to\infty} u(t) = 0\,. \end{cases}$$

In [29], we have shown that the solutions of a periodic Dirichlet problem for the Laplace equation in $\mathbb{S}[\Omega_{p,\epsilon}]^-$ depend analytically upon ϵ and upon (a rescaling of) the Dirichlet datum. By Lemma 3.1 (ii), we know that (a rescaling of) the Dirichlet datum of the auxiliary problem (3.5) depends analytically upon (ϵ, g, f). Then we deduce that the solution of problem (3.5) depends analytically on (ϵ, g, f), and we have the following.

Proposition 3.3. *Let* $m \in \mathbb{N} \setminus \{0\}$, $\alpha \in]0,1[$. *Let* $\rho > 0$. *Let* (3.1)–(3.4) *hold. Then there exist* $\epsilon_1 \in]0, \epsilon_0]$, *an open neighborhood* $\mathcal{U}$ *of* g_0 *in* $C^{m,\alpha}(\partial\Omega)$, *an open neighborhood* $\mathcal{V}$ *of* f_0 *in* $C^0_{q,\omega,\rho}(\mathbb{R}^n)_0$, *and a real analytic map* $(\Theta[\cdot,\cdot,\cdot], \Xi[\cdot,\cdot,\cdot])$ *from* $]-\epsilon_1, \epsilon_1[\times\mathcal{U}\times\mathcal{V}$ *to* $C^{m,\alpha}(\partial\Omega)_0 \times \mathbb{R}$ *such that*

$$u_\#[\epsilon, g, f] = w_q^-[\partial\Omega_{p,\epsilon}, \Theta[\epsilon, g, f]((\cdot - p)/\epsilon)] + \Xi[\epsilon, g, f] \qquad \text{on } \mathrm{cl}\mathbb{S}[\Omega_{p,\epsilon}]^-\,,$$

for all $(\epsilon, g, f) \in]0, \epsilon_1[\times\mathcal{U}\times\mathcal{V}$. *Moreover,* $(\Theta[0, g_0, f_0], \Xi[0, g_0, f_0]) = (\tilde{\theta}, \tilde{\xi})$, *where* $\tilde{\theta}$, $\tilde{\xi}$ *are as in Lemma* 3.2.

Then we have the following representation theorem for $u_\#[\cdot,\cdot,\cdot]$.

Theorem 3.4. *Let* $m \in \mathbb{N} \setminus \{0\}$, $\alpha \in]0,1[$. *Let* $\rho > 0$. *Let* (3.1)–(3.4) *hold. Let* ϵ_1, $\mathcal{U}$, $\mathcal{V}$, Ξ *be as in Proposition* 3.3. *Then the following statements hold.*

(i) *Let* V *be a bounded open subset of* $\mathbb{R}^n$ *such that* $\mathrm{cl}V \subseteq \mathbb{R}^n \setminus (p + q\mathbb{Z}^n)$. *Let* $r \in \mathbb{N}$. *Then there exist* $\epsilon_{\#,2} \in]0, \epsilon_1]$ *and a real analytic map* $U_\#$ *from* $]-\epsilon_{\#,2}, \epsilon_{\#,2}[\times\mathcal{U}\times\mathcal{V}$ *to* $C^r(\mathrm{cl}V)$ *such that the following statements hold.*
   (j) $\mathrm{cl}V \subseteq \mathbb{S}[\Omega_{p,\epsilon}]^-$ *for all* $\epsilon \in]-\epsilon_{\#,2}, \epsilon_{\#,2}[$.

(jj) $u_{\#}[\epsilon, g, f](x) = \epsilon^{n-1} U_{\#}[\epsilon, g, f](x) + \Xi[\epsilon, g, f] \qquad \forall x \in \mathrm{cl}V$,

for all $(\epsilon, g, f) \in]0, \epsilon_{\#,2}[\times \mathcal{U} \times \mathcal{V}$. *Moreover,*

$$\begin{aligned} U_{\#}[0, g_0, f_0](x) = & - \int_{\partial\Omega} (DS_{q,n}(x-p))\nu_\Omega(s)\tilde{\theta}(s)\, d\sigma_s \\ = & DS_{q,n}(x-p) \int_{\partial\Omega} \nu_\Omega(s)\tilde{u}(s)\, d\sigma_s \\ & - DS_{q,n}(x-p) \int_{\partial\Omega} s \frac{\partial \tilde{u}}{\partial \nu_\Omega}(s)\, d\sigma_s \qquad \forall x \in \mathrm{cl}V, \end{aligned}$$

where $\tilde{\theta}$, $\tilde{u}$ *are as in Lemma* 3.2.

(ii) *Let* $\widetilde{V}$ *be a bounded open subset of* $\mathbb{R}^n \setminus \mathrm{cl}\Omega$. *Then there exist* $\tilde{\epsilon}_{\#,2} \in]0, \epsilon_1]$ *and a real analytic map* $\widetilde{U}_{\#}$ *from* $]-\tilde{\epsilon}_{\#,2}, \tilde{\epsilon}_{\#,2}[\times \mathcal{U} \times \mathcal{V}$ *to* $C^{m,\alpha}(\mathrm{cl}\widetilde{V})$ *such that the following statements hold.*

(j′) $p + \epsilon \mathrm{cl}\widetilde{V} \subseteq Q \setminus \Omega_{p,\epsilon}$ *for all* $\epsilon \in]-\tilde{\epsilon}_{\#,2}, \tilde{\epsilon}_{\#,2}[\setminus\{0\}$.

(jj′) $u_{\#}[\epsilon, g, f](p+\epsilon t) = \widetilde{U}_{\#}[\epsilon, g, f](t) + \Xi[\epsilon, g, f] \qquad \forall t \in \mathrm{cl}\widetilde{V}$,

for all $(\epsilon, g, f) \in]0, \tilde{\epsilon}_{\#,2}[\times \mathcal{U} \times \mathcal{V}$. *Moreover,*

$$\widetilde{U}_{\#}[0, g_0, f_0](t) = \tilde{u}(t) \qquad \forall t \in \mathrm{cl}\widetilde{V}.$$

where $\tilde{u}$ *is as in Lemma* 3.2.

Proof. We follow the argument of the proof of [29, Thm. 4.1]. We first consider statement (i). By taking $\epsilon_{\#,2} \in]0, \epsilon_1]$ small enough, we can clearly assume that (j) holds. Consider now (jj). By Proposition 3.3, if $(\epsilon, g, f) \in]0, \epsilon_{\#,2}[\times \mathcal{U} \times \mathcal{V}$, we have

$$u_{\#}[\epsilon, g, f](x) = -\epsilon^{n-1} \int_{\partial\Omega} (DS_{q,n}(x-p-\epsilon s))\nu_\Omega(s)\Theta[\epsilon, g, f](s)\, d\sigma_s + \Xi[\epsilon, g, f] \qquad \forall x \in \mathrm{cl}V.$$

Thus it is natural to set

$$U_{\#}[\epsilon, g, f](x) \equiv - \int_{\partial\Omega} (DS_{q,n}(x-p-\epsilon s))\nu_\Omega(s)\Theta[\epsilon, g, f](s)\, d\sigma_s \qquad \forall x \in \mathrm{cl}V,$$

for all $(\epsilon, g, f) \in]-\epsilon_{\#,2}, \epsilon_{\#,2}[\times \mathcal{U} \times \mathcal{V}$. Then Proposition 3.3, standard properties of integral operators with real analytic kernels and with no singularity (cf., *e.g.*, [23, §4]), and classical potential theory (cf., *e.g.*, Miranda [27], Lanza and Rossi [25, Thm. 3.1]) imply that $U_{\#}$ is a real analytic map from $]-\epsilon_{\#,2}, \epsilon_{\#,2}[\times \mathcal{U} \times \mathcal{V}$ to $C^r(\mathrm{cl}V)$ such that (jj) holds (see also the proof of [29, Thm. 4.1].)

Consider now (ii). Let $R > 0$ be such that $(\mathrm{cl}\widetilde{V} \cup \mathrm{cl}\Omega) \subseteq \mathbb{B}_n(0, R)$. By the continuity of the restriction operator from $C^{m,\alpha}(\mathrm{cl}\mathbb{B}_n(0,R) \setminus \Omega)$ to $C^{m,\alpha}(\mathrm{cl}\widetilde{V})$, it suffices to prove statement (ii) with $\widetilde{V}$ replaced by $\mathbb{B}_n(0,R) \setminus \mathrm{cl}\Omega$. By taking $\tilde{\epsilon}_{\#,2} \in]0, \epsilon_1]$ small enough, we can assume that

$$p + \epsilon \mathrm{cl}\mathbb{B}_n(0, R) \subseteq Q \qquad \forall \epsilon \in]-\tilde{\epsilon}_{\#,2}, \tilde{\epsilon}_{\#,2}[\,.$$

If $(\epsilon, g, f) \in]0, \tilde{\epsilon}_{\#,2}[\times \mathcal{U} \times \mathcal{V}$, a simple computation based on the theorem of change of variables in integrals shows that

$$u_{\#}[\epsilon, g, f](p+\epsilon t) = -\int_{\partial\Omega} (DS_n(t-s))\nu_{\Omega}(s)\Theta[\epsilon, g, f](s)\, d\sigma_s \\ - \epsilon^{n-1}\int_{\partial\Omega} (DR_n(\epsilon(t-s)))\nu_{\Omega}(s)\Theta[\epsilon, g, f](s)\, d\sigma_s + \Xi[\epsilon, g, f]$$

for all $t \in \mathrm{cl}\mathbb{B}_n(0,R) \setminus \mathrm{cl}\Omega$. If $(\epsilon, g, f) \in]-\tilde{\epsilon}_{\#,2}, \tilde{\epsilon}_{\#,2}[\times \mathcal{U} \times \mathcal{V}$, classical potential theory implies that the function

$$-\int_{\partial\Omega} (DS_n(t-s))\nu_{\Omega}(s)\Theta[\epsilon, g, f](s)\, d\sigma_s$$

of the variable $t \in \mathrm{cl}\mathbb{B}_n(0,R) \setminus \mathrm{cl}\Omega$ admits an extension to $\mathrm{cl}\mathbb{B}_n(0,R) \setminus \Omega$ of class $C^{m,\alpha}(\mathrm{cl}\mathbb{B}_n(0,R) \setminus \Omega)$, which we denote by $w^-[\partial\Omega, \Theta[\epsilon, g, f]]_{|\mathrm{cl}\mathbb{B}_n(0,R)\setminus\Omega}$ (cf., *e.g.*, Miranda [27], Lanza and Rossi [25, Thm. 3.1].) Then classical potential theory and Proposition 3.3 imply that the map from $]-\tilde{\epsilon}_{\#,2}, \tilde{\epsilon}_{\#,2}[\times \mathcal{U} \times \mathcal{V}$ to $C^{m,\alpha}(\mathrm{cl}\mathbb{B}_n(0,R) \setminus \Omega)$ which takes (ϵ, g, f) to $w^-[\partial\Omega, \Theta[\epsilon, g, f]]_{|\mathrm{cl}\mathbb{B}_n(0,R)\setminus\Omega}$ is real analytic (cf., *e.g.*, Miranda [27], Lanza and Rossi [25, Thm. 3.1].) Therefore, if we set

$$\widetilde{U}_{\#}[\epsilon, g, f](t) \equiv w^-[\partial\Omega, \Theta[\epsilon, g, f]]_{|\mathrm{cl}\mathbb{B}_n(0,R)\setminus\Omega}(t) \\ - \epsilon^{n-1}\int_{\partial\Omega} (DR_n(\epsilon(t-s)))\nu_{\Omega}(s)\Theta[\epsilon, g, f](s)\, d\sigma_s \quad \forall t \in \mathrm{cl}\mathbb{B}_n(0,R) \setminus \Omega\,,$$

for all $(\epsilon, g, f) \in]-\tilde{\epsilon}_{\#,2}, \tilde{\epsilon}_{\#,2}[\times \mathcal{U} \times \mathcal{V}$, Proposition 3.3, standard properties of integral operators with real analytic kernels and with no singularity (cf., *e.g.*, [23, §4]), and classical potential theory imply that $\widetilde{U}_{\#}$ is a real analytic map from $]-\tilde{\epsilon}_{\#,2}, \tilde{\epsilon}_{\#,2}[\times \mathcal{U} \times \mathcal{V}$ to $C^{m,\alpha}(\mathrm{cl}\mathbb{B}_n(0,R) \setminus \Omega)$ such that (jj′) holds with $\widetilde{V}$ replaced by $\mathbb{B}_n(0,R) \setminus \mathrm{cl}\Omega$ (see also the proof of [29, Thm. 4.1].) Thus the proof is complete. □

Then we analyze the behaviour of the energy integral by means of the following.

Theorem 3.5. *Let $m \in \mathbb{N} \setminus \{0\}$, $\alpha \in]0,1[$. Let* (3.1)–(3.4) *hold. Let ϵ_1, $\mathcal{U}$, $\mathcal{V}$ be as in Proposition* 3.3. *Then there exist $\epsilon_{\#,3} \in]0, \epsilon_1]$ and a real analytic map $G_{\#}$ from $]-\epsilon_{\#,3}, \epsilon_{\#,3}[\times \mathcal{U} \times \mathcal{V}$ to $\mathbb{R}$, such that*

$$\int_{Q\setminus\mathrm{cl}\Omega_{p,\epsilon}} |D_x u_{\#}[\epsilon, g, f](x)|^2\, dx = \epsilon^{n-2} G_{\#}[\epsilon, g, f]\,,$$

for all $(\epsilon, g, f) \in]0, \epsilon_{\#,3}[\times \mathcal{U} \times \mathcal{V}$. Moreover,

$$G_{\#}[0, g_0, f_0] = \int_{\mathbb{R}^n\setminus\mathrm{cl}\Omega} |D\tilde{u}(t)|^2\, dt\,,$$

where $\tilde{u}$ is as in Lemma 3.2.

Proof. We follow the argument of the proof of [29, Thm. 4.6]. Let $(\epsilon, g, f) \in]0, \epsilon_1[\times \mathcal{U} \times \mathcal{V}$. By the Green Formula and by the periodicity of $u_\#[\epsilon, g, f](\cdot)$, we have

$$\int_{Q\setminus \mathrm{cl}\Omega_{p,\epsilon}} |D_x u_\#[\epsilon, g, f](x)|^2 \, dx \tag{3.7}$$

$$= -\epsilon^{n-1} \int_{\partial\Omega} D_x u_\#[\epsilon, g, f](p+\epsilon t)\nu_\Omega(t) u_\#[\epsilon, g, f](p+\epsilon t) \, d\sigma_t$$

$$= -\epsilon^{n-2} \int_{\partial\Omega} D\big(u_\#[\epsilon, g, f] \circ (p + \epsilon \mathrm{id}_n)\big)(t)\nu_\Omega(t)\big(g(t) - P_q[f](p+\epsilon t)\big) \, d\sigma_t \,,$$

where id_n denotes the identity in $\mathbb{R}^n$. Let $R > 0$ be such that $\mathrm{cl}\Omega \subseteq \mathbb{B}_n(0, R)$. By Proposition 3.3 and Theorem 3.4 (ii), there exist $\epsilon_{\#,3} \in]0, \epsilon_1]$ and a real analytic map $\widehat{U}_\#$ from $]-\epsilon_{\#,3}, \epsilon_{\#,3}[\times \mathcal{U} \times \mathcal{V}$ to $C^{m,\alpha}(\mathrm{cl}\mathbb{B}_n(0, R) \setminus \Omega)$, such that

$$p + \epsilon \mathrm{cl}(\mathbb{B}_n(0, R) \setminus \mathrm{cl}\Omega) \subseteq Q \setminus \Omega_{p,\epsilon} \qquad \forall \epsilon \in]-\epsilon_{\#,3}, \epsilon_{\#,3}[\setminus\{0\} \,,$$

and that

$$\widehat{U}_\#[\epsilon, g, f](t) = u_\#[\epsilon, g, f] \circ (p + \epsilon \mathrm{id}_n)(t) \quad \forall t \in \mathrm{cl}\mathbb{B}_n(0, R) \setminus \Omega \,,$$

for all $(\epsilon, g, f) \in]0, \epsilon_{\#,3}[\times \mathcal{U} \times \mathcal{V}$, and that

$$\widehat{U}_\#[0, g_0, f_0](t) = \tilde{u}(t) + \tilde{\xi} \qquad \forall t \in \mathrm{cl}\mathbb{B}_n(0, R) \setminus \Omega \,,$$

where $\tilde{u}$, $\tilde{\xi}$ are as in Lemma 3.2. By equality (3.7), we have

$$\int_{Q\setminus \mathrm{cl}\Omega_{p,\epsilon}} |D_x u_\#[\epsilon, g, f](x)|^2 \, dx$$

$$= -\epsilon^{n-2} \int_{\partial\Omega} D_t \widehat{U}_\#[\epsilon, g, f](t)\nu_\Omega(t)\big(g(t) - P_q[f](p+\epsilon t)\big) \, d\sigma_t \,,$$

for all $(\epsilon, g, f) \in]0, \epsilon_{\#,3}[\times \mathcal{U} \times \mathcal{V}$. Thus it is natural to set

$$G_\#[\epsilon, g, f] \equiv -\int_{\partial\Omega} D_t \widehat{U}_\#[\epsilon, g, f](t)\nu_\Omega(t)\big(g(t) - P_q[f](p+\epsilon t)\big) \, d\sigma_t \,,$$

for all $(\epsilon, g, f) \in]-\epsilon_{\#,3}, \epsilon_{\#,3}[\times \mathcal{U} \times \mathcal{V}$. Then by continuity of the partial derivatives from $C^{m,\alpha}(\mathrm{cl}\mathbb{B}_n(0, R) \setminus \Omega)$ to $C^{m-1,\alpha}(\mathrm{cl}\mathbb{B}_n(0, R) \setminus \Omega)$, and by continuity of the trace operator on $\partial\Omega$ from $C^{m-1,\alpha}(\mathrm{cl}\mathbb{B}_n(0, R) \setminus \Omega)$ to $C^{m-1,\alpha}(\partial\Omega)$, and by the continuity of the pointwise product in Schauder spaces, and by Lemma 3.1 (ii), and by classical potential theory, we conclude that $G_\#$ is a real analytic map from $]-\epsilon_{\#,3}, \epsilon_{\#,3}[\times \mathcal{U} \times \mathcal{V}$ to $\mathbb{R}$ and that the Theorem holds (see also the proof of [29, Thm. 4.6].) □

Finally, we consider the integral of $u_\#[\cdot, \cdot, \cdot]$, and we prove the following.

Theorem 3.6. *Let $m \in \mathbb{N} \setminus \{0\}$, $\alpha \in]0, 1[$. Let (3.1)–(3.4) hold. Let ϵ_1, $\mathcal{U}$, $\mathcal{V}$ be as in Proposition 3.3. Then there exists a real analytic map $J_\#$ from $]-\epsilon_1, \epsilon_1[\times \mathcal{U} \times \mathcal{V}$ to $\mathbb{R}$, such that*

$$\int_{Q\setminus \mathrm{cl}\Omega_{p,\epsilon}} u_\#[\epsilon, g, f](x) \, dx = J_\#[\epsilon, g, f] \,, \tag{3.8}$$

for all $(\epsilon, g, f) \in]0, \epsilon_1[\times \mathcal{U} \times \mathcal{V}$. *Moreover,*

$$J_{\#}[0, g_0, f_0] = \tilde{\xi}\mathrm{meas}(Q)\,, \tag{3.9}$$

where $\tilde{\xi}$ *is as in Lemma* 3.2.

Proof. Let $(\epsilon, g, f) \in]0, \epsilon_1[\times \mathcal{U} \times \mathcal{V}$. Clearly,

$$\begin{aligned}\int_{Q\setminus \mathrm{cl}\Omega_{p,\epsilon}} u_{\#}[\epsilon, g, f](x)\,dx = &\int_{Q\setminus \mathrm{cl}\Omega_{p,\epsilon}} w_q^-\big[\partial\Omega_{p,\epsilon}, \Theta[\epsilon, g, f]((\cdot - p)/\epsilon)\big](x)\,dx \\ &+ \Xi[\epsilon, g, f]\big(\mathrm{meas}(Q) - \epsilon^n \mathrm{meas}(\Omega)\big)\,,\end{aligned}$$

where $\mathrm{meas}(Q)$ and $\mathrm{meas}(\Omega)$ denote the n-dimensional measure of Q and of Ω, respectively. By equality (2.2), we have

$$\begin{aligned}&w_q^-\big[\partial\Omega_{p,\epsilon}, \Theta[\epsilon, g, f]((\cdot - p)/\epsilon)\big](x) \\ &= -\sum_{j=1}^{n} \frac{\partial}{\partial x_j} v_q^-\big[\partial\Omega_{p,\epsilon}, \Theta[\epsilon, g, f]((\cdot - p)/\epsilon)(\nu_{\Omega_{p,\epsilon}}(\cdot))_j\big](x) \quad \forall x \in \mathrm{cl}Q \setminus \mathrm{cl}\Omega_{p,\epsilon}\,.\end{aligned}$$

Let $j \in \{1, \dots, n\}$. By the Divergence Theorem and the periodicity of the periodic simple layer potential, we have

$$\begin{aligned}&\int_{Q\setminus \mathrm{cl}\Omega_{p,\epsilon}} \frac{\partial}{\partial x_j} v_q^-\big[\partial\Omega_{p,\epsilon}, \Theta[\epsilon, g, f]((\cdot - p)/\epsilon)(\nu_{\Omega_{p,\epsilon}}(\cdot))_j\big](x)\,dx \\ &= \int_{\partial Q} v_q^-\big[\partial\Omega_{p,\epsilon}, \Theta[\epsilon, g, f]((\cdot - p)/\epsilon)(\nu_{\Omega_{p,\epsilon}}(\cdot))_j\big](x)(\nu_Q(x))_j\,d\sigma_x \\ &\quad - \int_{\partial\Omega_{p,\epsilon}} v_q^-\big[\partial\Omega_{p,\epsilon}, \Theta[\epsilon, g, f]((\cdot - p)/\epsilon)(\nu_{\Omega_{p,\epsilon}}(\cdot))_j\big](x)(\nu_{\Omega_{p,\epsilon}}(x))_j\,d\sigma_x \\ &= -\,\epsilon^{n-1} \int_{\partial\Omega} v_q^-\big[\partial\Omega_{p,\epsilon}, \Theta[\epsilon, g, f]((\cdot - p)/\epsilon)(\nu_{\Omega_{p,\epsilon}}(\cdot))_j\big](p + \epsilon t)(\nu_\Omega(t))_j\,d\sigma_t\,.\end{aligned}$$

Then we note that

$$\begin{aligned}&v_q^-\big[\partial\Omega_{p,\epsilon}, \Theta[\epsilon, g, f]((\cdot - p)/\epsilon)(\nu_{\Omega_{p,\epsilon}}(\cdot))_j\big](p + \epsilon t) \\ &\quad = \epsilon^{n-1} \int_{\partial\Omega} S_n(\epsilon(t - s))\Theta[\epsilon, g, f](s)(\nu_\Omega(s))_j\,d\sigma_s \\ &\qquad + \epsilon^{n-1} \int_{\partial\Omega} R_n(\epsilon(t - s))\Theta[\epsilon, g, f](s)(\nu_\Omega(s))_j\,d\sigma_s \qquad \forall t \in \partial\Omega\,.\end{aligned}$$

We now observe that if $\epsilon > 0$ and $x \in \mathbb{R}^n \setminus \{0\}$ then we have

$$S_n(\epsilon x) = \epsilon^{2-n} S_n(x) + \delta_{2,n} \frac{1}{2\pi} \log \epsilon\,. \tag{3.10}$$

Moreover, by the Divergence Theorem, it is immediate to see that

$$\begin{aligned}&\int_{\partial\Omega} \Big(\int_{\partial\Omega} \Theta[\epsilon, g, f](s)(\nu_\Omega(s))_j\,d\sigma_s\Big)(\nu_\Omega(t))_j\,d\sigma_t \\ &\quad = \Big(\int_{\partial\Omega} \Theta[\epsilon, g, f](s)(\nu_\Omega(s))_j\,d\sigma_s\Big)\Big(\int_{\partial\Omega} (\nu_\Omega(t))_j\,d\sigma_t\Big) = 0\,.\end{aligned} \tag{3.11}$$

Hence, by equalities (3.10) and (3.11), if $(\epsilon, g, f) \in]0, \epsilon_1[\times \mathcal{U} \times \mathcal{V}$, we have

$$
\begin{aligned}
\int_{Q\setminus \mathrm{cl}\Omega_{p,\epsilon}} & w_q^-\left[\partial\Omega_{p,\epsilon}, \Theta[\epsilon, g, f]((\cdot - p)/\epsilon)\right](x)\, dx \\
&= \sum_{j=1}^{n} \epsilon^n \Bigg[\int_{\partial\Omega}\Big(\int_{\partial\Omega} S_n(t-s)\Theta[\epsilon, g, f](s)(\nu_\Omega(s))_j\, d\sigma_s\Big)(\nu_\Omega(t))_j\, d\sigma_t \\
&\quad + \epsilon^{n-2}\int_{\partial\Omega}\Big(\int_{\partial\Omega} R_n(\epsilon(t-s))\Theta[\epsilon, g, f](s)(\nu_\Omega(s))_j\, d\sigma_s\Big)(\nu_\Omega(t))_j\, d\sigma_t\Bigg].
\end{aligned}
$$

Thus we set

$$
\begin{aligned}
\tilde{J}_{\#}[\epsilon, g, f] \equiv \sum_{j=1}^{n} & \Bigg[\int_{\partial\Omega}\Big(\int_{\partial\Omega} S_n(t-s)\Theta[\epsilon, g, f](s)(\nu_\Omega(s))_j\, d\sigma_s\Big)(\nu_\Omega(t))_j\, d\sigma_t \\
& + \epsilon^{n-2}\int_{\partial\Omega}\Big(\int_{\partial\Omega} R_n(\epsilon(t-s))\Theta[\epsilon, g, f](s)(\nu_\Omega(s))_j\, d\sigma_s\Big)(\nu_\Omega(t))_j\, d\sigma_t\Bigg],
\end{aligned}
$$

for all $(\epsilon, g, f) \in]-\epsilon_1, \epsilon_1[\times \mathcal{U} \times \mathcal{V}$. Clearly, if $(\epsilon, g, f) \in]0, \epsilon_1[\times \mathcal{U} \times \mathcal{V}$, then

$$
\int_{Q\setminus \mathrm{cl}\Omega_{p,\epsilon}} w_q^-\left[\partial\Omega_{p,\epsilon}, \Theta[\epsilon, g, f]((\cdot - p)/\epsilon)\right](x)\, dx = \epsilon^n \tilde{J}_{\#}[\epsilon, g, f].
$$

Then the analyticity of Θ, the continuity of the linear map from $C^{m-1,\alpha}(\partial\Omega)$ to $C^{m,\alpha}(\partial\Omega)$ which takes f to the function $\int_{\partial\Omega} S_n(t-s)f(s)\, d\sigma_s$ of the variable $t \in \partial\Omega$ (cf., *e.g.*, Miranda [27], Lanza and Rossi [25, Thm. 3.1]), the continuity of the pointwise product in Schauder spaces, standard properties of integral operators with real analytic kernels and with no singularity (cf., *e.g.*, [23, §4]), and standard calculus in Banach spaces imply that the map $\tilde{J}_{\#}$ is real analytic from $]-\epsilon_1, \epsilon_1[\times\mathcal{U}\times \mathcal{V}$ to $\mathbb{R}$. Hence, if we set

$$
J_{\#}[\epsilon, g, f] \equiv \epsilon^n \tilde{J}_{\#}[\epsilon, g, f] + \Xi[\epsilon, g, f]\big(\mathrm{meas}(Q) - \epsilon^n \mathrm{meas}(\Omega)\big)
$$

for all $(\epsilon, g, f) \in]-\epsilon_1, \epsilon_1[\times\mathcal{U}\times\mathcal{V}$, we immediately deduce that $J_{\#}$ is a real analytic map from $]-\epsilon_1, \epsilon_1[\times\mathcal{U} \times \mathcal{V}$ to $\mathbb{R}$ such that equalities (3.8), (3.9) hold, and thus the proof is complete. □

4. A functional analytic representation theorem for the solution of problem (1.1)

In this Section, we deduce by the results of Section 3 for $u_{\#}[\cdot,\cdot,\cdot]$ the corresponding results for $u[\cdot,\cdot,\cdot]$. By formula (3.6) and by Theorem 3.4, we immediately deduce the following.

Theorem 4.1. *Let* $m \in \mathbb{N}\setminus\{0\}$, $\alpha \in]0,1[$. *Let* $\rho > 0$. *Let* (3.1)–(3.4) *hold. Let* ϵ_1, $\mathcal{U}$, $\mathcal{V}$ *be as in Proposition* 3.3. *Then the following statements hold.*

(i) *Let* V *be a bounded open subset of* $\mathbb{R}^n$ *such that* $\mathrm{cl}V \subseteq \mathbb{R}^n\setminus(p+q\mathbb{Z}^n)$. *Let* $r \in \mathbb{N}$. *Then there exist* $\epsilon_2 \in]0, \epsilon_1]$ *and a real analytic map* U *from* $]-\epsilon_2, \epsilon_2[\times\mathcal{U}\times\mathcal{V}$ *to* $C^r(\mathrm{cl}V)$ *such that the following statements hold.*

   (j) $\mathrm{cl}V \subseteq \mathbb{S}[\Omega_{p,\epsilon}]^-$ *for all* $\epsilon \in]-\epsilon_2, \epsilon_2[$.

(jj) $u[\epsilon, g, f](x) = U[\epsilon, g, f](x) + P_q[f](x) \qquad \forall x \in \mathrm{cl}V$,

for all $(\epsilon, g, f) \in]0, \epsilon_2[\times \mathcal{U} \times \mathcal{V}$. *Moreover,*

$$U[0, g_0, f_0](x) = \tilde{\xi} \qquad \forall x \in \mathrm{cl}V,$$

where $\tilde{\xi}$ *is as in Lemma* 3.2.

(ii) *Let* $\widetilde{V}$ *be a bounded open subset of* $\mathbb{R}^n \setminus \mathrm{cl}\Omega$. *Then there exist* $\tilde{\epsilon}_2 \in]0, \epsilon_1]$ *and a real analytic map* $\widetilde{U}$ *from* $]-\tilde{\epsilon}_2, \tilde{\epsilon}_2[\times \mathcal{U} \times \mathcal{V}$ *to* $C^{m,\alpha}(\mathrm{cl}\widetilde{V})$ *such that the following statements hold.*

(j′) $p + \epsilon \mathrm{cl}\widetilde{V} \subseteq Q \setminus \Omega_{p,\epsilon}$ *for all* $\epsilon \in]-\tilde{\epsilon}_2, \tilde{\epsilon}_2[\setminus\{0\}$.

(jj′) $u[\epsilon, g, f](p + \epsilon t) = \widetilde{U}[\epsilon, g, f](t) + P_q[f](p + \epsilon t) \qquad \forall t \in \mathrm{cl}\widetilde{V}$,

for all $(\epsilon, g, f) \in]0, \tilde{\epsilon}_2[\times \mathcal{U} \times \mathcal{V}$. *Moreover,*

$$\widetilde{U}[0, g_0, f_0](t) = \tilde{u}(t) + \tilde{\xi} \qquad \forall t \in \mathrm{cl}\widetilde{V}.$$

where $\tilde{u}$, $\tilde{\xi}$ *are as in Lemma* 3.2.

As far as the energy integral of the solution is concerned, we have the following.

Theorem 4.2. *Let* $m \in \mathbb{N} \setminus \{0\}$, $\alpha \in]0, 1[$. *Let* (3.1)–(3.4) *hold. Let* ϵ_1, $\mathcal{U}$, $\mathcal{V}$ *be as in Proposition* 3.3. *Then there exist* $\epsilon_3 \in]0, \epsilon_1]$ *and a real analytic map G from* $]-\epsilon_3, \epsilon_3[\times \mathcal{U} \times \mathcal{V}$ *to* $\mathbb{R}$, *such that*

$$\int_{Q\setminus\mathrm{cl}\Omega_{p,\epsilon}} |D_x u[\epsilon, g, f](x)|^2 \, dx = \epsilon^{n-2} G[\epsilon, g, f] + \int_Q |D_x P_q[f](x)|^2 \, dx, \tag{4.1}$$

for all $(\epsilon, g, f) \in]0, \epsilon_3[\times \mathcal{U} \times \mathcal{V}$. *Moreover,*

$$G[0, g_0, f_0] = \int_{\mathbb{R}^n\setminus\mathrm{cl}\Omega} |D\tilde{u}(t)|^2 \, dt, \tag{4.2}$$

where $\tilde{u}$ *is as in Lemma* 3.2.

Proof. Let $\epsilon_{\#,3}$, $G_\#$ be as in Theorem 3.5. If $(\epsilon, g, f) \in]0, \epsilon_{\#,3}[\times \mathcal{U} \times \mathcal{V}$, then we have

$$\begin{aligned}\int_{Q\setminus\mathrm{cl}\Omega_{p,\epsilon}} |D_x u[\epsilon, g, f](x)|^2 \, dx &= \int_{Q\setminus\mathrm{cl}\Omega_{p,\epsilon}} |D_x u_\#[\epsilon, g, f](x)|^2 \, dx \\ &\quad + 2\int_{Q\setminus\mathrm{cl}\Omega_{p,\epsilon}} D_x u_\#[\epsilon, g, f](x) \cdot D_x P_q[f](x) \, dx + \int_{Q\setminus\mathrm{cl}\Omega_{p,\epsilon}} |D_x P_q[f](x)|^2 \, dx.\end{aligned}$$

By the Divergence Theorem, by the harmonicity of $u_\#[\epsilon, g, f]$, and by the periodicity of $u_\#[\epsilon, g, f]$ and $P_q[f]$, we have

$$\begin{aligned}&2\int_{Q\setminus\mathrm{cl}\Omega_{p,\epsilon}} D_x u_\#[\epsilon, g, f](x) \cdot D_x P_q[f](x) \, dx \\ &\qquad = -2\epsilon^{n-1} \int_{\partial\Omega} P_q[f](p + \epsilon t) \Big(\frac{\partial u_\#[\epsilon, g, f]}{\partial \nu_{\Omega_{p,\epsilon}}}\Big)(p + \epsilon t) \, d\sigma_t \\ &\qquad = -2\epsilon^{n-2} \int_{\partial\Omega} P_q[f](p + \epsilon t) D\big(u_\#[\epsilon, g, f] \circ (p + \epsilon \mathrm{id}_n)\big)(t) \nu_\Omega(t) \, d\sigma_t,\end{aligned}$$

where id_n denotes the identity map in $\mathbb{R}^n$. Let $R>0$ be such that $\mathrm{cl}\Omega \subseteq \mathbb{B}_n(0,R)$. By Proposition 3.3 and Theorem 3.4 (ii), there exist $\epsilon_3 \in]0, \epsilon_{\#,3}]$ and a real analytic map $\widehat{U}_\#$ from $]-\epsilon_3, \epsilon_3[\times \mathcal{U} \times \mathcal{V}$ to $C^{m,\alpha}(\mathrm{cl}\mathbb{B}_n(0,R)\setminus\Omega)$, such that

$$p+\epsilon\mathrm{cl}(\mathbb{B}_n(0,R)\setminus \mathrm{cl}\Omega) \subseteq Q\setminus \Omega_{p,\epsilon} \qquad \forall \epsilon \in]-\epsilon_3,\epsilon_3[\setminus\{0\}\,,$$

and that

$$\widehat{U}_\#[\epsilon,g,f](t) = u_\#[\epsilon,g,f]\circ(p+\epsilon\mathrm{id}_n)(t) \qquad \forall t\in \mathrm{cl}\mathbb{B}_n(0,R)\setminus\Omega\,,$$

for all $(\epsilon,g,f)\in]0,\epsilon_3[\times\mathcal{U}\times\mathcal{V}$, and that

$$\widehat{U}_\#[0,g_0,f_0](t) = \tilde{u}(t)+\tilde{\xi} \qquad \forall t\in \mathrm{cl}\mathbb{B}_n(0,R)\setminus\Omega\,,$$

where $\tilde{u}$, $\tilde{\xi}$ are as in Lemma 3.2. Then we have

$$2\int_{Q\setminus \mathrm{cl}\Omega_{p,\epsilon}} D_x u_\#[\epsilon,g,f](x)\cdot D_x P_q[f](x)\,dx$$
$$= -2\epsilon^{n-2}\int_{\partial\Omega} P_q[f](p+\epsilon t) D_t\widehat{U}_\#[\epsilon,g,f](t)\nu_\Omega(t)\,d\sigma_t\,.$$

for all $(\epsilon,g,f)\in]0,\epsilon_3[\times\mathcal{U}\times\mathcal{V}$. Thus it is natural to set

$$G_1[\epsilon,g,f] \equiv -2\int_{\partial\Omega} P_q[f](p+\epsilon t) D_t\widehat{U}_\#[\epsilon,g,f](t)\nu_\Omega(t)\,d\sigma_t\,,$$

for all $(\epsilon,g,f)\in]-\epsilon_3,\epsilon_3[\times\mathcal{U}\times\mathcal{V}$. Then by continuity of the partial derivatives from $C^{m,\alpha}(\mathrm{cl}\mathbb{B}_n(0,R)\setminus\Omega)$ to $C^{m-1,\alpha}(\mathrm{cl}\mathbb{B}_n(0,R)\setminus\Omega)$, and by continuity of the trace operator on $\partial\Omega$ from $C^{m-1,\alpha}(\mathrm{cl}\mathbb{B}_n(0,R)\setminus\Omega)$ to $C^{m-1,\alpha}(\partial\Omega)$, and by the continuity of the pointwise product in Schauder spaces, and by Lemma 3.1 (ii), we conclude that $G_1[\cdot,\cdot,\cdot]$ is a real analytic map from $]-\epsilon_3,\epsilon_3[\times\mathcal{U}\times\mathcal{V}$ to $\mathbb{R}$. Moreover, by classical potential theory, we have

$$G_1[0,g_0,f_0] = -2\int_{\partial\Omega} P_q[f_0](p)\frac{\partial\tilde{u}}{\partial\nu_\Omega}(t)\,d\sigma_t = 0$$

(see also the proof of [29, Thm. 4.6].) If $(\epsilon,f)\in]0,\epsilon_0[\times\mathcal{V}$, then clearly

$$\int_{Q\setminus\mathrm{cl}\Omega_{p,\epsilon}} |D_xP_q[f](x)|^2\,dx = \int_Q |D_xP_q[f](x)|^2\,dx - \epsilon^n\int_\Omega |D_xP_q[f](p+\epsilon t)|^2\,dt\,.$$

By standard properties of functions in the Roumieu class, there exists $\rho'\in]0,\rho]$ such that the map from $C^0_{q,\omega,\rho}(\mathbb{R}^n)$ to $C^0_{q,\omega,\rho'}(\mathbb{R}^n)$ which takes f to $|D_xP_q[f](\cdot)|^2$ is real analytic (cf., *e.g.*, the proof of Lanza [18, Prop. 2.25].) By arguing as in the proof of Lemma 3.1, one can show that the map from $]-\epsilon_0,\epsilon_0[\times\mathcal{V}$ to $C^{m,\alpha}(\mathrm{cl}\Omega)$ which takes (ϵ,f) to the function $|D_xP_q[f](p+\epsilon t)|^2$ of the variable $t\in\mathrm{cl}\Omega$ is real analytic. Then, by the continuity of the linear operator from $C^{m,\alpha}(\mathrm{cl}\Omega)$ to $\mathbb{R}$ which takes h to $\int_\Omega h(y)\,dy$, we immediately deduce that the map from $]-\epsilon_0,\epsilon_0[\times\mathcal{V}$ to $\mathbb{R}$ which takes (ϵ,f) to $\int_\Omega |D_xP_q[f](p+\epsilon t)|^2\,dt$ is real analytic. Thus if we set

$$G[\epsilon,g,f] \equiv G_\#[\epsilon,g,f]+G_1[\epsilon,g,f]-\epsilon^2\int_\Omega |D_xP_q[f](p+\epsilon t)|^2\,dt$$

for all $(\epsilon, g, f) \in]-\epsilon_3, \epsilon_3[\times \mathcal{U} \times \mathcal{V}$, we deduce that G is a real analytic map from $]-\epsilon_3, \epsilon_3[\times \mathcal{U} \times \mathcal{V}$ to $\mathbb{R}$ such that equalities (4.1), (4.2) hold. □

Remark 4.3. We note that the map from $C^0_{q,\omega,\rho}(\mathbb{R}^n)$ to $\mathbb{R}$ which takes f to $\int_Q |D_x P_q[f](x)|^2 \, dx$ is real analytic.

Finally, we consider the integral of $u[\cdot, \cdot, \cdot]$ and we prove the following.

Theorem 4.4. *Let $m \in \mathbb{N} \setminus \{0\}$, $\alpha \in]0,1[$. Let (3.1)–(3.4) hold. Let ϵ_1, $\mathcal{U}$, $\mathcal{V}$ be as in Proposition 3.3. Then there exists a real analytic map J from $]-\epsilon_1, \epsilon_1[\times \mathcal{U} \times \mathcal{V}$ to $\mathbb{R}$, such that*

$$\int_{Q \setminus \mathrm{cl}\Omega_{p,\epsilon}} u[\epsilon, g, f](x) \, dx = J[\epsilon, g, f] + \int_Q P_q[f](x) \, dx ,$$

for all $(\epsilon, g, f) \in]0, \epsilon_1[\times \mathcal{U} \times \mathcal{V}$. Moreover,

$$J[0, g_0, f_0] = \tilde{\xi} \mathrm{meas}(Q) , \tag{4.3}$$

where $\tilde{\xi}$ is as in Lemma 3.2.

Proof. Let $J_\#$ be as in Theorem 3.6. If we set

$$J[\epsilon, g, f] \equiv J_\#[\epsilon, g, f] - \epsilon^n \int_\Omega P_q[f](p + \epsilon t) \, dt$$

for all $(\epsilon, g, f) \in]-\epsilon_1, \epsilon_1[\times \mathcal{U} \times \mathcal{V}$, then clearly

$$\int_{Q \setminus \mathrm{cl}\Omega_{p,\epsilon}} u[\epsilon, g, f](x) \, dx = J[\epsilon, g, f] + \int_Q P_q[f](x) \, dx ,$$

for all $(\epsilon, g, f) \in]0, \epsilon_1[\times \mathcal{U} \times \mathcal{V}$. By Lemma 3.1 (i) and by the continuity of the linear operator from $C^{m,\alpha}(\mathrm{cl}\Omega)$ to $\mathbb{R}$ which takes h to $\int_\Omega h(y) \, dy$, we immediately deduce that J is a real analytic map from $]-\epsilon_1, \epsilon_1[\times \mathcal{U} \times \mathcal{V}$ to $\mathbb{R}$ and that (4.3) holds. □

Remark 4.5. We note that the map from $C^0_{q,\omega,\rho}(\mathbb{R}^n)$ to $\mathbb{R}$ which takes f to $\int_Q P_q[f](x) \, dx$ is linear and continuous.

Remark 4.6. Let the assumptions of Proposition 3.3 hold. Let $\tilde{u}$, $\tilde{\xi}$ be as in Lemma 3.2. We observe that if V is a bounded open subset of $\mathbb{R}^n$ such that $\mathrm{cl}V \subseteq \mathbb{R}^n \setminus (p + q\mathbb{Z}^n)$ and if $r \in \mathbb{N}$, then Theorem 4.1 (i) implies that

$$\lim_{(\epsilon,g,f)\to(0,g_0,f_0)} u[\epsilon, g, f] = \tilde{\xi} + P_q[f_0] \qquad \text{in } C^r(\mathrm{cl}V) ,$$

for all $n \in \mathbb{N} \setminus \{0, 1\}$. Similarly, by Theorem 4.4, we have

$$\lim_{(\epsilon,g,f)\to(0,g_0,f_0)} \int_{Q \setminus \mathrm{cl}\Omega_{p,\epsilon}} u[\epsilon, g, f](x) \, dx = \tilde{\xi} \mathrm{meas}(Q) + \int_Q P_q[f_0](x) \, dx ,$$

for all $n \in \mathbb{N} \setminus \{0, 1\}$. Instead, Theorem 4.2 implies that

$$\lim_{(\epsilon,g,f)\to(0,g_0,f_0)} \int_{Q \setminus \mathrm{cl}\Omega_{p,\epsilon}} |D_x u[\epsilon, g, f](x)|^2 \, dx = \int_Q |D_x P_q[f_0](x)|^2 \, dx ,$$

if $n \in \mathbb{N} \setminus \{0,1,2\}$, whereas

$$\lim_{(\epsilon,g,f)\to(0,g_0,f_0)} \int_{Q\setminus \mathrm{cl}\Omega_{p,\epsilon}} |D_x u[\epsilon,g,f](x)|^2\,dx = \int_{\mathbb{R}^n\setminus \mathrm{cl}\Omega} |D\tilde{u}(t)|^2\,dt + \int_Q |D_x P_q[f_0](x)|^2\,dx\,,$$

if $n = 2$.

Appendix A. A real analyticity result for a composition operator

In this Appendix, we introduce a slight variant of Preciso [31, Prop. 4.2.16, p. 51], Preciso [32, Prop. 1.1, p. 101] on the real analyticity of a composition operator. See also Lanza [15, Prop. 2.17, Rem. 2.19] and the slight variant of the argument of Preciso of the proof of Lanza [19, Prop. 9, p. 214].

Proposition A.1. *Let $m, h, k \in \mathbb{N}$, $h, k \geq 1$. Let $\alpha \in]0,1]$, $\rho > 0$. Let Ω, Ω' be bounded open connected subsets of $\mathbb{R}^h$, $\mathbb{R}^k$, respectively. Let Ω' be of class C^1. Then the operator T defined by*

$$T[u,v] \equiv u \circ v$$

for all $(u,v) \in C^0_{\omega,\rho}(\mathrm{cl}\Omega) \times C^{m,\alpha}(\mathrm{cl}\Omega',\Omega)$ is real analytic from the open subset $C^0_{\omega,\rho}(\mathrm{cl}\Omega) \times C^{m,\alpha}(\mathrm{cl}\Omega',\Omega)$ of $C^0_{\omega,\rho}(\mathrm{cl}\Omega) \times C^{m,\alpha}(\mathrm{cl}\Omega',\mathbb{R}^h)$ to $C^{m,\alpha}(\mathrm{cl}\Omega')$.

Acknowledgment

The author is indebted to Prof. M. Lanza de Cristoforis for proposing him the problem and for a number of comments which have improved the quality of the paper.

References

[1] H. Ammari and H. Kang, *Polarization and moment tensors*, volume 162 of *Applied Mathematical Sciences*. Springer, New York, 2007.

[2] L.P. Castro and E. Pesetskaya, *A transmission problem with imperfect contact for an unbounded multiply connected domain.* Math. Methods Appl. Sci. **33** (2010), 517–526.

[3] L.P. Castro, E. Pesetskaya, and S.V. Rogosin, *Effective conductivity of a composite material with non-ideal contact conditions.* Complex Var. Elliptic Equ. **54** (2009), 1085–1100.

[4] D. Cioranescu and F. Murat, *Un terme étrange venu d'ailleurs.* In *Nonlinear partial differential equations and their applications. Collège de France Seminar, Vol. II (Paris,* 1979/1980), volume 60 of *Res. Notes in Math.*, pages 98–138, 389–390. Pitman, Boston, Mass., 1982.

[5] D. Cioranescu and F. Murat, *Un terme étrange venu d'ailleurs. II.* In *Nonlinear partial differential equations and their applications. Collège de France Seminar, Vol. III (Paris,* 1980/1981), volume 70 of *Res. Notes in Math.*, pages 154–178, 425–426. Pitman, Boston, Mass., 1982.

[6] M. Dalla Riva, *Stokes flow in a singularly perturbed exterior domain.* Complex Var. Elliptic Equ., (to appear).

[7] M. Dalla Riva and M. Lanza de Cristoforis, *Microscopically weakly singularly perturbed loads for a nonlinear traction boundary value problem: a functional analytic approach.* Complex Var. Elliptic Equ. **55** (2010), 771–794.

[8] M. Dalla Riva and M. Lanza de Cristoforis, *A singularly perturbed nonlinear traction boundary value problem for linearized elastostatics. A functional analytic approach.* Analysis (Munich) **30** (2010), 67–92.

[9] M. Dalla Riva, M. Lanza de Cristoforis, and P. Musolino, *Analytic dependence of volume potentials corresponding to parametric kernels in Roumieu classes.* Typewritten manuscript, 2012.

[10] P. Drygas and V. Mityushev, *Effective conductivity of unidirectional cylinders with interfacial resistance*, Quart. J. Mech. Appl. Math. **62** (2009), 235–262.

[11] D. Gilbarg and N.S. Trudinger, *Elliptic partial differential equations of second order*, volume 224 of *Grundlehren der Mathematischen Wissenschaften* [*Fundamental Principles of Mathematical Sciences*]. Springer-Verlag, Berlin, second edition, 1983.

[12] Eh.I. Grigolyuk and L.A. Fil'shtinskij, *Periodic piecewise homogeneous elastic structures*, (in Russian), Nauka, Moscow, 1992.

[13] V. Kozlov, V. Maz'ya, and A. Movchan, *Asymptotic analysis of fields in multistructures.* Oxford Mathematical Monographs. The Clarendon Press Oxford University Press, New York, 1999.

[14] M. Lanza de Cristoforis, *Properties and pathologies of the composition and inversion operators in Schauder spaces.* Rend. Accad. Naz. Sci. XL Mem. Mat. **15** (1991), 93–109.

[15] M. Lanza de Cristoforis, *Differentiability properties of a composition operator.* Rend. Circ. Mat. Palermo (2) Suppl. **56** (1998), 157–165.

[16] M. Lanza de Cristoforis, *Asymptotic behaviour of the conformal representation of a Jordan domain with a small hole in Schauder spaces.* Comput. Methods Funct. Theory **2** (2002), 1–27.

[17] M. Lanza de Cristoforis, *Asymptotic behaviour of the conformal representation of a Jordan domain with a small hole, and relative capacity.* In *Complex analysis and dynamical systems, Proc. Conf. Karmiel, June* 19–22, 2001, volume 364 of *Contemp. Math.*, pages 155–167. Amer. Math. Soc., Providence, RI, 2004.

[18] M. Lanza de Cristoforis, *A domain perturbation problem for the Poisson equation.* Complex Var. Theory Appl. **50** (2005), 851–867.

[19] M. Lanza de Cristoforis, *Perturbation problems in potential theory, a functional analytic approach.* J. Appl. Funct. Anal. **2** (2007), 197–222.

[20] M. Lanza de Cristoforis, *Asymptotic behavior of the solutions of the Dirichlet problem for the Laplace operator in a domain with a small hole. A functional analytic approach.* Analysis (Munich) **28** (2008), 63–93.

[21] M. Lanza de Cristoforis, *A singular domain perturbation problem for the Poisson equation.* In *More progresses in analysis, Proceedings of the* 5th *international ISAAC congress, Catania, Italy, July* 25–30, 2005, pages 955–965. World Scientific, Hackensack, NJ, 2009.

[22] M. Lanza de Cristoforis and P. Musolino, *A perturbation result for periodic layer potentials of general second-order differential operators with constant coefficients.* Far East J. Math. Sci. (FJMS) **52** (2011), 75–120.

[23] M. Lanza de Cristoforis and P. Musolino, *A real analyticity result for a nonlinear integral operator.* J. Integral Equations Appl., (to appear).

[24] M. Lanza de Cristoforis and P. Musolino, *A singularly perturbed nonlinear Robin problem in a periodically perforated domain: a functional analytic approach.* Complex Var. Elliptic Equ., (to appear).

[25] M. Lanza de Cristoforis and L. Rossi, *Real analytic dependence of simple and double layer potentials upon perturbation of the support and of the density.* J. Integral Equations Appl. **16** (2004), 137–174.

[26] V. Maz'ya, S. Nazarov, and B. Plamenevskij, *Asymptotic theory of elliptic boundary value problems in singularly perturbed domains. Vols. I, II*, volumes 111, 112 of *Operator Theory: Advances and Applications.* Birkhäuser Verlag, Basel, 2000.

[27] C. Miranda, *Sulle proprietà di regolarità di certe trasformazioni integrali.* Atti Accad. Naz. Lincei Mem. Cl. Sci. Fis. Mat. Natur. Sez. I **7** (1965), 303–336.

[28] V. Mityushev and P.M. Adler, *Longitudinal permeability of spatially periodic rectangular arrays of circular cylinders. I. A single cylinder in the unit cell.* ZAMM Z. Angew. Math. Mech. **82** (2002), 335–345.

[29] P. Musolino, *A singularly perturbed Dirichlet problem for the Laplace operator in a periodically perforated domain. A functional analytic approach.* Math. Methods Appl. Sci. **35** (2012), 334–349.

[30] S. Ozawa, *Electrostatic capacity and eigenvalues of the Laplacian.* J. Fac. Sci. Univ. Tokyo Sect. IA Math. **30** (1983), 53–62.

[31] L. Preciso, *Perturbation analysis of the conformal sewing problem and related problems.* PhD Dissertation, University of Padova, 1998.

[32] L. Preciso, *Regularity of the composition and of the inversion operator and perturbation analysis of the conformal sewing problem in Roumieu type spaces.* Tr. Inst. Mat. Minsk **5** (2000), 99–104.

[33] G. Prodi and A. Ambrosetti, *Analisi non lineare. I quaderno.* Editrice Tecnico Scientifica, Pisa, 1973.

[34] S. Rogosin, M. Dubatovskaya, and E. Pesetskaya, *Eisenstein sums and functions and their application at the study of heat conduction in composites.* Siauliai Math. Semin. **4** (2009), 167–187.

[35] G.M. Troianiello, *Elliptic differential equations and obstacle problems.* The University Series in Mathematics. Plenum Press, New York, 1987.

[36] M.J. Ward and J.B. Keller, *Strong localized perturbations of eigenvalue problems.* SIAM J. Appl. Math. **53** (1993), 770–798.

Paolo Musolino
Dipartimento di Matematica Pura ed Applicata
Università degli Studi di Padova
Via Trieste 63
I-35121 Padova, Italy
e-mail: `musolino@math.unipd.it`

Operator Theory:
Advances and Applications, Vol. 229, 291–301

Fractional Variational Calculus of Variable Order

T. Odzijewicz, A.B. Malinowska and D.F.M. Torres

To Professor Stefan Samko on the occasion of his 70th birthday

Abstract. We study the fundamental problem of the calculus of variations with variable order fractional operators. Fractional integrals are considered in the sense of Riemann–Liouville while derivatives are of Caputo type.

Mathematics Subject Classification (2010). 26A33; 34A08; 49K05.

Keywords. Fractional operators; fractional integration and differentiation of variable order; fractional variational analysis; Euler–Lagrange equations.

1. Introduction

Fractional calculus is a discipline that studies integrals and derivatives of non-integer (real or complex) order [13, 18, 27, 34]. The subject is nowadays very active due to its many applications in mechanics, chemistry, biology, economics, and control theory [36]. In 1993, Samko and Ross proposed an interesting generalization to fractional calculus [35] (see also Samko's paper [33] of 1995). They introduced the study of fractional integration and differentiation when the order is not a constant but a function. More precisely, they considered an extension of Riemann–Liouville and Fourier definitions [32, 33, 35]. Afterwards, several works were dedicated to variable order fractional operators, their applications and interpretations (see, e.g., [1, 7, 19]). In particular, Samko's variable order fractional calculus turns out to be very useful in mechanics and in the theory of viscous flows [7, 9, 19, 26, 28, 29]. Indeed, many physical processes exhibit fractional-order behavior that may vary with time or space [19]. The paper [7] is devoted to the study of a variable-order fractional differential equation that characterizes some problems in the theory of viscoelasticity. In [9] the authors analyze the dynamics and control of a nonlinear variable viscoelasticity oscillator, and two controllers

---

Part of the first author's Ph.D., which is carried out at the University of Aveiro under the *Doctoral Programme in Mathematics and Applications* of Universities of Aveiro and Minho.

are proposed for the variable order differential equations that track an arbitrary reference function. The work [26] investigates the drag force acting on a particle due to the oscillatory flow of a viscous fluid. The drag force is determined using the variable order fractional calculus, where the order of derivative vary according to the dynamics of the flow. In [29] a variable order differential equation for a particle in a quiescent viscous liquid is developed. For more on the application of variable order fractional operators to the modeling of dynamic systems, we refer the reader to the recent review article [28].

In this note we develop the fractional calculus of variations via a variable order approach. The fractional variational calculus was born in 1996-1997 with the works of Riewe in mechanics [30, 31], and is now under strong current research (see [2, 3, 5, 6, 8, 10, 11, 14–17, 20–25] and references therein). However, to the best of the authors knowledge, results for the variable order case are a rarity and reduce to those in [4]. Motivated by the advancements of [3, 23], and in contrast with [4], we consider here fractional problems of the calculus of variations where the Lagrangian depends on classical integer order derivatives and both on variable order fractional derivatives and integrals.

The paper is organized as follows. In §2 a brief review to the variable order fractional calculus is given. Our results are then formulated and proved in §3: we show the boundedness of the variable order Riemann–Liouville fractional integral in the space $L_1[a,b]$ (Theorem 3.1); integration by parts formulas for variable order fractional operators (Theorems 3.2 and 3.3); and a necessary optimality condition for our general fundamental problem of the variable order fractional variational calculus (Theorem 3.5). Two illustrative examples are discussed in §4.

2. Preliminaries

The reference book for fractional analysis and its applications is [34]. Here we recall the necessary definitions for the variable order fractional calculus (see, e.g., [19]).

Definition 2.1 (Left and right Riemann–Liouville integrals of variable order). Let $0<\alpha(t,\tau)<1$ for all $t,\tau\in[a,b]$ and $f\in L_1[a,b]$. Then,

$$ {}_aI_t^{\alpha(\cdot,\cdot)}f(t)=\int\limits_a^t \frac{1}{\Gamma(\alpha(t,\tau))}(t-\tau)^{\alpha(t,\tau)-1}f(\tau)d\tau \quad (t>a)$$

is called the left Riemann–Liouville integral of variable fractional order $\alpha(\cdot,\cdot)$, while

$$ {}_tI_b^{\alpha(\cdot,\cdot)}f(t)=\int\limits_t^b \frac{1}{\Gamma(\alpha(\tau,t))}(\tau-t)^{\alpha(\tau,t)-1}f(\tau)d\tau \quad (t<b)$$

denotes the right Riemann–Liouville integral of variable fractional order $\alpha(\cdot,\cdot)$.

Definition 2.2 (Left and right Riemann–Liouville derivatives of variable order). Let $0<\alpha(t,\tau)<1$ for all $t,\tau\in[a,b]$. If ${}_aI_t^{1-\alpha(\cdot,\cdot)}f\in AC[a,b]$, then the left

Riemann–Liouville derivative of variable fractional order $\alpha(\cdot,\cdot)$ is defined by

$$ {}_aD_t^{\alpha(\cdot,\cdot)}f(t) = \frac{d}{dt}{}_aI_t^{1-\alpha(\cdot,\cdot)}f(t) = \frac{d}{dt}\int_a^t \frac{1}{\Gamma(1-\alpha(t,\tau))}(t-\tau)^{-\alpha(t,\tau)}f(\tau)d\tau \quad (t>a)$$

while the right Riemann–Liouville derivative of variable fractional order $\alpha(\cdot,\cdot)$ is defined for functions f such that ${}_tI_b^{1-\alpha(\cdot,\cdot)}f \in AC[a,b]$ by

$$ {}_tD_b^{\alpha(\cdot,\cdot)}f(t) = -\frac{d}{dt}{}_tI_b^{1-\alpha(\cdot,\cdot)}f(t) = \frac{d}{dt}\int_t^b \frac{-1}{\Gamma(1-\alpha(\tau,t))}(\tau-t)^{-\alpha(\tau,t)}f(\tau)d\tau \quad (t<b).$$

Definition 2.3 (Left and right Caputo derivatives of variable fractional order). Let $0<\alpha(t,\tau)<1$ for all $t,\tau\in[a,b]$. If $f\in AC[a,b]$, then the left Caputo derivative of variable fractional order $\alpha(\cdot,\cdot)$ is defined by

$$ {}_a^CD_t^{\alpha(\cdot,\cdot)}f(t) = \int_a^t \frac{1}{\Gamma(1-\alpha(t,\tau))}(t-\tau)^{-\alpha(t,\tau)}\frac{d}{d\tau}f(\tau)d\tau \quad (t>a)$$

while the right Caputo derivative of variable fractional order $\alpha(\cdot,\cdot)$ is given by

$$ {}_t^CD_b^{\alpha(\cdot,\cdot)}f(t) = \int_t^b \frac{-1}{\Gamma(1-\alpha(\tau,t))}(\tau-t)^{-\alpha(\tau,t)}\frac{d}{d\tau}f(\tau)d\tau \quad (t<b).$$

The following result will be useful in the proof of Theorems 3.1 and 3.2.

Theorem 2.4 (cf. [12]). *Let $x\in[0,1]$. The Gamma function satisfies the inequalities*

$$\frac{x^2+1}{x+1} \le \Gamma(x+1) \le \frac{x^2+2}{x+2}. \tag{2.1}$$

3. Main results

In §3.1 we prove that the variable fractional order Riemann–Liouville integral ${}_aI_t^{\alpha(\cdot,\cdot)}$ is bounded on the space $L_1[a,b]$ (Theorem 3.1). In §3.2 we obtain a formula of integration by parts for Riemann–Liouville integrals of variable order (Theorem 3.2) and a formula of integration by parts for derivatives of variable fractional order (Theorem 3.3). Finally, in §3.3 we use the obtained formulas of integration by parts to derive a necessary optimality condition for a general problem of the calculus of variations involving variable order fractional operators (Theorem 3.5).

3.1. Boundedness

The following result allow us to consider problems of the calculus of variations with a Lagrangian depending on left Riemann–Liouville integrals of variable order.

Theorem 3.1. *Let $\frac{1}{n} < \alpha(t,\tau) < 1$ for all $t,\tau \in [a,b]$ and a certain $n \in \mathbb{N}$ greater or equal than two. The Riemann–Liouville integral ${}_aI_t^{\alpha(\cdot,\cdot)} : L_1[a,b] \to L_1[a,b]$ of variable fractional order $\alpha(\cdot,\cdot)$ is a linear and bounded operator.*

Proof. The operator is obviously linear. Let $\frac{1}{n} < \alpha(t,\tau) < 1$, $f \in L_1[a,b]$, and define

$$F(\tau,t) := \begin{cases} \left|\frac{1}{\Gamma(\alpha(t,\tau))}(t-\tau)^{\alpha(t,\tau)-1}\right| \cdot |f(\tau)| & \text{if } \tau < t; \\ 0 & \text{if } \tau \geq t \end{cases}$$

for all $(\tau,t) \in \Delta = [a,b] \times [a,b]$. Since $\frac{1}{n} < \alpha(t,\tau) < 1$, then

1. for $\tau + 1 \leq t$ we have $\ln(t-\tau) \geq 0$ and $(t-\tau)^{\alpha(t,\tau)-1} < 1$;
2. for $\tau < t < \tau + 1$ we have $\ln(t-\tau) < 0$ and $(t-\tau)^{\alpha(t,\tau)-1} < (t-\tau)^{\frac{1}{n}-1}$.

Therefore,

$$\begin{aligned} \int_a^b \left(\int_a^b F(\tau,t)dt\right) d\tau &= \int_a^b |f(\tau)| \left(\int_\tau^b \left|\frac{1}{\Gamma(\alpha(t,\tau))}(t-\tau)^{\alpha(t,\tau)-1}\right| dt\right) d\tau \\ &= \int_a^b |f(\tau)| \left(\int_\tau^{\tau+1} \left|\frac{(t-\tau)^{\alpha(t,\tau)-1}}{\Gamma(\alpha(t,\tau))}\right| dt + \int_{\tau+1}^b \left|\frac{(t-\tau)^{\alpha(t,\tau)-1}}{\Gamma(\alpha(t,\tau))}\right| dt\right) d\tau \\ &< \int_a^b |f(\tau)| \left(\int_\tau^{\tau+1} \left|\frac{(t-\tau)^{\frac{1}{n}-1}}{\Gamma(\alpha(t,\tau))}\right| dt + \int_{\tau+1}^b \left|\frac{1}{\Gamma(\alpha(t,\tau))}\right| dt\right) d\tau. \end{aligned}$$

Moreover, by inequality (2.1), one has

$$\begin{aligned} &\int_a^b |f(\tau)| \left(\int_\tau^{\tau+1} \left|\frac{1}{\Gamma(\alpha(t,\tau))}(t-\tau)^{\frac{1}{n}-1}\right| dt + \int_{\tau+1}^b \left|\frac{1}{\Gamma(\alpha(t,\tau))}\right| dt\right) d\tau \\ &< \int_a^b |f(\tau)| \left(\int_\tau^{\tau+1} \frac{\alpha^2(t,\tau)+\alpha(t,\tau)}{\alpha^2(t,\tau)+1}(t-\tau)^{\frac{1}{n}-1} dt + \int_{\tau+1}^b \frac{\alpha^2(t,\tau)+\alpha(t,\tau)}{\alpha^2(t,\tau)+1} dt\right) d\tau \\ &< \int_a^b |f(\tau)| \left(\int_\tau^{\tau+1} (t-\tau)^{\frac{1}{n}-1} dt + b - \tau - 1\right) d\tau = \int_a^b |f(\tau)|(n+b-\tau-1)d\tau \\ &< (n+b-a)\|f\| < \infty. \end{aligned}$$

It follows from Fubini's theorem that F is integrable in the square Δ and

$$\begin{aligned} \left\|{}_aI_t^{\alpha(\cdot,\cdot)} f\right\| &= \int_a^b \left|\int_a^t \frac{1}{\Gamma(\alpha(t,\tau))}(t-\tau)^{\alpha(t,\tau)-1} f(\tau)d\tau\right| dt \\ &\leq \int_a^b \left(\int_a^t \left|\frac{1}{\Gamma(\alpha(t,\tau))}(t-\tau)^{\alpha(t,\tau)-1} f(\tau)\right| d\tau\right) dt \\ &= \int_a^b \left(\int_a^b F(\tau,t)d\tau\right) dt < (n+b-a)\|f\|. \end{aligned}$$

Therefore, ${}_aI_t^{\alpha(\cdot,\cdot)} : L_1[a,b] \to L_1[a,b]$ and $\left\|{}_aI_t^{\alpha(\cdot,\cdot)}\right\| < n+b-a$. □

3.2. Integration by parts formulas

The integration by parts formulas, we now obtain, have an important role in the proof of the generalized Euler–Lagrange equation (3.5). We note that in Theorem 3.2 the left-hand side of (3.1) involves a left integral of variable order while on the right-hand side it appears a right integral.

Theorem 3.2. *Let $\frac{1}{n} < \alpha(t,\tau) < 1$ for all $t,\tau \in [a,b]$ and a certain $n \in \mathbb{N}$ greater or equal than two, and $f,g \in C([a,b];\mathbb{R})$. Then,*

$$\int_a^b g(t)\,{}_aI_t^{\alpha(\cdot,\cdot)} f(t)dt = \int_a^b f(t)\,{}_tI_b^{\alpha(\cdot,\cdot)} g(t)dt. \tag{3.1}$$

Proof. Define

$$F(\tau,t) := \begin{cases} \left|\frac{1}{\Gamma(\alpha(t,\tau))}(t-\tau)^{\alpha(t,\tau)-1} g(t) f(\tau)\right| & \text{if } \tau < t; \\ 0 & \text{if } \tau \geq t \end{cases}$$

for all $(\tau,t) \in \Delta = [a,b] \times [a,b]$. Since f and g are continuous functions on $[a,b]$, they are bounded on $[a,b]$, i.e., there exist $C_1, C_2 > 0$ such that $|g(t)| \leq C_1$ and $|f(t)| \leq C_2$, $t \in [a,b]$. Therefore,

$$\begin{aligned} \int_a^b \left(\int_a^b F(\tau,t)d\tau\right) dt &= \int_a^b \left(\int_a^t \left|\frac{1}{\Gamma(\alpha(t,\tau))}(t-\tau)^{\alpha(t,\tau)-1} g(t) f(\tau)\right| d\tau\right) dt \\ &\leq C_1 C_2 \int_a^b \left(\int_a^t \left|\frac{1}{\Gamma(\alpha(t,\tau))}(t-\tau)^{\alpha(t,\tau)-1}\right| d\tau\right) dt \\ &= C_1 C_2 \int_a^b \left(\int_a^t \frac{1}{\Gamma(\alpha(t,\tau))}(t-\tau)^{\alpha(t,\tau)-1} d\tau\right) dt. \end{aligned}$$

Since $\frac{1}{n} < \alpha(t,\tau) < 1$, then

1. for $1 \leq t-\tau$ we have $\ln(t-\tau) \geq 0$ and $(t-\tau)^{\alpha(t,\tau)-1} < 1$;
2. for $1 > t-\tau$ we have $\ln(t-\tau) < 0$ and $(t-\tau)^{\alpha(t,\tau)-1} < (t-\tau)^{\frac{1}{n}-1}$.

Therefore,

$$\begin{aligned} &C_1 C_2 \int_a^b \left(\int_a^t \frac{1}{\Gamma(\alpha(t,\tau))}(t-\tau)^{\alpha(t,\tau)-1} d\tau\right) dt \\ &\quad < C_1 C_2 \int_a^b \left(\int_a^{t-1} \frac{1}{\Gamma(\alpha(t,\tau))} d\tau + \int_{t-1}^t \frac{1}{\Gamma(\alpha(t,\tau))}(t-\tau)^{\frac{1}{n}-1} d\tau\right) dt. \end{aligned}$$

Moreover, by (2.1), one has

$$\begin{aligned} &C_1 C_2 \int_a^b \left(\int_a^{t-1} \frac{1}{\Gamma(\alpha(t,\tau))} d\tau + \int_{t-1}^t \frac{1}{\Gamma(\alpha(t,\tau))}(t-\tau)^{\frac{1}{n}-1} d\tau\right) dt \\ &\leq C_1 C_2 \int_a^b \left(\int_a^{t-1} \frac{\alpha^2(t,\tau)+\alpha(t,\tau)}{\alpha^2(t,\tau)+1} d\tau + \int_{t-1}^t \frac{\alpha^2(t,\tau)+\alpha(t,\tau)}{\alpha^2(t,\tau)+1}(t-\tau)^{\frac{1}{n}-1} d\tau\right) dt \end{aligned}$$

$$< C_1C_2 \int_a^b \left(\int_a^{t-1} d\tau + \int_{t-1}^t (t-\tau)^{\frac{1}{n}-1} d\tau \right) dt$$

$$= C_1C_2(b-a)\left(\frac{b+a}{2} - 1 + n - a\right) < \infty.$$

Hence, one can use the Fubini theorem to change the order of integration:

$$\int_a^b g(t)_a I_t^{\alpha(\cdot,\cdot)} f(t)dt = \int_a^b \left(\int_a^t g(t)f(\tau)\frac{1}{\Gamma(\alpha(t,\tau))}(t-\tau)^{\alpha(t,\tau)-1} d\tau \right) dt$$

$$= \int_a^b \left(\int_\tau^b g(t)f(\tau)\frac{1}{\Gamma(\alpha(t,\tau))}(t-\tau)^{\alpha(t,\tau)-1} dt \right) d\tau = \int_a^b f(\tau)_\tau I_b^{\alpha(\cdot,\cdot)} g(\tau)d\tau.$$

□

In our second formula (3.2) of fractional integration by parts, the left-hand side contains a left Caputo derivative of variable fractional order $\alpha(\cdot,\cdot)$, while on the right-hand side it appears a right Riemann–Liouville integral of variable order $1-\alpha(\cdot,\cdot)$ and a right Riemann–Liouville derivative of variable order $\alpha(\cdot,\cdot)$.

Theorem 3.3. *Let $0 < \alpha(t,\tau) < 1-\frac{1}{n}$ for all $t,\tau \in [a,b]$ and a certain $n \in \mathbb{N}$ greater or equal than two. If $f \in C^1([a,b];\mathbb{R})$, $g \in C([a,b];\mathbb{R})$, and ${}_tI_b^{1-\alpha(\cdot,\cdot)} g \in AC[a,b]$, then*

$$\int_a^b g(t)_a^C D_t^{\alpha(\cdot,\cdot)} f(t)dt = f(t)_t I_b^{1-\alpha(\cdot,\cdot)} g(t)\Big|_a^b + \int_a^b f(t)_t D_b^{\alpha(\cdot,\cdot)} g(t)dt. \tag{3.2}$$

Proof. By Definition 2.3, it follows that ${}_a^C D_t^{\alpha(\cdot,\cdot)} f(t) = {}_aI_t^{1-\alpha(\cdot,\cdot)} \frac{d}{dt} f(t)$. Applying Theorem 3.2 and integration by parts for classical (integer) derivatives, we obtain

$$\int_a^b g(t)_a^C D_t^{\alpha(\cdot,\cdot)} f(t)dt = \int_a^b g(t)_a I_t^{1-\alpha(\cdot,\cdot)} \frac{d}{dt} f(t)dt = \int_a^b \frac{d}{dt} f(t)_t I_b^{1-\alpha(\cdot,\cdot)} g(t)dt$$

$$= f(t)_t I_b^{1-\alpha(\cdot,\cdot)} g(t)\Big|_a^b - \int_a^b f(t)\frac{d}{dt}{}_t I_b^{1-\alpha(\cdot,\cdot)} g(t)dt$$

$$= f(t)_t I_b^{1-\alpha(\cdot,\cdot)} g(t)\Big|_a^b + \int_a^b f(t)_t D_b^{\alpha(\cdot,\cdot)} g(t)dt.$$

□

3.3. A fundamental variational problem of variable fractional order

We consider the problem of extremizing (minimizing or maximizing) a functional

$$\mathcal{J}[y(\cdot)] = \int_a^b F\left(t, y(t), y'(t), {}_a^C D_t^{\alpha(\cdot,\cdot)} y(t), {}_a I_t^{\beta(\cdot,\cdot)} y(t)\right) dt \tag{3.3}$$

subject to boundary conditions

$$y(a) = y_a, \quad y(b) = y_b, \tag{3.4}$$

where $\alpha, \beta : [a,b] \times [a,b] \to \mathbb{R}$ are given functions taking values in $(0, 1-\frac{1}{n})$ and $(\frac{1}{n}, 1)$, respectively, with $n \in \mathbb{N}$ greater or equal than two. For simplicity of notation, we introduce the operator $\{\cdot,\cdot,\cdot\}$ defined by

$$\{y, \alpha, \beta\}(t) = \left(t, y(t), y'(t), {}_a^C D_t^{\alpha(\cdot,\cdot)} y(t), {}_a I_t^{\beta(\cdot,\cdot)} y(t)\right).$$

We assume that $F \in C^1\left([a,b] \times \mathbb{R}^4; \mathbb{R}\right)$; $t \mapsto \partial_4 F\{y,\alpha,\beta\}(t)$ is continuous and has absolutely continuous integral ${}_t I_b^{1-\alpha(\cdot,\cdot)}$ and continuous derivative ${}_t D_b^{\alpha(\cdot,\cdot)}$; $t \mapsto \partial_5 F\{y,\alpha,\beta\}(t)$ is continuous and has continuous variable order fractional integral ${}_t I_b^{\beta(\cdot,\cdot)}$; and $t \mapsto \partial_3 F\{y,\alpha,\beta\}(t)$ has continuous usual derivative $\frac{d}{dt}$.

Definition 3.4. A continuously differentiable function $y \in C^1([a,b];\mathbb{R})$ is said to be admissible for the variational problem (3.3)–(3.4), if ${}_a^C D_t^{\alpha(\cdot,\cdot)} y$ and ${}_a I_t^{\beta(\cdot,\cdot)} y$ exist and are continuous on the interval $[a,b]$, and y satisfies the given boundary conditions (3.4).

Theorem 3.5. *Let y be a solution to problem* (3.3)–(3.4). *Then, y satisfies the generalized Euler–Lagrange equation*

$$\begin{aligned}\partial_2 F\{y,\alpha,\beta\}(t) - \frac{d}{dt}\partial_3 F\{y,\alpha,\beta\}(t) + {}_t I_b^{\beta(\cdot,\cdot)} \partial_5 F\{y,\alpha,\beta\}(t) \\ + {}_t D_b^{\alpha(\cdot,\cdot)} \partial_4 F\{y,\alpha,\beta\}(t) = 0\end{aligned} \tag{3.5}$$

for all $t \in [a,b]$.

Proof. Suppose that y is an extremizer of $\mathcal{J}$. Consider the value of $\mathcal{J}$ at a nearby function $\hat{y}(t) = y(t) + \varepsilon\eta(t)$, where $\varepsilon \in \mathbb{R}$ is a small parameter and $\eta \in C^1([a,b];\mathbb{R})$ is an arbitrary function satisfying $\eta(a) = \eta(b) = 0$ and such that ${}_a^C D_t^{\alpha(\cdot,\cdot)} \hat{y}(t)$ and ${}_a I_t^{\beta(\cdot,\cdot)} \hat{y}(t)$ are continuous. Let

$$J(\varepsilon) = \mathcal{J}[\hat{y}(\cdot)] = \int_a^b F\{\hat{y},\alpha,\beta\}(t)dt.$$

A necessary condition for y to be an extremizer is given by

$$\begin{aligned}\left.\frac{dJ}{d\varepsilon}\right|_{\varepsilon=0} = 0 \Leftrightarrow \int_a^b \Bigg(\partial_2 F\{y,\alpha,\beta\}(t)\cdot\eta(t) + \partial_3 F\{y,\alpha,\beta\}(t)\frac{d}{dt}\eta(t) \\ + \partial_4 F\{y,\alpha,\beta\}(t) {}_a^C D_t^{\alpha(\cdot,\cdot)} \eta(t) + \partial_5 F\{y,\alpha,\beta\}(t)\cdot {}_a I_t^{\beta(\cdot,\cdot)} \eta(t)\Bigg)dt = 0.\end{aligned} \tag{3.6}$$

Using the classical and the generalized fractional integration by parts formulas of Theorems 3.2 and 3.3, we obtain

$$\int_a^b \partial_3 F \frac{d\eta}{dt} dt = \partial_3 F \eta|_a^b - \int_a^b \left(\eta \frac{d}{dt}\partial_3 F\right) dt,$$

$$\int_a^b \partial_4 F {}_a^C D_t^{\alpha(\cdot,\cdot)} \eta dt = \eta_t I_b^{1-\alpha(\cdot,\cdot)} \partial_4 F \Big|_a^b + \int_a^b \eta_t D_b^{\alpha(\cdot,\cdot)} \partial_4 F dt,$$

and

$$\int_a^b \partial_5 F {}_a I_t^{\beta(\cdot,\cdot)} \eta dt = \int_a^b \eta_t I_b^{\beta(\cdot,\cdot)} \partial_5 F dt.$$

Because $\eta(a) = \eta(b) = 0$, (3.6) simplifies to

$$\int_a^b \eta(t) \Big(\partial_2 F\{y, \alpha, \beta\}(t) - \frac{d}{dt} \partial_3 F\{y, \alpha, \beta\}(t) + {}_t D_b^{\alpha(\cdot,\cdot)} \partial_4 F\{y, \alpha, \beta\}(t) \\ + {}_t I_b^{\beta(\cdot,\cdot)} \partial_5 F\{y, \alpha, \beta\}(t) \Big) dt = 0.$$

One obtains (3.5) by the fundamental lemma of the calculus of variations (see, e.g., [37]). □

4. Illustrative examples

Let $\beta(t,\tau) = \beta(t)$ be a function depending only on variable t, $\frac{1}{n} < \beta(t) < 1$ for all $t \in [a,b]$ and a certain $n \in \mathbb{N}$ greater or equal than two, and $\gamma > -1$. We make use of the identity

$$ {}_a I_t^{\beta(\cdot)} (t-a)^\gamma = \frac{\Gamma(\gamma+1)(t-a)^{\gamma+\beta(t)}}{\Gamma(\gamma+\beta(t)+1)} \tag{4.1}$$

that one can find in the Samko and Ross paper [35].

Example 1. Let $\mathcal{J}$ be the functional defined by

$$\mathcal{J}[y(\cdot)] = \int_a^b \sqrt{1 + \frac{\Gamma(\beta(t)+3)}{2\Gamma(3)(t-a)^{2+\beta(t)}} \left({}_a I_t^{\beta(\cdot)} y(t)\right)^2 - {}_a I_t^{\beta(\cdot)} y(t)}\, dt$$

with endpoint conditions $y(a) = 0$ and $y(b) = (b-a)^2$. If y is an extremizer for $\mathcal{J}$, then the necessary optimality condition of Theorem 3.5 gives

$$ {}_t I_b^{\beta(\cdot)} \left(\frac{\frac{\Gamma(\beta(t)+3)}{\Gamma(3)(t-a)^{2+\beta(t)}} {}_a I_t^{\beta(\cdot)} y(t) - 1}{2\sqrt{1 + \frac{\Gamma(\beta(t)+3)}{2\Gamma(3)(t-a)^{2+\beta(t)}} \left({}_a I_t^{\beta(\cdot)} y(t)\right)^2 - {}_a I_t^{\beta(\cdot)} y(t)}} \right) = 0. \tag{4.2}$$

By identity (4.1), function

$$y(t) = (t-a)^2 \tag{4.3}$$

is a solution to the variable order fractional differential equation

$$ {}_a I_t^{\beta(\cdot)} y(t) = \frac{\Gamma(3)(t-a)^{2+\beta(t)}}{\Gamma(\beta(t)+3)}.$$

Therefore, (4.3) is a solution to the Euler–Lagrange equation (4.2) and an extremal of $\mathcal{J}$.

In the next example $\alpha(t,\tau)$ is a function taking values in the set $(0, 1-\frac{1}{n})$.

Example 2. Consider the following problem:

$$\mathcal{J}[y(\cdot)] = \int_a^b \left({}_a^C D_t^{\alpha(\cdot,\cdot)} y(t)\right)^2 + \left({}_a I_t^{\beta(\cdot)} y(t) - \frac{\xi(t-\tau)^{\beta(t)}}{\Gamma(\beta(t)+1)}\right)^2 dt \longrightarrow \min,$$

$$y(a) = \xi, \quad y(b) = \xi,$$

for a given real ξ. Because $\mathcal{J}[y(\cdot)] \geq 0$ for any function y and $\mathcal{J}[\tilde{y}(\cdot)] = 0$ for the admissible function $\tilde{y} = \xi$ (use relation (4.1) for $\gamma = 0$, the linearity of operator ${}_a I_t^{\beta(\cdot)}$, and the definition of left Caputo derivative of a variable fractional order), we conclude that $\tilde{y}$ is the global minimizer to the problem. It is straightforward to check that $\tilde{y}$ satisfies our variable order fractional Euler–Lagrange equation (3.5).

Acknowledgment

Work supported by *FEDER* funds through *COMPETE* – Operational Programme Factors of Competitiveness ("Programa Operacional Factores de Competitividade") and by Portuguese funds through the *Center for Research and Development in Mathematics and Applications* (University of Aveiro) and the Portuguese Foundation for Science and Technology ("FCT – Fundação para a Ciência e a Tecnologia"), within project PEst-C/MAT/UI4106/2011 with COMPETE number FCOMP-01-0124-FEDER-022690. Odzijewicz was also supported by FCT through the Ph.D. fellowship SFRH/BD/33865/2009; Malinowska by Białystok University of Technology grant S/WI/02/2011; and Torres by FCT through the project PTDC/MAT/113470/2009.

References

[1] A. Almeida, S. Samko, *Fractional and hypersingular operators in variable exponent spaces on metric measure spaces.* Mediterr. J. Math. **6** (2009), 215–232.

[2] R. Almeida, A.B. Malinowska, D.F.M. Torres, *A fractional calculus of variations for multiple integrals with application to vibrating string.* J. Math. Phys. **51** (2010), 033503, 12 pp.

[3] R. Almeida, D.F.M. Torres, *Calculus of variations with fractional derivatives and fractional integrals.* Appl. Math. Lett. **22** (2009), 1816–1820.

[4] T.M. Atanackovic, S. Pilipovic, *Hamilton's principle with variable order fractional derivatives.* Fract. Calc. Appl. Anal. **14** (2011), 94–109.

[5] N.R.O. Bastos, R.A.C. Ferreira, D.F.M. Torres, *Necessary optimality conditions for fractional difference problems of the calculus of variations.* Discrete Contin. Dyn. Syst. **29** (2011), 417–437.

[6] N.R.O. Bastos, R.A.C. Ferreira, D.F.M. Torres, *Discrete-time fractional variational problems.* Signal Process. **91** (2011), 513–524.

[7] C.F.M. Coimbra, *Mechanics with variable-order differential operators.* Ann. Phys. **12** (2003), 692–703.

[8] J. Cresson, *Fractional embedding of differential operators and Lagrangian systems.* J. Math. Phys. **48** (2007), 033504, 34 pp.

[9] G. Diaz, C.F.M. Coimbra, *Nonlinear dynamics and control of a variable order oscillator with application to the van der Pol equation.* Nonlinear Dynam. **56** (2009), 145–157.

[10] G.S.F. Frederico, D.F.M. Torres, *Fractional conservation laws in optimal control theory.* Nonlinear Dynam. **53** (2008), 215–222.

[11] G.S.F. Frederico, D.F.M. Torres, *Fractional Noether's theorem in the Riesz-Caputo sense.* Appl. Math. Comput. **217** (2010), 1023–1033.

[12] P. Ivady, *A note on a gamma function inequality.* J. Math. Inequal. **3** (2009), 227–236.

[13] A.A. Kilbas, H.M. Srivastava, J.J. Trujillo, *Theory and applications of fractional differential equations*, North-Holland Mathematics Studies, 204, Elsevier, Amsterdam, 2006.

[14] M. Klimek, *Fractional sequential mechanics – models with symmetric fractional derivative*, Czechoslovak J. Phys. **51** (2001), no. 12, 1348–1354.

[15] M. Klimek, *Lagrangian and Hamiltonian fractional sequential mechanics*, Czechoslovak J. Phys. **52** (2002), no. 11, 1247–1253.

[16] M. Klimek, *Stationarity-conservation laws for fractional differential equations with variable coefficients.* J. Phys. A **35** (2002), 6675–6693.

[17] M. Klimek, *Lagrangian fractional mechanics – a noncommutative approach.* Czechoslovak J. Phys. **55** (2005), 1447–1453.

[18] M. Klimek, *On solutions of linear fractional differential equations of a variational type.* The Publishing Office of Czenstochowa University of Technology, Czestochowa, 2009.

[19] C.F. Lorenzo, T.T. Hartley, *Variable order and distributed order fractional operators.* Nonlinear Dynam. **29** (2002), 57–98.

[20] A.B. Malinowska, D.F.M. Torres, *Generalized natural boundary conditions for fractional variational problems in terms of the Caputo derivative.* Comput. Math. Appl. **59** (2010), 3110–3116.

[21] D. Mozyrska, D.F.M. Torres, *Minimal modified energy control for fractional linear control systems with the Caputo derivative.* Carpathian J. Math. **26** (2010), 210–221.

[22] D. Mozyrska, D.F.M. Torres, *Modified optimal energy and initial memory of fractional continuous-time linear systems.* Signal Process. **91** (2011), 379–385.

[23] T. Odzijewicz, A.B. Malinowska, D.F.M. Torres, *Fractional variational calculus with classical and combined Caputo derivatives.* Nonlinear Anal. **75** (2012), 1507–1515.

[24] T. Odzijewicz, A.B. Malinowska, D.F.M. Torres, *Generalized fractional calculus with applications to the calculus of variations*, Comput. Math. Appl. **64** (2012), 3351–3366.

[25] T. Odzijewicz, A.B. Malinowska, D.F.M. Torres, *Fractional calculus of variations in terms of a generalized fractional integral with applications to physics*, Abstr. Appl. Anal. **2012** (2012), Art. ID 871912, 24 pp.

[26] H.T.C. Pedro, M.H. Kobayashi, J.M.C. Pereira, C.F.M. Coimbra, *Variable order modeling of diffusive-convective effects on the oscillatory flow past a sphere.* J. Vib. Control **14** (2008), 1569–1672.

[27] I. Podlubny, *Fractional differential equations.* Mathematics in Science and Engineering, 198, Academic Press, San Diego, CA, 1999.

[28] L.E.S. Ramirez, C.F.M. Coimbra, *On the selection and meaning of variable order operators for dynamic modeling.* Int. J. Differ. Equ. **2010** (2010), Art. ID 846107, 16 pp.

[29] L.E.S. Ramirez and C.F.M. Coimbra, *On the variable order dynamics of the nonlinear wake caused by a sedimenting particle.* Phys. D **240** (2011), 1111–1118.

[30] F. Riewe, *Nonconservative Lagrangian and Hamiltonian mechanics.* Phys. Rev. E (3) **53** (1996), 1890–1899.

[31] F. Riewe, *Mechanics with fractional derivatives.* Phys. Rev. E (3) **55** (1997), 3581–3592.

[32] B. Ross, S.G. Samko, *Fractional integration operator of a variable order in the Hölder spaces* $H^{\lambda(x)}$. Internat. J. Math. Math. Sci. **18** (1995), 777–788.

[33] S.G. Samko, *Fractional integration and differentiation of variable order.* Anal. Math. **21** (1995), 213–236.

[34] S.G. Samko, A.A. Kilbas, O.I. Marichev, *Fractional integrals and derivatives*, translated from the 1987 Russian original, Gordon and Breach, Yverdon, 1993.

[35] S.G. Samko, B. Ross, *Integration and differentiation to a variable fractional order.* Integral Transform. Spec. Funct. **1** (1993), 277–300.

[36] J.A. Tenreiro Machado, V. Kiryakova, F. Mainardi, *Recent history of fractional calculus.* Commun. Nonlinear Sci. Numer. Simul. **16** (2011), 1140–1153.

[37] B. van Brunt, *The calculus of variations.* Universitext, Springer, New York, 2004.

T. Odzijewicz and D.F.M. Torres
Center for Research and Development
in Mathematics and Applications
Department of Mathematics
University of Aveiro
P-3810-193 Aveiro, Portugal
e-mail: `tatianao@ua.pt`
`delfim@ua.pt`

A.B. Malinowska
Faculty of Computer Science, Białystok University of Technology
PL-15-351 Białystok, Poland
e-mail: `a.malinowska@pb.edu.pl`

Operator Theory:
Advances and Applications, Vol. 229, 303–321

Improving Bounds for Singular Operators via Sharp Reverse Hölder Inequality for A_∞

Carmen Ortiz-Caraballo, Carlos Pérez and Ezequiel Rela

Dedicated to Prof. Samko

Abstract. In this expository article we collect and discuss some recent results on different consequences of a Sharp Reverse Hölder Inequality for A_∞ weights. For two given operators T and S, we study $L^p(w)$ bounds of Coifman-Fefferman type:

$$\|Tf\|_{L^p(w)} \leq c_{n,w,p}\|Sf\|_{L^p(w)},$$

that can be understood as a way to control T by S.

We will focus on a *quantitative* analysis of the constants involved and show that we can improve classical results regarding the dependence on the weight w in terms of Wilson's A_∞ constant

$$[w]_{A_\infty} := \sup_Q \frac{1}{w(Q)} \int_Q M(w\chi_Q).$$

We will also exhibit recent improvements on the problem of finding sharp constants for weighted norm inequalities involving several singular operators. In the same spirit as in [10], we obtain mixed A_1–A_∞ estimates for the commutator $[b,T]$ and for its higher-order analogue T_b^k. A common ingredient in the proofs presented here is a recent improvement of the Reverse Hölder Inequality for A_∞ weights involving Wilson's constant from [10].

Mathematics Subject Classification (2010). Primary 42B20, 42B25. Secondary 46B70, 47B38.

Keywords. Weighted norm inequalities, Reverse Hölder Inequality, maximal operators, singular integrals, Calderón-Zygmund theory, commutators.

The second author is supported by the Spanish Ministry of Science and Innovation grant MTM2009-08934, the second and third authors are also supported by the Junta de Andalucía, grant FQM-4745.

1. Introduction

The purpose of this survey is to exhibit and discuss some recent results involving improvements in two closely related topics in weight theory: Coifman–Fefferman type inequalities for weighted L^p spaces and boundedness of Calderón–Zygmund (C–Z) operators and their commutators with BMO functions. More generally, we present several examples of the so-called "Calderón–Zygmund Principle" which says, essentially, that any C–Z operator is controlled (in some sense) by an adequate maximal operator. Typically, those inequalities provide control over an operator T with some singularity, like a singular integral operator, by means of a "better" operator S, like a maximal operator. As a model example of this phenomenon, we can take the classical Coifman–Fefferman inequality involving a C–Z operator and the usual Hardy-Littlewood maximal function M (see [5], [6]). Throughout this paper, we will denote $L^p(\mathbb{R}^n, w)$ as $L^p(w)$. As usual, for $1 < p < \infty$, we will denote with p' the dual index of p, defined by the equation $\frac{1}{p} + \frac{1}{p'} = 1$.

Theorem 1.1 (Coifman–Fefferman). *For any weight w in the Muckenhoupt class A_∞, the following norm inequality holds:*

$$\|Tf\|_{L^p(w)} \leq c\,\|Mf\|_{L^p(w)}, \tag{1.1}$$

where $0 < p < \infty$ and $c = c_{n,w,p}$ is a positive constant depending on the dimension n, the exponent p and on the weight w.

Let us recall that a weight w is a non-negative measurable function. Such a function w belongs to the Muckenhoupt class A_p, $1 < p < \infty$ if

$$[w]_{A_p} := \sup_Q \left(\frac{1}{|Q|}\int_Q w(y)\,dy\right)\left(\frac{1}{|Q|}\int_Q w(y)^{1-p'}\,dy\right)^{p-1} < \infty,$$

where the supremum is taken over all the cubes in $\mathbb{R}^n$. This number is called the A_p constant or characteristic of the weight w. For $p = 1$, the condition is that there exists a constant $c > 0$ such that the Hardy–Littlewood maximal function M satisfies the bound

$$Mw(x) \leq c\,w(x) \text{ a.e. } x \in \mathbb{R}^n.$$

In that case, $[w]_{A_1}$ will denote the smallest of these constants. Since the A_p classes are increasing with respect to p, we can define the A_∞ class in the natural way by

$$A_\infty := \bigcup_{p>1} A_p.$$

For any weight in this larger class, the A_∞ constant can be defined as follows:

$$\|w\|_{A_\infty} := \sup_Q \left(\frac{1}{|Q|}\int_Q w\right)\exp\left(\frac{1}{|Q|}\int_Q \log w^{-1}\right).$$

This constant was introduced by Hruščev [9] (see also [8]) and has been the standard A_∞ constant until very recently when a "new" A_∞ constant was found to be better suited. This new constant is defined as

$$[w]_{A_\infty} := \sup_Q \frac{1}{w(Q)} \int_Q M(w\chi_Q).$$

However, this constant was introduced by M. Wilson a long time ago (see [26, 27, 28]) with a different notation. This constant is more relevant since there are examples of weights $w \in A_\infty$ so that $[w]_{A_\infty}$ is much smaller than $\|w\|_{A_\infty}$. Indeed, first it can be shown that

$$c_n\,[w]_{A_\infty} \le \|w\|_{A_\infty} \le [w]_{A_p}, \quad 1 < p < \infty, \tag{1.2}$$

where c_n is a constant depending only on the dimension. The first inequality is the only nontrivial part and can be found in [10] where a more interesting fact is also shown, namely that this inequality can be strict. More precisely, the authors exhibit a family of weights $\{w_t\}$ such that $[w_t]_{A_\infty} \le 4\log(t)$ and $\|w_t\|_{A_\infty} \sim t/\log(t)$ for $t \gg 1$.

It should be mentioned that this constant has also been used by Lerner [14, Section 5.5] (see also [12]) where the term "A_∞ constant" was coined.

Going back to Theorem 1.1, we can see that the result is, in the way that it is stated, a *qualitative* result. It says that the maximal operator M acts as a "control operator" for C–Z operators, but the dependence of the constant c on both w and p is not precise enough for some applications. For instance, let us mention that a precise knowledge of the behavior of this constant was crucial in the proof of the following theorems from [16] where the authors address a problem due to Muckenhoupt and Wheeden.

Theorem 1.2. *Let T be a C–Z operator and let $1 < p < \infty$. Let w be a weight in A_1. Then*

$$\|T\|_{L^p(w)} \le c\,pp'\,[w]_{A_1},$$

where $c = c_{n,T}$.

As an application of this result, the following sharp endpoint estimate can be proven.

Theorem 1.3. *Let T be a C–Z operator. Let w be a weight in A_1. Then*

$$\|T\|_{L^{1,\infty}(w)} \le c\,[w]_{A_1}(1 + \log[w]_{A_1}),$$

where $c = c_{n,T}$.

An important point in the proof of these theorems was the use of a special instance of (1.1) with a gain in the constant $c_{n,w,p}$. To be more precise the following L^1 estimate was needed:

$$\|Tf\|_{L^1(w)} \le c_{n,T}\,[w]_{A_q}\,\|Mf\|_{L^1(w)} \quad w \in A_q \tag{1.3}$$

where there is an improvement in the constant appearing in (1.1), because $c_{n,T}$ is a structural constant.

The same kind of qualitative and quantitative Coifman–Fefferman type results are known or can be proved for a variety of singular operators, namely commutators of C–Z operators with BMO functions, vector-valued extensions and square functions. In addition, there are also weak type estimates. In that direction, there is also a notion of singularity that can be assigned to these operators. Roughly, these commutators are "more singular" than C–Z operators, and these are "more singular" than, for example, square functions. This notion of singularity is reflected in the kind of maximal operators involved in the norm inequalities. For the commutator, we have from [24] (see also [23]) that, for any weight $w \in A_\infty$,

$$\|[b,T]f\|_{L^p(w)} \leq c_{n,w,p}\, \|b\|_{BMO} \|M^2 f\|_{L^p(w)},$$

where $M^2 = M \circ M$. This result is sharp on the BMO norm of b but not on the weight w. More generally, for the iterated commutator T_b^k, the control operator is M^{k+1}. We will show that in this case the constant depends on a $(k+1)$th power of $[w]_{A_\infty}$. Therefore, as our first main purpose, we will show how to obtain the above-mentioned Coifman–Fefferman type inequalities involving Wilson's A_∞ constant $[w]_{A_\infty}$, which is, in light of the chain of inequalities (1.2), the good one.

We will also address the problem of finding sharp operator bounds for a variety of singular operators. The improvement here will be reflected in a *mixed* bound in terms of the A_1 and A_∞ constant of the weight $w \in A_1$. This approach is taken from [10], where the authors prove a new sharp Reverse Hölder Inequality (RHI) for A_∞ weights with the novelty of involving Wilson's constant. It is important to note that this RHI for A_q weights, $1 \leq q \leq \infty$, was already known. Moreover, the same proof as in the case of A_1, which can be found in [15] with minor modifications, also works for A_∞ weights but with the "older" $\|w\|_{A_\infty}$ constant. However, the proof of the same property for Wilson's constant $[w]_{A_\infty}$ is more difficult and requires a different approach.

Recall that if $w \in A_p$ there are constants $r > 1$ and $c \geq 1$ such that for any cube Q, the RHI holds:

$$\left(\frac{1}{|Q|}\int_Q w^r\,dx\right)^{1/r} \leq \frac{c}{|Q|}\int_Q w. \tag{1.4}$$

In the standard proofs both constants c, r depend upon the A_p constant of the weight. A more precise version of (1.4) is the following result that can be found, for instance, in [21].

Lemma 1.4. *Let $w \in A_p$, $1 < p < \infty$ and let*

$$r_w = 1 + \frac{1}{2^{2p+n+1}[w]_{A_p}}.$$

Then for any cube Q,

$$\left(\frac{1}{|Q|}\int_Q w^{r_w}\,dx\right)^{1/r_w} \leq \frac{2}{|Q|}\int_Q w.$$

Here we present the new result of T. Hytönen and the second author (see [10]). This sharper version of the RHI plays a central role in the proofs of all the results presented in this article.

Theorem 1.5 (A new sharp reverse Hölder inequality). *Define* $r_w := 1+\dfrac{1}{\tau_n\,[w]_{A_\infty}}$, *where* τ_n *is a dimensional constant that we may take to be* $\tau_n = 2^{11+n}$. *Note that* $r'_w \approx [w]_{A_\infty}$.

a) *If* $w \in A_\infty$, *then*

$$\left(\frac{1}{|Q|}\int_Q w^{r_w}\right)^{1/r_w} \leq 2\frac{1}{|Q|}\int_Q w.$$

b) *Furthermore, the result is optimal up to a dimensional factor: There exists a dimensional constant* $c = c_n$ *such that, if a weight* w *satisfies the RHI, i.e., there exists a constant* K *such that*

$$\left(\frac{1}{|Q|}\int_Q w^{r}\right)^{1/r} \leq K\frac{1}{|Q|}\int_Q w,$$

for all cubes Q, *then* $[w]_{A_\infty} \leq c_n\,K\,r'$.

Among other important results, in [10] the authors derive mixed A_1–A_∞ type results of A. Lerner, S. Ombrosi and the second author in [16] of the form:

$$\|Tf\|_{L^p(w)} \leq c\,pp'\,[w]_{A_1}^{1/p}[w]_{A_\infty}^{1/p'}, \qquad w\in A_1, \quad 1<p<\infty,$$

which is an improvement of Theorem 1.2. They also derive weak type estimates like

$$\|Tf\|_{L^{1,\infty}(w)} \leq c[w]_{A_1}\log(e+[w]_{A_\infty})\|f\|_{L^1(w)}.$$

As another main purpose in the present work, we will show how to extend this results on mixed A_1–A_∞ bounds to the case of the commutator and its higher-order analogue. The analogues of Theorem 1.2 and Theorem 1.3 for the commutators we consider were already proved by the first author in [18]:

Theorem 1.6 (Quadratic A_1 bound for commutators). *Let* T *be a Calderón–Zygmund operator and let* b *be in* BMO. *Also let* $1<p,r<\infty$. *Then there exists a constant* $c = c_{n,T}$ *such that for any weight* w, *the following inequality holds*

$$\|[b,T]f\|_{L^p(w)} \leq c\,\|b\|_{BMO}\,(pp')^2\,(r')^{1+\frac{1}{p'}}\,\|f\|_{L^p(M_rw)}. \tag{1.5}$$

In particular if $w\in A_1$, *we have*

$$\|[b,T]f\|_{L^p(w)} \leq c\,\|b\|_{BMO}(pp')^2[w]_{A_1}^2\,\|f\|_{L^p(M_rw)}. \tag{1.6}$$

Furthermore this result is sharp in both p *and the exponent of* $[w]_{A_1}$.

Theorem 1.7. *Let T and b as above. Then there exists a constant $c = c_{n,T}$ such that for any weight*

$$w(\{x \in \mathbb{R}^n : |[b,T]f(x)| > \lambda\}) \leq c\,(pp')^{2p}(r')^{2p-1} \int_{\mathbb{R}^n} \Phi\left(\|b\|_{BMO}\frac{|f|}{\lambda}\right) M_r w dx, \tag{1.7}$$

where $\Phi(t) = t(1+\log^+ t)$.

As a consequence, if $w \in A_1$

$$w(\{x \in \mathbb{R}^n : |[b,T]f(x)| > \lambda\}) \leq c\,\Phi([w]_{A_1})^2 \int_{\mathbb{R}^n} \Phi\left(\frac{|f(x)|}{\lambda}\right) w(x)\,dx \tag{1.8}$$

where $\Phi(t) = t(1+\log^+ t)$.

In this article we present an improvement of these theorems in terms of mixed A_1–A_∞ norms for the commutator and for its iterations, proved by the first author in her dissertation (2011). For any $k \in \mathbb{N}$, the kth iterated commutator T_b^k of a BMO function b and a C–Z operator T is defined by

$$T_b^k := [b, T_b^{k-1}].$$

This paper is organized as follows. In Section 2 we formulate the precise statements of the results announced in the introduction. In Section 3 we present some background and auxiliary results for the proofs and finally, in Section 4, we describe the main features of the proofs.

2. Main results

In this section we will present the precise statements of the results discussed in the introduction. First, we present several Coifman–Fefferman inequalities for a variety of singular operators. Our purpose is to emphasize that there is a notion of order of singularity that allows us to distinguish them. This higher or lower singularity can be seen in the power of the maximal functions involved and also in the dependence of the constant of the weight. Next, we present our results on strong and weak norm inequalities for commutator and for its iterations. In this latter case, we study in detail how the iterations affects on each part of the weight, namely the A_1 and the A_∞ fractions of the constant.

2.1. Coifman–Fefferman inequalities

Theorem 2.1 (C–Z operators). *Let T be a C–Z operator and let $w \in A_\infty$. Then there is constant $c = c_{n,T}$ such that, for any $0 < p < \infty$,*

$$\|Tf\|_{L^p(w)} \leq c \max\{1,p\}[w]_{A_\infty}\|Mf\|_{L^p(w)}$$

whenever f is a function satisfying the condition $|\{x : |Tf(x)| > t\}| < \infty$ for all $t > 0$.

We need some additional notation for the following theorem. Let $1 < q < \infty$ and define, for any sequence of functions $f = \{f_j\}_{j\in\mathbb{N}}$, $|f|_q := \left(\sum_j |f_j|^q\right)^{\frac{1}{q}}$. Also define, for a C–Z operator T, the vector-valued extension $\overline{T}f = \{Tf_j\}_{j\in\mathbb{N}}$.

Theorem 2.2 (C–Z operators – vector-valued extensions). *Let T be a C–Z operator and let $w \in A_\infty$. Then there is constant $c = c_{n,T}$ such that, for any $0 < p < \infty$, $1 < q < \infty$ and for any sequence of compactly supported functions $\{f_j\}_{j\in\mathbb{N}}$,*

$$\left\| |\overline{T}f|_q \right\|_{L^p(w)} \le c \max\{1,p\}[w]_{A_\infty} \left\| M\left(|f|_q\right) \right\|_{L^p(w)} .$$

We also have the following result on multilinear C–Z operators. For the precise definitions and properties, see [17]. Let T be an m-linear C–Z operator acting on a vector $\vec{f}$ of m functions $\vec{f} = (f_1, \dots, f_m)$. Define also the maximal function $\mathcal{M}$ by

$$\mathcal{M}(\vec{f})(x) = \sup_{Q\ni x} \prod_{i=1}^{m} \frac{1}{|Q|} \int_Q |f_i(y_i)|\, dy_i.$$

The following theorem is a refinement of [17, Corollary 3.8].

Theorem 2.3 (Multilinear C–Z operators). *Let T be an m-linear C–Z operator, let $w \in A_\infty$ and let $p > 0$. Then there exists a constant $c = c_{n,m,T}$ such that*

$$\|T(\vec{f})\|_{L^p(w)} \le c \max\{1,p\}[w]_{A_\infty} \|\mathcal{M}(\vec{f})\|_{L^p(w)},$$

whenever $\vec{f}$ is a vector of compactly supported functions.

Theorem 2.4 (Commutators). *Let T be a C–Z operator, $b \in BMO$ and let $w \in A_\infty$. Then there is a constant $c = c_{n,T}$ such that for $0 < p < \infty$,*

$$\|[b,T]\|_{L^p(w)} \le c \, \|b\|_{BMO} \max\{1,p^2\}[w]^2_{A_\infty} \|M^2 f\|_{L^p(w)}$$

More generally, we have the following result for the iterated commutator.

Theorem 2.5 (kth iterated commutator). *Let T be a C–Z operator, $b \in BMO$ and let $w \in A_\infty$. For the kth ($k \ge 2$) commutator T^k_b, there is a constant $c = c_{n,T}$ such that, for $0 < p < \infty$,*

$$\|T^k_b f\|_{L^p(w)} \le c\, 2^k \max\{1,p^{k+1}\}[w]^{k+1}_{A_\infty} \|b\|^k_{BMO} \|M^{k+1} f\|_{L^p(w)}.$$

2.2. Mixed A_1–A_∞ strong and weak norm inequalities for commutators

Theorem 2.6. *Let T be a C–Z operator, $b \in BMO$ and let $1 < p < \infty$. Then there is a constant $c = c_{n,T}$ such that, for any $w \in A_1$,*

$$\|[b,T]\|_{L^p(w)} \le c \, \|b\|_{BMO} (pp')^2 [w]^{1/p}_{A_1} [w]^{1+1/p'}_{A_\infty} .$$

Theorem 2.7. *Let T and b as above. Then there exists a constant $c = c_{n,T}$ such that for any weight $w \in A_1$*

$$w(\{x \in \mathbb{R}^n : |[b,T]f| > \lambda\}) \le c\,\beta \int_{\mathbb{R}^n} \Phi\left(\|b\|_{BMO} \frac{|f|}{\lambda} \right) w(x) dx,$$

where $\beta = [w]_{A_1}[w]_{A_\infty}(1 + \log^+[w]_{A_\infty})^2$ and $\Phi(t) = t(1 + \log^+ t)$.

More generally, we have the following generalization for the kth iterated commutator.

Theorem 2.8. *Let T be a C–Z operator, $b \in BMO$ and $1 < p, r < \infty$. Consider the higher-order commutators T_b^k, $k = 1, 2, \dots$. Then there exists a constant $c = c_{n,T}$ such that for any weight w the following inequality holds*

$$\|T_b^k f\|_{L^p(w)} \le c\, \|b\|_{BMO}^k\, (pp')^{k+1}\, (r')^{k+1/p'}\, \|f\|_{L^p(M_r w)}. \tag{2.1}$$

In particular if $w \in A_1$, we have that

$$\|T_b^k\|_{L^p(w)} \le c\, \|b\|_{BMO}^k (pp')^{k+1} [w]_{A_1}^{1/p} [w]_{A_\infty}^{k+1/p'}.$$

Theorem 2.9. *Let T and b as above, and let $1 < p, r < \infty$. Then there exists a constant $c = c_{n,T}$ such that for any w*

$$w(\{x \in \mathbb{R}^n : |T_b^k f| > \lambda\}) \le c\, (pp')^{(k+1)p} (r')^{(k+1)p-1} \int_{\mathbb{R}^n} \Phi\left(\|b\|_{BMO} \frac{|f|}{\lambda}\right) M_r w\, dx, \tag{2.2}$$

where $\Phi(t) = t(1 + \log^+ t)^k$.

If $w \in A_1$ we obtain that

$$w(\{x \in \mathbb{R}^n : |T_b^k f(x)| > \lambda\}) \le c_n\, \beta \int_{\mathbb{R}^n} \Phi\left(\|b\|_{BMO} \frac{|f(x)|}{\lambda}\right) w(x) dx,$$

where $\beta = [w]_{A_1} [w]_{A_\infty}^k (1 + \log^+ [w]_{A_\infty})^{k+1}$ and $\Phi(t) = t(1 + \log^+ t)^k$.

3. Background and preliminaries

3.1. Rearrangement type estimates

In this section we present an important tool based on rearrangements of functions. The main lemma is the following about the control of L^p norms by a sharp maximal function. We start by recalling some standard definitions. Given a locally integrable function f on $\mathbb{R}^n$, the *Hardy–Littlewood maximal operator* M is defined by

$$Mf(x) = \sup_{Q \ni x} \frac{1}{|Q|} \int_Q f(y), dy,$$

where the supremum is taken over all cubes Q containing the point x. We will also use the following operator:

$$M_\delta f(x) = (M(|f|^\delta)(x))^{1/\delta}, \qquad \delta > 0.$$

Recall that the Fefferman–Stein sharp maximal function is defined as

$$M^\#(f)(x) = \sup_{Q \ni x} \frac{1}{|Q|} \int_Q |f(y) - f_Q|\, dy,$$

where

$$f_Q = \frac{1}{|Q|} \int_Q f(y)\, dy.$$

If we only consider dyadic cubes, we obtain the *dyadic* sharp maximal function, denoted by $M^{\#,d}$.

Define also, for $0 < \delta < 1$,

$$M_\delta^{\#} f(x) = M^{\#}(|f|^\delta)(x)^{1/\delta}.$$

and

$$M_\delta^{\#,d} f(x) = M^{\#,d}(|f|^\delta)(x)^{1/\delta}.$$

Lemma 3.1. *Let $0 < p < \infty$, $0 < \delta < 1$ and let $w \in A_\infty$. Then*

$$\|f\|_{L^p(w)} \leq c \max\{1, p\} [w]_{A_\infty} \|M_\delta^{\#,d}(f)\|_{L^p(w)} \tag{3.1}$$

for any function f such that $|\{x : |f(x)| > t\}| < \infty$ for all $t > 0$.

The proof of Lemma 3.1 will follow from an analogous inequality for the non-increasing rearrangements of f and $M_\delta^{\#,d}$ with respect to the weight w. Recall that the non-increasing rearrangement f_w^* of a measurable function f with respect to a weight w is defined by

$$f_w^*(t) = \inf \{\lambda > 0 : w_f(\lambda) < t\}, \qquad t > 0,$$

where

$$w_f(\lambda) = w(\{x \in \mathbb{R}^n : |f(x)| > \lambda\}), \ \lambda > 0,$$

is the distribution function of f associated to w. An important fact is that

$$\int_{\mathbb{R}^n} |f|^p \, w dx = \int_0^\infty f_w^*(t)^p \, dt.$$

The key rearrangement lemma is the following.

Lemma 3.2. *Let and $w \in A_\infty$, $0 < \delta < 1$ and $0 < \gamma < 1$. There is a constant $c = c_{n,\gamma,\delta}$, such that for any measurable function:*

$$f_w^*(t) \leq c [w]_{A_\infty} \left(M_\delta^{\#,d} f\right)_w^*(\gamma t) + f_w^*(2t) \qquad t > 0. \tag{3.2}$$

This sort of estimates goes back to the work of R. Bagby and D. Kurtz in the middle of the 80s (see [2] and [3]). The proof of Lemma 3.2 can be found in [21] but with A_q instead of A_∞ weights. This improvement to A_∞ weights and, moreover, with the smaller $[w]_\infty$ constant, is a consequence of the sharp RHI for A_∞ weights from [10]. We will sketch the proof in the next section. With this lemma, we can prove Lemma 3.1. If we iterate (3.2) we have:

$$\begin{aligned} f_w^*(t) &\leq c [w]_{A_\infty} \sum_{k=0}^{\infty} (M_\delta^{\#} f)_w^*(2^k \gamma t) + f_w^*(+\infty) \\ &\leq \frac{[w]_{A_\infty}}{\log 2} \int_{t\gamma/2}^{\infty} (M_\delta^{\#} f)_w^*(s) \, \frac{ds}{s} + f_w^*(+\infty). \end{aligned}$$

Hence if we assume that

$$f_w^*(+\infty) = 0,$$

the inequality we obtain is

$$f_w^*(t) \leq c\,[w]_{A_\infty} \int_{t\gamma/2}^{\infty} (M_\delta^\# f)_w^*(s)\,\frac{ds}{s}.$$

We continue using the Hardy operator. Recall that if $f : (0,\infty) \to [0,\infty)$

$$Af(x) = \frac{1}{x}\int_0^x f(t)dt, \qquad x > 0$$

is called the Hardy operator. The dual operator is given by

$$Sf(x) = \int_x^\infty f(s)\frac{ds}{s}.$$

Hence the above estimate can be expressed as:

$$f_w^*(t) \leq c\,[w]_{A_\infty}\, S((M_\delta^\# f)_w^*)(t\gamma/2).$$

Finally since it is well known that these operators are bounded on $L^p(0,\infty)$ and furthermore, it is known that if $p \geq 1$, then $\|S\|_{L^p(0,\infty)} = p$, we have that

$$\begin{aligned}
\|f\|_{L^p(w)} &= \|f_w^*\|_{L^p(0,\infty)} \\
&\leq c\,[w]_{A_\infty}\, \|S((M_\delta^\# f)_w^*)\|_{L^p(0,\infty)} \\
&\leq c\,p\,[w]_{A_\infty}\, \|(M_\delta^\# f)_w^*\|_{L^p(0,\infty)} \\
&= c\,p\,[w]_{A_\infty}\, \|M_\delta^\# f\|_{L^p(w)}.
\end{aligned}$$

This concludes the strong estimate in the case $p \geq 1$. For $0 < p < 1$, use the triangle inequality and L^1 boundedness of the operator S (we get no p here, so we just get $\max\{1,p\}$ for general p). This concludes the proof of Lemma 3.1.

3.2. Pointwise inequalities

Once we have the result of the previous subsection, i.e., the Fefferman–Stein inequality involving $M_\delta^{\#,d}$ with a sharp dependence on the weight, we can then deduce sharp Coifman–Fefferman weighted inequalities for any pair of operators T and S satisfying a pointwise inequality like

$$M_\delta^\#(Tf)(x) \leq c_\delta Sf(x) \qquad \text{a.e. } x \in \mathbb{R}^n.$$

We will elaborate on this in the next section. Here we want to collect those inequalities that involve the operators under study (we refer the interested reader to [7, Chapter 9] for a deeper treatment of this subject). The inequalities that are going to be used are:

- For $0 < \delta < 1$ and T any C–Z operator (see [1]),
$$M_\delta^\#(Tf)(x) \leq c_\delta\, Mf(x). \tag{3.3}$$
- We have the following vector-valued extension from [25]: Let $1 < q < \infty$ and $0 < \delta < 1$. Let T be a C–Z operator and consider the vector-valued extension $\overline{T}$. There exists a constant c_δ such that
$$M_\delta^\#\left(|\overline{T}f|_q\right)(x) \leq c_\delta\, M(|f|_q)(x).$$

- We also have a pointwise inequality for multilinear C–Z operators. Let T be an m-linear C–Z operator. Then for all $\vec{f}$ in any product of $L^{q_j}(\mathbb{R}^n)$ spaces, with $1 \le q_j < \infty$,

$$M_\delta^\#(T(\vec{f}))(x) \le c_\delta\, \mathcal{M}(\vec{f})(x) \qquad \text{for } 0 < \delta < 1/m.$$

- The following result is from [18]. If $0 < \delta < \varepsilon < 1$, there is a constant $c = c_{\varepsilon,\delta}$ such that

$$M_\delta^{\#,d}(M_\varepsilon^d(f))(x) \le c\, M_\varepsilon^{\#,d} f(x).$$

What we really need, for the proof of the Coifman–Fefferman inequality in the case of the commutator, is a consequence of this inequality, which is a subtle improvement of Lemma 3.1: for $0 < p < \infty$ and $0 < \delta < 1$,

$$\|M_\delta^d f\|_{L^p(w)} \le c \max\{1,p\}[w]_{A_\infty} \|M_\delta^{\#,d}(f)\|_{L^p(w)}. \tag{3.4}$$

- Another related and very important pointwise inequality for iterated commutators is the following result from [24]. For each $b \in BMO$, $0 < \delta < \varepsilon < 1$, there exists $c = c_{\delta,\varepsilon}$ such that, for all smooth functions f, we have that

$$M_\delta^\#(T_b^k f)(x) \le c\,\|b\|_{BMO} \sum_{j=0}^{k-1} M_\varepsilon^d(T_b^j f)(x) + \|b\|_{BMO}^k M^{k+1} f(x). \tag{3.5}$$

3.3. Building A_1 weights from duality

The following lemma gives a way to produce A_1 weights with special control on the constant. It is based on the so-called Rubio de Francia iteration scheme or algorithm.

Lemma 3.3. [16] *Let $1 < s < \infty$, and let v be a weight. Then, there exists a non-negative sublinear operator R bounded in $L^s(v)$ satisfying the following properties:*

(i) $h \le R(h)$;
(ii) $\|Rh\|_{L^s(v)} \le 2\|h\|_{L^s(v)}$;
(iii) $Rh\, v^{1/s} \in A_1$ *with* $[Rh\, v^{1/s}]_{A_1} \le cs'$.

4. About the proofs

4.1. Coifman–Fefferman inequalities

We begin this section with a brief description of how to obtain Coifman–Fefferman inequalities for all the operators presented in the previous section. These inequalities will follow from a general scheme involving pointwise control of the singular operator by an appropriate maximal function and the key inequality from Lemma 3.1. Suppose that we have, for a pair of operators T and S, a pointwise inequality like

$$M_\delta^\#(Tf)(x) \le c_\delta\, Sf(x) \qquad \text{a.e. } x \in \mathbb{R}^n.$$

Then, for any f such that $|\{x : |Tf(x)| > t\}| < \infty$ for all $t > 0$ and any $w \in A_\infty$, we have that

$$\begin{aligned}\|T(f)\|_{L^p(w)} &\le c \max\{1,p\}[w]_{A_\infty} \|M_\delta^{\#,d}(T(f))\|_{L^p(w)} \\ &\le c_\delta \max\{1,p\}[w]_{A_\infty} \|S(f)\|_{L^p(w)}.\end{aligned}$$

This argument completes the proof of Theorem 2.1, Theorem 2.2 and Theorem 2.3.

For the case of the commutator and its iterations, we need an extra step for each iteration. More precisely, for the commutator we have the following estimate. Let w be any A_∞ weight and let $p > 0$. For $0 < \varepsilon < \delta < 1$, we have that

$$\begin{aligned}\|[b,T]f\|_{L^p(w)} &\le c \max\{1,p\}[w]_{A_\infty} \|M_\varepsilon^{\#,d}([b,T]f)\|_{L^p(w)} \\ &\le c \max\{1,p\}[w]_{A_\infty} \|b\|_{BMO} \left(\|M_\delta^d(Tf)\|_{L^p(w)} + \|M^2 f\|_{L^p(w)}\right).\end{aligned}$$

We start with (3.1) applied to $[b,T]f$, and then we use (3.5). Now we can combine (3.4) with (3.3) to control the first term and therefore obtain that

$$\|[b,T]f\|_{L^p(w)} \le c\, \|b\|_{BMO} \max\{1,p^2\}[w]_{A_\infty}^2 \|b\|_{BMO}\|M^2 f\|_{L^p(w)}.$$

This final estimate follows from the fact that $[w]_{A_\infty} \ge 1$ and by dominating M by M^2. For the iterated commutator, we proceed in the same way as before. The relevant inequality, which can be easily proved by induction, is contained in the following lemma.

Lemma 4.1. *Let $0 < \delta < 1$. Then there exists η such that $0 < \delta < \eta < 1$ and a universal constant $c > 0$ such that*

$$\|M_\varepsilon^{\#,d}(T_b^k f)\|_{L^p(w)} \le 2^k\, c \max\{1,p\}\|b\|_{BMO}\|M_\eta^{k+1} f\|_{L^p(w)}.$$

Following the preceding scheme and applying this lemma, we obtain the result stated in Theorem 2.5. This will conclude the proofs of all the announced Coifman–Fefferman inequalities, if we prove the essential inequality (3.2). In fact, it is in this inequality where the improvement on the dependence on the weight appears. We will present here a sketch of the proof from [21]. The idea is to show that the key ingredient for the improvement is an exponential decay lemma and the new Reverse Hölder Property for A_∞ weights.

4.1.1. Proof of the key rearrangement estimate. Fix $t > 0$ and let $A = [w]_{A_\infty}$. By definition of rearrangement, (3.2) will follow if we prove that for each $t > 0$,

$$w\Big\{x \in \mathbb{R}^n : f(x) > cA\left(M_\delta^{\#} f\right)_w^*(\gamma t) + f_w^*(2t)\Big\} \le t.$$

We split the left-hand side L as follows:

$$\begin{aligned}L &\le w\{x \in \mathbb{R}^n : cA\, M_\delta^{\#} f(x) > cA\left(M_\delta^{\#} f\right)_w^*(\gamma t) \\ &\quad + w\{x \in \mathbb{R}^n : f(x) > cA\, M_\delta^{\#} f(x) + f_w^*(2t)\} \\ &= I + II.\end{aligned}$$

Observe that

$$I = w\Big\{x \in \mathbb{R}^n : M_\delta^{\#} f(x) > \left(M_\delta^{\#} f\right)_w^*(\gamma t)\Big\} \le \gamma t$$

by definition of rearrangement, and hence the heart of the matter is to prove that

$$II = w\{x \in \mathbb{R}^n : f(x) > cA\, M_\delta^{\#} f(x) + f_w^*(2t)\} \le (1-\gamma)\, t.$$

The set $\{x \in \mathbb{R}^n : f(x) > cA\, M_\delta^{\#} f(x) + f_w^*(2t)\}$ is contained in the set $E = \{x \in \mathbb{R}^n : f(x) > f_w^*(2t)\}$ which has w-measure at most $2t$ by definition of rearrangement. Now, by the regularity of the measure we can find an open set Ω containing E such that $w(\Omega) < 3t$. The remainder of the proof from [21] consists in proving that there exists a constant c depending only on δ and the dimension such that

$$II = w\{x \in \Omega : f(x) > cA\, M_\delta^{\#} f(x) + f_w^*(2t)\} \le (1-\gamma)\, t.$$

By applying an appropriate Calderón–Zygmund decomposition of the set Ω into a union of dyadic cubes $\{Q_j\}$, it can be proved that, for $c > 1 + 2^{1/\delta}$, we have the following inclusion for any j. If we denote

$$E_j = \big\{x \in Q_j : |f(x) - m_f(Q_j)| > cA\, M_\delta^{\#} f(x)\big\}, \tag{4.1}$$

then

$$\{x \in Q_j : f(x) > cA\, M_\delta^{\#} f(x) + f_w^*(2t)\} \subset E_j,$$

where $m_f(Q)$ is the median value of f over Q (see [13]). The key ingredient now is an "exponential decay" Lemma from [21] adapted to E_j.

Lemma 4.2. *Let $f \in L^\delta_{loc}$. For $t > 0$ we define*

$$\varphi(t) = \sup_{Q\in\mathcal{D}} \frac{1}{|Q|}\Big|\{x \in Q : |f(x) - m_f(Q)| > t\, M_\delta^{\#} f(x)\}\Big|.$$

There are dimensional constants c_1, c_2 such that $\varphi(t) \le \dfrac{c_1}{e^{c_2 t}}$.

Applying this lemma to (4.1), we obtain that

$$\frac{|E_j|}{|Q_j|} \le c_1 e^{-c_2\, cA}.$$

Now we apply Hölder's inequality to obtain that

$$\begin{aligned} w(E_j) &= \frac{1}{|Q_j|}\int_{Q_j} \chi_{E_j}(x) w(x)\, dx\, |Q_j| \\ &\le \left(\frac{|E_j|}{|Q_j|}\right)^{1/r'} \left(\frac{1}{|Q_j|}\int_{Q_j} w(x)^r\, dx\right)^{1/r} |Q_j|. \end{aligned} \tag{4.2}$$

We conclude by choosing $r = r_w$, the sharp reverse exponent for w, (see Theorem 1.5), $r_w = 1 + \dfrac{1}{2^{11+d}[w]_{A_\infty}}$. We obtain

$$\left(\frac{1}{|Q_j|}\int_{Q_j} w^{r_w} dx\right)^{1/r_w} \le \frac{2}{|Q_j|}\int_{Q_j} w$$

and therefore

$$w(E_j) \le 2\left(\frac{|E_j|}{|Q_j|}\right)^{1/r'} w(Q_j) \le c_1 e^{-c\,c_2} w(Q_j)$$

since $r' \approx [w]_{A_\infty} = A$. Here c and c_2 depend on the dimension and we still can choose c as big as needed. If we finally we choose c such that $c_1 e^{-c_2\,c} < \frac{1-\gamma}{3}$, then we have that

$$II \le \sum_j w(E_j) \le \frac{1-\gamma}{3} \sum_j w(Q_j) = \frac{1-\gamma}{3}\, w(\Omega) < (1-\gamma)\,t,$$

since $w(\Omega) < 3t$.

As a final remark on Coifman–Fefferman type inequalities, we want to mention that for the case of C–Z operators, Theorem 2.1 can be deduced from a combination of standard good-λ techniques together with the following local exponential estimate from [20], which is an improvement of the classical result of Buckley [4] (see also [11]).

Theorem 4.3. *Let T be a Calderón–Zygmund operator. Let Q be a cube and let $f \in L^\infty_c(\mathbb{R}^n)$ such that* $\operatorname{supp}(f) = Q$. *Then there exist $c > 0$ and $k > 0$ such that*

$$|\{x \in Q : |Tf(x)| > tMf(x)\}| \le k e^{-ct}|Q|, \qquad t > 0.$$

This theorem is essentially all we need for the proof of Theorem 2.1 since, by standard truncation arguments, we can restrict ourselves to look at only the local part. Now, if we put $E_j := \{x \in Q_j : T(f) > \lambda, Mf < \gamma\lambda\}$, we have that, for some constant c_1,

$$\frac{|E_j|}{|Q_j|} \le c_1 e^{-c_2/\gamma}, \qquad \lambda > 0, \gamma_0 > \gamma > 0.$$

Here we are in the exact same setting as in (4.2), and therefore we can obtain that

$$w(E_j) \le 2\left(\frac{|E_j|}{|Q_j|}\right)^{1/r'} w(Q_j) \le c_1 e^{-c_2/\gamma r'} w(Q_j),$$

where $r' \approx [w]_{A_\infty}$. Now we sum over j to obtain for any constant $B > 0$ the following good-λ inequality:

$$w\left\{x \in \mathbb{R}^n : T(f) > 3\lambda, Mf < \frac{B}{[w]_\infty}\lambda\right\} \le c_1 e^{-c_2/B} w\left\{x \in \mathbb{R}^n : T(f) > \lambda\right\}.$$

From this inequality, we can apply standard good-λ arguments to obtain a sharp Coifman–Fefferman inequality with $[w]_\infty$ bound.

We also remark that in [20] a similar local *sub*exponential decay is proved for the commutator $[b, T]$ above. It is natural to ask if it is possible to prove a Coifman–Fefferman inequality for $[b, T]$ via truncation, exponential decay and sharp RHI. Unfortunately, we do not know how to control the part of the truncated operator "away" from some fixed cube.

4.2. Mixed A_1–A_∞ strong and weak norm inequalities for commutators

In this section we present the proofs of the results announced in Section 2.2. Those results actually follow from the two weight bounds for commutators (1.5) proved in [18] for $k = 1$, and similarly (2.1) (see[19]).

Let us briefly describe how to derive this inequality in the case $k = 1$. The aim is to prove that

$$\|[b,T]f\|_{L^p(w)} \leq c\,\|b\|_{BMO}\,(pp')^2\,(r')^{1+\frac{1}{p'}}\,\|f\|_{L^p(M_r w)}. \tag{4.3}$$

By duality, it is equivalent to prove that

$$\left\|\frac{([b,T])^* f}{M_r w}\right\|_{L^{p'}(M_r w)} \leq c\,(pp')^2\,(r')^{1+\frac{1}{p'}}\,\left\|\frac{f}{w}\right\|_{L^{p'}(w)},$$

for the adjoint operator $[b,T]^*$. To prove this last inequality, we write the norm as

$$\left\|\frac{([b,T])^* f}{M_r w}\right\|_{L^{p'}(M_r w)} = \sup_{\|h\|_{L^p(M_r w)}=1}\left|\int_{\mathbb{R}^n}([b,T])^* f(x)h(x)\,dx\right|.$$

Now we use Rubio de Francia's algorithm and Lemma 3.1 to obtain that

$$\left|\int_{\mathbb{R}^n}([b,T])^* f(x)h(x)\,dx\right| \leq c_\delta\,[Rh]_{A_3}\int_{\mathbb{R}^n} M_\delta^{\#}(([b,T])^* f)(x)\,Rh(x)\,dx.$$

The key here is to use that $[Rh]_{A_3} \leq c_n\,p'$. Also note that for our purposes here it is sufficient to use (3.1) with the A_3 constant of the weight. To handle the integral we use the pointwise inequality (3.5), and therefore we obtain that

$$\left|\int_{\mathbb{R}^n}([b,T])^* f(x)h(x)\,dx\right| \leq c_{\delta,n,\varepsilon}\,p'\|b\|_{BMO}\int_{\mathbb{R}^n} M_\varepsilon^d(T^* f)(x) + M^2 f\,Rh(x)\,dx.$$

For the second term we only need to use Hölder and the properties of the weight Rh. For the first term, we use (3.4) and then (3.3). We obtain that

$$\left|\int_{\mathbb{R}^n}([b,T])^* f(x)h(x)\,dx\right| \leq c_{\delta,n,\varepsilon}\,(p')^2\|b\|_{BMO}\|M^2 f\|_{L^{p'}((M_r w)^{1-p'})}.$$

Finally, we use (see [18]) that

$$\|M^2 f\|_{L^{p'}((M_r w)^{1-p'})} \leq cp^2 (r')^{1+1/p'}\,\|f\|_{L^p(w^{1-p'})},$$

which implies the desired result.

The case $k > 1$ is technically more difficult, but the ideas are similar. The key estimate is the pointwise inequality (3.5) and an induction argument (see [19] for details).

Once we have proved (2.1), we can derive the announced mixed bound as follows. First, note that it is easy to verify that for *any* $r > 1$,

$$M_r w(x) \leq 2[w]_{A_1} w(x),$$

which implies in a standard way a weighted estimate with sharp dependence on $[w]_{A_1}$. But we are able to obtain an improvement by choosing the best RHI at our disposal. Choose then $r = r_w$ as the sharp exponent in the RHI for A_∞ weights, i.e., $r = r_w = 1 + \frac{1}{c_n[w]_{A_\infty}}$. In this case we get a fraction of the weight from the pointwise inequality (4.2) applied to the integrand and, in addition, now r'_w is comparable with $[w]_{A_\infty}$, and therefore from (2.1) we conclude that

$$\|T_b^k f\|_{L^p(w)} \le c\,\|b\|_{BMO}^k\,(pp')^{k+1}\,[w]_{A_1}^{\frac{1}{p}}\,[w]_{A_\infty}^{k+\frac{1}{p'}}\,\|f\|_{L^p(w)}.$$

Now we present the weak-type inequalities. In the same way as before, the mixed bounds will follow from a two weight result (1.7) from [18] for $k = 1$ and the analog (2.2) for $k > 1$ from [19].

Once again, we only sketch the proof for $k = 1$ and refer the reader to the original paper [18] for the details. Let us assume (by homogeneity) that $\|b\|_{BMO} = 1$. For any $f \in C_0^\infty(\mathbb{R}^n)$ consider the Calderón–Zygmund decomposition of f at level λ. We obtain a collection of pairwise disjoint dyadic cubes $\{Q_j\} = Q_j(x_{Q_j}, r_j)$ such that

$$\lambda < \frac{1}{|Q_j|}\int_{Q_j} |f(x)|\,dx \le 2^n\lambda.$$

Define $\Omega = \bigcup_j Q_j$ and write $f = g + b$, where

$$g(x) = \begin{cases} f(x), & x \in \mathbb{R}^n \setminus \Omega \\ f_{Q_j}, & x \in Q_j, \end{cases}$$

Consider also $h = \sum_j h_j$ with $h_j(x) = (f(x) - f_{Q_j})\chi_{Q_j}(x)$ and define, for each j,

$$\widetilde{Q_j} = 3Q_j, \quad \widetilde{\Omega} = \cup_j \widetilde{Q_j} \quad \text{and} \quad w_j(x) = w(x)\chi_{\mathbb{R}^n \setminus 3Q_j}.$$

Therefore,

$$\begin{aligned} w(\{x \in \mathbb{R}^n : |[b,T]f(x)| > \lambda\}) &\le w(\{x \in \mathbb{R}^n \setminus \widetilde{\Omega} : |[b,T]g(x)| > \lambda/2\}) \\ &\quad + w(\widetilde{\Omega}) \\ &\quad + w(\{x \in \mathbb{R}^n \setminus \widetilde{\Omega} : |[b,T]h(x)| > \lambda/2\}) \\ &= I + II + III. \end{aligned}$$

Now define $\tilde{w}(x) := w(x)\chi_{\mathbb{R}^n \setminus \widetilde{\Omega}}$. We can deal with the first term by using Chebyshev and the strong result, obtaining that

$$I \le \frac{c}{\lambda}(pp')^{2p}(r')^{2p-1}\left(\int_{\mathbb{R}^n \setminus \Omega} |f(x)|\,M_r\tilde{w}(x)\,dx + \int_\Omega |g(x)|\,M_r\tilde{w}(x)\,dx\right).$$

Now, by definition of $\tilde{w}$, away from each Q_j the maximal function $M_r\tilde{w}$ is almost constant, and hence we obtain that

$$I \le \frac{c}{\lambda}(pp')^{2p}(r')^{2p-1}\int_{\mathbb{R}^n} |f(x)|\,M_r w(x)\,dx.$$

The second term is the easy one. It is straightforward to show that

$$II = w(\tilde{\Omega}) \leq \frac{c}{\lambda}\int_{\mathbb{R}^n} |f(x)| Mw(x)\, dx.$$

Finally, for the third term, we expand the commutator:

$$\begin{aligned} III &\leq w\left(\left\{y \in \mathbb{R}^n \setminus \tilde{\Omega} : |\sum_j (b(y) - b_{Q_j}) Th_j(y)| > \frac{\lambda}{4}\right\}\right) \\ &\quad + w\left(\left\{y \in \mathbb{R}^n \setminus \tilde{\Omega} : |\sum_j T((b - b_{Q_j})h_j)(y)| > \frac{\lambda}{4}\right\}\right) \\ &= A + B. \end{aligned}$$

The A term can be bounded by using the properties of the kernel, cancellation of the h_j's and standard BMO estimates. We obtain that

$$\begin{aligned} A &\leq \frac{c}{\lambda}\int_{\mathbb{R}^n} |f(x)|\, M_{L(\log L)}w(x)dx \\ &\leq cr' \int_{\mathbb{R}^n} \frac{|f(x)|}{\lambda} M_r w(x)dx. \end{aligned}$$

For the second inequality, we have used that $M_{L(\log L)}w(x) \leq cr' M_r w(x)$. Now, for the B term, we can use the known results about weak type boundedness of T. It can be proved then (see [18] for details) that

$$B \leq c\,(pp')^p (r')^{p-1} \int_{\mathbb{R}^n} \Phi\left(\frac{|f(x)|}{\lambda}\right) M_r w(x)dx.$$

for the function $\Phi(t) = t(1 + \log^+ t)$. Combining all estimates together, we obtain the desired result:

$$w(x \in \mathbb{R}^n : |[b, T]f(x)| > \lambda) \leq c\,(pp')^{2p} (r')^{2p-1} \int_{\mathbb{R}^n} \Phi\left(\frac{|f(x)|}{\lambda}\right) M_r w(x)dx.$$

As in the strong case, we refer the reader to [19] for the details about the analogous result for iterated commutators.

Now we want to optimize inequality (2.2) by choosing p and r. In the same way as in the strong case, we choose $r = r_w = 1 + \frac{1}{c_n[w]_{A_\infty}}$. This is the best that we can do to control the part with r'. In addition, take $p = 1 + \frac{1}{\log([w]_{A_\infty})}$. We finish using that $pp' \approx \log([w]_{A_\infty})$ and that $A^{1/A}$ is bounded for $A > 1$. Define $\beta = [w]_{A_1}[w]_{A_\infty}^k (1+\log^+[w]_{A_\infty})^{k+1}$ and recall that $\Phi(t) = t(1+\log^+ t)^k$. Inequality (2.2) then becomes

$$w(\{x \in \mathbb{R}^n : |T_b^k f(x)| > \lambda\}) \leq c_n\, \beta \int_{\mathbb{R}^n} \Phi\left(\|b\|_{BMO}\frac{|f(x)|}{\lambda}\right) w(x)dx.$$

Acknowledgement

The second author is supported by the Spanish Ministry of Science and Innovation grant MTM2009-08934, the second and third authors are also supported by the Junta de Andalucía, grant FQM-4745.

References

[1] J. Alvarez and C. Pérez, *Estimates with A_∞ weights for various singular integral operators*, Bollettino U.M.I. **7** 8-A (1994), 123–133.

[2] R.J. Bagby and D.S. Kurtz, *Covering lemmas and the sharp function,* Proc. Amer. Math. Soc. **93** (1985), 291–296.

[3] R.J. Bagby and D.S. Kurtz, *A rearranged good-λ inequality,* Trans. Amer. Math. Soc. **293** (1986), 71–81.

[4] S.M. Buckley, *Estimates for operator norms on weighted spaces and reverse Jensen inequalities*, Trans. Amer. Math. Soc. **340** (1993), no. 1, 253–272.

[5] R.R. Coifman, *Distribution function inequalities for singular integrals*, Proc. Nat. Acad. Sci. U.S.A. **69** (1972), 2838–2839.

[6] R.R. Coifman and C. Fefferman, *Weighted norm inequalities for maximal functions and singular integrals*, Studia Math. **51** (1974), 241–250.

[7] D. Cruz-Uribe, SFO, J.M. Martell and C. Pérez, *Weights, Extrapolation and the Theory of Rubio de Francia*, Operator Theory: Advances and Applications **215**, Birkhäuser/Springer Basel AG, Basel, 2011.

[8] J. García-Cuerva and J.L. Rubio de Francia, *Weighted Norm Inequalities and Related Topics*, North-Holland Math. Studies **116**, North-Holland, Amsterdam, 1985.

[9] S. Hruščev, *A description of weights satisfying the A_∞ condition of Muckenhoupt, Proc. Amer. Math. Soc.* **90**(2), 253–257, (1984).

[10] T. Hytönen and C. Pérez, *Sharp weighted bounds involving A_∞*, Journal of Analysis and Partial Differential Equations, (2011), (to appear).

[11] G.A., Karagulyan, *Exponential estimates for the Calderón-Zygmund operator and related problems of Fourier series*, Mat. Zametki 3 **71**, 398–41, (2002).

[12] M. Lacey, *An A_p–A_∞ inequality for the Hilbert Transform,* preprint. Available at http://arxiv.org/abs/1104.2199

[13] A.K. Lerner, *Weighted rearrangements inequalities for local sharp maximal functions,* Trans. Amer. Math. Soc. (2004), **357**, (6), 2445–2465.

[14] A.K. Lerner, *Sharp weighted norm inequalities for Littlewood-Paley operators and singular integrals*, Advances in Mathematics (2011), **226**, (5), 3912–3926.

[15] A.K. Lerner, S. Ombrosi and C. Pérez, *Sharp A_1 bounds for Calderón-Zygmund operators and the relationship with a problem of Muckenhoupt and Wheeden,* International Mathematics Research Notices, 2008, **no. 6**, Art. ID rnm161, 11 pp. 42B20.

[16] A. Lerner, S. Ombrosi and C. Pérez, *A_1 bounds for Calderón-Zygmund operators related to a problem of Muckenhoupt and Wheeden,* Mathematical Research Letters (2009), **16**, 149–156.

[17] A. Lerner, S. Ombrosi, C. Pérez, R. Torres and R. Trujillo-González, *New maximal functions and multiple weights for the multilinear Calderón-Zygmund theory,* Advances in Mathematics (2009), **220**, 1222–1264.

[18] C. Ortiz-Caraballo, *Quadratic* A_1 *bounds for commutators of singular integrals with BMO functions*, to appear in Indiana Univ. Math. J. (2012).

[19] C. Ortiz-Caraballo, *Conmutadores de integrales singulares y pesos* A_1, Ph. D. Dissertation, (2011), Universidad de Sevilla.

[20] C. Ortiz-Caraballo, C. Pérez and E. Rela, *Local subexponential estimates for classical operators*, Preprint 2011.

[21] C. Pérez, *A course on Singular Integrals and weights*, to appear in Advanced Courses in Mathematics, CRM Barcelona, Birkhäuser editors.

[22] C. Pérez, *Weighted norm inequalities for singular integral operators*, J. London Math. Soc., **49** (1994), 296–308.

[23] C. Pérez, *Endpoint Estimates for Commutators of Singular Integral Operators*, Journal of Functional Analysis, **(1) 128** (1995), 163–185.

[24] C. Pérez, *Sharp estimates for commutators of singular integrals via iterations of the Hardy-Littlewood maximal function*, J. Fourier Anal. Appl. **3**, 743–756, (1997).

[25] C. Pérez and R. Trujillo-González, *Sharp weighted estimates for vector-valued singular integrals operators and commutators*, Tohoku Math. J. **55**, 109–129, (2003).

[26] J.M. Wilson, *Weighted inequalities for the dyadic square function without dyadic* A_∞, Duke Math. J., **55**(1), 19–50, (1987).

[27] J.M. Wilson, *Weighted inequalities for the continuous square function*, Trans. Amer. Math. Soc., **314**(2), 661–692, (1989).

[28] J.M. Wilson, *Weighted Littlewood-Paley theory and exponential-square integrability*, volume 1924 of *Lecture Notes in Mathematics.* Springer, Berlin, (2008).

Carmen Ortiz-Caraballo
Departamento de Matemáticas, Escuela Politécnica
Universidad de Extremadura
Avda. Universidad, s/n
E-10003 Cáceres, Spain
e-mail: `carortiz@unex.es`

Carlos Pérez and Ezequiel Rela
Departamento De Análisis Matemático
Facultad de Matemáticas
Universidad De Sevilla
E-41080 Sevilla, Spain
e-mail: `carlosperez@us.es`
`erela@us.es`

Operator Theory:
Advances and Applications, Vol. 229, 323–347

Potential Type Operators on Weighted Variable Exponent Lebesgue Spaces

Vladimir Rabinovich

Dedicated to my friend and colleague Stefan Samko on the occasion of his 70th birthday

Abstract. We consider double-layer potential type operators acting in weighted variable exponent Lebesgue space $L^{p(\cdot)}(\Gamma, w)$ on some composed curves with oscillating singularities. We obtain a Fredholm criterion for operators $A = aI + bD_{g,\Gamma} : L^{p(\cdot)}(\Gamma, w) \to L^{p(\cdot)}(\Gamma, w)$ where $D_{g,\Gamma}$ is the operator of the form

$$D_{g,\Gamma}u(t) = \frac{1}{\pi}\int_{\Gamma} \frac{g(t,\tau)\,(\nu(\tau), \tau - t)\,u(\tau)dl_\tau}{|t-\tau|^2}, t \in \Gamma,$$

$\nu(\tau)$ is the inward unit normal vector to Γ at the point $\tau \in \Gamma\backslash\mathcal{F}$, dl_τ is the oriented Lebesgue measure on $\Gamma, \mathcal{F}$ is the set of the nodes, $a, b : \Gamma \to \mathbb{C}, g : \Gamma \times \Gamma \to \mathbb{C}$ are a bounded functions with oscillating discontinuities at the nodes only.

We give applications of such operators to the Dirichlet and Neumann problems with boundary function in $L^{p(\cdot)}(\Gamma, w)$ for domains with boundaries having a finite set of oscillating singularities.

Mathematics Subject Classification (2010). 31A10, 31A25.

1. Introduction

We consider the operators of double-layer potentials acting in weighted variable exponent Lebesgue spaces $L^{p(\cdot)}(\Gamma, w)$ on some composed curves Γ. The simple curves $\Gamma_{t_0,j}$, $j = 1, \dots, n(t_0)$ which form a node t_0 are slowly oscillating near the point t_0. It means that the curves $\Gamma_{t_0,j}$ have the parametrization near t_0 of the form

$$\Gamma_{t_0,j} = \left\{t = t_0 + re^{i\varphi_{t_0,j}(r)}, r \in (0,\varepsilon)\right\},$$

The work is partially supported by the Mexico CONACYT Project 81615.

where $\varphi_{t_0,j} \in C^\infty((0,\varepsilon))$,

$$\sup_{r\in(0,\varepsilon)} \left| (r\frac{d}{dr})^\alpha \varphi_{t_0,j}(r) \right| < \infty \tag{1}$$

for every $\alpha = 0, 1, 2, \ldots$, and

$$\lim_{r\to 0} r\frac{d\varphi_{t_0,j}(r)}{dr} = 0.$$

A typical example of $\Gamma_{t_0,j}$ is the curve with

$$\varphi_{t_0,j}(r) = \theta_j + \gamma_j \sin(-\log r)^{\alpha_j}, 0 < \alpha_j < 1, 0 < \gamma_j < \theta_j < 2\pi.$$

One can see that the slope of the tangent to the curve $\Gamma_{t_0,j}$ near the point t_0 is an oscillating function which does not have a limit at the point t_0.

Remark 1. Note that the above-introduced composed curves belong to the class of composed Carleson curves (see for instance [1], p. 2, and Chap. 4.5).

Let $D_{g,\Gamma}$ be the double-layer type potential operator of the form

$$D_{g,\Gamma}u(t) = \frac{1}{\pi}\int_\Gamma \frac{g(t,\tau)\,(\nu(\tau), \tau - t)\,u(\tau)dl_\tau}{|t-\tau|^2}, t \in \Gamma,$$

where $\nu(\tau)$ is the inward unit normal vector to Γ at the point $\tau \in \Gamma \backslash \mathcal{F}$, $\mathcal{F}$ is a finite set of nodes of Γ, dl_τ is the oriented Lebesgue measure on Γ, $g : \Gamma \times \Gamma \to \mathbb{C}$ is a bounded function with discontinuities at the nodes. If Γ is a Radon curve without peaks and a and b are piece-wise continuous functions the operators $A_\Gamma = aI + bD_{g,\Gamma}$ have been studied by many authors (see the classic Radon paper [29], papers [17], [34], the book [6], and the survey [18], and references cited there). Note also the recent monograph [19] where the boundary integral equations on contours with peaks have been considered.

We study here the Fredholm properties of the operators $A = aI + bD_{g,\Gamma}$ acting on the spaces $L^{p(\cdot)}(\Gamma, w)$ with coefficients a, b having slowly oscillating discontinuities at the nodes of Γ. Applying the Simonenko local principle (see the original works [31], [32], [33], and the recent monograph [30]) we reduce the investigation of Fredholmness of the operator A to the investigation of the local invertibility of A at every point of the curve Γ. The operator A is realized at the points $\Gamma \backslash \mathcal{F}$ as a usual pseudodifferential operator, and at the nodes as a Mellin pseudodifferential operator. This approach arises to the author papers [20], [23], [22], [21] (see also [5], [2], [3], [4]) devoted to singular integral operators acting in the spaces $L^p(\Gamma, w)$ on composed Carleson curves. In the paper [24] the potential type operators were considered as acting on $L^p(\Gamma, w)$, $p \in (1, \infty)$, where Γ is a closed simple oscillating curve with a finite set of nodes. We extend here the results of [24] on the potential type operators acting on the *variable exponent Lebesgue spaces* $L^{p(\cdot)}(\Gamma, w)$ on some *composed* curves Γ.

In the paper [28] we studied the Fredholm property of singular integral operators acting on the weighted variable exponent Lebesgue spaces $L^{p(\cdot)}(\Gamma, w)$ on some composed Carleson curves. This investigation was based on the works [27], [26] devoted to the problems of local invertibility and Fredholmness of pseudodifferential operators on $\mathbb{R}^n$ acting in weighted variable exponent Lebesgue spaces.

Note that in the last time there is a big interest to investigations of main operators of Analysis, i.e., singular and maximal operators, Hardy operators, pseudodifferential operators in the $L^{p(\cdot)}$-spaces with variable exponent $p(\cdot)$, see for instance [12], [7], [13], [15], [10], [11], [14], [26], [27], [28], and references cited there.

The paper is organized as follows. In Section 2 we give an auxiliary material concerning of pseudodifferential operators on $\mathbb{R}$ and Mellin pseudodifferential operators on $\mathbb{R}_+$ acting on the Lebesgue spaces with variable exponents. We follow the papers [27], [28]. Section 3 is devoted to definitions of composed slowly oscillating curves, weighted variable exponent spaces $L^{p(\cdot)}(\Gamma, w)$ on such curves, and the Simonenko local principle for operators acting on $L^{p(\cdot)}(\Gamma, w)$.

In Section 4 we consider double-layer potential type operators with coefficients having slowly oscillating discontinuity acting on the spaces $L^{p(\cdot)}(\Gamma, w)$ on composed slowly oscillating curves Γ. We obtain a criterion of Fredholmness that takes into account the oscillations of weights and the curve and values $p(t)$ at nodes $t \in \mathcal{F}$.

More detail we investigate the local spectrum and the Fredholm index of the Dirichlet problem in a bounded domain Ω with boundary Γ having above-described singular points (nodes) and with a boundary function $f \in L^{p(\cdot)}(\Gamma, w)$. We obtain the exact estimates for values of $p(t)$ at the nodes for which the problem is Fredholm and calculate its Fredholm index. We also consider here the integral operator of the Neumann problem.

Remark 2. There are deep results concerning the Dirichlet and Neumann problems in domains Ω with Lipschitz boundary (see for instance the monograph [9] and references given there). The Harmonic Analyses technique allows one to obtain the condition of solvability of the Dirichlet and Neumann problems for boundary functions in $L^p(\partial\Omega)$, where the exponent p for which the problem is solvable globally depends on the Lipschitz constants of the boundary $\partial\Omega$. The applications of the local technique and variable exponent Lebesgue spaces allow us to obtain the exact estimates of $p(t)$ for which the problems are solvable locally at an every node $t \in \mathcal{F}$. These estimates depend on a local behavior of a curve and weight at the node t only.

2. Pseudodifferential operators

2.1. Pseudodifferential operators on $\mathbb{R}$

We will use the following notations:

- for a Banach spaces X, the notation $\mathcal{B}(X)$ stands for the space of all bounded linear operators acting in X, $\mathcal{K}(X)$ is the space of all compact operators acting in X.

Let $\mathbb{M}$ be a smooth manifold. Then we denote by:

- $C^\infty(\mathbb{M})$ the linear space of infinitely differentiable complex-valued functions on $\mathbb{M}$,
- $C_0^\infty(\mathbb{M})$ the subspace of $C^\infty(\mathbb{M})$ of complex-valued functions with compact supports,
- $C_b^\infty(\mathbb{M})$ the subspace of $C^\infty(\mathbb{M})$ of complex-valued functions bounded with all derivatives.

Definition 3. We say that a function $a \in C^\infty(\mathbb{R}\times\mathbb{R})$ is a symbol of the class $S_{1,0}^m$ if

$$\sum_{\alpha\leq l_1,\beta\leq l_2}\sup_{(x,\xi)\in\mathbb{R}^2}\left|\partial_x^\beta\partial_\xi^\alpha a(x,\xi)\right|\langle\xi\rangle^{-m+\alpha}<\infty,\ \langle\xi\rangle=\left(1+\xi^2\right)^{1/2}$$

for every $l_1, l_2 \in \mathbb{N}_0 = \{0\}\cup\mathbb{N}$; we correspond to a symbol a the pseudodifferential operator

$$Op(a)u(x)=\frac{1}{2\pi}\int_{\mathbb{R}}d\xi\int_{\mathbb{R}}a(x,\xi)u(y)e^{i(x-y)\xi}dy,\ u\in C_0^\infty(\mathbb{R})$$

and denote the class of pseudodifferential operators with symbols in $S_{1,0}^m$ by $OPS_{1,0}^m(\mathbb{R})$.

Let p be a measurable function on $\mathbb{R}$ such that $p:\mathbb{R}\to(1,\infty)$ and let

$$I^p(f):=\int_{\mathbb{R}}|f(x)|^{p(x)}\,dx$$

be the modular of f. We denote by $L^{p(\cdot)}(\mathbb{R})$ the space of measurable functions on $\mathbb{R}$ such that

$$\|f\|_{L^{p(\cdot)}(\mathbb{R})}=\inf\left\{\lambda>0: I^p\left(\frac{f}{\lambda}\right)\leq 1\right\}<\infty.$$

We assume that p satisfies conditions:

$$1<p_-\leq p(x)\leq p_+<\infty, x\in\mathbb{R}, \tag{2}$$

$$|p(x)-p(y)|\leq\frac{C}{\log\frac{1}{|x-y|}},\qquad x,y\in\mathbb{R},\qquad |x-y|\leq\frac{1}{2}, \tag{3}$$

there exists $\lim_{x\to\infty}p(x)=p(\infty)$ and

$$|p(x)-p(\infty)|\leq\frac{C}{\log(2+|x|)},\quad C>0,\quad x\in\mathbb{R}. \tag{4}$$

The class of $p(\cdot)$ satisfying (2), (3) and (4) is denoted by $\mathcal{P}(\mathbb{R})$.

We denote by aI the operator of multiplication by a function a. Note that under condition (2) we have

$$\|aI\|_{\mathcal{B}(L^{p(\cdot)}(\mathbb{R}))}\leq\|a\|_{L^\infty(\mathbb{R})} \tag{5}$$

if $a\in L^\infty(\mathbb{R})$.

Below we need in the following results from [27] on boundedness and compactness of pseudodifferential operators.

Theorem 4. *Let $p(\cdot) \in \mathcal{P}(\mathbb{R})$. Then an operator $Op(a) \in OPS^0_{1,0}(\mathbb{R})$ is bounded in $L^{p(\cdot)}(\mathbb{R})$.*

Proposition 5. *Let $p(\cdot) \in \mathcal{P}(\mathbb{R})$, $Op(a) \in OPS^{-\varepsilon}_{1,0}(\mathbb{R}), \varepsilon > 0$ and*

$$\lim_{x \to \infty} \sup_{\xi \in \mathbb{R}} |a(x, \xi)| = 0.$$

Then $Op(a) \in \mathcal{K}(L^{p(\cdot)}(\mathbb{R}))$.

Proposition 5 implies that an operator $Op(a) \in OPS^{-\varepsilon}_{1,0}(\mathbb{R}), \varepsilon > 0$ is locally compact in $L^{p(\cdot)}(\mathbb{R})$, that is if $U \subset \mathbb{R}$ is an open set with a compact closure, $\varphi, \psi \in C_0^\infty(U)$ then the operator $\varphi Op(a)\psi I \in \mathcal{K}(L^{p(\cdot)}(\mathbb{R}))$.

Definition 6. We say that $A \in \mathcal{B}(L^{p(\cdot)}(\mathbb{R}))$ is locally invertible at the point $x_0 \in \mathbb{R}$, if there exists an interval $\mathcal{I}_\varepsilon(x_0) = (x_0 - \varepsilon, x_0 + \varepsilon\ )$ and operators $\mathcal{L}_{x_0,\varepsilon}, \mathcal{R}_{x_0,\varepsilon} \in \mathcal{B}(L^{p(\cdot)}(\mathbb{R}))$ such that

$$\mathcal{L}_{x_0,\varepsilon} A \chi_{\mathcal{I}_\varepsilon(x_0)} I = \chi_{\mathcal{I}_\varepsilon(x_0)} I, \chi_{\mathcal{I}_\varepsilon(x_0)} A \mathcal{R}_{x_0,\varepsilon} = \chi_{\mathcal{I}_\varepsilon(x_0)} I.$$

Theorem 7 ([28, Theorem 23]). *Let $Op(a) \in OPS^0_{1,0}(\mathbb{R}), p(\cdot) \in \mathcal{P}(\mathbb{R})$ and there exist uniform with respect to $x \in \mathbb{R}$ limits*

$$\lim_{\xi \to \pm\infty} a(x, \xi) = a_\pm(x).$$

Then $Op(a) : L^{p(\cdot)}(\mathbb{R}) \to L^{p(\cdot)}(\mathbb{R})$ is locally invertible at a point $x_0 \in \mathbb{R}$ if and only if $a_\pm(x_0) \neq 0$.

2.2. Mellin pseudodifferential operators

We give some facts on the calculus of the Mellin pseudodifferential operators following [22], [28].

We say that a matrix-function $\mathbb{R}_+ \times \mathbb{R} \ni (r, \xi) \to (a_{ij}(r, \xi))_{i,j=1}^n$ belongs to $\mathcal{E}^m(n)$ if $a_{ij} \in C^\infty(\mathbb{R}_+ \times \mathbb{R})$ and

$$\sum_{\alpha \le l_1, \beta \le l_2} \sup_{(r,\xi) \in \mathbb{R}_+ \times \mathbb{R}} \left|(r\partial_r)^\beta \partial_\xi^\alpha a_{ij}(r, \xi)\right| \langle \xi \rangle^{-m+\alpha} < \infty$$

for all $l_1, l_2 \in \mathbb{N}_0$.

Let $a \in \mathcal{E}^m(n)$. The operator

$$Op_M(a)u(r) = (2\pi)^{-1} \int_{\mathbb{R}} d\xi \int_{\mathbb{R}_+} a(r, \xi) \left(r\rho^{-1}\right)^{i\xi} u(\rho) \frac{d\rho}{\rho}, \tag{6}$$

where $u \in C_0^\infty(\mathbb{R}_+, \mathbb{C}^n) = C_0^\infty(\mathbb{R}_+) \otimes \mathbb{C}^n$ is called the Mellin pseudodifferential operator with symbol $a \in \mathcal{E}^m(n)$. The class of all such operators is denoted by $OP\mathcal{E}^m(n)$.

We say that a matrix-function $a = (a_{ij})_{i,j=1}^n$ $(\in \mathcal{E}^m(n))$ is slowly oscillating at the point 0 and belongs to $\mathcal{E}^m_{sl}(n)$, if for all $i, j = 1, \dots, n$

$$\lim_{r \to +0} \sup_{\xi \in \mathbb{R}} |(r\partial_r)^\beta \partial_\xi^\alpha a_{ij}(r, \xi)| \langle \xi \rangle^{-m+\alpha} = 0 \tag{7}$$

for all $\alpha \in \mathbb{N}_0$ and $\beta \in \mathbb{N}$. By $\mathcal{E}_0^m(n)$ we denote the set of matrix-functions satisfying condition (7) for all $\alpha, \beta \in \mathbb{N}_0$.

We denote by $\mathcal{E}_d^m(n)$ the class of matrix-functions $\mathbb{R}_+ \times \mathbb{R}_+ \times \mathbb{R} \ni (r, \rho, \xi) \to a(r, \rho, \xi) = (a_{i,j}(r, \rho, \xi))_{(i,j)=1}^n$ such that for all $i, j \in \{1, 2, \dots, n\}$

$$\sup_{(r,\rho,\xi)\in\mathbb{R}_+\times\mathbb{R}_+\times\mathbb{R}} \sum_{\alpha\le l_1,\beta\le l_2,\gamma\le l_3} \left|(r\partial_r)^\beta(\rho\partial_\rho)^\gamma\partial_\xi^\alpha a_{ij}(r,\rho,\xi)\right| \langle\xi\rangle^{-m+\alpha} < \infty$$

for all $l_1, l_2, l_3 \in \mathbb{N}_0$. An operator defined by (6) with $a(r, \xi)$ replaced by $a(r, \rho, \xi)$ is called the Mellin pseudodifferential operator $Op_{M,d}(a)$ with the double symbol a and $OP\mathcal{E}_d^m(n)$ denotes the class of such operators. We will say that a double symbol $a \in \mathcal{E}_d^m(n)$ is slowly oscillating at the point 0 if

$$\lim_{r\to 0} \sup_{\rho\in\mathbb{R}_+,\xi\in\mathbb{R}_+} |(r\partial_r)^\beta(\rho\partial_\rho)^\gamma\partial_\xi^\alpha a_{ij}(r,\rho,\xi)|\langle\xi\rangle^{-m+\alpha} = 0,$$

for all $\beta \in \mathbb{N}$; $\alpha, \gamma \in \mathbb{N}_0$, $i, j = 1, \dots, n$, and

$$\lim_{\rho\to 0} \sup_{r\in\mathbb{R}_+,\xi\in\mathbb{R}} |(r\partial_r)^\beta(\rho\partial_\rho)^\gamma\partial_\lambda^\alpha a_{ij}(r,\rho,\xi)|\langle\xi\rangle^{-m+\alpha} = 0$$

for all $\gamma \in \mathbb{N}; \alpha, \beta \in \mathbb{N}_0$. We denote by $\mathcal{E}_{sl,d}^m(n)$ the class of slowly oscillating double symbols and by $OP\mathcal{E}_{sl,d}^m(n)$ the corresponding class of operators. For $m = 0$ we use the notations $\mathcal{E}(n), \mathcal{E}_{sl}(n), \mathcal{E}_{sl,d}(n)$ and corresponding notations for the operators.

Proposition 8. *Let $a \in \mathcal{E}_{sl,d}^m(n)$. Then $Op_{M,d}(a) = Op_M(a^\#) \in OP\mathcal{E}_{sl}^m(n)$ where*

$$a^\#(r,\xi) = a(r,r,\xi) + q(r,\xi),$$

with $q \in \mathcal{E}_0^{m-1}(n)$.

Let p be a measurable function on $\mathbb{R}_+$ such that $p : \mathbb{R}_+ \to (1, \infty)$. We assume that p satisfies the conditions

$$1 < p_- \le p(r) \le p_+ < \infty, \tag{8}$$

$$|p(r) - p(\rho)| \le \frac{A}{\log \frac{1}{\left|\log \frac{r}{\rho}\right|}}, \qquad r, \rho \in \mathbb{R}_+, \qquad \frac{1}{\sqrt{e}} \le \frac{r}{\rho} \le \sqrt{e}, \tag{9}$$

and

$$|p(r) - p(0)| \le \frac{A}{\log\left(2 + |\log r|\right)}, \qquad r \in \mathbb{R}_+, \tag{10}$$

$$|p(r) - p(+\infty)| \le \frac{A}{\log\left(2 + |\log r|\right)}, \qquad r \in \mathbb{R}_+, \tag{11}$$

where $p(0) = \lim_{r\to+0} p(r), p(+\infty) = \lim_{r\to+\infty} p(r)$, and $p(0) = p(+\infty)$. The class of functions satisfying conditions (8), (9), (10), (11) is denoted by $\mathcal{P}(\mathbb{R}_+)$.

The variable exponent Lebesgue space $L_n^{p(\cdot)}(\mathbb{R}_+, \frac{dr}{r})$ with $p(\cdot) \in \mathcal{P}(\mathbb{R}_+)$ is defined as the space of measurable vector-functions on $\mathbb{R}_+$ with a finite modular

$$I_n^p(f) := \int_{\mathbb{R}_+} \|f(r)\|_{\mathbb{C}^n}^{p(r)} \frac{dr}{r}.$$

The norm in $L_n^{p(\cdot)}(\mathbb{R}_+, \frac{dr}{r})$ is introduced as

$$\|f\|_{L_n^{p(\cdot)}(\mathbb{R}_+,d\mu)} = \inf\left\{\gamma > 0 : I_n^p\left(\frac{f}{\gamma}\right) \leq 1\right\}.$$

Note that conditions (8), (9), (10), (11) follow from conditions (2), (3), (4) since the mapping $\mathbb{R} \ni x \to \exp x \in \mathbb{R}_+$ generates the isomorphism of spaces $L^{p(\cdot)}(\mathbb{R}_+, \frac{dr}{r})$ and $L^{\tilde{p}(\cdot)}(\mathbb{R})$ where $p(r) = \tilde{p}(\log r)$.

Theorem 9 ([28, Proposition 27]). *Let $a \in \mathcal{E}(n), p(\cdot) \in \mathcal{P}(\mathbb{R}_+)$. Then $Op_M(a)$ is bounded in $L_n^{p(\cdot)}(\mathbb{R}_+, \frac{dr}{r})$.*

Definition 10. We say that $A \in \mathcal{B}(L_n^{p(\cdot)}(\mathbb{R}_+, \frac{dr}{r}))$ is locally invertible at the point 0 if there exist $R > 0$ and operators $\mathcal{L}_R, \mathcal{R}_R \in \mathcal{B}(L_n^{p(\cdot)}(\mathbb{R}_+, \frac{dr}{r}))$ such that

$$\mathcal{L}_R A \chi_{[0,R]} I = \chi_{[0,R]} I, \chi_{[0,R]} A \mathcal{R}_R = \chi_{[0,R]} I$$

where $\chi_{[0,R]}$ is the characteristic function of $[0, R]$.

Theorem 11 ([28, Theorem 34]). *Let $Op_M(a) \in OP\mathcal{E}_{sl}(n)$. Then*

$$Op_M(a) : L_n^{p(\cdot)}(\mathbb{R}_+, \frac{dr}{r}) \to L_n^{p(\cdot)}(\mathbb{R}_+, \frac{dr}{r})$$

is locally invertible at the point 0 if and only if

$$\liminf_{r \to +0} \inf_{\xi \in \mathbb{R}} |\det a(r, \xi)| > 0. \tag{12}$$

- We say that $\lambda \in \mathbb{C}$ is a point of the local spectrum of $A \in \mathcal{B}(L_n^{p(\cdot)}(\mathbb{R}_+, \frac{dr}{r}))$ at the point 0 if the operator $A - \lambda I$ is not locally invertible at the point 0. We denote the local spectrum of A at the point 0 as $sp_0 A$.
- Let $a \in \mathcal{E}_{sl}(n)$. We say that $a \in \mathcal{E}'_{sl}(n)$ if there exist $\varepsilon > 0$ small enough such that there exist uniform with respect to $r \in (0, \varepsilon)$ limits

$$\lim_{\xi \to \pm\infty} a(r, \xi) = a_\pm(r).$$

Let $a \in \mathcal{E}'_{sl}(n), \mathbb{R}_+ \ni h_m \to 0$. We consider the functional sequence $a(h_m, \xi)$. It follows from Arzelà-Ascoli Theorem that there exists a subsequence g_m of h_m such that uniformly for $\xi \in [-\infty, +\infty]$ there exists a limit

$$\lim_{m \to \infty} a(g_m, \xi) = a^g(\xi). \tag{13}$$

The following proposition follows from Theorem 11.

Proposition 12. *Let $a \in \mathcal{E}'_{sl}(n)$. Then*

$$sp_0 Op_M(a) = \bigcup_{g_m} \bigcup_{\xi \in [-\infty, +\infty]} sp(a^g(\xi)) \tag{14}$$

where the union $\cup_{g_m}$ is taken with respect to all sequences for which limits (13) *exist, and $sp(a^g(\xi))$ is the spectrum of the $n \times n$ matrix $a^g(\xi)$.*

3. Operators on slowly oscillating curves

3.1. Curves, weights, coefficients

We say that a complex-valued function $a \in \mathcal{C}^{\infty}((0,\varepsilon)), \varepsilon > 0$ if for every $j \in \mathbb{N}_0$

$$\sup_{r\in(0,\varepsilon)} \left| \left(r\frac{d}{dr} \right)^j a(r) \right| < \infty,$$

and we say that $a \in \tilde{\mathcal{C}}^{\infty}((0,\varepsilon))$ if

$$\varkappa_a(r) = r\frac{da}{dr} \in \mathcal{C}^{\infty}((0,\varepsilon)).$$

A function $a \in \mathcal{C}^{\infty}((0,\varepsilon))$ is said to be *slowly oscillating* at the point 0 and belongs to $\mathcal{C}^{\infty}_{sl}((0,\varepsilon))$ if

$$\lim_{r\to 0} \varkappa_a(r) = \lim_{r\to+0} r\frac{da}{dr} = 0.$$

We denote by $\tilde{\mathcal{C}}^{\infty}_{sl}((0,\varepsilon))$ the class of functions a such that $\varkappa_a \in \mathcal{C}^{\infty}_{sl}((0,\varepsilon))$.

A set $\gamma \subset \mathbb{C}$ is called a *smooth arc* if there exists an homeomorphism $\varphi : [0,1] \to \gamma$ such that $\varphi \in C^{\infty}((0,1))$, and $\varphi'(r) \neq 0$ for all $r \in (0,1)$. The points $\varphi(0)$ and $\varphi(1)$ are called the *endpoints* of γ. We refer to a set $\Gamma \subset \mathbb{C}$ as a *composed curve* if $\Gamma = \cup_{k=1}^{K} \Gamma_k$ and $\Gamma_1, \dots, \Gamma_K$ are oriented and rectifiable smooth arcs, each pair of which has at most endpoints in common. A *node* of Γ is a point which is endpoint of at least one of the arcs $\Gamma_1, \dots, \Gamma_K$. The set of all the nodes is denoted by $\mathcal{F}$.

Let $t_0 \in \mathcal{F}$. We suppose that there exists an $\varepsilon > 0$ such that the portion

$$\Gamma(t_0,\varepsilon) = \{t \in \Gamma : |t_0 - t| < \varepsilon\}$$

is of the form

$$\Gamma(t_0,\varepsilon) = \{t_0\} \cup \Gamma^1_{t_0,\varepsilon} \cup \dots \cup \Gamma^{n(t_0)}_{t_0,\varepsilon} \tag{15}$$

where

$$\Gamma^j_{t_0,\varepsilon} = \left\{ z \in \mathbb{C} : z = t_0 + re^{i\varphi_{t_0,j}(r)} : r \in (0,\varepsilon), (j = 1,\dots,n(t_0)) \right\}, \tag{16}$$

$\varphi_{t_0,j}$ are real-valued functions such that $\varphi_{t_0,j} \in \mathcal{C}^{\infty}_{sl}((0,\varepsilon))$ and

$$\begin{aligned} 0 \le m_1 < \varphi_{t_0,1}(r) < M_1 < m_2 < \varphi_{t_0,2}(r) < M_2 \\ \dots < m_{n(t_0)} < \varphi_{t_0,n(t_0)}(r) < M_{n(t_0)} < 2\pi \end{aligned} \tag{17}$$

for all $r \in (0,\varepsilon)$ with certain constants m_j, M_j. We denote by $\mathcal{L}_{sl}$ the class of compact curves satisfying the above-given conditions.

Important subclass of $\mathcal{L}_{sl}$ is the class of piecewise smooth composed curves without cusps.

Let $\Gamma \in \mathcal{L}_{sl}$, $p : \Gamma \to (1,\infty)$ be a measurable function and

$$|p(t) - p(\tau)| \le \frac{C}{\log \frac{1}{|t-\tau|}}, C > 0$$

for all $t, \tau \in \Gamma \backslash \mathcal{F}$ such that $|t-\tau| < 1/2$. For $t_0 \in \mathcal{F}$ we suppose that there exists an $\varepsilon > 0$ such that the functions

$$p_{t_0,j}(r) := p(t_0 + re^{i\varphi_{t_0,j}(r)}) \equiv p_{t_0}(r), r \in [0, \varepsilon) \tag{18}$$

are independent on j, and $p_{t_0} = p_{t_0,j} \in C^\infty([0,\varepsilon))$. We denote such class by $\mathcal{P}(\Gamma)$.

Let $w : \Gamma \to [0, \infty]$ be a measurable function such that $mes\ (w^{-1}\{0,\infty\}) = 0$ referred to us as a weight. The weighted variable exponent Lebesgue space $L^{p(\cdot)}(\Gamma, w)$ is defined as a class of measurable on Γ functions such that

$$I^p\,(w,f) := \int_\Gamma |w(\tau)f(\tau)|^{p(\tau)}\ |d\tau| < \infty$$

with the norm

$$\|f\|_{L^{p(\cdot)}(\Gamma,w)} = \inf\left\{\lambda > 0 : I^p\left(w, \frac{f}{\lambda}\right) \leq 1\right\}.$$

We will write $L^{p(\cdot)}(\Gamma) := L^{p(\cdot)}(\Gamma, w)$ if $w \equiv 1$.

We consider the weights on $(0,\varepsilon)$ of the form $w(r) = \exp v(r)$ where $v(r) \in \tilde{C}_{sl}^\infty\,((0,\varepsilon))$ and real valued and we denote this class of weights by $\mathcal{R}_{sl}\,((0,\varepsilon))$. Let $\varkappa_v(r) = rv'(r)$ and

$$\varkappa_v^+ = \limsup_{r\to 0} \varkappa_v(r) = \limsup_{r\to 0} \frac{rw'(r)}{w(r)}, \tag{19}$$

$$\varkappa_v^- = \liminf_{r\to 0} \varkappa_v(r) = \liminf_{r\to 0} \frac{rw'(r)}{w(r)}. \tag{20}$$

Note that the weight $w(r) = \exp v(r)$ with

$$v(r) = f(\log(-\log r))\log r, \qquad r \in (0,\varepsilon)$$

and $f \in C_b^\infty((0,\varepsilon))$ belongs to $\mathcal{R}_{sl}\,((0,\varepsilon))$. For example, if $f(x) = \sin x$ then

$$\begin{aligned}\varkappa_v(r) &= \cos(\log(-\log r)) + \sin(\log(-\log r))\\ &= \sqrt{2}\cos\left(\log(-\log r) - \frac{\pi}{2}\right),\end{aligned}$$

and $\varkappa_w^+ = \sqrt{2}, \varkappa_w^- = -\sqrt{2}$.

Let $w : \Gamma \to [0, +\infty]$ be a weight on the curve Γ. We suppose that for every point $t_j \in \mathcal{F}$ there exists a neighborhood U_j such that w and w^{-1} belong $L^\infty(\Gamma \backslash \cup_{t_j\in\mathcal{F}} (\Gamma \cap U_j))$. We say that $w \in \mathcal{R}_{sl}(\Gamma)$ if for every point $t_k \in \mathcal{F}$ and for every $j \in \{1,\ldots,n(t_k)\}$ the function

$$w_{t_k}(r) = w(t_k + re^{i\varphi_{t_k,j}(r)}) = e^{v_{t_k}(r)}, \qquad r \in (0,\varepsilon) \tag{21}$$

is independent of j and $w_{t_k} \in \mathcal{R}_{sl}\,((0,\varepsilon))$. By $\mathcal{A}_{p(\cdot)}(\Gamma)$ we denote a class of weights in $\mathcal{R}_{sl}(\Gamma)$ such that

$$-\frac{1}{p(t_k)} < \liminf_{r\to 0} \varkappa_{v_{t_k}}(r) \leq \limsup_{r\to 0} \varkappa_{v_{t_k}}(r) < 1 - \frac{1}{p(t_k)}, \tag{22}$$

for every node $t_k \in \mathcal{F}$.

Proposition 13 ([28, Proposition 36]). *If $w \in \mathcal{A}_{p(\cdot)}(\Gamma)$ then $w \in L^{p(\cdot)}(\Gamma)$ and $w^{-1} \in L^{p'(\cdot)}(\Gamma)$, where $\frac{1}{p(t)} + \frac{1}{p'(t)} = 1$ for all $t \in \Gamma$.*

3.2. Simonenko local principle

Let X be a Hausdorff compact topological space, μ be a σ-additive measure on X, $p(\cdot)$ be a measurable function on X such that

$$1 < p_1 \leq p(x) \leq p_2 < \infty, x \in X.$$

As above we introduce the Banach space $L^{p(\cdot)}(X, d\mu)$ of measurable on X functions f such that

$$I_{p(\cdot)}(f) = \int_X |f(x)|^{p(x)}\, d\mu < \infty$$

with the norm

$$\|f\|_{L^{p(\cdot)}(X,d\sigma)} = \inf\left\{\lambda > 0 : I_{p(\cdot)}(\frac{f}{\lambda}) \leq 1\right\}.$$

Definition 14. An operator $A \in \mathcal{B}(L^{p(\cdot)}(X, d\mu))$ is called the local type operator if for every two closed sets $F_1, F_2 \subset X$ such that $F_1 \cap F_2 = \varnothing$ the operator $\chi_{F_1} A \chi_{F_2} I \in \mathcal{K}(L^{p(\cdot)}(X))$.

Definition 15. An operator $A \in \mathcal{B}(L^{p(\cdot)}(X))$ is called locally Fredholm at a point $x_0 \in X$ if there exists a neighborhood U of the point x_0 and operators $L^{x_0}, R^{x_0} \in \mathcal{B}(L^{p(\cdot)}(X))$ such that

$$L^{x_0} A \chi_U I = \chi_U I + T_1, \chi_U A R^{x_0} = \chi_U I + T_2, \tag{23}$$

where $T_1, T_2 \in \mathcal{K}(L^{p(\cdot)}(X))$. If in (23) T_1 and T_2 are 0-operators A is called the locally invertible operator at the point x_0.

Theorem 16 (Simonenko's local principle, [31], [32], [33], see also [30, Chap 2]). *Let $A \in \mathcal{B}(L^{p(\cdot)}(X, d\mu))$ be an operator of the local type on X. Then A is a Fredholm operator if and only if A is a locally Fredholm operator at every point $x \in X$.*

The proof of Theorem 16 for variable $p(\cdot)$ repeats word for word the Simonenko's proof for a constant p. (See for instance [33], pp. 21–24.)

- Let $A \in \mathcal{B}(E)$ where E is a Banach space. We denote by $sp_{ess}A$ the *essential spectrum* of A, that is the set of $\lambda \in \mathbb{C}$ such that $A - \lambda I$ is not Fredholm operator on E.
- Let $A \in \mathcal{B}(L^{p(\cdot)}(X))$. We denote by $sp_x A$ the *local spectrum* of A at the point $x \in X$, that is the set of points $\lambda \in \mathbb{C}$ such that $A - \lambda I$ is not locally invertible operator at the point x.

The next Theorem is a simple corollary of the Simonenko local principle.

Theorem 17. *Let for every point $x \in X$ there exists a sequence of neighborhoods $\{U_j^x\}_{j=1}^{\infty}$ such that $U_1^x \supset U_2^x \supset \cdots \supset U_m^x \supset \cdots$ and $\lim_{j\to\infty} \mu(U_j^x) = 0$. If $A \in \mathcal{B}(L^{p(\cdot)}(X, d\sigma))$ is the local type operator then*

$$sp_{ess}A = \bigcup_{x \in X} sp_x A.$$

4. Potential type operators in $L^{p(\cdot)}(\Gamma, w)$

4.1. Operators of the double-layer potentials type

Here we consider the operator of the double-layer potential type

$$D_{g,\Gamma}u(t) = \frac{1}{\pi}\int_\Gamma g(t,\tau)\frac{\partial \log(|t-\tau|)}{\partial \nu_\tau}u(\tau)dl_\tau$$
$$= \frac{1}{\pi}\int_\Gamma \frac{g(t,\tau)\,(\nu(\tau),\tau-t)}{|t-\tau|^2}u(\tau)dl_\tau, \quad t\in\Gamma$$

where $\Gamma\subset\mathbb{R}^2$ is an oriented curve, dl_τ is the oriented Lebesgue measure on Γ, $\Gamma\ni\tau\to\nu(\tau)=(\nu_1(\tau),\nu_2(\tau))\in\mathbb{R}^2$ is the inward unit normal vector, $(a,b)=a_1b_1+a_2b_2$ is the scalar product in $\mathbb{R}^2$, u is a complex-valued function on Γ.

In what follows we consider the curve Γ as embedded in $\mathbb{C}$, and we suppose that $\Gamma\in\mathcal{L}_{sl}$.

Let $t_0\in\mathcal{F}$ and the portion $\Gamma(t_0,\varepsilon)$ of the curve Γ be of the form (15)–(17). Let $g:\Gamma\times\Gamma\to\mathbb{C}$. We set

$$g^{t_0}(r,\rho) = (g^{t_0}_{jk}(r,\rho))_{j,k=1}^{n(t_0)}$$

where

$$g^{t_0}_{jk}(r,\rho) = g(t_0+re^{i\varphi_{t_0,j}(r)}, t_0+\rho e^{i\varphi_{t_0,k}(\rho)}), \quad r,\rho\in(0,\varepsilon).$$

We suppose that

$$\sup_{r,\rho\in(0,\varepsilon)}\left|\left(r\frac{\partial}{\partial r}\right)^\alpha\left(\rho\frac{\partial}{\partial\rho}\right)^\beta g^{t_0}_{jk}(r,\rho)\right|\le C_{\alpha\beta} \tag{24}$$

for all $\alpha,\beta\in\mathbb{N}_0$ and for all $j,k\in\{1,\dots,n(t_0)\}$,

$$\lim_{r\to0}\sup_{\rho\in(0,\varepsilon)}\left|r\frac{\partial g^{t_0}_{jk}(r,\rho)}{\partial r}\right| = 0, \tag{25}$$

and

$$\lim_{\rho\to0}\sup_{r\in(0,\varepsilon)}\left|\rho\frac{\partial g^{t_0}_{jk}(r,\rho)}{\partial \rho}\right| = 0. \tag{26}$$

We denote by $SO^\infty(\Gamma\times\Gamma,\mathcal{F})$ the class of functions $g:\Gamma\times\Gamma\to\mathbb{C}$ such that $g\,|_{(\Gamma\backslash\mathcal{F})\times(\Gamma\backslash\mathcal{F})}\in C^\infty((\Gamma\backslash\mathcal{F})\times(\Gamma\backslash\mathcal{F}))$, and for the every node t_0 conditions (24), (25), (26) hold.

We associate a function $f:\Gamma\to\mathbb{C}$ and a point $t_0\in\mathcal{F}$ the vector-function

$$(0,\varepsilon)\ni r\to(f_{1,t_0}(r),\dots,f_{n(t_0),t_0}(r))\in\mathbb{C}^{n(t_0)}$$

where

$$f_{j,t_0}(r) = f(t_0+re^{i\varphi_{t_0,j}(r)}), r\in(0,\varepsilon).$$

We say that a function $a:\Gamma\to\mathbb{C}$ belongs to the class $SO^\infty(\Gamma,\mathcal{F})$ if $a\,|_{\Gamma\backslash\mathcal{F}}\in C^\infty(\Gamma\backslash\mathcal{F})$ and $a_{j,t_0}\in C^\infty_{sl}((0,\varepsilon))$ for every $t_0\in\mathcal{F}$ and $j=1,\dots,n(t_0)$.

Let ϕ_{t_0} be a function defined on $\Gamma(t_0,\varepsilon)$. We suppose that $\tilde\phi_{t_0}(\rho)=\phi_{t_0}(t_0+\rho e^{i\varphi_{t_0,k}(\rho)})\in C^\infty(\overline{\mathbb{R}_+})$, $\tilde\phi_{t_0}$ is independent of $k\in\{1,\dots,n(t_0)\}$, and $\tilde\phi_{t_0}$ has a support on $[0,\varepsilon)$ and equals 1 on $[0,\varepsilon'),\varepsilon'<\varepsilon$. We say that ϕ_{t_0} is a cut-off function of $\Gamma(t_0,\varepsilon)$.

We will consider the operator $D_{g,\Gamma}$ as acting on the space $L^{p(\cdot)}(\Gamma,w)$ where $p(\cdot)\in\mathcal{P}(\Gamma)$ and the weight $w\in\mathcal{A}_{p(\cdot)}(\Gamma)$. We set $r^{\frac{1}{p_{t_0}(r)}}w_{t_0}(r)=\tilde{w}_{t_0}(r)=e^{\tilde{v}_{t_0}(r)}$.

We consider now the structure of the operator $D_{g,\Gamma}$ in a neighborhood of a node t_0. Let

$$\Phi_{t_0}:L^{p(\cdot)}(\Gamma(t_0,\varepsilon),w)\to L^{p_{t_0}(\cdot)}_{n(t_0)}\left((0,\varepsilon),\frac{dr}{r}\right)$$

be an isomorphism ([28], Proposition 37) defined as

$$(\Phi_{t_0}u)(r)=\begin{pmatrix}\tilde{w}_{t_0}(r)u_{1,t_0}(r)\\ \cdot\\ \cdot\\ \cdot\\ \tilde{w}_{t_0}(r)u_{n(t_0),t_0}(r)\end{pmatrix},\quad r\in(0,\varepsilon).$$

We set

$$\theta^{t_0}_{jk}(r,\rho)=\varphi_{j,t_0}(r)-\varphi_{k,t_0}(\rho).$$

Proposition 18. *Let $\Gamma\in\mathcal{L}_{sl}$, $p(\cdot)\in\mathcal{P}(\Gamma),w\in\mathcal{A}_{p(\cdot)}(\Gamma)$, $t_0\in\mathcal{F}$, $g\in SO^{\infty}(\Gamma\times\Gamma,\mathcal{F})$, and ϕ_{t_0} be a cut-off function of $\Gamma(t_0,\varepsilon)$. Then the operator*

$$D^{t_0}_{g,\Gamma}=\Phi_{t_0}\phi_{t_0}D_{g,\Gamma}\phi_{t_0}\Phi^{-1}_{t_0}$$

is a Mellin pseudodifferential operator of the class $OP\mathcal{E}_{sl}(n(t_0))$ with the symbol $\tilde{\sigma}_{t_0}(D_{g,\Gamma})=\sigma_{t_0}(D_{g,\Gamma})+q$ where $\sigma_{t_0}(D_{g,\Gamma})=\left(\sigma^{jk}_{t_0}(D_{g,\Gamma})\right)^{n(t_0)}_{j,k=1},q\in\mathcal{E}^{-1}_0(n(t_0))$, and

$$\sigma^{jk}_{t_0}(D_{g,\Gamma})(r,\xi)=\begin{cases}\frac{\varepsilon_k g_{jk}(r,r)\sinh(\pi-\theta^{t_0}_{jk}(r,r))(\xi+i(\frac{1}{p(t_0)}+rv'_{t_0}(r)))}{\sinh\pi(\xi+i(\frac{1}{p(t_0)}+rv'_{t_0}(r)))}, & k<j,\\ \frac{\varepsilon_k g_{jk}(r,r)\sinh(\theta^{t_0}_{jk}(r,r)-\pi)(\xi+i(\frac{1}{p(t_0)}+rv'_{t_0}(r)))}{\sinh\pi(\xi+i(\frac{1}{p(t_0)}+rv'_{t_0}(r)))}, & k>j,\\ 0,k=j.\end{cases}$$

where $\varepsilon_k=1$ if t_0 is the starting point of the curve $\Gamma^{t_0}_k$, and $\varepsilon_k=-1$ if t_0 is the end point of the curve $\Gamma^{t_0}_k$.

Proof. The operator $D_{g,\Gamma}$ can be written in the complex variables as

$$D_{g,\Gamma}u(t)=\frac{1}{\pi}\int_\Gamma\frac{g(t,\tau)\Re(\nu(\tau)\overline{(\tau-t)})u(\tau)dl_\tau}{|t-\tau|^2},\quad t\in\Gamma.$$

Then

$$D_{g,\Gamma}=\frac{1}{2}\left(D^{(1)}_{g,\Gamma}+D^{(2)}_{\Gamma}\right)$$

where

$$D_{g,\Gamma}^{(1)}u(t) = \frac{1}{\pi}\int_\Gamma \frac{g(t,\tau)\nu(\tau)u(\tau)dl_\tau}{\tau - t},$$

$$D_{g,\Gamma}^{(2)}u(t) = \frac{1}{\pi}\int_\Gamma \frac{g(t,\tau)\overline{\nu(\tau)}u(\tau)dl_\tau}{\bar{\tau} - \bar{t}}.$$

Hence $D_{g,\Gamma}^{t_0} = \frac{1}{2}(D_{g,\Gamma}^{(1)t_0} + D_{g,\Gamma}^{(2)t_0})$ where $D_{g,\Gamma}^{(l)t_0} = \Phi_{t_0}\phi_{t_0}D_{g,\Gamma}^{(l)}\phi_{t_0}\Phi_{t_0}^{-1}, l = 1,2$. The operators $D_{g,\Gamma}^{(l)t_0}, l = 1,2$ have matrix representations of the form

$$\left(D_{g,\Gamma}^{(l)t_0}u\right)_{j,t_0}(r) = \sum_{k=1}^{n(t_0)}\left(K_{jk}^{(l)t_0}u_{k,t_0}\right)(r)$$

where

$$\left(K_{jk}^{(l)t_0}u_{l,t_0}\right)(r) = \int_0^\varepsilon k_{jk}^{(l)t_0}(r,\rho)u_{k,t_0}(\rho)\frac{d\rho}{\rho}, r \in (0,\varepsilon).$$

First we consider $K_{jk}^{(1)t_0}$. The function $k_{jk}^{(1)t_0}$ is of the form:

$$k_{jk}^{(1)t_0}(r,\rho) = \frac{g_{jk}^{t_0}(r,\rho)\tilde{\phi}_{t_0}(r)\tilde{\phi}_{t_0}(\rho)e^{(\tilde{v}_{t_0}(r)-\tilde{v}_{t_0}(\rho))}e^{-i\theta_{jk}^{t_0}(r,\rho)}}{\pi i(r - \rho e^{-i\theta_{jk}^{t_0}(r,\rho)})} \times (1 + i\rho\varphi'_{t_0,k}(\rho))\sqrt{1 + \left(\rho\varphi'_{t_0,k}(\rho)\right)^2}.$$

Taking into account that

$$\frac{1}{\pi i}\int_0^\infty \frac{t^{is-1}dt}{t - e^{i\theta}} = \begin{cases} \frac{e^{(\pi-\theta)s}e^{-i\theta}}{\sinh \pi s}, 0 < \theta < 2\pi, \\ \frac{e^{(\theta-\pi)s}e^{-i\theta}}{\sinh \pi s}, -2\pi < \theta < 0, \end{cases} \quad 0 < \Im(s) < 1, \tag{27}$$

$$\frac{1}{\pi i}v.p.\int_0^\infty \frac{t^{is-1}dt}{t-1} = \coth \pi s, 0 < \Im(s) < 1, \tag{28}$$

(see [8], formulas 3.238.1 and 3.238.2), and

$$\tilde{v}_{t_0}(r) - \tilde{v}_{t_0}(\rho) = \gamma_{\tilde{v}_{t_0}}(r,\rho)\frac{r}{\rho}, \quad \text{where} \quad \gamma_{\tilde{v}_{t_0}}(r,\rho) = \int_0^1 \varkappa_{\tilde{v}}(r^\theta\rho^{1-\theta})d\theta$$

we can write the operator $K_{jk}^{(1)t_0}$ as a Mellin pseudodifferential operator with a double symbol

$$\sigma(K_{jk}^{(1)t_0})(r,\rho,\xi) = \begin{cases} \varepsilon_k\tilde{\phi}_{t_0}(r)\tilde{\phi}_{t_0}(\rho)g_{jk}^{t_0}(r,\rho)\frac{\exp\left[\left(\theta_{jk}^{t_0}(r,\rho)-\pi\right)\left(\xi+i\gamma_{\tilde{v}_{t_0}}(r,\rho)\right)\right]}{\sinh\left(\xi+i\gamma_{\tilde{v}_{t_0}}(r,\rho)\right)}, k<j, \\ \varepsilon_k\tilde{\phi}_{t_0}(r)\tilde{\phi}_{t_0}(\rho)g_{jk}^{t_0}(r,\rho)\coth\pi\left(\xi+i(\frac{1}{p(t_0)}+rv'_{t_0}(r))\right), j=k, \\ \varepsilon_k\tilde{\phi}_{t_0}(r)\tilde{\phi}_{t_0}(\rho)g_{jk}^{t_0}(r,\rho)\frac{\exp\left[(\pi-\theta_{jk}^{t_0}(r,\rho))(\xi+i\gamma_{\tilde{v}_{t_0}}(r,\rho))\right]}{\sinh\left(\xi+i\gamma_{\tilde{v}_{t_0}}(r,\rho))\right)}, k>j. \end{cases}$$

(See details of the calculations in the paper [22].) Applying Proposition 8 we obtain that

$$D_{g,\Gamma}^{(1)t_0} = Op_M(\sigma_{t_0}(D_{g,\Gamma}^{(1)}) + q^{(1)}), \tag{29}$$

where $q^{(1)} \in \mathcal{E}_0^{-1}(n(t_0))$,

$$\sigma_{t_0}(D_{g,\Gamma}^{(1)}) = \left(\sigma_{t_0}^{jk}(D_{g,\Gamma}^{(1)})\right)_{j,k=1}^{n(t_0)},$$

and

$$\sigma_{t_0}^{jk}(D_{g,\Gamma}^{(1)})(r,\xi) = \begin{cases} \varepsilon_k g_{jk}^{t_0}(r,r)\frac{\exp\left[\left(\theta_{jk}^{t_0}(r,r)-\pi\right)\left(\xi+i(\frac{1}{p(t_0)}+rv_{t_0}'(r))\right)\right]}{\sinh\left(\xi+i(\frac{1}{p(t_0)}+rv_{t_0}'(r))\right)}, & k<j, \\ \varepsilon_k g_{jk}^{t_0}(r,r)\coth\pi\left(\xi+i(\frac{1}{p(t_0)}+rv_{t_0}'(r))\right), & j=k, \\ \varepsilon_k g_{jk}^{t_0}(r,r)\frac{\exp\left[\left(\pi-(\theta_{jk}^{t_0}(r,r))\right)\left(\xi+i(\frac{1}{p(t_0)}+rv_{t_0}'(r))\right)\right]}{\sinh\left(\xi+i(\frac{1}{p(t_0)}+rv_{t_0}'(r)))\right)}, & k>j. \end{cases} \tag{30}$$

The operator $K_{jk}^{(2)t_0}$ has a similar representation as a Mellin pseudodifferential operator with double symbol and as above we obtain

$$D_{g,\Gamma}^{(2)t_0} = Op_M(\sigma_{t_0}(D_{g,\Gamma}^{(2)})) + Op_M(q^{(2)}),$$

where $q^{(2)} \in \mathcal{E}_0^{-1}(n(t_0))$ and

$$\sigma_{t_0}^{jk}(D_{g,\Gamma}^{(2)})(r,\xi) = \begin{cases} -\varepsilon_k g_{jk}^{t_0}(r,r)\frac{\exp\left[-\left(\theta_{jk}^{t_0}(r,r)-\pi\right)\left(\xi+i(\frac{1}{p(t_0)}+rv_{t_0}'(r))\right)\right]}{\sinh\left(\xi+i(\frac{1}{p(t_0)}+rv_{t_0}'(r))\right)}, & j>k, \\ -\varepsilon_k g_{jk}^{t_0}(r,r)\coth\pi\left(\xi+i(\frac{1}{p(t_0)}+rv_{t_0}'(r))\right), & j=k, \\ -\varepsilon_k g_{jk}^{t_0}(r,r)\frac{\exp\left[-\left(\pi-\theta_{jk}^{t_0}(r,r)\right)\left(\xi+i(\frac{1}{p(t_0)}+rv_{t_0}'(r))\right)\right]}{\sinh\left(\xi+i(\frac{1}{p(t_0)}+rv_{t_0}'(r))\right)}, & j<k. \end{cases} \tag{31}$$

Taking into account (30) and (31) we obtain that

$$\begin{aligned} D_{g,\Gamma}^{t_0} &= Op_M(\frac{1}{2}(\sigma_{t_0}(D_{g,\Gamma}^{(1)}) + \sigma_{t_0}(D_{g,\Gamma}^{(2)})) + Op_M\frac{1}{2}(q^{(1)} + q^2) \\ &= Op_M(\sigma_{t_0}(D_{g,\Gamma})) + Op_M(q) \end{aligned}$$

where $q \in \mathcal{E}_0^{-1}(n(t_0))$. □

- In what follows we say that $\sigma_{t_0}(D_{g,\Gamma}) = \left(\sigma_{t_0}^{jk}(D_{g,\Gamma})\right)_{j,k=1}^{n(t_0)}$ is the local symbol of $D_{g,\Gamma}$ at the point $t_0 \in \mathcal{F}$.

The following proposition follows easily from the well-known result that the operator of the double-layer potential on a smooth manifold is a pseudodifferential operator in $OPS_{1,0}^{-1}$ (see for instance [37], p. 361). (For the definition of pseudodifferential operators on a smooth manifold see, for instance, [35], Chap. 1.)

Proposition 19. *Let* $\mathcal{X}$ *be a smooth one-dimensional manifold,* $g \in C^\infty(\mathcal{X} \times \mathcal{X})$, $\varphi, \psi \in C_0^\infty(\Omega)$. *Then* $\phi D_{g,\Omega}\psi I$ *is a pseudodifferential operator on* $\mathcal{X}$ *of the class* $OPS_{1,0}^{-1}$.

Proposition 20. *Let* $\Gamma \in \mathcal{L}_{sl}$, $p(\cdot) \in \mathcal{P}(\Gamma)$, $g \in SO^\infty(\Gamma \times \Gamma, \mathcal{F})$. *Then the operator* $D_{g,\Gamma}$ *is locally compact on* $\Gamma\backslash\mathcal{F}$, *that is for an arbitrary open set* U *such that* $\bar{U} \subset \Gamma\backslash\mathcal{F}$ *and functions* $\phi, \psi \in C_0^\infty(U)$ *the operator* $\phi D_{g,\Gamma}\psi I \in \mathcal{K}(L^{p(\cdot)}(\Gamma))$.

Proof. Let $\kappa : U \to V \subset \mathbb{R}$ be a diffeomorphism, $\kappa^* : C_0^\infty(V) \to C_0^\infty(U)$ be an invertible mapping defined as

$$\kappa^* f(t) = f(\kappa^{-1}(t)), t \in U,$$

and let $\kappa_* : C_0^\infty(U) \to C_0^\infty(V)$ be the inverse mapping for κ^*. By Proposition 19 $\kappa_*\phi D_{g,\Gamma}\psi\kappa^* \in OPS_{1,0}^{-1}(\mathbb{R})$. According to Theorem 4 $\kappa_*\phi D_{g,\Gamma}\psi\kappa^*$ is bounded in all $L^{p(\cdot)}(\mathbb{R}), p(\cdot) \in \mathcal{P}(\mathbb{R})$. Moreover this operator is compact in $L^2(\mathbb{R})$ (see for instance [35], Chapter 1). Taking into account Proposition 2.2, [27] on the interpolation of compactness in $L^{p(\cdot)}(\mathbb{R})$ we obtain that $\kappa_*\phi D_{g,\Gamma}\psi\kappa^* \in \mathcal{K}(L^{p(\cdot)}(\mathbb{R}))$. It implies that $\phi D_{g,\Gamma}\psi I \in \mathcal{K}(L^{p(\cdot)}(\Gamma))$. □

Theorem 21. *Let* $\Gamma \in \mathcal{L}_{sl}$, $p(\cdot) \in \mathcal{P}(\Gamma)$, $w \in \mathcal{A}_{p(\cdot)}(\Gamma)$, $g \in SO^\infty(\Gamma \times \Gamma, \mathcal{F})$. *Then* $D_{g,\Gamma} : L^{p(\cdot)}(\Gamma, w) \to L^{p(\cdot)}(\Gamma, w)$ *is a bounded operator.*

Proof. The proof follows the proof of Theorem 40 in [28] and uses an admissible partition of unity, Proposition 13, boundedness of Mellin pseudodifferential operators in $L^{p(\cdot)}(\mathbb{R}_+)$, Proposition 18, boundedness of pseudodifferential operators of the class $OPS_{1,0}^0$ on $L^{p(\cdot)}(\mathbb{R})$. □

Theorem 22. *Let* $\Gamma \in \mathcal{L}_{sl}$, $p(\cdot) \in \mathcal{P}(\Gamma)$, $w \in \mathcal{A}_{p(\cdot)}(\Gamma)$, $g \in SO^\infty(\Gamma \times \Gamma, \mathcal{F})$. *Then* $D_{g,\Gamma} : L^{p(\cdot)}(\Gamma, w) \to L^{p(\cdot)}(\Gamma, w)$ *is a local type operator.*

Proof. The proof follows to the proof of Theorem 47 in [28]. It based on Proposition 13, compactness of the operator $\phi_1 D_{g,\Gamma}\phi_2 I$ acting from $L^1(\Gamma)$ into $L^\infty(\Gamma)$ for all $\phi_1, \phi_2 \in C(\Gamma)$ such that $\operatorname{supp}\phi_1 \cap \mathit{supp}\phi_2 = \varnothing$. □

4.2. Fredholm property and essential spectrum of operators of double-layer potentials type

4.2.1. Fredholm property. We consider here the operator

$$A_{g,\Gamma} = aI + bD_{g,\Gamma} : L^{p(\cdot)}(\Gamma, w) \to L^{p(\cdot)}(\Gamma, w)$$

under assumptions:

$$\Gamma \in \mathcal{L}_{sl}, w \in \mathcal{A}_{p(\cdot)}(\Gamma), p(\cdot) \in \mathcal{P}(\Gamma), g \in SO^\infty(\Gamma \times \Gamma\mathcal{F}), a, b \in SO^\infty(\Gamma \times \Gamma\mathcal{F}). \quad (32)$$

- We define the local symbol of $A_{g,\Gamma}$ at a point $t \in \mathcal{F}$ as

$$\sigma_t(A_{g,\Gamma}) = \left(\sigma_t^{jk}(A_{g,\Gamma})\right)_{j,k=1}^{n(t)}$$

where

$$\sigma_t^{jk}(A_{g,\Gamma})(r, \xi) = a_{j,t}(r) + b_{j,t}(r)\sigma_t^{jk}(D_{g,\Gamma})(r, \xi).$$

Theorem 23. *Let assumptions* (32) *hold. Then* $A_{g,\Gamma} : L^{p(\cdot)}(\Gamma, w) \to L^{p(\cdot)}(\Gamma, w)$ *is locally invertible:*

(i) *at the point* $t \in \Gamma \backslash \mathcal{F}$ *if and only if* $a(t) \neq 0$;

(ii) *at the point* $t \in \mathcal{F}$ *if and only if*

$$\liminf_{r \to +0} \inf_{\xi \in \mathbb{R}} |\det \sigma_t(A_{g,\Gamma})(r, \xi)| > 0.$$

Proof. (i) Let $t_0 \in \Gamma \backslash \mathcal{F}$, U be a small neighborhood of the point t_0, $\varphi, \psi \in C_0^\infty(U), \varphi\psi = \varphi$, and $\varphi(t_0) = \psi(t_0) = 1$. Applying standard arguments of the pseudodifferential operators theory (see for instance [35], Chapter 1) one can prove that $A_{g,\Gamma} : L^{p(\cdot)}(\Gamma, w) \to L^{p(\cdot)}(\Gamma, w)$ is locally invertible at the point t_0 if and only if $\varphi A_{g,\Gamma} \psi I : L^{p(\cdot)}(U) \to L^{p(\cdot)}(U)$ is locally invertible at this point. In turn $\varphi A_{g,\Gamma} \psi I : L^{p(\cdot)}(U) \to L^{p(\cdot)}(U)$ is locally invertible at the point t_0 if and only if $\kappa_* \phi A_{g,\Gamma} \psi \kappa^* : L^{\tilde{p}(\cdot)}(V) \to L^{\tilde{p}(\cdot)}(V)$ is locally invertible at the point $x_0 = \varkappa(t_0)$ where $\tilde{p}(x) = p(\varkappa^{-1}(x))$, and $\kappa, \kappa_*, \kappa^*$ are defined in the proof of Proposition 20. Hence statement (i) follows from Theorem 7.

(ii) It follows from Proposition 18 that $A_{g,\Gamma} : L^{p(\cdot)}(\Gamma, w) \to L^{p(\cdot)}(\Gamma, w)$ is locally invertible at a point $t \in \mathcal{F}$ if and only if the Mellin pseudodifferential operator

$$Op_M(\sigma_t(A_{g,\Gamma})) : L^{p_t(\cdot)}(\mathbb{R}_+, \frac{dr}{r}) \to L^{p_t(\cdot)}(\mathbb{R}_+, \frac{dr}{r})$$

is locally invertible at the point 0. Hence statement (ii) follows from Theorem 11. □

Remark 24. Note that condition (ii) implies that

$$\liminf_{r \to 0} |a_{j,t}(r)| > 0, j = 1, \dots, n(t),$$

because

$$\lim_{\xi \to \infty} \sigma_{t_0}^{jk}(D_{g,\Gamma})(r, \xi) = 0,$$

uniformly with respect to $r \in (0, \varepsilon)$, for $\varepsilon > 0$ small enough.

Theorem 25. *Let assumptions* (32) *hold. Then the operator* $A_{g,\Gamma} : L^{p(\cdot)}(\Gamma, w) \to L^{p(\cdot)}(\Gamma, w)$ *is a Fredholm operator if and only if conditions* (i) *and* (ii) *of Theorem* 23 *hold for every point* $t \in \Gamma \backslash \mathcal{F}$, *and* $t \in \mathcal{F}$, *respectively. The Fredholm index of* $A_{g,\Gamma}$ *is defined by the formula*

$$\operatorname{Ind} A_{g,\Gamma} = -\frac{1}{2\pi} \sum_{t \in \mathcal{F}} \lim_{r \to 0} \left[\arg \det(\hat{\sigma}_t(A_{g,\Gamma})(r, \xi))\right]_{\xi=-\infty}^{+\infty}, \tag{33}$$

where

$$\hat{\sigma}_t(A_{g,\Gamma})(r, \xi) = \left(\hat{\sigma}_t^{jk}(A_{g,\Gamma})(r, \xi)\right)_{j,k=1}^{n(t)}$$

and $\hat{\sigma}_t^{jk}(A_{g,\Gamma})(r, \xi)$ *are defined as*

$$(\hat{\sigma}_t^{jk}(A_{g,\Gamma})(r, \xi) = a_{j,t}^{-1}(r) \sigma_t^{jk}(A_{g,\Gamma})(r, \xi)$$

for $r > 0$ *small enough.*

Proof. Because $A_{g,\Gamma} : L^{p(\cdot)}(\Gamma, w) \to L^{p(\cdot)}(\Gamma, w)$ is a local type operator the statement on the Fredholmness follows from Theorem 23 and Theorem 16. The proof of the index formula is standard and based on a separation of singularities and it is the same as in the case of the constant $p \in (1, \infty)$ (see for instance [22]). □

4.2.2. Essential spectrum of the operator $D_{g,\Gamma}$.

Theorem 26. *Let assumptions* (32) *hold. Then the essential spectrum of* $D_{g,\Gamma} : L^{p(\cdot)}(\Gamma, w) \to L^{p(\cdot)}(\Gamma, w)$ *is given by the formula*

$$sp_{ess} D_{g,\Gamma} = \bigcup_{t\in\mathcal{F}} \bigcup_{h} \bigcup_{\xi\in[-\infty,\infty]} sp\, \sigma_t^h(D_{g,\Gamma})(\xi) \tag{34}$$

where $\bigcup_h$ *is taken with respect to the all sequences* $h = (h_m)$ $(\mathbb{R}_+ \ni h_m \to 0)$ *for which there exist the uniform with respect to* $\xi \in [-\infty, +\infty]$ *limits*

$$\lim_{m\to\infty} \sigma_t(A_{g,\Gamma})(h_m, \xi) = \sigma_t^h(D_{g,\Gamma})(\xi), \tag{35}$$

$sp\, \sigma_t^h(D_{g,\Gamma})(\xi)$ *denotes the spectrum of* $n(t) \times n(t)$ *matrix.*

Proof. It follows from Theorem 17 that the essential spectrum of $D_{g,\Gamma}$ is the union of the local spectra of $D_{g,\Gamma}$ for all points $t \in \Gamma$. By Proposition 20 the operator $D_{g,\Gamma}$ is locally compact at $t \in \Gamma\backslash\mathcal{F}$. It implies that the local spectrum of $D_{g,\Gamma}$ at the every point $t \in \Gamma\backslash\mathcal{F}$ is $\{0\}$. But 0 belongs to the local spectrum of the operator $D_{g,\Gamma}$ at a point $t \in \mathcal{F}$ because $\lim_{\xi\to\infty} \sigma_t^{jk}(A_{g,\Gamma})(r, \xi) = 0$. Hence

$$sp_{ess} D_{g,\Gamma} = \bigcup_{t\in\mathcal{F}} sp_t D_{g,\Gamma}. \tag{36}$$

Applying formula (36) and Proposition 12 we obtain formula (34). □

4.3. Integral operators of the Dirichlet problem

Let Ω be a simply connected bounded domain with a boundary $\Gamma \in \mathcal{L}_{sl}$ oriented by the standard way and let

$$D_\Gamma u(t) = D_{1,\Gamma}(t)u(t) = \frac{1}{\pi} \int_\Gamma \frac{(\nu(\tau), \tau - t)}{|t - \tau|^2} u(\tau) dl_\tau, t \in \Gamma,$$

be the double-layer potential on the curve Γ.

Proposition 27. *Let* $\Lambda \subset \Gamma$ *be an open connected set such that* $\bar{\Lambda} \subset \Gamma\backslash\mathcal{F}$ *and* $u \in L^{p(\cdot)}(\Gamma, w), w \in \mathcal{A}_{p(\cdot)}(\Gamma)$. *Then for almost all points* $t \in \Lambda$ *there exist nontangential limits*

$$\begin{aligned} \lim_{\Omega\ni x\to t\in\Lambda} (D_\Gamma u)(x) &= \frac{1}{\pi} \lim_{\Omega\ni x\to t\in\Lambda} \int_\Gamma \frac{(\nu(\tau), \tau - t)}{|x - \tau|^2} u(\tau) dl_\tau \\ &= u(t) + (D_\Gamma u)(t). \end{aligned} \tag{37}$$

Proof. Let $\Lambda'(\supset \bar{\Lambda})$ be a neighborhood of Λ such that $\overline{\Lambda'} \subset \Gamma\backslash\mathcal{F}$

$$D'_\Gamma u(x) = \frac{1}{\pi}\int_{\Lambda'} \frac{(\nu(\tau), \tau - t)}{|x-\tau|^2} u(\tau) dl_\tau, \quad x \in \Omega,$$

$$D''_\Gamma u(x) = \frac{1}{\pi}\int_{\Gamma\backslash\Lambda'} \frac{(\nu(\tau), \tau - t)}{|x-\tau|^2} u(\tau) dl_\tau, \ x \in \Omega.$$

Applying the embedding $L^{p(\cdot)}(\Lambda) \subset L^{p_-}(\Lambda)$ where $p_- = \min_{t\in\Gamma} p(t) \in (1, \infty)$ (see [7], p. 83), and well-known results (see [9], [18], p. 199, [36]) with respect to the limit values of the double-layered potentials with density in L^p, $p \in (1, \infty)$ we obtain that for almost all $t \in \Lambda$

$$\lim_{\Omega\ni x\to t\in\Lambda} D'_\Gamma u(x) = u(t) + \frac{1}{\pi}\int_{\Lambda'} \frac{(\nu(\tau), \tau - t)}{|t-\tau|^2} u(\tau) dl_\tau. \tag{38}$$

By Proposition 13 $w^{-1} \in L^{p'(\cdot)}(\Gamma)$ then applying the Hölder inequality for variable exponent Lebesgue spaces (see [7], p. 82) we obtain that $u = w^{-1}(wu) \in L^1(\Gamma)$. Then the Lebesgue dominated convergence theorem yields that

$$\lim_{\mathcal{D}\ni x\to t\in\Lambda} D''_\Gamma u(x) = \frac{1}{\pi}\int_{\Gamma\backslash\Lambda'} \frac{(\nu(\tau), \tau - t)}{|t-\tau|^2} u(\tau) dl_\tau \tag{39}$$

for all $t \in \Lambda$. Formulas (38) and (39) imply formula (37). □

Corollary 28. *Let $\Gamma \in \mathcal{L}_{sl}, w \in \mathcal{A}_{p(\cdot)}(\Gamma)$, $u \in L^{p(\cdot)}(\Gamma, w)$. Then for almost all $t \in \Gamma$ there exist non tangential limits*

$$\lim_{\mathcal{D}\ni x\to t\in\Gamma} D_\Gamma u(x) = u(t) + \frac{1}{\pi}\int_{\Gamma} \frac{(\nu(\tau), \tau - t)}{|t-\tau|^2} u(\tau) dl. \tag{40}$$

We consider the Dirichlet problem in the domain Ω

$$\Delta\psi(x) = 0, x \in \Omega, \psi\,|_\Gamma = f \tag{41}$$

where ψ is a distribution in $\mathcal{D}'(\Omega)$, $f \in L^{p(\cdot)}(\Gamma, w)$, $w \in \mathcal{A}_{p(\cdot)}(\Gamma)$,and $\psi\,|_\Gamma$ means the non-tangential limit

$$(\psi\,|_\Gamma)(t) = \lim_{\Omega\ni x\to t\in\Gamma} \psi(x)$$

which exists for almost all points of Γ.

The solution of the problem (41) is sought of the double-layer potential form

$$\psi(x) = D_\Gamma u(x) = \frac{1}{\pi}\int_{\Gamma} \frac{(\nu(\tau), \tau - x)}{|x-\tau|^2} u(\tau) dl_\tau, \quad x \in \Omega$$

with unknown density $u \in L^{p(\cdot)}(\Gamma, w)$. Applying Corollary 28 we obtain that u have to satisfy the equation

$$u + D_\Gamma u = f \tag{42}$$

in $L^{p(\cdot)}(\Gamma, w)$. Because $D_\Gamma \in \mathcal{B}(L^{p(\cdot)}(\Gamma, w))$ the equation (42) has sense. Note if $u \in L^{p(\cdot)}(\Gamma, w)$ is a solution of equation (42) then $\psi = D_\Gamma u$ is a solution of the Dirichlet problem (41).

Let $t_0 \in \mathcal{F}$. Then

$$\Gamma(t_0,\varepsilon) = \Gamma_1(t_0,\varepsilon) \cup \Gamma_2(t_0,\varepsilon),$$

where $\Gamma_j(t_0,\varepsilon) = \left\{t \in \mathbb{C} : t = t_0 + re^{i\varphi_{t_0,j}(r)}, r \in (0,\varepsilon)\right\}$. We set

$$\theta^{t_0}(r) = \varphi_{t_0,2}(r) - \varphi_{t_0,1}(r).$$

Applying Proposition 18 we obtain the local symbol of D_Γ at the point $t_0 \in \mathcal{F}$

$$\sigma_{t_0}(D_\Gamma)(r,\xi) = \begin{pmatrix} 0 & \omega_{t_0}(r,\xi) \\ \omega_{t_0}(r,\xi) & 0 \end{pmatrix}$$

where

$$\omega_{t_0}(r,\xi) = \frac{\sinh(\pi - \theta^{t_0}(r))(\xi + i(\frac{1}{p(t_0)} + rv'_{t_0}(r)))}{\sinh \pi(\xi + i(\frac{1}{p(t_0)} + rv'_{t_0}(r)))}.$$

- Let $A \in \mathcal{B}(L^{p(\cdot)}(\Gamma))$. Then the local spectral radius of A is defined as

$$r_t(A) = \sup_{\lambda \in sp_t A} |\lambda|.$$

- Let $A \in \mathcal{B}(E)$ where E is a Banach space. Then the essential spectral radius of A is defined as

$$r_{ess}(A : E \to E) = \sup_{\lambda \in sp_{ess} A} |\lambda|.$$

Theorem 29. *Let $\Gamma \in \mathcal{L}_{sl}$, $p(\cdot) \in \mathcal{P}(\Gamma), w \in \mathcal{A}_p(\Gamma), t \in \mathcal{F}$. Then*

$$r_t(D_\Gamma) = \limsup_{r\to 0} \frac{\sin|\pi - \theta^t(r)|\,(\frac{1}{p(t)} + rv'_t(r))}{\sin \pi(\frac{1}{p(t)} + rv'_t(r))}. \tag{43}$$

Proof. We have

$$\sigma_t(D_\Gamma - \lambda I)(r,\xi) = \begin{pmatrix} -\lambda & \omega_t(r,\xi) \\ \omega_t(r,\xi) & -\lambda \end{pmatrix}. \tag{44}$$

Hence

$$\det \sigma_t(D_\Gamma - \lambda I)(r,\xi) = \lambda^2 - \omega_t(r,\xi)^2. \tag{45}$$

It follows from Theorem 23 (ii) and (45) that

$$r_t(D_\Gamma) = \limsup_{r\to 0} \sup_{\xi\in\mathbb{R}} |\omega_t(r,\xi)|, t \in \mathcal{F}. \tag{46}$$

One can prove that

$$\sup_{\xi\in\mathbb{R}} \left|\frac{\sinh \phi(\xi + i\gamma)}{\sinh \pi(\xi + i\gamma)}\right| = \frac{\sin \phi\gamma}{\sin \pi\gamma}, \phi \in (0,\pi), \gamma \in (0,1). \tag{47}$$

Hence (46), (47) imply formula (43). ∎

Remark 30. It follows from Theorem 29 that the local spectral radius $r_t(D_\Gamma)$ of the operator $D_\Gamma : L^{p(\cdot)}(\Gamma, w)$ depends on the value $p(t)$ at the point t only.

Corollary 31. *Let $\Gamma \in \mathcal{L}_{sl}$, $p(\cdot) \in \mathcal{P}(\Gamma), w \in \mathcal{A}_{p(\cdot)}(\Gamma)$. Then*

$$r_{ess}(D_\Gamma) = \max_{t \in \mathcal{F}} r_t(D_\Gamma).$$

This corollary follows from Theorem 29 and Simonenko's local principle, because $r_t(D_\Gamma) = 0$ for every $t \in \Gamma \backslash \mathcal{F}$.

Theorem 32. *Let $\Gamma \in \mathcal{L}_{sl}$, $p(\cdot) \in \mathcal{P}(\Gamma), w \in \mathcal{A}_{p(\cdot)}(\Gamma)$. Then the integral operator $I + D_\Gamma : L^{p(\cdot)}(\Gamma, w) \to L^{p(\cdot)}(\Gamma, w)$ of Dirichlet problem (41) is locally invertible at a node $t \in \mathcal{F}$ if and only if*

$$\frac{1}{p(t)} \notin [\mu_-^t, \mu_+^t] \tag{48}$$

where

$$\mu_-^t = \liminf_{r \to 0} \left(\frac{\pi}{\pi + |\theta^t(r) - \pi|} - r v_t'(r) \right),$$

$$\mu_+^t = \limsup_{r \to 0} \left(\frac{\pi}{\pi + |\theta^t(r) - \pi|} - r v_t'(r) \right).$$

Proof. We have

$$\det \sigma_t(D_\Gamma + I)(r, \xi) = 1 - \left(\frac{\sinh |\pi - \theta^t(r)| \, (\xi + i(\frac{1}{p(t)} + r v_t'(r)))}{\sinh \pi(\xi + i(\frac{1}{p(t)} + r v_t'(r)))} \right)^2.$$

The equation

$$1 - \frac{\sinh^2 |\pi - \theta| \, \zeta}{\sinh^2 \pi\zeta} = 0, \zeta \in \mathbb{R} + i(0,1), \theta \in (0, \pi)$$

has the unique solution

$$\zeta = \frac{i\pi}{\pi + |\theta - \pi|}.$$

Hence if $\frac{1}{p(t)} \notin [\mu_-^t, \mu_+^t]$

$$\liminf_{r \to 0} \inf_{\xi \in \mathbb{R}} |\det \sigma_t(I + D_\Gamma)(r, \xi)| > 0$$

and by Theorem 23 the operator $I + D_\Gamma : L^{p(\cdot)}(\Gamma, w) \to L^{p(\cdot)}(\Gamma, w)$ is locally invertible at the point $t \in \mathcal{F}$. If $\frac{1}{p(t)} \in [\mu_-^t, \mu_+^t]$ then there exists a sequence $r_m \to 0$ such that

$$\lim_{m \to \infty} \det \sigma_t(I + D_\Gamma)(r_m, 0) = 0. \tag{49}$$

But (49) contradicts necessary and sufficient condition (ii) of Theorem 23 for local invertibility at the point $t \in \mathcal{F}$. □

Remark 33. The segment $[\mu_-^t, \mu_+^t]$ of the prohibited values $p(t)$ for the local invertibility of $I + D_\Gamma$ at a point $t \in \mathcal{F}$ depends on the oscillation of the curve Γ and the weight w at this point. In the case of an angular point of a curve and a power weight the prohibited value is unique.

Theorem 34. *Let $\Gamma \in \mathcal{L}_{sl}$, $p(\cdot) \in \mathcal{P}(\Gamma), w \in \mathcal{A}_{p(\cdot)}(\Gamma)$. Then the operator $I + D_\Gamma : L^{p(\cdot)}(\Gamma, w) \to L^{p(\cdot)}(\Gamma, w)$ is Fredholm if and only if condition (48) holds for every point $t \in \mathcal{F}$. In this case the Fredholm index is given by the formula*

$$\operatorname{Ind}(I + D_\Gamma) = \sum_{t \in \mathcal{F}} \beta_t, \tag{50}$$

where

$$\beta_t = \begin{cases} 1, \frac{1}{p(t)} > \mu_+^t, \\ 0, \frac{1}{p(t)} < \mu_-^t \end{cases}.$$

Proof. The Fredholm condition follows from Theorem 25. For the calculation of the index we use formula (33). The function

$$\Phi(\zeta) = 1 - \frac{\sinh^2 |\pi - \theta| \, \zeta}{\sinh^2 \pi\zeta}, \theta \in (0, \pi)$$

is analytical in the strip $\mathbb{R} + i(0,1)$ and continuously extended on the real axis. Moreover

$$\lim_{\xi \to \infty} \Phi(\xi + i\eta) = 1$$

uniformly with respect to $\eta \in [0,1)$, and Φ has only one zero $\frac{i\pi}{\pi + |\theta - \pi|}$ in $\mathbb{R} + i[0,1)$. Moreover one can see that

$$[\arg \Phi(\xi)]_{\xi=-\infty}^{+\infty} = 0.$$

Hence

$$-\frac{1}{2\pi} [\arg \Phi(\xi + i\eta)]_{\xi=-\infty}^{+\infty} = \begin{cases} 1, \eta \in \left(\frac{\pi}{\pi + |\theta - \pi|}, 1\right) \\ 0, \eta \in \left(0, \frac{\pi}{\pi + |\theta - \pi|}\right) \end{cases}. \tag{51}$$

Hence equality (51) and formula (33) imply formula (50). □

4.4. Integral operators of the Neumann problem

Let Ω be a simply connected domain in $\mathbb{R}^2$ with a standard oriented boundary $\Gamma \in \mathcal{L}_{sl}$. We consider the Neumann problem in Ω

$$\begin{aligned} \Delta\psi(x) &= 0, x \in \Omega, \\ \frac{d\psi}{d\nu} &\mid {}_\Gamma = f \end{aligned} \tag{52}$$

where $f \in L^{p(\cdot)}(\Gamma, w), w \in \mathcal{A}_{p(\cdot)}(\Gamma), \nu(t)$ is the unit inward normal vector to the boundary Γ at the point $t \in \Gamma \backslash \mathcal{F}$.

We consider a solution of the problem (52) represented by a logarithmic potential

$$\psi(x) = \frac{1}{\pi} \int_\Gamma \log(|x - \tau|) u(\tau) dl_\tau, x \in \Omega \tag{53}$$

with density $u \in L^{p(\cdot)}(\Gamma, w)$. We set

$$N_\Gamma u(t) = \frac{1}{\pi} \int_\Gamma \frac{d \log(|t - \tau|)}{d\nu(t)} dl_\tau = \frac{1}{\pi} \int_\Gamma \frac{(\nu(t), t - \tau) \, u(\tau) dl_\tau}{|t - \tau|^2}, t \in \Gamma.$$

Proposition 35. *Let $t_0 \in \mathcal{F}$. Then the operator*

$$N_\Gamma^{t_0} = \Phi_{t_0}\phi_{t_0}N_\Gamma\phi_{t_0}\Phi_{t_0}^{-1}$$

is a Mellin pseudodifferential operator with symbol $\sigma_{t_0}(N_\Gamma) + q \in \mathcal{E}_{sl}(2)$ where $q \in \mathcal{E}_0(2)$ and

$$\sigma_{t_0}(N_\Gamma)(r,\xi) = \begin{pmatrix} 0 & e^{i\theta^{t_0}(r)}\omega_{t_0}(r,\xi) \\ e^{-i\theta^{t_0}(r)}\omega_{t_0}(r,\xi) & 0 \end{pmatrix}.$$

This proposition is proved by calculations similar to those used in the proof of Proposition 18. Note that the operator N_Γ is a pseudodifferential operator in the class $OPS_{1,0}^{-1}$ on the one-dimensional manifold $\Gamma\backslash\mathcal{F}$ (see for instance [37], p. 361). It implies that the operator N_Γ is a bounded local type operator on $L^{p(\cdot)}(\Gamma, w)$. Moreover N_Γ is a locally compact operator on $\Gamma\backslash\mathcal{F}$.

Proposition 36. *Let $\Gamma\in\mathcal{L}_{sl}$, $p(\cdot)\in\mathcal{P}(\Gamma)$, $w\in\mathcal{A}_{p(\cdot)}(\Gamma)$. Then $N_\Gamma : L^{p'(\cdot)}(\Gamma, w^{-1}) \to L^{p'(\cdot)}(\Gamma, w^{-1})$ is an adjoint operator to $D_\Gamma : L^{p(\cdot)}(\Gamma, w) \to L^{p(\cdot)}(\Gamma, w)$.*

Proof. Note if $p(\cdot) \in \mathcal{P}(\Gamma)$, and $w \in \mathcal{A}_{p(\cdot)}(\Gamma)$ the space $L^{p(\cdot)}(\Gamma, w)$ is reflexive and $\left[L^{p(\cdot)}(\Gamma, w)\right]^* = L^{p'(\cdot)}(\Gamma, w^{-1})$ (see for instance [7], p. 383). The general form of a linear functional on $L^{p(\cdot)}(\Gamma, w)$ is

$$f(\varphi) = \int_\Gamma \bar{f}(t)\varphi(t)dl_t$$

where $f \in L^{p'(\cdot)}(\Gamma, w^{-1})$. Moreover if $w \in \mathcal{A}_{p(\cdot)}(\Gamma)$ then $w^{-1} \in \mathcal{A}_{p'(\cdot)}(\Gamma)$. The kernels $\Gamma\times\Gamma \in (t,\tau) \to k_{D_\Gamma}(t,\tau)$, $\Gamma\times\Gamma \in (t,\tau) \to k_{N_\Gamma}(t,\tau)$ of operators D_Γ, N_Γ are real-valued functions and $k_{N_\Gamma}(t,\tau) = k_{D_\Gamma}(\tau, t)$. It implies that $N_\Gamma = D_\Gamma^*$. □

Proposition 37. *Let $u \in L^{p(\cdot)}(\Gamma, w)$, $\Gamma \in \mathcal{L}_{sl}, w \in \mathcal{A}_{p(\cdot)}(\Gamma)$. Then for almost all $t \in \Gamma$ there exist non-tangent limits*

$$\lim_{\Omega\ni x\to t\in\Gamma} \frac{d\psi(x)}{d\nu(t)} = -u(t) + N_\Gamma u(t).$$

Proof. The proof is similar to the proofs of Proposition 27 and Corollary 28. □

Applying Proposition 37 we obtain that the density u in (53) satisfies the integral equation on $L^{p(\cdot)}(\Gamma, w)$

$$(I - N_\Gamma)\, u = -f \in L^{p(\cdot)}(\Gamma, w).$$

If this equation has a solution $u \in L^{p(\cdot)}(\Gamma, w)$ then the function ψ defined by (53) is a solution of the Neumann problem (52).

Theorem 38. *Let $p(\cdot) \in \mathcal{P}(\Gamma), w \in \mathcal{A}_{p(\cdot)}(\Gamma)$. Then*

$$r_{ess}(N_\Gamma : L^{p'(\cdot)}(\Gamma, w^{-1}) \to L^{p'(\cdot)}(\Gamma, w^{-1})) = \max_{t\in\mathcal{F}} \limsup_{r\to 0} \sup_{\xi\in\mathbb{R}} |\omega_t(r,\xi)|\,, t \in \mathcal{F}.$$

Proof. Note that if $A \in \mathcal{B}(E)$. Then

$$r_{ess}(A^* : E^* \to E^*) = r_{ess}(A : E \to E).$$

Hence Theorem 38 follows from Proposition 36 and Corollary 31. □

Corollary 39. *Let* $p(\cdot) \in \mathcal{P}(\Gamma), w \in \mathcal{A}_{p(\cdot)}(\Gamma)$ *be such that* $r_{ess}(N_\Gamma : L^{p'(\cdot)}(\Gamma, w^{-1}) \to L^{p'(\cdot)}(\Gamma, w^{-1})) < 1$. *Then the operator* $I - N_\Gamma : L^{p'(\cdot)}(\Gamma, w^{-1}) \to L^{p'(\cdot)}(\Gamma, w^{-1})$ *of the Neumann problem is Fredholm.*

References

[1] A. Böttcher and Yu.I. Karlovich. *Carleson Curves, Muckenhoupt Weights, and Toeplitz Operators.* Birkhäuser Verlag, Basel, 1997.

[2] A. Böttcher, Yu.I. Karlovich and V.S. Rabinovich, *Emergence, persistence, and disappearance of logarithmic spirals in the spectra of singular integral operators.* Integr. Equ. Oper. Theory **25** (1996), 405–444.

[3] A. Böttcher, Yu.I. Karlovich and V.S. Rabinovich, *Mellin pseudodifferential operators with slowly varying symbols and singular integrals on Carleson curves with Muckenhoupt weights.* Manuscripta Mathematica **95** (1998), 363–376.

[4] A. Böttcher, Yu.I. Karlovich and V.S. Rabinovich, *The method of limit operators for one-dimensional singular integrals with oscillating data.* J. Operator Theory **43** (2000), 171–198.

[5] A. Böttcher, Yu.I. Karlovich and V.S. Rabinovich, *Singular integral operators with complex conjugation from the viewpoint of pseudodifferential operators.* In: Problems and Methods in Mathematical Physics, Operator Theory: Advances and Applications **121**, Birkhäuser, Basel, 2001, 36–59.

[6] I.I. Daniluk, *Nonregular Boundary Value Problems on the Plane.* M. Nauka, 1975 (in Russian).

[7] L. Diening, P. Harjulehto, P. Hästö and M. Růžička, *Lebesgue and Sobolev Spaces with Variable Exponents.* Lecture Notes in Mathematics, Springer, New York, 2011.

[8] I.S. Gradstein and I.M. Ryzhik, *Tables of Integrals, Series, and Products.* Fourth edition. Academic Press, New York, 1965.

[9] C.E. Kenig, *Harmonic Analysis Techniques for Second Order Elliptic Boundary Value Problem.* Regional Conference Series in Mathematics **83**, American Mathematical Society, Providence, RI, 1994.

[10] V. Kokilashvili, V. Paatashvili and S. Samko, *Boundedness in Lebesgue spaces with variable exponent of the Cauchy singular operators on Carleson curves.* In: Operator Theory: Advances and Applications **170**, Birkhäuser, Basel, 2006, 167–186.

[11] V. Kokilashvili, N. Samko and S. Samko, *Singular operators in variable spaces* $L_{p(\cdot)}$ *with oscillating weights.* Math. Nachr. **280** (2007), 1145–1156.

[12] V. Kokilashvili and S. Samko, *Singular integrals in weighted Lebesgue spaces with variable exponent.* Georgian Math. J. **10**(1) (2003), 145–156.

[13] V. Kokilashvili and S. Samko, *Singular integral equations in the Lebesgue spaces with variable exponent.* Proc. of A.Razmadze Math. Inst. **131** (2003), 61–78.

[14] V. Kokilashvili and S. Samko, *Weighted boundedness in Lebesgue spaces with variable exponents of classical operators on Carleson curves.* Proc. A. Razmadze Math.Inst. **138** (2005), 106–110.

[15] V. Kokilashvili and S. Samko, *Boundedness of maximal operators and potential operators on Carleson curves in Lebesgue spaces with variable exponent.* Acta Mathematica Sinica **23**(6) (2007), 965–972.

[16] V. Kokilashvili, S. Samko, *Sobolev theorem for potentials on Carleson curves in variable Lebesgue spaces.* Mem. Differential Equations Math. Phys. **33** (2004), 157–158.

[17] Ya.B. Lopatinskiy, *On a type of singular integral equations.* Theor. and Appl. Math. Lvov State Univ. **2** (1963), 53–57 (in Ukrainian).

[18] V.G. Maz'ya, *Boundary integral equations.* Itogi Nauki i Techniki. Sovremennie Problemi Matematiki. Fundamentalnie Napravlenia **27**, Moskva, 1988, 131–237 (in Russian).

[19] V. Maz'ya and A. Soloviev, *Boundary Integral Equations on Contours with Peaks.* Operator Theory: Advances and Applications **196**, Birkhäuser, Basel, 2010.

[20] V.S. Rabinovich, *Singular integral operators on a composed contour with oscillating tangent and pseudodifferential Mellin operators.* Soviet Math. Dokl. **44** (1992), 791–796.

[21] V.S. Rabinovich, *Singular integral operators on composed contours and pseudodifferential operators.* Math. Notes **58** (1995), 722–734.

[22] V.S. Rabinovich, *Algebras of singular integral operators on complicated contours with nodes being of logarithmic whirl points.* Izvestia AN Rossii, ser. mathem. **60**(6) (1996), 169–200 (in Russian); English transl.: Izvestia: Mathematics **60**(6) (1996), 1261–1292.

[23] V.S. Rabinovich, *Mellin pseudodifferential operators technique in the theory of singular integral operators on some Carleson curves.* In: Operator Theory: Advances and Applications **102** (1998), 201–218.

[24] V.S. Rabinovich, *Potential type operators on curves with Vorticity Points.* Zeitschrift für Analysis und ihre Anwendungen **18**(4) (1999), 1065–1081.

[25] V. Rabinovich, S. Roch and B. Silbermann, *Limit Operators and their Applications in Operator Theory.* Operator Theory: Advances and Applications **150**, Birkhäuser Verlag, Basel, 2004.

[26] V.S. Rabinovich and S.G. Samko, *Essential spectra of pseudodifferential operators in Sobolev spaces with variable smoothness and variable Lebesgue indices.* Doklady Mathematics **76**(3) (2007), 835–838; Dokl. AN Rossii **417**(2) (2007), 167–170 (in Russian).

[27] V. Rabinovich and S. Samko, *Boundedness and Fredholmness of pseudodifferential operators in variable exponent spaces.* Integr. Equ. Oper. Theory **60**(4) (2008), 507–537.

[28] V. Rabinovich and S. Samko, *Pseudodifferential operators approach to singular integral operators in weighted variable exponent Lebesgue spaces on Carleson curves.* Integr. Equ. Oper. Theory **69**(3) (2011), 405–444.

[29] I. Radon, *On boundary problems for the logarithmic potential.* Uspehi Math. Nauk **1**(3-4) (1946), 96–124 (Russian).

[30] S. Roch, P.A. Santos and B. Silbermann, *Noncommutative Gelfand Theory*. Springer, Berlin, 2011.

[31] I.B. Simonenko, *A new general method of investigating linear operator equations of singular integral equations type, I, II*. Izv. Acad. Nauk SSSR, Ser. Mat. **29**(3-4) (1965), 567–586; 757–782.

[32] I.B. Simonenko and Chin Ngok Min, *The local principle in the theory of one-dimensional singular integral operators with piece-wise continuous coefficients*. Rostov State University, Rostov-na-Donu, 1986, (Russian).

[33] I.B. Simonenko, *The local method in the theory of invariant with respect to shifts operators and their enveloping*. Rostov State University, Rostov-na-Donu, 2007, (Russian).

[34] V.Yu. Shelepov, *On index and spectrum of integral operators of potential type along the Radon curves*. Mathem. Sbornik **181**(6) (1990), 751–778 (in Russian).

[35] M.A. Shubin, *Pseudodifferential Operators and Spectral Theory*. Second edition. Springer, New York, 2001.

[36] G. Verchota, *Layers potentials and regularity for the Dirichlet problem for Laplace equations in Lipschitz domains*. J. Funct. Anal. **59**(3) (1984), 572–611.

[37] N.N. Vojtovich, B.Z. Katsenelenbaum and A.N. Sivov (M.S. Agranovich), *The Generalized Method of Eigenoscillations in the Diffraction Theory*. With a supplement by M.S. Agranovich: *Spectral Properties of Diffraction Problems*, Izdatel'stvo "Nauka", Moscow, 1977 (in Russian).

Vladimir Rabinovich
National Polytechnic Institute of Mexico
Av. IPN, ESIME Zacatenco, 07738
DF, MEXICO
e-mail: `vladimir.rabinovich@gmail.com`

Operator Theory:
Advances and Applications, Vol. 229, 349–356

A Note on Boundedness of Operators in Grand Grand Morrey Spaces

Humberto Rafeiro

Dedicated with great pleasure to Stefan Samko on the occasion of his 70th birthday

Abstract. In this note we introduce grand grand Morrey spaces, in the spirit of the grand Lebesgue spaces. We prove a kind of *reduction lemma* which is applicable to a variety of operators to reduce their boundedness in grand grand Morrey spaces to the corresponding boundedness in Morrey spaces. As a result of this application, we obtain the boundedness of the Hardy-Littlewood maximal operator and Calderón-Zygmund operators in the framework of grand grand Morrey spaces.

Mathematics Subject Classification (2010). Primary 46E30; Secondary 42B20, 42B25.

Keywords. Morrey spaces, maximal operator, Hardy-Littlewood maximal operator, Calderón-Zygmund operator.

1. Introduction

In 1992 T. Iwaniec and C. Sbordone [12], in their studies related with the integrability properties of the Jacobian in a bounded open set Ω, introduced a new type of function spaces $L^{p)}(\Omega)$, called *grand Lebesgue spaces*. A generalized version of them, $L^{p),\theta}(\Omega)$ appeared in L. Greco, T. Iwaniec and C. Sbordone [11]. Harmonic analysis related to these spaces and their associate spaces (called *small Lebesgue spaces*), was intensively studied during last years due to various applications, we mention, e.g., [2, 4, 6, 7, 8, 9, 13, 14].

Recently in [21] there was introduced a version of weighted grand Lebesgue spaces adjusted for sets $\Omega \subseteq \mathbb{R}^n$ of infinite measure, where the integrability of $|f(x)|^{p-\varepsilon}$ at infinity was controlled by means of a weight, and there generalized grand Lebesgue spaces were also considered, together with the study of classical op-

H.R. was supported in part by *Fundaçâo para a Ciência e a Tecnologia* (FCT), Grant SFRH/BPD/63085/2009, Portugal, and by Pontificia Universidad Javeriana, under the research project ID 5118.

erators of harmonic analysis in such spaces. Another idea of introducing "bilateral" grand Lebesgue spaces on sets of infinite measure was suggested in [16], where the structure of such spaces was investigated, not operators; the spaces in [16] are two parametrical with respect to the exponent p, with the norm involving $\sup_{p_1<p<p_2}$.

Morrey spaces $L^{p,\lambda}$ were introduced in 1938 by C. Morrey [17] in relation to regularity problems of solutions to partial differential equations, and provided a useful tool in the regularity theory of PDE's (for Morrey spaces we refer to books [10, 15], see also [20] where an overview of various generalizations may be found).

Recently, in the spirit of grand Lebesgue spaces, A. Meskhi [18, 19] introduced *grand Morrey spaces* (in [18] it was already defined on quasi-metric measure spaces with doubling measure) and obtained results on the boundedness of the maximal operator, Calděron-Zygmund singular operators and Riesz potentials. Note that the "*grandification* procedure" was applied only to the parameter p.

In this paper we make a further step and apply the "grandification procedure" to both the parameters, p and λ, obtaining *grand grand Morrey spaces* $L^{p),\lambda)}_{\theta,\alpha}(\Omega)$. In this new framework we obtain a reduction boundedness theorem, which reduces the boundedness of operators (not necessarily linear ones) in grand grand Morrey spaces to the corresponding boundedness in classical Morrey spaces, as well as giving an upper estimate for the norm of the operator.

Notation

Throughout the text we use the following notation:
c and C denote various absolute positive constants, which may have different values even in the same line,
Ω stands for a bounded open set in $\mathbb{R}^n$,
$|A|$ denote the Lebesgue measure of a measurable set $A \subset \Omega$,
$B(x,r) = \{y \in \mathbb{R}^n : |y-x| < r\}$,
$\widetilde{B}(x,r) = B(x,r) \cap \Omega$,
$\mathsf{d} := \operatorname{diam} \Omega$,
$⨍_B f(x)\,\mathrm{d}x$ denote the integral average of the function f, i.e., $⨍_B f(x)\,\mathrm{d}x := |B|^{-1}\int_B f(x)\,\mathrm{d}x$,
$\hookrightarrow$ means continuous embedding.

2. Preliminaries

Everywhere in the sequel, Ω is supposed to be a bounded open set.

2.1. Grand Lebesgue spaces

For $1 < p < \infty$ and $\theta > 0$ the *grand Lebesgue space* (less known as *Iwaniec-Sbordone space*) is the set of all real-valued measurable functions for which

$$\|f\|_{L^{p),\theta}(\Omega)} := \sup_{0<\varepsilon<p-1} \varepsilon^{\frac{\theta}{p-\varepsilon}} \|f\|_{L^{p-\varepsilon}(\Omega)} < \infty. \tag{1}$$

In the case $\theta = 1$, we simply denote $L^{p),\theta}(\Omega) := L^{p)}(\Omega)$.

Remark 2.1. We emphasize that, for our purposes, the definition given in (1) is different from the one existing in literature, namely we don't normalize the $L^{p-\varepsilon}$ norm. Since we assume Ω to be a bounded open set, the two definitions give equivalent norms.

We note that, for all $0 < \varepsilon \leqslant p-1$, we have

$$L^p(\Omega) \hookrightarrow L^{p)}(\Omega) \hookrightarrow L^{p-\varepsilon}(\Omega).$$

For more properties of grand Lebesgue spaces, see [13].

2.2. Morrey spaces

For $1 \leqslant p < \infty$ and $0 \leqslant \lambda < 1$, the usual Morrey space $L^{p,\lambda}(\Omega)$ is introduced as the set of all real-valued measurable functions such that

$$\|f\|_{L^{p,\lambda}(\Omega)} := \sup_{\substack{x\in\Omega \\ 0<r\leqslant \mathsf{d}}} \left(\frac{1}{|B(x,r)|^\lambda} \int_{\widetilde{B}(x,r)} |f(y)|^p \, \mathrm{d}y \right)^{\frac{1}{p}} < \infty$$

where $\mathsf{d} := \operatorname{diam} \Omega$.

3. Grand grand Morrey spaces and the reduction lemma

For $\theta > 0$, $\alpha \geqslant 0$, $1 < p < \infty$ and $0 \leqslant \lambda < 1$, we consider the functional

$$\Phi^{p,\lambda}_{\theta,\alpha}(f,s) := \sup_{0<\varepsilon\leqslant s} \varepsilon^{\frac{\theta}{p-\varepsilon}} \|f\|_{L^{p-\varepsilon,\lambda-\alpha\varepsilon}(\Omega)}, \tag{2}$$

where $0 < s < \min\{p-1, \lambda/\alpha\}$.

We make a convention that the quotient λ/α when $\alpha = 0$ is always $\lambda/\alpha := \infty$ even if $\lambda = 0$.

Definition 3.1 (Grand grand Morrey spaces). Let $1 < p < \infty$, $\theta > 0$, $\alpha \geqslant 0$ and $0 \leqslant \lambda < 1$. By $L^{p),\lambda)}_{\theta,\alpha}(\Omega)$ we denote the space of real-valued measurable functions having the finite norm

$$\|f\|_{L^{p),\lambda)}_{\theta,\alpha}(\Omega)} := \Phi^{p,\lambda}_{\theta,\alpha}(f, s_{\max}), \qquad s_{\max} = \min\left\{p-1, \frac{\lambda}{\alpha}\right\}. \tag{3}$$

Remark 3.2. In the case $\alpha = 0, \lambda > 0$ we recover the grand Morrey space introduced in [19], and when $\lambda = \alpha = 0$, in view of the convention made above, we have the grand Lebesgue space introduced in [11] (and in [12] in the case $\theta = 1$).

For fixed $p, \theta, \alpha, \lambda, f$ we have that $s \mapsto \Phi^{p,\lambda}_{\theta,\alpha}(f,s)$ is a non-decreasing function, but it is possible to estimate $\Phi^{p,\lambda}_{\theta,\alpha}(f,s)$ via $\Phi^{p,\lambda}_{\theta,\alpha}(f,\sigma)$ with $\sigma < s$ as follows.

Lemma 3.3. *For $0 < \sigma < s < \min\left\{p-1, \frac{\lambda}{\alpha}\right\}$ we have that*

$$\Phi^{p,\lambda}_{\theta,\alpha}(f,s) \leqslant C s^{\frac{\theta}{p-s}} \sigma^{-\frac{\theta}{p-\sigma}} \Phi^{p,\lambda}_{\theta,\alpha}(f,\sigma), \tag{4}$$

where C depends on n, the parameters $p, \theta, \alpha, \lambda$ and the diameter d, but does not depend on f, s and σ.

Proof. For $0<\sigma<s<\min\{p-1,\frac{\lambda}{\alpha}\}$ we have

$$\Phi^{p,\lambda}_{\theta,\alpha}(f,s)=\max\Big\{\Phi^{p,\lambda}_{\theta,\alpha}(f,\sigma),\underbrace{\sup_{\sigma<\varepsilon\leqslant s}\varepsilon^{\frac{\theta}{p-\varepsilon}}\|f\|_{L^{p-\varepsilon,\lambda-\alpha\varepsilon}(\Omega)}}_{I}\Big\}. \tag{5}$$

To estimate

$$I=\sup_{\sigma<\varepsilon\leqslant s}\varepsilon^{\frac{\theta}{p-\varepsilon}}\sup_{\substack{x\in\Omega\\0<r\leqslant \mathsf{d}}}|B(x,r)|^{\frac{\alpha\varepsilon-\lambda}{p-\varepsilon}}\|f\|_{L^{p-\varepsilon}(\widetilde{B}(x,r))},$$

note that the function $g(\varepsilon):=\varepsilon^{\frac{\theta}{p-\varepsilon}}$ is increasing in $0<\varepsilon\leqslant\min\{p-1,\frac{\lambda}{\alpha}\}$, so that

$$\begin{aligned}I&\leqslant s^{\frac{\theta}{p-s}}\sup_{\sigma<\varepsilon\leqslant s_{\max}}\sup_{\substack{x\in\Omega\\0<r\leqslant \mathsf{d}}}|B(x,r)|^{\frac{1+\alpha\varepsilon-\lambda}{p-\varepsilon}}\left(\fint_{B(x,r)}|f(y)\chi_{\widetilde{B}(x,r)}(y)|^{p-\sigma}\,\mathrm{d}y\right)^{\frac{1}{p-\sigma}}\\&\leqslant s^{\frac{\theta}{p-s}}\sup_{\sigma<\varepsilon\leqslant s_{\max}}\sup_{\substack{x\in\Omega\\0<r\leqslant \mathsf{d}}}|B(x,r)|^{\Delta(\varepsilon)}\left(\frac{\sigma^\theta\sigma^{-\theta}}{|B(x,r)|^{\lambda-\alpha\sigma}}\int_{\widetilde{B}(x,r)}|f(y)|^{p-\sigma}\,\mathrm{d}y\right)^{\frac{1}{p-\sigma}}\end{aligned}$$

where $s_{\max}$ is defined in (3) and $\Delta(\varepsilon):=\frac{1+\alpha\varepsilon-\lambda}{p-\varepsilon}-\frac{1+\alpha\sigma-\lambda}{p-\sigma}$.

Observe that

$$\Delta(\varepsilon)=\frac{1+\alpha\varepsilon-\lambda}{p-\varepsilon}-\frac{1+\alpha\sigma-\lambda}{p-\sigma}=\frac{(1-\lambda+\alpha p)(\varepsilon-\sigma)}{(p-\sigma)(p-\varepsilon)}\geqslant 0,$$

and for $0\leqslant\varepsilon<\min\{p-1,\lambda/\alpha\}$ we have $\frac{1-\lambda}{p}\leqslant\frac{1+\alpha\varepsilon-\lambda}{p-\varepsilon}\leqslant 1$, so that $0\leqslant\Delta(\varepsilon)\leqslant 1$. Then

$$|B(x,r)|^{\frac{1+\alpha\varepsilon-\lambda}{p-\varepsilon}-\frac{1+\alpha\sigma-\lambda}{p-\sigma}}\leqslant C\max\{1,\mathsf{d}^n\}$$

and we obtain

$$\begin{aligned}I&\leqslant Cs^{\frac{\theta}{p-s}}\cdot\sup_{\substack{x\in\Omega\\0<r\leqslant \mathsf{d}}}\sigma^{-\frac{\theta}{p-\sigma}}\left(\frac{\sigma^\theta}{|B(x,r)|^{\lambda-\alpha\sigma}}\int_{\widetilde{B}(x,r)}|f(y)|^{p-\sigma}\,\mathrm{d}y\right)^{\frac{1}{p-\sigma}}\\&\leqslant Cs^{\frac{\theta}{p-s}}\cdot\sigma^{-\frac{\theta}{p-\sigma}}\sup_{0<\varepsilon\leqslant\sigma}\varepsilon^{\frac{\theta}{p-\varepsilon}}\|f\|_{L^{p-\varepsilon,\lambda-\alpha\varepsilon}(\Omega)}\\&=Cs^{\frac{\theta}{p-s}}\cdot\sigma^{-\frac{\theta}{p-\sigma}}\cdot\Phi^{p,\lambda}_{\theta,\alpha}(f,\sigma).\end{aligned}$$

□

From Lemma 3.3 we immediately have

Lemma 3.4. *For $0<\sigma<\min\{p-1,\frac{\lambda}{\alpha}\}$, the norm defined in (3) has the following dominant*

$$\|f\|_{L^{p),\lambda)}_{\theta,\alpha}(\Omega)}\leqslant C\frac{\Phi^{p,\lambda}_{\theta,\alpha}(f,\sigma)}{\sigma^{\frac{\theta}{p-\sigma}}}, \tag{6}$$

where C depends on $n,p,\theta,\alpha,\lambda$ and d, but does not depend on f and σ.

Lemma 3.5 (Reduction lemma). *Let U be an operator (not necessarily linear) bounded in the usual Morrey spaces $L^{p-\varepsilon,\lambda-\alpha\varepsilon}(\Omega)$:*

$$\|Uf\|_{L^{p-\varepsilon,\lambda-\alpha\varepsilon}(\Omega)} \leqslant C_{p-\varepsilon,\lambda-\alpha\varepsilon}\|f\|_{L^{p-\varepsilon,\lambda-\alpha\varepsilon}(\Omega)} \tag{7}$$

for all sufficiently small $\varepsilon \in [0,\sigma]$, where $0 < \sigma < \min\left\{p-1, \frac{\lambda}{\alpha}\right\}$. If we have $\sup_{0<\varepsilon<\sigma} C_{p-\varepsilon,\lambda-\alpha\varepsilon} < \infty$, then it is also bounded in the grand grand Morrey space $L^{p),\lambda)}_{\theta,\alpha}(\Omega)$:

$$\|Uf\|_{L^{p),\lambda)}_{\theta,\alpha}(\Omega)} \leqslant C\|f\|_{L^{p),\lambda)}_{\theta,\alpha}(\Omega)} \tag{8}$$

with

$$C = \frac{C_0}{\sigma^{\frac{\theta}{p-\sigma}}} \sup_{0<\varepsilon\leqslant\sigma} C_{p-\varepsilon,\lambda-\alpha\varepsilon},$$

where C_0 may depend on $n, p, \theta, \alpha, \lambda$ and d, but does not depend on σ.

Proof. By (6), we have

$$\|Uf\|_{L^{p),\lambda)}_{\theta,\alpha}(\Omega)} \leqslant \frac{C}{\sigma^{\frac{\theta}{p-\sigma}}} \Phi^{p,\lambda}_{\theta,\alpha}(Uf,\sigma). \tag{9}$$

The estimation of $\Phi^{p,\lambda}_{\theta,\alpha}(Uf,\sigma)$ by $\|f\|_{L^{p),\lambda)}_{\theta,\alpha}(\Omega)}$ is direct:

$$\begin{aligned}\Phi^{p,\lambda}_{\theta,\alpha}(Uf,\sigma) &= \sup_{0<\varepsilon\leqslant\sigma} \varepsilon^{\frac{\theta}{p-\varepsilon}}\|Uf\|_{L^{p-\varepsilon,\lambda-\alpha\varepsilon}(\Omega)} \\ &\leqslant \sup_{0<\varepsilon\leqslant\sigma} \varepsilon^{\frac{\theta}{p-\varepsilon}} \cdot C_{p-\varepsilon,\lambda-\alpha\varepsilon} \cdot \|f\|_{L^{p-\varepsilon,\lambda-\alpha\varepsilon}(\Omega)} \\ &\leqslant \sup_{0<\varepsilon\leqslant\sigma} C_{p-\varepsilon,\lambda-\alpha\varepsilon} \cdot \|f\|_{L^{p),\lambda)}_{\theta,\alpha}(\Omega)}\end{aligned} \tag{10}$$

which completes the proof. □

4. On boundedness of operators in the grand grand Morrey spaces

4.1. Maximal operator in grand grand Morrey spaces

Let

$$Mf(x) = \sup_{0<r<\mathsf{d}} \fint_{\widetilde{B}(x,r)} |f(y)|\,dy, \qquad x \in \Omega \tag{11}$$

be the usual centered maximal operator. The Hardy-Littlewood-Wiener theorem regarding the boundedness of the maximal operator in Lebesgue spaces is a well-known result, see, e.g., [5]. A similar result is valid in the framework of Morrey spaces, namely

Lemma 4.1. *Let $1 < p < \infty$ and $0 \leqslant \lambda < 1$. Then*

$$\|Mf\|_{L^{p,\lambda}(\Omega)} \leqslant \left(2^{\frac{n\lambda}{p}} c\left(\frac{p}{p-1}\right)^{\frac{1}{p}} + 1\right)\|f\|_{L^{p,\lambda}(\Omega)}. \tag{12}$$

where c is a positive constant non-depending on p.

Remark 4.2. The above explicit evaluation of the constant in Lemma 4.1 is the one obtained in [19]. For another approach with slightly different evaluation of the constant, see [1, 3].

Theorem 4.3. *Let* $1<p<\infty$, $\theta>0$, $\alpha\geqslant 0$ *and* $0\leqslant\lambda<1$. *Then the Hardy-Littlewood maximal operator* (11) *is bounded in grand grand Morrey spaces* $L^{p),\lambda)}_{\theta,\alpha}(\Omega)$.

Proof. By the reduction lemma 3.5 and (12), we only need to show the finiteness of

$$\sup_{0<\varepsilon\leqslant\sigma} C_{p-\varepsilon,\lambda-\alpha\varepsilon} = \sup_{0<\varepsilon\leqslant\sigma}\left(2^{\frac{n(\lambda-\alpha\varepsilon)}{p-\varepsilon}}c_0\left(\frac{p-\varepsilon}{p-\varepsilon-1}\right)^{\frac{1}{p-\varepsilon}}+1\right)$$

which holds if we choose $\sigma<p-1$. Note that the use of the reduction lemma in this proof is not necessary when the grand grand space $L^{p),\lambda)}_{\theta,\alpha}(\Omega)$ is considered with $\alpha>\frac{\lambda}{p-1}$ since, in that case, the constant in (12) is uniformly bounded for $0<\varepsilon\leqslant s_{\max}$. □

4.2. Singular integral operators in grand grand Morrey spaces

We follow [19] in this section, in particular, making use of the following definition of the Calderón-Zygmund singular operators. Namely, the Calderón-Zygmund operator is treated as the integral operator

$$Tf(x) = \text{p.v.}\int_\Omega K(x,y)f(y)\,\mathrm{d}y$$

with the kernel $K:\Omega\times\Omega\backslash\{(x,x):x\in\Omega\}\to\mathbb{R}$ satisfying the conditions:

$$|K(x,y)|\leqslant\frac{C}{|x-y|^n},\quad x,y\in\Omega,\quad x\neq y;$$

$$|K(x_1,y)-K(x_2,y)|+|K(y,x_1)-K(y,x_2)|\leqslant Cw\left(\frac{|x_2-x_1|}{|x_2-y|}\right)\frac{1}{|x_2-y|^n}$$

for all x_1, x_2 and y with $|x_2-y|>C|x-x_2|$, where w is a positive non-decreasing function on $(0,\infty)$ which satisfies the doubling condition $w(2t)\leqslant cw(t)$ and the Dini condition $\int_0^1 w(t)/t\,\mathrm{d}t<\infty$. In the case where w is a power function, this goes back to Coifman-Meyers version of singular operators with *standard kernel*. We also assume that Tf exists almost everywhere on Ω in the principal value sense for all $f\in L^2(\Omega)$ and that T is bounded in $L^2(\Omega)$.

The boundedness of such Calderón-Zygmund operators in Morrey spaces is valid, as can be seen in the following Proposition, proved in [19]

Proposition 4.4. *Let* $1<p<\infty$ *and* $0\leqslant\lambda<1$. *Then*

$$\|Tf\|_{L^{p,\lambda}(\Omega)}\leqslant C_{p,\lambda}\|f\|_{L^{p,\lambda}(\Omega)}$$

where

$$C_{p,\lambda}\leqslant c\begin{cases}\frac{p}{p-1}+\frac{p}{2-p}+\frac{p-\lambda+1}{1-\lambda} & \text{if } 1<p<2;\\ p+\frac{p}{p-2}+\frac{p-\lambda+1}{1-\lambda} & \text{if } p>2.\end{cases}\tag{13}$$

with c *not depending on* p *and* λ.

Theorem 4.5. *Let $1 < p < \infty$, $\theta > 0$, $\alpha \geqslant 0$ and $0 < \lambda < 1$. Then the Calderón-Zygmund operator T is bounded in grand grand Morrey spaces $L^{p),\lambda)}_{\theta,\alpha}(\Omega)$.*

Proof. Keeping in mind that by the reduction lemma 3.5 we are interested only in small values of ε, from (13), we deduce that

$$C_{p-\varepsilon,\lambda-\alpha\varepsilon} \leqslant c \begin{cases} \frac{p}{p-\varepsilon-1} + \frac{p-\varepsilon}{2-p+\varepsilon} + \frac{p-\varepsilon-\lambda+\alpha\varepsilon+1}{1-\lambda+\alpha\varepsilon} & \text{if } p \leqslant 2 \text{ and } 0 < \varepsilon < p-1; \\ p-\varepsilon + \frac{p-\varepsilon}{p-\varepsilon-2} + \frac{p-\varepsilon-\lambda+\alpha\varepsilon+1}{1-\lambda+\alpha\varepsilon} & \text{if } p > 2 \text{ and } 0 < \varepsilon < p-2, \end{cases}$$

so that when applying the reduction lemma, it suffices to take $\sigma < \min\left\{p-1, \frac{\lambda}{\alpha}\right\}$ when $p \leqslant 2$ and $\sigma < \min\left\{p-2, \frac{\lambda}{\alpha}\right\}$ when $p > 2$. □

Acknowledgment

The author wish to thank Stefan Samko for pointing out this topic as well as fruitful discussions and useful comments on this paper. The author is also thankful to the referees for valuable comments.

References

[1] A. Almeida, J. Hasanov, S. Samko, *Maximal and potential operators in variable exponent Morrey spaces.* Georgian Math. J. **15**(2) (2008), 195–208.

[2] C. Capone, A. Fiorenza, *On small Lebesgue spaces.* J. Function Spaces and Applications, **3** (2005), 73–89.

[3] F. Chiarenza, M. Frasca, *Morrey spaces and Hardy-Littlewood maximal function.* Rend. Mat. Appl. **7** (1987), 273–279.

[4] G. Di Fratta, A. Fiorenza, *A direct approach to the duality of grand and small Lebesgue spaces.* Nonlinear Analysis: Theory, Methods and Applications, **70**(7) (2009), 2582–2592.

[5] J. Duoandikoetxea, *Fourier Analysis.* Graduate Studies, Amer. Math. Soc., 2001.

[6] A. Fiorenza, *Duality and reflexivity in grand Lebesgue spaces.* Collect. Math. **51**(2) (2000), 131–148.

[7] A. Fiorenza, B. Gupta, P. Jain, *The maximal theorem in weighted grand Lebesgue spaces.* Studia Math. **188**(2) (2008), 123–133.

[8] A. Fiorenza, G.E. Karadzhov, *Grand and small Lebesgue spaces and their analogs.* Z. Anal. Anwend., **23**(4) (2004), 657–681.

[9] A. Fiorenza, J.M. Rakotoson, *Petits espaces de Lebesgue et leurs applications.* C.R.A.S. t, **333** (2001), 1–4.

[10] M. Giaquinta, *Multiple integrals in the calculus of variations and non-linear elliptic systems.* Princeton Univ. Press, 1983.

[11] L. Greco, T. Iwaniec, C. Sbordone, *Inverting the p-harmonic operator.* Manuscripta Math. **92** (1997), 249–258.

[12] T. Iwaniec, C. Sbordone, *On the integrability of the Jacobian under minimal hypotheses.* Arch. Rational Mech. Anal., **119** (1992), 129–143.

[13] V. Kokilashvili, *Weighted problems for operators of harmonic analysis in some Banach function spaces.* Lecture course of Summer School and Workshop "Harmonic Analysis and Related Topics" (HART2010), Lisbon, June 21–25, `http://www.math.ist.utl.pt/~hart2010/kokilashvili.pdf`, 2010.

[14] V. Kokilashvili, *Singular integrals and strong maximal functions in weighted grand Lebesgue spaces.* Jiří Rákosník (ed.): Nonlinear Analysis, Function Spaces and Applications, Proceedings of the Spring School held in Třešt', September 11–17, 2010, Vol. 9. Czech Academy of Sciences, Mathematical Institute, Praha, 2011. pp. 261–269.

[15] A. Kufner, O. John, S. Fučik, *Function spaces.* Noordhoff International Publishing, Leyden; Academia, Prague, 1977.

[16] E. Liflyand, E. Ostrovsky, L. Sirota, *Structural properties of Bilateral Grand Lebesgue Spaces.* Turk. J. Math. **34** (2010), 207–219.

[17] C. B. Morrey, *On the solutions of quasi-linear elliptic partial differential equations.* Trans. Amer. Math. Soc. **43**(1) (1938), 126–166.

[18] A. Meskhi, *Maximal functions and singular integrals in Morrey spaces associated with grand Lebesgue spaces.* Proc. A. Razmadze Math. Inst. **151** (2009), 139–143.

[19] A. Meskhi, *Maximal functions, potentials and singular integrals in grand Morrey spaces.* Complex Var. Elliptic Equ., 56:10-11 (2011), 1003–1019.

[20] H. Rafeiro, N. Samko, S. Samko, *Morrey-Campanato spaces: an overview*, Operator Theory: Advances and Applications (Birkhäuser), Vol. Operator Theory, Pseudo-Differential Equations, and Mathematical Physics. (The Vladimir Rabinovich Anniversary Volume), to appear.

[21] S.G. Samko, S.M. Umarkhadzhiev, *On Iwaniec-Sbordone spaces on sets which may have infinite measure.* Azerb. J. Math. **1**(1) (2010) 67–84.

Humberto Rafeiro
Instituto Superior Técnico
Departamento de Matemática
Centro CEAF, Av. Rovisco Pais
P-1049–001 Lisboa, Portugal

and

Pontificia Universidad Javeriana
Cra 7a No 43-82 Ed. Carlos Ortiz
Departamento de Matemáticas
Bogotá, Colombia
e-mail: `hrafeiro@math.ist.utl.pt`
`silva-h@javeriana.edu.co`

Operator Theory:
Advances and Applications, Vol. 229, 357–370

Operational Calculus for Bessel's Fractional Equation

M.M. Rodrigues, N. Vieira and S. Yakubovich

Abstract. This paper is intended to investigate a fractional differential Bessel's equation of order 2α with $\alpha \in]0,1]$ involving the Riemann-Liouville derivative. We seek a possible solution in terms of power series by using operational approach for the Laplace and Mellin transform. A recurrence relation for coefficients is obtained. The existence and uniqueness of solutions is discussed via Banach fixed point theorem.

Mathematics Subject Classification (2010). Primary 35R11; Secondary 34B30, 42A38, 47H10.

Keywords. Fractional differential equations, Riemann Liouville derivative, Mellin transform, Laplace transform, Bessel equation.

1. Introduction

As it is known, one of the most important equations in physical sciences is the Bessel equation, which occurs in many problems of potential theory for cylindrical domains. Its solutions are called cylinder functions. Certain special kind of these functions are known in the literature as Bessel functions, and this term is sometimes applied to the whole class of cylinder functions (for more details see [5, Ch. 5]). As far as the authors are aware, there were no attempts to study the corresponding fractional Bessel equation (see below). Probably its solutions can be also applied in several areas of mathematical-physics, as for example, in analysis of electromagnetic waves in cylindrical domains, examining the fractional heat equation and radial-fractional Schrödinger equation, a study of fractional diffusion problems on a lattice, etc.

Fractional differential equations are widely used for modeling anomalous relaxation and diffusion phenomena (see [2, Ch. 5], [4, Ch. 2]). A systematic development of the analytic theory of fractional differential equations with variable coefficients can be found, for instance, in the books of Samko, Kilbas and Marichev [11, Ch. 3] and Yakubovich and Luchko [13, Ch. 20].

Among analytical methods, which are widely used to solve precisely fractional equations, are methods of integral transforms (see, e.g., [6, 7]). As for different problems, the Laplace, Fourier, Mellin transforms are applicable. Solutions of fractional equations are usually given in the forms of special functions ([4, Ch. 2], [9, Ch. 3]), such as the Mittag-Leffler, Wright's function, the Fox H-function and the Meijer G-function. However, since some useful properties of the standard calculus cannot carry over analogously to the case of fractional calculus, the analytical solutions of fractional differential equations under certain initial and boundary conditions are difficult to obtain.

Recently, (see [14]) by using Laplace and Fourier transforms methods for a symbolic operational form of solutions in terms of the Mittag-Leffler functions multidimensional space-time fractional Schrödinger equation was obtained.

The aim of the paper is to study the following fractional Bessel equation

$$x^{2\alpha}(D_{0+}^{2\alpha}y)(x) + x^{\alpha}(D_{0+}^{\alpha}y)(x) + (x^{2\alpha} - \lambda^2)y(x) = 0,$$

where $x > 0$, $\alpha \in]0,1]$ and $\lambda \in \mathbb{C}$. In particular, the case $\alpha = 1$ derives us to the classical Bessel's equation (see [5, Ch. 5]), whose solutions are, correspondingly, the Bessel functions. The paper is structured as follows: in the Preliminaries we will recall basic properties of the Mellin transform and necessary elements of fractional calculus. Section 3 deals with a recurrence relation of the coefficients for the series solution associated to the considered fractional differential equation. Besides the Mellin transform method for solving fractional Bessel equations is presented. Finally, in Section 4 we will study the existence and uniqueness of solutions appealing to the Banach fix point theorem.

2. Preliminaries

2.1. The Mellin transform of fractional derivatives

The resolution of differential equations with polynomial coefficients becomes more efficient considering the Mellin transform. Usually, this integral transform is defined by [12, Sec. 2.7], [8, Ch. 10]

$$M\{f(x);s\} = \int_0^\infty f(x)x^{s-1}dx, \tag{2.1}$$

where s is complex, such as $\gamma_1 < \Re(s) < \gamma_2$. The Mellin transform exists if $f(x)$ is piecewise continuous in every closed interval $[a,b] \subset]0,+\infty[$ and

$$\int_0^1 |f(x)|x^{\gamma_1-1}dx < \infty, \qquad \int_1^\infty |f(x)|x^{\gamma_2-1}dx < \infty.$$

If the function satisfies Dirichlet's condition in every closed interval $[a,b] \subset]0,+\infty[$, then it can be restored using the inverse Mellin transform formula

$$f(x) = \frac{1}{2\pi i}\int_{\gamma-i\infty}^{\gamma+i\infty} M\{f(x);s\}x^{-s}\,ds, \qquad 0 < x < \infty, \tag{2.2}$$

where $\gamma_1 < \gamma < \gamma_2$. Moreover, we recall the following elementary properties (see [8, Ch. 10], [10, Ch. 8])

$$M\{x^\beta f(x);s\} = M\{f(x);s+\beta\} \equiv F(s+\beta),\ \gamma_1 - \Re(\beta) < \Re(s) < \gamma_2 - \Re(\beta); \tag{2.3}$$

$$M\{f(\beta x);s\} = \beta^{-s} M\{f(x);s\} \equiv \beta^{-s} F(s),\ \gamma_1 < \Re(s) < \gamma_2,\ \beta > 0; \tag{2.4}$$

$$M\{f(x^\beta);s\} = \frac{1}{|\beta|} M\left\{f(x);\frac{s}{\beta}\right\} \equiv \frac{1}{|\beta|}\ F\left(\frac{s}{\beta}\right),\ \left\{\begin{array}{ll} \beta\gamma_1 < \Re(s) < \beta\gamma_2, & \Leftarrow \beta > 0 \\ \beta\gamma_2 < \Re(s) < \beta\gamma_1, & \Leftarrow \beta < 0 \end{array}\right. . \tag{2.5}$$

Further, if additionally $f \in C^n(\mathbb{R}^+), n \in \mathbb{N}$, then

$$M\{f^{(n)}(x);s\} = \frac{\Gamma(1+n-s)}{\Gamma(1-s)}\ F(s-n). \tag{2.6}$$

In the sequel, we will appeal to the Mellin transform to examine the fractional Bessel equation. Namely, we will consider the following differential properties of this integral transform (see [10, Ch. 8], [11, Ch. 3])

$$M\{x^{2\beta}(D_{0+}^{2\beta} y)(x);s\} = \frac{\Gamma(1-s)}{\Gamma(1-s-2\beta)}\ Y(s), \tag{2.7}$$

$$M\{x^{\beta}(D_{0+}^{\beta} y)(x);s\} = \frac{\Gamma(1-s)}{\Gamma(1-s-\beta)}\ Y(s), \tag{2.8}$$

$$M\{x^{2\beta} y(x);s\} = Y(s+2\beta). \tag{2.9}$$

Theorem 2.1 (see [12, Sec. 2.7]). *Let $x^{\beta-1}f(x)$ and $x^{\beta-1}g(x)$ belong to $L(0,+\infty)$, and let*

$$h(x) = (f *_M g)(x) = \int_0^\infty f(y)\, g\left(\frac{x}{y}\right)\ \frac{dy}{y}.$$

Then $x^{\beta-1}h(x)$ belongs to $L(0,+\infty)$, and its Mellin transform is $F(s)G(s)$, where $F(s) = M\{f(x);s\}$ and $G(s) = M\{g(x);s\}$.

Let us take $0 \leq n-1 < \beta < n$. According to the definition of Riemann-Liouville fractional derivative (see [4, Ch. 2]), we can write

$$(D_{0+}^\beta f)(x) = \left(\frac{d}{dx}\right)^n \frac{1}{\Gamma(n-\beta)} \int_0^x \frac{f(t)}{(x-t)^{\beta-n+1}}\ dt,\ n = [\beta]+1, \tag{2.10}$$

where $[\beta]$ is the integer part of β. Note that the Riemann-Liouville derivative is defined for some functions with a singularity at the origin. For example, if $f(x) = x^d, d > -1$, then

$$(D_{0+}^\beta f)(x) = \frac{\Gamma(d+1)}{\Gamma(d+1-\beta)}\ x^{d-\beta} \tag{2.11}$$

so that $D_{0+}^\beta f = 0$ if $f(x) = x^{\beta-1}$.

The Caputo derivative of fractional order β of a function $f(x)$ is defined, in turn, as (see [4, Ch. 2])

$$ {}^{C}D^{\beta}_{0+} f(x) = \frac{1}{\Gamma(n-\beta)} \int_0^x (x-\tau)^{n-\beta-1} f^{(n)}(\tau)\, d\tau, \tag{2.12}$$

where $n-1 < \beta < n$ and $n \in \mathbb{N}$.

We have that the Mellin transform (2.1) of the Riemann-Liouville derivative (2.10) is equal to the following formula (see [4, Ch. 2])

$$M\{(D^{\beta}_{0+} f)(x); s\} = \sum_{k=0}^{n-1} \frac{\Gamma(1+k-s)}{\Gamma(1-s)} (D^{\beta-n}_{0+} f)(x) x^{s-k-1} \Big|_0^{+\infty} + \frac{\Gamma(1-s+\beta)}{\Gamma(1-s)} F(s-\beta).$$

If f is such that all out integrand terms are vanished, then it takes more simple form

$$M\{(D^{\beta}_{0+} f)(x); s\} = \frac{\Gamma(1-s+\beta)}{\Gamma(1-s)} F(s-\beta). \tag{2.13}$$

2.2. Fractional calculus

In this section we recall some results about fractional calculus which will be used below. In what follows $n = [\beta]+1$ for $\beta \notin \mathbb{N}_0$ and $n = \beta$ for $\beta \in \mathbb{N}_0$.

Theorem 2.2 (see [4, Ch. 2]). *Let $\beta \geq 0$ and $v(x) \in AC^n([a,b])$. Then $D^{\beta}_{a+} v$ exists almost everywhere and may be represented in the form*

$$D^{\beta}_{a+} v = \sum_{k=0}^{n-1} \frac{v^{(k)}(a)}{\Gamma(1+k-\beta)} (x-a)^{k-\beta} + \frac{1}{\Gamma(n-\beta)} \int_a^x \frac{v^{(n)}(t)}{(x-t)^{\beta-n+1}}\, dt. \tag{2.14}$$

We remark that fractional derivatives (2.10) verify the following relation

$$ {}^{C}(D^{\beta}_{a+} v)(x) = (D^{\beta}_{a+} v)(x) - \sum_{k=0}^{n-1} \frac{v^{(k)}(a)}{\Gamma(k-\beta+1)} (x-a)^{k-\beta}. \tag{2.15}$$

Theorem 2.3 (see [4, Ch. 2]). *Let $\beta \geq 0$. If $v(x) \in AC^n([a,b])$, then the Caputo fractional derivative ${}^{C}D^{\beta}_{a+} v$ exists almost everywhere on $[a,b]$, and if $\beta \notin \mathbb{N}_0$, ${}^{C}D^{\beta}_{a+} v$ is represented by*

$$ {}^{C}D^{\beta}_{a+} v(x) = \frac{1}{\Gamma(n-\beta)} \int_a^x \frac{v^{(n)}(t)}{(x-t)^{\beta-n+1}}\, dt := (I^{n-\beta}_{a+} D^n v)(x), \tag{2.16}$$

where $D = \frac{d}{dx}$ is the ordinary derivative.

If $\beta \notin \mathbb{N}_0$ and $n = [\beta]+1$, then [4, Ch. 2]

$$\left| (I^{n-\beta}_{a+} D^n v)(x) \right| \leq \frac{\|v^{(n)}\|_C}{\Gamma(n-\beta)\,(n-\beta+1)} (x-a)^{n-\beta}. \tag{2.17}$$

Lemma 2.4 (see [4, Ch. 2]). *Let $\beta > 0$. If $v(x) \in AC^n([a,b])$ or $v(x) \in C^n([a,b])$, then*

$$\left(I_{a+}^{\alpha}\ {}^{C}D_{a+}^{\beta}v\right)(x) = v(x) - \sum_{k=0}^{n-1} \frac{v^{(k)}(a)}{k!}(x-a)^k.$$

Lemma 2.5 (see [11, Ch. 3]). *If the series $f(x) = \sum_{n=0}^{\infty} f_n(x)$, $f_n(x) \in C([a,b])$, is uniformly convergent on $[a,b]$, then its termwise fractional integration is admissible:*

$$\left(I_{a+}^{\beta} \sum_{n=0}^{\infty} f_n\right)(x) = \sum_{n=0}^{\infty} \left(I_{a+}^{\beta} f_n\right)(x),\ \beta > 0,\ a < x < b, \tag{2.18}$$

the series on the right-hand side being also uniformly convergent on $[a,b]$.

Lemma 2.6 (see [11, Ch. 3]). *Let the fractional derivatives $D_{a+}^{\beta} f_n$ exist for all $n = 0, 1, 2, \ldots$ and let the series $\sum_{n=0}^{\infty} f_n$ and $\sum_{n=0}^{\infty} D_{a+}^{\beta} f_n$ be uniformly convergent on every sub-interval $[a+\epsilon, b]$, $\epsilon > 0$. Then, the former series admit termwise fractional differentiation using the formula*

$$\left(D_{a+}^{\beta} \sum_{n=0}^{\infty} f_n\right)(x) = \left(\sum_{n=0}^{\infty} D_{a+}^{\beta} f_n\right)(x),\ \beta > 0,\ a < x < b, \tag{2.19}$$

3. Fractional Bessel equation

The aim of this section is to obtain particular solutions for the following fractional Bessel equation

$$x^{2\alpha}(D_{0+}^{2\alpha}y)(x) + x^{\alpha}(D_{0+}^{\alpha}y)(x) + (x^{2\alpha} - \lambda^2)y(x) = 0,\ \alpha \in]0,1],\ \lambda \in \mathbb{C},\ x \in [x_0, X_0], \tag{3.1}$$

where $x_0, X_0 \in \mathbb{R}_+$ and D_{0+}^{α} is the operator of the Riemann fractional derivative (2.10).

3.1. Recurrence relation for the coefficients of the series solution

Here, we derive a recurrence relation for coefficients of the series solution associated to (3.1). We seek a solution of (3.1) in the form of a generalized power series in increasing powers of argument x

$$y(x) = \sum_{n=0}^{+\infty} a_n x^{\alpha n}, \tag{3.2}$$

such that the following condition holds

$$\sum_{n=0}^{\infty} |a_n| n^2 X_0^n < +\infty. \tag{3.3}$$

In fact, since by Stirling asymptotic formula for Gamma function [11, Ch. 1]

$$\frac{\Gamma(\alpha n+1)}{\Gamma(\alpha(n-2)+1)} = O(n^{2\alpha}), \quad n \to \infty,$$
$$\frac{\Gamma(\alpha n+1)}{\Gamma(\alpha(n-1)+1)} = O(n^{\alpha}), \quad n \to \infty.$$

Therefore by relation (2.11), condition (3.3) and Lemma 2.4 we can guarantee the absolute and uniform on $[x_0, X_0]$ convergence of the corresponding series in Lemma 2.6. Hence we get

$$(D^{2\alpha}_{0+}y)(x) = \sum_{n=0}^{+\infty} a_n \frac{\Gamma(\alpha n+1)}{\Gamma(\alpha n+1-2\alpha)}\, x^{\alpha n-2\alpha}, \tag{3.4}$$

$$(D^{\alpha}_{0+}y)(x) = \sum_{n=0}^{+\infty} a_n \frac{\Gamma(\alpha n+1)}{\Gamma(\alpha n+1-\alpha)}\, x^{\alpha n-\alpha}. \tag{3.5}$$

Substituting these expressions into (3.1) and collecting terms containing equal powers of x, we derive

$$\begin{aligned}\sum_{n=0}^{+\infty} a_n \frac{\Gamma(\alpha n+1)}{\Gamma(\alpha n+1-2\alpha)}\, x^{\alpha n} &+ \sum_{n=0}^{+\infty} a_n \frac{\Gamma(\alpha n+1)}{\Gamma(\alpha n+1-\alpha)}\, x^{\alpha n} \\ &+ \sum_{n=2}^{+\infty} a_{n-2}\, x^{\alpha n} - \lambda^2 \sum_{n=0}^{+\infty} a_n\, x^{\alpha n} = 0.\end{aligned} \tag{3.6}$$

Evidently all coefficients of $x^{\alpha n}$ should be equal to zero. Hence

$$\begin{cases} a_0 \left(\dfrac{1}{\Gamma(1-2\alpha)} + \dfrac{1}{\Gamma(1-\alpha)} - \lambda^2 \right) = 0 \\ a_1 \left[\Gamma(1+\alpha) \left(\dfrac{1}{\Gamma(1-\alpha)} + 1 \right) - \lambda^2 \right] = 0 \\ a_n + \left[\Gamma(\alpha(n+2)+1) \left(\dfrac{1}{\Gamma(\alpha n+1)} \right.\right. \\ \qquad \left.\left. + \dfrac{1}{\Gamma(\alpha(n+1)+1)} \right) - \lambda^2 \right] a_{n+2} = 0, \quad n \in \mathbb{N}_0. \end{cases} \tag{3.7}$$

Suppose that

$$\lambda^2 \neq \Gamma(\alpha(n+2)+1) \left(\frac{1}{\Gamma(\alpha n+1)} + \frac{1}{\Gamma(\alpha(n+1)+1)} \right), \quad \alpha \in]0,1], \; n \in \mathbb{N}.$$

This means, from the previous system, that the case $a_0 = a_1 = 0$ drives to the trivial solution of the equation (3.1). Moreover, at least numerically it is possible to verify that

$$\Gamma(1+\alpha) \left(\frac{1}{\Gamma(1-\alpha)} + 1 \right) - \frac{1}{\Gamma(1-2\alpha)} - \frac{1}{\Gamma(1-\alpha)} \neq 0, \; \alpha \in]0,1].$$

This implies that a_0, a_1 cannot be nonzero simultaneously as well. So, let, for instance, $a_0 \neq 0$. Consequently,

$$\lambda^2 = \frac{1}{\Gamma(1-2\alpha)} + \frac{1}{\Gamma(1-\alpha)}. \tag{3.8}$$

Hence the third equation of (3.7) becomes

$$a_n + \left[\Gamma(\alpha(n+2)+1)\left(\frac{1}{\Gamma(\alpha n+1)} + \frac{1}{\Gamma(\alpha(n+1)+1)}\right) - \frac{1}{\Gamma(1-2\alpha)} - \frac{1}{\Gamma(1-\alpha)}\right] a_{n+2} = 0, \ n \in \mathbb{N}_0.$$

Furthermore, numerical simulations confirm

$$\Gamma(\alpha(n+2)+1)\left(\frac{1}{\Gamma(\alpha n+1)} + \frac{1}{\Gamma(\alpha(n+1)+1)}\right) - \frac{1}{\Gamma(1-2\alpha)} - \frac{1}{\Gamma(1-\alpha)} \neq 0$$

for all $\alpha \in]0,1]$ and $n \in \mathbb{N}_0$. Therefore one can express coefficients a_n with even indices by the relation

$$a_{2n} = a_0 \prod_{k=1}^{n} C_2(k,\alpha), \quad n \in \mathbb{N}_0, \tag{3.9}$$

where

$$C_2(k,\alpha) = \left[\frac{1}{\Gamma(1-2\alpha)} + \frac{1}{\Gamma(1-\alpha)} - \frac{\Gamma(\alpha(k+2)+1)}{\Gamma(\alpha k+1)} - \frac{\Gamma(2\alpha(k+1)+1)}{\Gamma(\alpha(2k+1)+1)}\right]^{-1}.$$

Since in this case $a_1 = 0$, we have, from the third equation of (3.7), that all odd coefficients are zero. We summarize the above discussion as

Theorem 3.1. *Let $a_0 \neq 0$, $\alpha \in]0,1]$, $x \in [x_0, X_0]$, and*

$$\lambda^2 = \frac{1}{\Gamma(1-2\alpha)} + \frac{1}{\Gamma(1-\alpha)}.$$

Then the fractional Bessel equation (3.1) *admits a particular solution in terms of the power series* (3.2), *with even coefficients satisfying conditions* (3.4), (3.5) *and given by formula* (3.9).

In the odd case, we presume $a_1 \neq 0$ and the determining equation for λ becomes

$$\lambda^2 = \Gamma(1+\alpha)\left(\frac{1}{\Gamma(1-\alpha)} + 1\right). \tag{3.10}$$

As above via numerical methods

$$\Gamma(\alpha(n+2)+1)\left(\frac{1}{\Gamma(\alpha n+1)} + \frac{1}{\Gamma(\alpha(n+1)+1)}\right) - \frac{\Gamma(1+\alpha)}{\Gamma(1-\alpha)} - \Gamma(1+\alpha) \neq 0, \ \alpha \in]0,1], \ n \in \mathbb{N}_0.$$

Hence in the same manner, we express the odd coefficients by the relation

$$a_{2n+1} = a_1 \prod_{k=1}^{n} C_1(k,\alpha), \quad n \in \mathbb{N}_0, \tag{3.11}$$

where

$$C_1(k,\alpha) = \left[\Gamma(1+\alpha) + \frac{\Gamma(1+\alpha)}{\Gamma(1-\alpha)} - \frac{\Gamma(\alpha(2k+3)+1)}{\Gamma(\alpha(2k+1)+1)} - \frac{\Gamma(\alpha(2k+1)+1)}{\Gamma(2k\alpha+1)}\right]^{-1}.$$

Since in this case $a_0 = 0$, we have, from the third equation of (3.7), that all even coefficients are zero. We summarize the above discussion as

Theorem 3.2. *Let $a_1 \neq 0$, $\alpha \in]0,1]$, $x \in [x_0, X_0]$, and*

$$\lambda^2 = \Gamma(1+\alpha)\left(\frac{1}{\Gamma(1-\alpha)} + 1\right).$$

Then the fractional Bessel equation (3.1) *admits a particular solution in terms of the power series* (3.2), *with odd coefficients satisfying conditions* (3.4), (3.5) *and given by formula* (3.11).

3.2. The Mellin transform method for solving the fractional Bessel equation

In this section we propose a method to obtain an approximate solution of the fractional Bessel equation by using the direct and inverse Mellin transforms M and M^{-1}, (2.1) and (2.2) respectively. Namely, applying Mellin's transform to (3.1) and taking into account properties (2.7), (2.8) and (2.9), we have

$$\frac{\Gamma(1-s)}{\Gamma(1-s-2\alpha)} Y(s) + \frac{\Gamma(1-s)}{\Gamma(1-s-\alpha)} Y(s) - \lambda^2 Y(s) + Y(s+2\alpha) = 0. \tag{3.12}$$

Denoting $H(s) = \Gamma(1-s)Y(s)$ and

$$h(x) = \int_0^{\infty} e^{-xt} y(t)\, dt, \tag{3.13}$$

we get owing to Theorem 2.1 that $h(x)$ is the inverse Mellin transform of $H(s)$. Thus, (3.12) became

$$H(s) + \frac{\Gamma(1-s-2\alpha)}{\Gamma(1-s-\alpha)} H(s) - \lambda^2 \frac{\Gamma(1-s-2\alpha)}{\Gamma(1-s)} H(s) + H(s+2\alpha) = 0. \tag{3.14}$$

Hence taking the inverse Mellin transform it gives correspondingly the equality

$$\left(x^{2\alpha}+1\right) h(x) - \lambda^2 \int_0^{\infty} h\left(\frac{x}{t}\right) k_1(t) \frac{dt}{t} + \int_0^{\infty} h\left(\frac{x}{t}\right) k_2(t) \frac{dt}{t} = 0, \tag{3.15}$$

where

$$\begin{aligned} k_1(t) &= M^{-1}\left\{\frac{\Gamma(1-s-2\alpha)}{\Gamma(1-s)}\right\} = \frac{1}{\Gamma(2\alpha)}\,(t-1)_+^{2\alpha-1}, \\ k_2(t) &= M^{-1}\left\{\frac{\Gamma(1-s-2\alpha)}{\Gamma(1-s-\alpha)}\right\} = \frac{t^{\alpha}}{\Gamma(\alpha)}\,(t-1)_+^{\alpha-1}. \end{aligned} \tag{3.16}$$

Let $h(x)$ admit a formal series representation

$$h(x) \sim \sum_{n=1}^{\infty} b_n x^{-\alpha n}, \tag{3.17}$$

i.e.,

$$h(x) = \sum_{n=1}^{N} b_n x^{-\alpha n} + O(x^{-\alpha N}), \quad x \to \infty, \ N \in \mathbb{N}.$$

Substituting this into (3.15) and using Lemma 2.5 with (3.16), we come out with asymptotic equality

$$\begin{aligned}(x^{2\alpha}+1)\sum_{n=1}^{N} b_n x^{-\alpha n} - \frac{\lambda^2}{\Gamma(2\alpha)}\sum_{n=1}^{N} b_n x^{-\alpha n}\int_1^{\infty} t^{\alpha n-1}(t-1)^{2\alpha-1}dt \\ + \frac{1}{\Gamma(\alpha)}\sum_{n=1}^{N} b_n x^{-\alpha n}\int_1^{\infty} t^{\alpha(1+n)-1}(t-1)^{\alpha-1}dt = O(x^{-\alpha N}).\end{aligned} \tag{3.18}$$

At the meantime, the elementary Beta-integrals are calculated explicitly under condition $\alpha < \frac{1}{N+2}$.

$$\begin{aligned}\int_1^{\infty} t^{\alpha n-1}(t-1)^{2\alpha-1}dt &= \frac{\Gamma(2\alpha)\,\Gamma(-\alpha(n+2)+1)}{\Gamma(-\alpha n+1)}, \quad n=0,\dots,N, \\ \int_1^{\infty} t^{\alpha(1+n)-1}(t-1)^{\alpha-1}dt &= \frac{\Gamma(\alpha)\,\Gamma(-\alpha(n+2)+1)}{\Gamma(-\alpha(n+1)+1)}, \quad n=0,\dots,N.\end{aligned} \tag{3.19}$$

Therefore, substituting (3.19) into (3.18), we arrive at the following truncated equation

$$\begin{aligned}(x^{2\alpha}+1)\sum_{n=1}^{N} b_n x^{-\alpha n} - \lambda^2\sum_{n=1}^{N} b_n \frac{\Gamma(-\alpha(n+2)+1)}{\Gamma(-\alpha n+1)} x^{-\alpha n} \\ + \sum_{n=1}^{N} b_n \frac{\Gamma(-\alpha(n+2)+1)}{\Gamma(-\alpha(n+1)+1)} x^{-\alpha n} = 0.\end{aligned}$$

Collecting the terms which contain equal powers of x and equating them to zero, we find

$$\begin{cases} b_1\left[1-\lambda^2\,\Gamma(1-3\alpha)\left(\dfrac{1}{\Gamma(1-\alpha)}+\dfrac{1}{\Gamma(1-2\alpha)}\right)\right]=0 \\ b_2\left[1-\lambda^2\,\Gamma(1-4\alpha)\left(\dfrac{1}{\Gamma(1-2\alpha)}+\dfrac{1}{\Gamma(1-3\alpha)}\right)\right]=0 \\ b_n+\left[1-\lambda^2\,\Gamma(1-\alpha(n+4))\left(\dfrac{1}{\Gamma(1-\alpha(n+2))}\right.\right. \\ \quad\left.\left.+\dfrac{1}{\Gamma(1-\alpha(n+3))}\right)\right]b_{n+2}=0, \quad n\in\mathbb{N},\ n\le N. \end{cases} \tag{3.20}$$

Suppose that

$$\lambda^2 \neq \left[\Gamma(1-\alpha(n+4))\left(\frac{1}{\Gamma(1-\alpha(n+2))}+\frac{1}{\Gamma(1-\alpha(n+3))}\right)\right]^{-1},$$

with $\alpha \in \left]0, \frac{1}{N+2}\right]$, $n \in \mathbb{N}$, $n \leq N$. This means, from the previous system, that the case $b_1 = b_2 = 0$ drives to the trivial solution of the equation (3.1). Moreover, at least numerically it is possible to verify that

$$1 - \frac{\Gamma(1-4\alpha)\left(\frac{1}{\Gamma(1-2\alpha)}+\frac{1}{\Gamma(1-3\alpha)}\right)}{\Gamma(1-3\alpha)\left(\frac{1}{\Gamma(1-\alpha)}+\frac{1}{\Gamma(1-2\alpha)}\right)} \neq 0, \ \alpha \in \left]0, \frac{1}{N+2}\right].$$

This implies that b_1, b_2 cannot be nonzero simultaneously as well. So, let, for instance, $b_1 \neq 0$. Consequently,

$$\lambda^2 = \left[\Gamma(1-3\alpha)\left(\frac{1}{\Gamma(1-\alpha)}+\frac{1}{\Gamma(1-2\alpha)}\right)\right]^{-1}. \tag{3.21}$$

Hence the third equation of (3.20) becomes

$$b_n + \left[1 - \frac{\Gamma(1-\alpha(n+4))\left(\frac{1}{\Gamma(1-\alpha(n+2))}+\frac{1}{\Gamma(1-\alpha(n+3))}\right)}{\Gamma(1-3\alpha)\left(\frac{1}{\Gamma(1-\alpha)}+\frac{1}{\Gamma(1-2\alpha)}\right)}\right] b_{n+2} = 0, \quad n \in \mathbb{N},\ n \leq N.$$

Furthermore, numerical simulations confirm

$$1 - \frac{\Gamma(1-\alpha(n+4))\left(\frac{1}{\Gamma(1-\alpha(n+2))}+\frac{1}{\Gamma(1-\alpha(n+3))}\right)}{\Gamma(1-3\alpha)\left(\frac{1}{\Gamma(1-\alpha)}+\frac{1}{\Gamma(1-2\alpha)}\right)} \neq 0, \ \alpha \in \left]0, \frac{1}{N+2}\right], \ n \in \mathbb{N},\ n \leq N.$$

Therefore one can express coefficients b_n with odd indices by the relation

$$b_{2n+1} = b_1 \prod_{k=1}^{n} C_3(k,\alpha), \quad n \in \mathbb{N}_0, \tag{3.22}$$

where

$$C_3(k,\alpha) = \left[\frac{\Gamma(1-\alpha(2k+3))\,\Gamma(1-\alpha)}{\Gamma(1-\alpha(2k+1))}\left(\frac{1}{\Gamma(1-3\alpha)} + \frac{1}{\Gamma(1-2\alpha)}\right) - \frac{\Gamma(1-\alpha(2k+3))}{\Gamma(1-2\alpha(k+1))} - 1\right]^{-1}.$$

Since in this case $b_2 = 0$, we have, from the third equation of (3.20), that all even coefficients are zero. We get an approximate solution of equation (3.18) in the form

$$h(x) \sim \sum_{n=1}^{\infty} b_{2n+1} x^{\alpha(2n+1)}.$$

Since $h(x)$ is approximated by the previous series, we obtain the corresponding expression for the solution $y(t)$, employing operational relation for the Laplace transform. Namely, equality (3.13) gives $y(t)$ as a formal series

$$y(t) \sim \sum_{n=0}^{\infty} \frac{b_{2n+1}}{\Gamma(n\alpha)}\, t^{\alpha n-1}. \tag{3.23}$$

We summarize the above discussion as

Theorem 3.3. *Let* $b_1 \neq 0$, $\alpha \in \left]0, \frac{1}{N+2}\right]$, $N \in \mathbb{N}_0$, $x \in [x_0, X_0]$, *and*

$$\lambda^2 = \left[\Gamma(1-3\alpha)\left(\frac{1}{\Gamma(1-\alpha)} + \frac{1}{\Gamma(1-2\alpha)}\right)\right]^{-1}.$$

Then the fractional Bessel equation (3.1) *admits a particular solution in terms of the power series* (3.23), *with odd coefficients given by formula* (3.22).

In the even case, we presume $b_2 \neq 0$ and the determining equation for λ becomes

$$\lambda^2 = \left[\Gamma(1-4\alpha)\left(\frac{1}{\Gamma(1-2\alpha)} + \frac{1}{\Gamma(1-3\alpha)}\right)\right]^{-1}. \tag{3.24}$$

As above via numerical methods

$$b_n + \left[1 - \frac{\Gamma(1-\alpha(n+4))\left(\frac{1}{\Gamma(1-\alpha(n+2))} + \frac{1}{\Gamma(1-\alpha(n+3))}\right)}{\Gamma(1-4\alpha)\left(\frac{1}{\Gamma(1-2\alpha)} + \frac{1}{\Gamma(1-3\alpha)}\right)}\right] b_{n+2} \neq 0.$$

for all $\alpha \in \left]0, \frac{1}{N+2}\right]$, $N \in \mathbb{N}$, $n \in \mathbb{N}$ and $n \leq N$. Hence in the same manner, we express the even coefficients by relation

$$b_{2(n+1)} = b_2 \prod_{k=1}^{n} C_4(k,\alpha), \quad n \in \mathbb{N}_0, \tag{3.25}$$

where

$$C_4(k,\alpha) = \left[\frac{\Gamma(1-2\alpha(k+1))\,\Gamma(1-2\alpha)}{\Gamma(1-2\alpha k)}\left(\frac{1}{\Gamma(1-4\alpha)} + \frac{1}{\Gamma(1-3\alpha)}\right) - \frac{\Gamma(1-2\alpha(k+1))}{\Gamma(1-\alpha(2k+1))} - 1\right]^{-1}.$$

Since in this case $b_1 = 0$, we have, from the third equation of (3.20), that all odd coefficients are zero. We get an approximate solution of equation (3.18) in the form

$$h(x) \sim \sum_{n=1}^{\infty} b_{2n} x^{2\alpha n}.$$

Further, since $h(x)$ is approximated by the series (3.17), we obtain the corresponding expression for the solution $y(t)$, employing operational relation for the Laplace transform. Namely, equality (3.13) gives $y(t)$ as a formal series

$$y(t) \sim \sum_{n=1}^{\infty} \frac{b_{2n}}{\Gamma(n\alpha)} t^{\alpha n-1}. \tag{3.26}$$

We summarize the above discussion as

Theorem 3.4. *Let $b_2 \neq 0$, $\alpha \in \left]0, \frac{1}{N+2}\right]$, $N \in \mathbb{N}_0$, $x \in [x_0, X_0]$, and*

$$\lambda^2 = \left[\Gamma(1-4\alpha)\left(\frac{1}{\Gamma(1-2\alpha)} + \frac{1}{\Gamma(1-3\alpha)}\right)\right]^{-1}.$$

Then the fractional Bessel equation (3.1) *admits a particular solution in terms of the power series* (3.26), *with even coefficients given by formula* (3.25).

4. Existence and uniqueness of solutions

Here we will use Banach's fixed point theorem to study the existence and uniqueness of equation (3.1)

$$x^{2\alpha}(D_{0+}^{2\alpha}y)(x) + x^{\alpha}(D_{0+}^{\alpha}y)(x) + (x^{2\alpha} - \lambda^2)y(x) = 0, \quad \alpha \in \left[\frac{1}{2}, 1\right], \ x \in [0, X_0], \tag{4.1}$$

with $X_0 \in \mathbb{R}_+$, $\lambda \in \mathbb{C}$ under initial conditions $y(0) = y_0$, $y'(0) = y_0^*$.

Let $I = [a, b]$ $(a < b,\ a, b \in \mathbb{R})$ and $m \in \mathbb{N}_0$. Denote by C^m the usual space of functions v which are m times continuously differentiable on I with the norm

$$\|v\|_{C^m} = \sum_{k=0}^{m} \|v^{(k)}\|_C = \sum_{k=0}^{m} \max_{x \in I} |v^{(k)}(x)|.$$

In particular, for $m = 0$, $C^0(I) \equiv C(I)$ is the space of continuous functions v on I with the norm $\|v\|_C = \max_{x \in I} |v(x)|$.

Theorem 4.1. *The fractional equation* (3.1) *for a rectangular domain $x \in [0, X_0]$ has a unique solution for every $\lambda \in \mathbb{C}$ such that*

$$|\lambda|^2 > X_0 \left(X_0 + \frac{1}{\Gamma(1-\alpha)(2-\alpha)} + \frac{X_0}{\Gamma(2(1-\alpha))\,(3-2\alpha)}\right) \tag{4.2}$$

with $\alpha \in \left[\frac{1}{2}, 1\right]$.

Proof. Consider the following Banach spaces

$$X = \{y : y(x) \in C^2([0, X_0])\}, \qquad Y = \{y : y(x) \in C([0, X_0])\}.$$

Putting $T : X \to Y$,

$$(Ty)(x) = \frac{1}{\lambda^2}\left[x^{2\alpha}y(x) + x^{\alpha}\left(D_{0+}^{\alpha}y\right)(x) + x^{2\alpha}\left(D_{0+}^{2\alpha}y\right)(x)\right],$$

we rewrite equation (3.1) in the form $y(x) = (Ty)(x)$. Taking into account Theorems 2.2, 2.3, Lemma 2.4 and relations (2.15), (2.17), we have

$$\begin{aligned}
&\|Ty_1 - Ty_2\|_Y \\
&= \frac{1}{|\lambda|^2} \left\| x^{2\alpha}(y_1(x) - y_2(x)) + x^{\alpha}(D^{\alpha}_{0+}(y_1 - y_2))(x) + x^{2\alpha}(D^{2\alpha}_{0+}(y_1 - y_2))(x) \right\|_Y \\
&\leq \frac{X_0^2}{|\lambda|^2}\|y_1 - y_2\|_Y + \frac{X_0}{|\lambda^2|\ \Gamma(1-\alpha)\ (2-\alpha)}\|D(y_1 - y_2)\|_Y \\
&\quad + \frac{X_0^2}{|\lambda|^2\ \Gamma(2(1-\alpha))\ (3-2\alpha)}\|D^2(y_1 - y_2)\|_Y \\
&\leq \frac{X_0}{|\lambda|^2}\left(X_0 + \frac{1}{\Gamma(1-\alpha)(2-\alpha)} + \frac{X_0}{\Gamma(2(1-\alpha))\ (3-2\alpha)} \right)\|y_1 - y_2\|_X.
\end{aligned}$$

Taking into account (4.2) we conclude that T is a contraction. Hence we can apply the Banach fix point theorem to complete the proof. □

Acknowledgment

The work of the first and second authors were supported by *Fundação para a Ciência e a Tecnologia* via grants `SFRH/BPD/73537/2010`, `SFRH/BPD/65043/2009`, respectively.

The first author was supported by *FEDER* funds through *COMPETE* – Operational Programme Factors of Competitiveness ("Programa Operacional Factores de Competitividade") and by Portuguese funds through the *Center for Research and Development in Mathematics and Applications* (University of Aveiro) and the Portuguese Foundation for Science and Technology ("FCT – Fundação para a Ciência e a Tecnologia"), within project PEst-C/MAT/UI4106/2011 with COMPETE number FCOMP-01-0124-FEDER-022690.

The work of the third author was supported by the European Regional Development Fund through the programme COMPETE and by the Portuguese Government through the FCT – Fundação para a Ciência e a Tecnologia under the project PEst-C/MAT/UI0144/2011.

References

[1] J. Butzer, *Mellin transform theory and the role of its differential and integral operators.* Transform Methods and Special Functions, Varna'96 (Conf. Proc.), Bulgarian Acad. Sci., (1998), 63–83.

[2] S.D. Eidelman, S.D. Ivasyshen and A.N. Kochubei, *Analytic methods in the theory of differential and pseudo-differential equations of parabolic type*, Operator Theory: Advances and Applications 152, Birkhäuser, Basel, 2004.

[3] A.A. Kilbas, M. Rivero, L. Rodríguez-Germá and J.J. Trujillo, *Analytic solutions of some linear fractional differential equations with variable coefficients* Appl. Math. Comput. **187** (2007), 239–249.

[4] A. Kilbas, H. Srivastava and J. Trujillo, *Theory and applications of fractional differential equations*, North-Holland Mathematics Studies 204, Elsevier, Amsterdam, 2006.

[5] N.N. Lebedev, *Special functions and their applications*, Translated from Russian by R.A. Silverman, N.J.: Prentice-Hall Inc., Englewood Cliffs, 1965.

[6] F. Mainardi and M. Tomirotti, *Seismic pulse propagation with constant Q and stable probability distributions.* Annali di Geofisica **40** (1997), 1311–1328.

[7] F. Mainardi, *Fractional diffusive waves in viscoelastic solids*, IUTAM Symposium – Nonlinear Waves in Solids (Conf. Proc.), J.L. Wagnern and F.R. Norwood (editors), (1995), 93–97.

[8] O.I. Marichev, *Handbook of integral transforms of higher transcendental functions: theory and algorithmic tables*, translated from Russian by L.W. Longdon, Ellis Horwood Limited, Chichester, 1983.

[9] I. Podlubny, *Fractional differential equations. An introduction to fractional derivatives, fractional differential equations, to methods of their solutions and some applications*, Mathematics in Science and Engineering 198, Academic Press, San Diego, 1999.

[10] A.P. Prudnikov, Yu.A. Brychkov and O.I. Marichev, *Integrals and Series, Volume 3: More special functions*, translated from Russian by G.G. Gould, Gordon and Breach Publisher, New York, 1990.

[11] S.G. Samko, A.A. Kilbas, and O.I. Marichev, *Fractional integrals and derivatives: theory and applications*, Gordon and Breach, New York, 1993, 277–284.

[12] E.C. Titchmarsh, *Introduction to the theory of Fourier integrals*, Clarendon Press, Oxford, 1948.

[13] S. Yakubovich and Y.F. Luchko, *The hypergeometric approach to integral transforms and convolutions*, Mathematics and Applications 287, Kluwer Academic Publishers, Dordrecht, 1994.

[14] S. Yakubovich, M.M. Rodrigues and N. Vieira, *Fractional two-parameter Schrödinger equation.* Preprint CMUP, No. 8, 2011, 12 pp.

M.M. Rodrigues
Center for Research and Development in Mathematics and Applications
Department of Mathematics, University of Aveiro
Campus Universitário de Santiago, 3810-193 Aveiro, Portugal
e-mail: `mrodrigues@ua.pt`

N. Vieira
Center of Mathematics of University of Porto
Faculty of Science, University of Porto
Rua do Campo Alegre 687, 4169-007 Porto, Portugal
e-mail: `nvieira@fc.up.pt`

S. Yakubovich
Department of Mathematics, Faculty of Science
University of Porto
Rua do Campo Alegre 687, 4169-007 Porto, Portugal
e-mail: `syakubov@fc.up.pt`

Operator Theory:
Advances and Applications, Vol. 229, 371–386

The Dirichlet Problem for Elliptic Equations with VMO Coefficients in Generalized Morrey Spaces

Lubomira G. Softova

Abstract. We consider the Dirichlet problem in a bounded smooth domain $\Omega \subset \mathbb{R}^n$ for linear uniformly elliptic equation $\mathcal{L}u(x) = f(x)$ with VMO principal coefficients. Its unique strong solvability is proved in [5] and [6]. Our aim is to show that for every f belonging to the generalized Morrey space $L^{p,\omega}(\Omega)$, $p \in (1,\infty)$, $\omega : \mathbb{R}^n \times \mathbb{R}_+ \to \mathbb{R}_+$ the operator $\mathcal{L} : W^{2,p,\omega} \cap W_0^{1,p}(\Omega) \to L^{p,\omega}(\Omega)$ is bijective and the estimate $\|D^2u\|_{L^{p,\omega}(\Omega)} \le C(\|f\|_{L^{p,\omega}(\Omega)} + \|u\|_{L^{p,\omega}(\Omega)})$ holds.

Mathematics Subject Classification (2010). Primary 35J25; Secondary 35D35, 35B45, 35R05, 47B38.

Keywords. Uniformly elliptic operators, Dirichlet problem, VMO, generalized Morrey spaces, a priori estimate.

1. Introduction

The spaces $L^{p,\lambda}$, $p \in (1,\infty)$, $\lambda \in (0,n)$ were introduced by Morrey in his celebrated work [17] in order to study the regularity of solutions to second-order elliptic partial differential equations. In fact, the embedding properties of the *Morrey spaces* permit to obtain better Hölder regularity of the strong solutions to different problems. In [16] Mizuhara introduced the *generalized Morrey spaces* taking a weight $\omega : \mathbb{R}^n \times \mathbb{R}_+ \to \mathbb{R}_+$ instead or r^λ and studied continuity in $L^{p,\omega}$ of some classical integral operators. Later, Nakai showed boundedness of the Hardy-Littlewood maximal operator and the Calderón-Zygmund integrals in $L^{p,\omega}$ (see [18]). In our work [27], we extended these results to singular integrals with variable kernels satisfying mixed homogeneity condition, expanding this way the classical Morrey spaces studies from [20, 22] to the settings of generalized Morrey spaces. We refer the reader to [9, 24] and the references therein for more recent results regarding continuity of singular integral operators in generalized Morrey and new functional spaces.

The reason to study continuity properties of these integrals in various functional spaces is that they permit to investigate the regularity of solutions to linear elliptic and parabolic partial differential equations and systems in terms of the data of the corresponding problems. The method, associated to the names of A. Calderón and A. Zygmund (see [3, 4]) uses explicit representation formula for the highest-order derivatives of the solution in terms of singular integrals acting on the known right-hand side plus another one acting on the very same derivatives. This last term appears in a commutator which norm can be made small enough if the coefficients have small oscillation over small balls. This way, suitable "integral continuity" of the principal coefficients ensure boundedness of the commutator and therefore validity of the corresponding a priori estimate. The Sarason class of functions with *vanishing mean oscillation* verifies this requirement although they could be discontinuous. Their good behavior on small balls allows to extend the classical theory of elliptic and parabolic equations and systems with continuous coefficients (see [8, 14, 13]) to operators with discontinuous coefficients (cf. [5, 6, 15, 12]). A vast number of works are dedicated to boundary value problems for linear elliptic and parabolic operators with VMO coefficients in the framework of Sobolev and Sobolev-Morrey spaces (see [7, 19, 20, 22, 21, 23, 26, 27, 28, 29]).

The main goal of the present paper is to extend the global Morrey regularity results from [7], regarding linear elliptic equations with VMO principal coefficients, to the settings of *generalized Morrey spaces* $L^{p,\omega}$ (see Definition 2.2). The approach adopted is that of [3, 6, 7] and relies on proving boundedness of suitable integral operators and their commutators, that appear at the representation formula for the highest-order derivatives of the solution. Even if standard in some sense, that method requires precise analysis due to the specifics of the considered generalized Morrey spaces, and we employ our results from [27] to get the desired $L^{p,\omega}$-boundedness of the singular integrals. Indeed, for particular choice of the weight ω, one gets the regularity in the classical Morrey spaces $L^{p,\lambda}$, as stated in [7], for the second-order derivatives of the strong solutions to the considered problem. However, the class $L^{p,\omega}$ is wider than $L^{p,\lambda}$ and therefore the regularity results here obtained generalize those from [7].

The article is organized as follows. In Section 2 we introduce the problem and give some basic notions. In this section we recall also continuity results regarding the *Calderón-Zygmund integrals* that appear in the *interior representation formula* of the derivatives $D_{ij}u$ of the solution. The corresponding *nonsingular integrals* are studied in Section 3. These results permit to obtain $L^{p,\omega}$-estimate of $D_{ij}u$ near the boundary. The a priori estimate is established in the last section.
Throughout this paper the following notations will be used

- $D_iu = \partial u/\partial x_i$, $Du = (D_1u, \dots, D_nu)$ stands for the gradient of u,
- $D_{ij}u = \partial^2 u/\partial x_i \partial x_j$, $D^2u = \{D_{ij}u\}_{ij=1}^n$ is the Hessian matrix of u,
- $\mathcal{B}_r(x_0) = \{x \in \mathbb{R}^n : |x - x_0| < r\}$ is a ball centered in a fixed point $x_0 \in \mathbb{R}^n$,
- $\mathcal{B}_r$ is a ball centered at a point $x \in \mathbb{R}^n$ with Lebesgue measure $|\mathcal{B}_r| = Cr^n$,
- $2\mathcal{B}_r$ is a ball centered at the same point as $\mathcal{B}_r$ and of radius $2r$,

- $\mathcal{B}_r^+ \equiv \mathcal{B}_r^+(x') = \mathcal{B}_r(x') \cap \{x_n > 0\}$, where $x' = (x_1, \dots, x_{n-1}, 0)$,
- $\mathbb{S}^n = \{x \in \mathbb{R}^n : |x - x_0| = 1\}$ is the unit sphere in $\mathbb{R}^n$.

The standard summation convention on repeated upper and lower indices is adopted. The letter C is used for various positive constants and may change from one occurrence to another.

2. Definitions and preliminary results

Let $\Omega \subset \mathbb{R}^n$, $n \geq 2$ be a bounded domain with $C^{1,1}$-boundary. We consider the Dirichlet problem

$$\begin{cases} \mathcal{L}u := a^{ij}(x)D_{ij}u = f(x) & \text{a.a. } x \in \Omega \\ u = 0 & \text{on } \partial\Omega. \end{cases} \tag{2.1}$$

The operator $\mathcal{L}$ is *uniformly elliptic*, i.e., there exists a constant $\Lambda > 0$ such that

$$\begin{cases} \Lambda^{-1}|\xi|^2 \leq a^{ij}(x)\xi_i\xi_j \leq \Lambda|\xi|^2 & \text{a.a. } x \in \Omega,\ \forall \xi \in \mathbb{R}^n \\ a^{ij}(x) = a^{ji}(x) & 1 \leq i,j \leq n. \end{cases} \tag{2.2}$$

The symmetry of the matrix $\mathbf{a}(x) = \{a^{ij}(x)\}_{ij=1}^n$ implies essential boundedness of a^{ij}'s and we set $\|\mathbf{a}\|_{\infty,\Omega} = \sum_{i,j=1}^n \|a^{ij}\|_{L^\infty(\Omega)}$.

To describe the regularity of the coefficients and the data of the problem we need of the following definitions.

Definition 2.1. Let $a \in L^1_{\text{loc}}(\mathbb{R}^n)$ we say that

1. $a \in BMO$ (*bounded mean oscillation*, [10]) if

$$\|a\|_* = \sup_{R>0}\sup_{\mathcal{B}_r, r\leq R} \frac{1}{|\mathcal{B}_r|}\int_{\mathcal{B}_r} |a(y) - a_{\mathcal{B}_r}|dy < +\infty \quad a_{\mathcal{B}_r} = \frac{1}{|\mathcal{B}_r|}\int_{\mathcal{B}_r} a(y)dy.$$

   The quantity $\|a\|_*$ is a norm in BMO modulo constant function under which BMO is a Banach space;
2. $a \in VMO$ (*vanishing mean oscillation*, [25]) if $a \in BMO$ and

$$\gamma_a(R) := \sup_{R>0}\sup_{\mathcal{B}_r, r\leq R} \frac{1}{|\mathcal{B}_r|}\int_{\mathcal{B}_r} |a(y) - a_{\mathcal{B}_r}|dy$$

   tends to zero as $R \to 0$. The quantity $\gamma_a(R)$ is called VMO-modulus of a.

For any bounded domain $\Omega \subset \mathbb{R}^n$ we define $BMO(\Omega)$ and $VMO(\Omega)$ taking $a \in L^1(\Omega)$ and $\mathcal{B}_r \cap \Omega$ instead of $\mathcal{B}_r$ in the above definition.

According to [1, 11], having a function $a \in BMO(\Omega)$ or $VMO(\Omega)$ it is possible to extend it in $\mathbb{R}^n$ preserving its BMO-norm or VMO-modulus, respectively. Any bounded uniformly continuous function is also VMO with a VMO-modulus coinciding with its modulus of continuity. The space VMO contains also discontinuous functions and the following example (cf. [15]) show the inclusion $W^{1,n}(\mathbb{R}^n) \subset VMO \subset BMO$.

Example. $f_\alpha = |\log|x||^\alpha \in VMO$ for any $\alpha \in (0,1)$, $f_\alpha \in W^{1,n}(\mathbb{R}^n)$ for $\alpha \in (0, 1-1/n)$, $f_\alpha \notin W^{1,n}(\mathbb{R}^n)$ for $\alpha \in [1-1/n, 1)$, $f = |\log|x|| \in BMO \setminus VMO$.

Definition 2.2. Let $\omega : \mathbb{R}^n \times \mathbb{R}_+ \to \mathbb{R}_+$, $1 \le p < \infty$ and $\omega(\mathcal{B}_r) := \omega(x,r)$. We say that a function $f \in L^p_{\mathrm{loc}}(\mathbb{R}^n)$ belongs to the generalized Morrey space $L^{p,\omega}(\mathbb{R}^n)$ if

$$\|f\|_{L^{p,\omega}(\mathbb{R}^n)} = \|f\|_{p,\omega} = \sup_{x\in\mathbb{R}^n, r>0} \left(\frac{1}{\omega(\mathcal{B}_r)} \int_{\mathcal{B}_r} |f(y)|^p dy \right)^{1/p} < +\infty$$

and the supremum is taken over all balls $\mathcal{B}_r$ centered in some point $x \in \mathbb{R}^n$ and of radius $r > 0$.

The space $L^{p,\omega}(\mathbb{R}^n_+)$ consists of all functions $f \in L^p_{\mathrm{loc}}(\mathbb{R}^n_+)$ for which

$$\|f\|_{L^{p,\omega}(\mathbb{R}^n_+)} = \|f\|_{p,\omega;\mathbb{R}^n_+} = \sup_{x'\in\mathbb{R}^{n-1}, r>0} \left(\frac{1}{\omega(\mathcal{B}_r(x'))} \int_{\mathcal{B}^+_r(x')} |f(y)|^p dy \right)^{1/p} < +\infty$$

and the supremum is taken over all balls centered at $x' = (x_1, \dots, x_{n-1}, 0)$ and of radius $r > 0$.

For any bounded domain Ω we define $L^{p,\omega}(\Omega)$ as the space of functions $f \in L^p(\Omega)$ for which the following norm is finite

$$\|f\|_{L^{p,\omega}(\Omega)} = \|f\|_{p,\omega;\Omega} = \sup_{x\in\mathbb{R}^n, r>0} \left(\frac{1}{\omega(\mathcal{B}_r)} \int_{\mathcal{B}_r\cap\Omega} |f(y)|^p dy \right)^{1/p}.$$

The generalized Sobolev-Morrey space $W^{2,p,\omega}(\Omega)$ consists of all Sobolev functions $u \in W^{2,p}(\Omega)$ with generalized derivatives $D^s u \in L^{p,\omega}(\Omega)$, $0 \le |s| \le 2$ endowed by the norm

$$\|u\|_{W^{2,p,\omega}(\Omega)} = \sum_{s=0}^{2} \|D^s u\|_{p,\omega;\Omega}.$$

If $\omega = r^\lambda$, $\lambda \in (0,n)$ then $L^{p,\omega} = L^{p,\lambda}$ if $\omega = r^n$ then $L^{p,\omega} = L^\infty$. However, there exist examples of weight functions of more general form.

Example. 1) Let u be an A_p weight of Muckenhoupt, then

$$\omega(\mathcal{B}_r) = \left(\int_{\mathcal{B}_r} u(y) dy \right)^\alpha, \quad 0 < \alpha < 1, \quad 1 < p < 1/\alpha$$

is a generalized Morrey weight (cf. [18]).

2) The function $f(x) = \chi_{[-1,1]}|x|^{-1/2}$ belongs to $L^{1,\omega}(\mathbb{R})$ with

$$\omega(\mathcal{B}_r) = \int_{\mathcal{B}_r} |y|^\alpha dy, \quad -1 < \alpha \le -\frac{1}{2}$$

where $\mathcal{B}_r$ is any interval with a length $2r$.

Definition 2.3. For a given function $f \in L^1_{\rm loc}(\mathbb{R}^n)$ we define the *Hardy-Littlewood maximal operator* Mf, the operator $M_s f$, $s \geq 1$ and the *sharp maximal operator* $f^\sharp$ as follows

$$Mf(x) := \sup_{\mathcal{B}_r \ni x} \frac{1}{|\mathcal{B}_r|} \int_{\mathcal{B}_r} |f(y)|dy, \quad M_s f(x) := (M|f|^s(x))^{1/s} \tag{2.3}$$

$$f^\sharp(x) := \sup_{\mathcal{B}_r \ni x} \frac{1}{|\mathcal{B}_r|} \int_{\mathcal{B}_r} |f(y) - f_{\mathcal{B}_r}|dy \quad \text{a.e. in } \mathbb{R}^n \tag{2.4}$$

where the supremum is taken over all balls $\mathcal{B}_r$ containing x.

The following results are weighted variants of the well-known maximal and sharp inequalities (see [18, 27] and the references therein).

Lemma 2.4. *Assume that there exist constants $\kappa_1, \kappa_2 > 0$ such that for any $x_0 \in \mathbb{R}^n$ and $r > 0$*

$$\forall t > 0 \quad r \leq t \leq 2r \Longrightarrow \kappa_1^{-1} \leq \frac{\omega(x_0, t)}{\omega(x_0, r)} \leq \kappa_1$$
$$\int_r^\infty \frac{\omega(x_0, t)}{t^{n+1}} dt \leq \kappa_2 \frac{\omega(x_0, r)}{r^n}. \tag{2.5}$$

Then for all $1 \leq s < p < \infty$ there is a constant $C = C(p, s, n)$ such that

$$\|M_s f\|_{p,\omega} \leq C\|f\|_{p,\omega} \qquad \forall\, f \in L^{p,\omega}(\mathbb{R}^n).$$

Let $\omega(x_0, r)$ satisfy (2.5) and suppose there is a constant $\kappa' > 0$ such that

$$\int_r^\infty \frac{\omega(x_0, t)}{t^{\sigma n+1}} dt \leq \kappa' \frac{\omega(x_0, r)}{r^{\sigma n}} \tag{2.6}$$

with $0 < \sigma < 1$. Then there is a constant $C = C(p, n, \sigma)$ such that

$$\|f\|_{p,\omega} \leq C\|f^\sharp\|_{p,\omega} \qquad \forall\, f \in L^{p,\omega}(\mathbb{R}^n).$$

The above assertion holds also in $\mathbb{R}^n_+$ and the proof follows the same line as in Lemma 2.4.

Lemma 2.5. *Let (2.5) holds, then there exist constants depending on n, p, s, and σ such that*

$$\|M_s f\|_{p,\omega;\mathbb{R}^n_+} \leq C\|f\|_{p,\omega;\mathbb{R}^n_+}, \quad \|f\|_{p,\omega;\mathbb{R}^n_+} \leq C\|f^\#\|_{p,\omega;\mathbb{R}^n_+} \tag{2.7}$$

for all $f \in L^{p,\omega}(\mathbb{R}^n_+)$.

In [27] we study continuity in $L^{p,\omega}(\mathbb{R}^n)$ of singular integrals with Calderón-Zygmund type kernel. These integrals play an essential role in obtaining the a priori estimate of the solution of (2.1). For the sake of completeness we present here these results.

Definition 2.6. A function $k(x;\xi) : \mathbb{R}^n \times \mathbb{R}^n \setminus \{0\} \to \mathbb{R}$ is said to be a variable Calderón-Zygmund kernel if:

i) for each fixed $x \in \mathbb{R}^n$, $k(x;\cdot)$ is a Calderón-Zygmund kernel, i.e.:

i$_a$) $k(x;\cdot) \in C^\infty(\mathbb{R}^n \setminus \{0\})$

i$_b$) $k(x;\mu\xi) = \mu^{-n} k(x;\xi) \ \ \forall\, \mu > 0$

i$_c$) $\displaystyle\int_{\mathbb{S}^n} k(x;\xi)\, d\sigma_\xi = 0 \qquad \int_{\mathbb{S}^n} |k(x;\xi)|\, d\sigma_\xi < +\infty$

ii) for every multi-index β : $\displaystyle\sup_{\xi\in\mathbb{S}^n} |D_\xi^\beta k(x;\xi)| \le C(\beta)$ independently of x

where $\mathbb{S}^n$ is the unite sphere in $\mathbb{R}^n$.

Theorem 2.7 ([27, Theorem 3.1]). *Let $k(x;\xi)$ be as in Definition 2.6, $a \in BMO$, $f \in L^{p,\omega}(\mathbb{R}^n)$, $p \in (1,\infty)$ with ω satisfying (2.5). Then the integrals*

$$\mathfrak{K} f(x) := P.V. \int_{\mathbb{R}^n} k(x, x-y) f(y) dy \tag{2.8}$$

$$\mathfrak{C}[a,f](x) := P.V. \int_{\mathbb{R}^n} k(x, x-y)[a(y) - a(x)] f(y) dy \tag{2.9}$$

are continuous from $L^{p,\omega}(\mathbb{R}^n)$ into itself and

$$\|\mathfrak{K} f\|_{p,\omega} \le C\|f\|_{p,\omega} \qquad \|\mathfrak{C}[a,f]\|_{p,\omega} \le C\|a\|_*\|f\|_{p,\omega} \tag{2.10}$$

where the constants depend on n, p, ω and k.

Corollary 2.8 ([27, Corollary 3.1]). *Let $a \in VMO$ with modulus γ_a and $f \in L^{p,\omega}_{\rm loc}$. Then for each $\varepsilon > 0$ there exists $r_0 = r_0(\varepsilon, \gamma_a) > 0$ such that for any $r \in (0, r_0)$ and any ball $\mathcal{B}_r$ it holds*

$$\|\mathfrak{C}[a,f]\|_{p,\omega;\mathcal{B}_r} < C\varepsilon \|f\|_{p,\omega;\mathcal{B}_r}\,. \tag{2.11}$$

In what follows the expression "known quantities" means the data of the problem. Namely, n, p, Ω, ω, the constants $\kappa_1, \kappa_2, \kappa'$ that appear in (2.5) and (2.6), the kernel k, the VMO-modulus $\gamma_a(r)$, the norm $\|\mathbf{a}\|_{\infty,\Omega}$ and the ellipticity constant Λ of the coefficients.

3. Nonsingular integrals in generalized Morrey spaces

For any $x, y \in \mathbb{R}^n_+$ define $\widetilde{x} = (x_1, \dots, -x_n) \in \mathbb{R}^n_-$ and the *generalized reflection* $\mathcal{T}(x;y)$

$$\mathcal{T}(x;y) = x - 2x_n \frac{\mathbf{a}^n(y)}{a^{nn}(y)} \qquad \mathcal{T}(x) = \mathcal{T}(x;x) : \mathbb{R}^n_+ \to \mathbb{R}^n_- \tag{3.1}$$

where $\mathbf{a}^n$ is the last row of the coefficients matrix $\mathbf{a}$. In case of the Laplace operator $\mathcal{T}(x) = \widetilde{x}$ while for a constant coefficient operator the point $\mathcal{T}(x)$ lies on the hyperplane $(x'_1, \dots, x'_{n-1}, -x_n)$ where $x'_i = x_i - 2x_n \frac{a^{ii}}{a^{nn}}$, $i = 1, \dots, n-1$. There exist constants C_1, C_2 depending on n and Λ, such that

$$C_1|\widetilde{x} - y| \le |\mathcal{T}(x) - y| \le C_2|\widetilde{x} - y| \qquad \forall\, x, y \in \mathbb{R}^n_+. \tag{3.2}$$

For any $f \in L^{p,\omega}(\mathbb{R}^n_+)$ and $a \in BMO(\mathbb{R}^n_+)$ define the operators

$$\begin{aligned}\widetilde{\mathfrak{K}}f(x) &:= \int_{\mathbb{R}^n_+} k(x, \mathcal{T}(x) - y) f(y) dy \\ \widetilde{\mathfrak{C}}[a, f](x) &:= \int_{\mathbb{R}^n_+} k(x, \mathcal{T}(x) - y)[a(y) - a(x)] f(y) dy\end{aligned} \tag{3.3}$$

with a nonsingular kernel satisfying the homogeneity condition i_b) from Definition 2.6. One more notion we need is that of *n-dimensional spherical harmonics* (see [3, 4, 5, 27] for details). Recall that the restriction on the unit sphere $\mathbb{S}^n$ of any homogeneous and harmonic polynomial is called *n-dimensional spherical harmonics*. Set $\mathcal{Y}_m$ for the space of all n-dimensional spherical harmonics of degree m that is

$$\mathcal{Y}_m = \{P_m(x)|_{\mathbb{S}^n} : \ P_m(\lambda x) = \lambda^m P_m(x), \ \Delta P_m(x) = 0\}$$

and $\mathcal{Y} = \{\mathcal{Y}_m\}_{m=0}^\infty$. The dimension of $\mathcal{Y}_m$ is g_m, where $g_0 = 1$, $g_1 = n$

$$g_m = \binom{m+n-1}{n-1} - \binom{m+n-3}{n-1} \le C(n)\, m^{n-2} \qquad m \ge 2, \tag{3.4}$$

Denote by $\{Y_{ms}\}_{s=1}^{g_m}$ an orthonormal base in $\mathcal{Y}_m$. The system of all basic functions in $\mathcal{Y}$ is $\{Y_{ms}\}_{m=0,s=1}^{\infty,\ g_m}$ and it is a complete orthonormal system in $L^2(\mathbb{S}^n)$. For each multi-index β the following estimate holds

$$\sup_{\xi \in \mathbb{S}^n} |D^\beta Y_{ms}(\xi)| \le C(n)\, m^{|\beta| + (n-2)/2} \qquad m = 1, 2, \dots. \tag{3.5}$$

For any function $\phi \in C^\infty(\mathbb{S}^n)$ consider its development in the Fourier series

$$\varphi(x) \sim \sum_{m=0}^\infty \sum_{s=1}^{g_m} b^{ms} Y_{ms}(x) \qquad b^{ms} = \int_{\mathbb{S}^n} \varphi(x) Y_{ms}(x) d\sigma.$$

Then for every $l > 1$

$$|b^{ms}| \le C(l)\, m^{-2l} \sup_{\substack{|\beta| = l \\ \xi \in \mathbb{S}^n}} |D^\beta \phi(\xi)|. \tag{3.6}$$

Because of the homogeneity property of the kernel we can take its expansion in Fourier series, shifting the dependence of k on x on the coefficients $b^{ms}(x)$. The last ones are essentially bounded with a bound independent of x. Precisely

$$\begin{aligned} k(x, \mathcal{T}(x) - y) &= \frac{k(x, \overline{\mathcal{T}(x) - y})}{|\mathcal{T}(x) - y|^n} = \sum_{m,s} b^{ms}(x) \frac{Y_{ms}(\overline{\mathcal{T}(x) - y})}{|\mathcal{T}(x) - y|^n} \\ &=: \sum_{m,s} b^{ms}(x) \mathcal{H}_{ms}(\mathcal{T}(x) - y) \end{aligned}$$

where $\|b^{ms}\|_\infty \le C(l)m^{-2l}$ for any integer $l > 1$ and $\sum_{m,s}$ stands for $\sum_{m=0}^{\infty}\sum_{s=1}^{g_m}$. Hence, by the dominated convergence theorem

$$\widetilde{\mathfrak{K}}f(x) = \sum_{m,s} b^{ms}(x)\int_{\mathbb{R}^n_+} \mathcal{H}_{ms}(\mathcal{T}(x)-y)f(y)dy =: \sum_{m,s} b^{ms}(x)\widetilde{\mathfrak{K}}_{ms}f(x)$$

$$\widetilde{\mathfrak{C}}[a,f](x) = \sum_{m,s} b^{ms}(x)\int_{\mathbb{R}^n_+} \mathcal{H}_{ms}(\mathcal{T}(x)-y)[a(x)-a(y)]f(y)dy \tag{3.7}$$

$$=: \sum_{m,s} b^{ms}(x)\widetilde{\mathfrak{C}}_{ms}[a,f](x) \tag{3.8}$$

with a kernel verifying

$$|\mathcal{H}_{ms}(\mathcal{T}(x)-y)| \le C\,\frac{|Y_{ms}(\mathcal{T}(x)-y)|}{|\widetilde{x}-y|^n} \le C(n)\,\frac{m^{(n-2)/2}}{|\widetilde{x}-y|^n}.$$

Proposition 3.1. *Let $f \in L^{p,\omega}(\mathbb{R}^n_+)$ with $p \in (1,\infty)$ and ω verifying (2.5) and (2.6), and $a \in BMO(\mathbb{R}^n_+)$. Then*

$$\mathcal{R}f(x) := \int_{\mathbb{R}^n_+} \frac{f(y)}{|\widetilde{x}-y|^n}\,dy, \qquad \mathcal{S}(a,f)(x) := \int_{\mathbb{R}^n_+} \frac{|a(y)-a(x)|f(y)}{|\widetilde{x}-y|^n}\,dy$$

are continuous operators acting from $L^{p,\omega}(\mathbb{R}^n_+)$ into itself and

$$\|\mathcal{R}f\|_{p,\omega;\mathbb{R}^n_+} \le C\|f\|_{p,\omega;\mathbb{R}^n_+} \tag{3.9}$$

$$\|\mathcal{S}(a,f)\|_{p,\omega;\mathbb{R}^n_+} \le C\|a\|_*\|f\|_{p,\omega;\mathbb{R}^n_+} \tag{3.10}$$

where the constants depend on known quantities only.

Proof. Take a semi-ball $\mathcal{B}^+_r(x')$ and construct a dyadic partition of $\mathbb{R}^n_+$ with respect to the system of semi-balls $\{2^k\mathcal{B}^+_r(x')\}_{k=0}^{\infty}$. For any function defined on $\mathbb{R}^n_+$ consider the decomposition

$$f(x) = f(x)\chi_{2\mathcal{B}^+_r}(x) + \sum_{k=1}^{\infty} f(x)\chi_{2^{k+1}\mathcal{B}^+_r\setminus 2^k\mathcal{B}^+_r}(x) = f_0(x) + \sum_{k=1}^{\infty} f_k(x).$$

Here χ is the characteristic function of the respective set. As it is shown in [6], the operator $\mathcal{R}f$ is continuous, acting from $L^p(\mathbb{R}^n_+)$ into itself, whence

$$\|\mathcal{R}f_0\|^p_{p,\mathcal{B}^+_r} \le \|\mathcal{R}f_0\|^p_{p,\mathbb{R}^n_+} \le C\|f_0\|^p_{p,\mathbb{R}^n_+} \le C(n,p)\omega(x',r)\|f\|^p_{p,\omega;\mathbb{R}^n_+}.$$

It is easy to see that for any $x \in \mathcal{B}^+_r$ and $y \in 2^{k+1}\mathcal{B}^+_r\setminus 2^k\mathcal{B}^+_r$ it holds $|\widetilde{x}-y| \ge 2^{k-1}r$. Thus

$$|\mathcal{R}f_k(x)|^p \le \left(\int_{2^{k+1}\mathcal{B}^+_r\setminus 2^k\mathcal{B}^+_r} \frac{|f(y)|}{|\widetilde{x}-y|^n}\,dy\right)^p$$

$$\le \frac{1}{(2^{k-1}r)^{pn}}\left(\int_{2^{k+1}\mathcal{B}^+_r\setminus 2^k\mathcal{B}^+_r} |f(y)|\,dy\right)^p$$

$$\le C\frac{(2^{k+1}r)^{n(p-1)}}{(2^{k-1}r)^{pn}}\left(\int_{2^{k+1}\mathcal{B}_r^+}|f(y)|^p\,dy\right)$$

$$\le C\frac{\omega(x',2^{k+1}r)}{(2^{k+1}r)^n}\|f\|^p_{p,\omega;\mathbb{R}^n_+}\le C(n,p,\kappa_1,\kappa_2)\|f\|^p_{p,\omega;\mathbb{R}^n_+}\int_{2^kr}^{2^{k+1}r}\frac{\omega(x',t)}{t^{n+1}}\,dt$$

and the last estimate holds because of the properties (2.5). Further on,

$$\begin{aligned}\int_{\mathcal{B}_r^+}|\mathcal{R}f(y)|^p dy &\le \sum_{k=0}^{\infty}\int_{\mathcal{B}_r^+}|\mathcal{R}f_k(y)|^p dy\\ &\le C\left(\omega(x',r)+\int_{\mathcal{B}_r^+}\sum_{k=1}^{\infty}\left(\int_{2^kr}^{2^{k+1}r}\frac{\omega(x',t)}{t^{n+1}}\,dt\right)dy\right)\|f\|^p_{p,\omega;\mathbb{R}^n_+}\\ &\le C\|f\|^p_{p,\omega;\mathbb{R}^n_+}\left(\omega(x',r)+r^n\int_{2r}^{\infty}\frac{\omega(x',t)}{t^{n+1}}\,dt\right)\\ &\le C(n,p,\omega)\omega(x',r)\|f\|^p_{p,\omega;\mathbb{R}^n_+}\,.\end{aligned}$$

Moving $\omega(x',r)$ on the left-hand side and taking the supremum over all semi-balls $\mathcal{B}_r^+(x')$ we get

$$\|\mathcal{R}f\|_{p,\omega;\mathbb{R}^n_+}\le C(n,p)\|f\|_{p,\omega;\mathbb{R}^n_+}. \tag{3.11}$$

To prove the second part of (3.10) we use the following inequality (see [2, Theorem 2.1])

$$|\mathcal{S}(a,f)^{\#}(x)|\le C\|a\|_*((M((\mathcal{R}|f|)^q)(x))^{1/q}+(M(|f|^q)(x))^{1/q}).$$

Thus, for any $q\in(1,p)$ and $f\in L^{p,\omega}(\mathbb{R}^n_+)$ we have

$$\begin{aligned}\int_{\mathcal{B}_r^+}|\mathcal{S}(a,f)^{\#}(y)|^p dy &\le C\|a\|_*^p\Big\{\int_{\mathcal{B}_r^+}\left(M(\mathcal{R}|f|)^q(y)\right)^{p/q}dy\\ &\quad+\int_{\mathcal{B}_r^+}\left(M(|f|^q)(y)\right)^{p/q}dy\Big\}=C\|a\|_*^p(J_1+J_2).\end{aligned}$$

Straightforward calculations using the Hölder inequality give

$$\begin{aligned}J_1&\le\int_{\mathcal{B}_r^+}|M(\mathcal{R}(|f|^q))(x)|^{p/q}\,dx\\ &\le\omega(x',r)\|M(\mathcal{R}(|f|^q))\|^{p/q}_{p/q,\omega;\mathbb{R}^n_+}\le C\omega(x',r)\|f\|^p_{p,\omega;\mathbb{R}^n_+}\\ J_2&\le\omega(x',r)\|M(|f|^q)\|^{p/q}_{p/q,\omega;\mathbb{R}^n_+}\le\omega(x',r)\|f\|^p_{p,\omega;\mathbb{R}^n_+}\end{aligned}$$

through Lemma 2.5 and (3.9). The assertion of the theorem follows by (2.7). □

Proposition 3.2. *Let the functions f and a be as above and $\widetilde{\mathfrak{K}}_{ms}f$, $\widetilde{\mathfrak{C}}_{ms}[a,f]$ be the operators defined by the series (3.7). Then there exist constants depending on n,p such that*

$$\|\widetilde{\mathfrak{K}}_{ms}f\|_{p,\omega;\mathbb{R}^n_+}\le C\,m^{(n-2)/2}\|f\|_{p,\omega;\mathbb{R}^n_+} \tag{3.12}$$

$$\|\widetilde{\mathfrak{C}}_{ms}[a,f]\|_{p,\omega;\mathbb{R}^n_+}\le C\,m^{(n-2)/2}\|a\|_*\|f\|_{p,\omega;\mathbb{R}^n_+}. \tag{3.13}$$

Proof. From the boundedness of the spherical harmonics (3.5) and (3.2) it follows

$$|\widetilde{\mathfrak{K}}_{ms}f(x)| \le Cm^{(n-2)/2}\mathcal{R}|f|(x).$$

Thus the estimate (3.9) gives

$$\|\widetilde{\mathfrak{K}}_{ms}f\|_{p,\omega;\mathbb{R}^n_+} \le Cm^{(n-2)/2}\|f\|_{p,\omega;\mathbb{R}^n_+}.$$

The estimate for $\widetilde{\mathfrak{C}}_{ms}[a,f]$ follows using the same arguments and (3.10). □

Theorem 3.3. *For any $a \in BMO(\mathbb{R}^n_+)$ and $f \in L^{p,\omega}(\mathbb{R}^n_+)$ with $p \in (1,\infty)$ and ω verifying (2.5) and (2.6) the operators $\widetilde{\mathfrak{K}}f$ and $\widetilde{\mathfrak{C}}[a,f]$ defined by (3.3) are continuous acting from $L^{p,\omega}(\mathbb{R}^n_+)$ into itself. Moreover there exist constants depending on n and p such that*

$$\|\widetilde{\mathfrak{K}}f\|_{p,\omega;\mathbb{R}^n_+} \le C\|f\|_{p,\omega;\mathbb{R}^n_+}, \qquad \|\widetilde{\mathfrak{C}}[a,f]\|_{p,\omega;\mathbb{R}^n_+} \le C\|a\|_*\|f\|_{p,\omega;\mathbb{R}^n_+}. \tag{3.14}$$

Proof. The estimates (3.4), (3.6) and (3.12) ensure total convergence in $L^{p,\omega}(\mathbb{R}^n_+)$ of the Fourier series for any $l > (3n-4)/4$

$$\begin{aligned}\|\widetilde{\mathfrak{K}}f\|_{p,\omega;\mathbb{R}^n_+} &\le \sum_{m,s} \|b^{ms}\|_{\infty;\mathbb{R}^n_+}\|\widetilde{\mathfrak{K}}_{ms}f\|_{p,\omega;\mathbb{R}^n_+}\\ &\le C\|f\|_{p,\omega;\mathbb{R}^n_+}\sum_{m=1}^{\infty} m^{(n-2)/2-2l+(n-2)}.\end{aligned}$$

The estimate of the commutator follows by analogous arguments. □

Corollary 3.4. *Let $a \in VMO$, then for every $\varepsilon > 0$ there exists a positive number $r_0(\varepsilon,\gamma_a)$, such that for every $f \in L^{p,\omega}_{\mathrm{loc}}(\mathbb{R}^n_+)$ and $\mathcal{B}^+_r$, $r < r_0$ it holds*

$$\|\widetilde{C}[a,f]\|_{p,\omega;\mathcal{B}^+_r} < C\varepsilon\|f\|_{p,\omega;\mathcal{B}^+_r}. \tag{3.15}$$

The proof is analogous to that of Corollary 2.8.

4. The Dirichlet problem

Theorem 4.1 (Interior estimate). *Suppose $a^{ij} \in VMO(\Omega)$ and $\mathcal{L}$ be a uniformly elliptic operator verifying (2.2). Let $u \in W^{2,p}_{\mathrm{loc}}(\Omega)$ and $\mathcal{L}u \in L^{p,\omega}_{\mathrm{loc}}(\Omega)$ with $p \in (1,\infty)$ and let ω satisfy (2.5). Then $D_{ij}u \in L^{p,\omega}(\Omega')$ for any $\Omega' \subset\subset \Omega'' \subset\subset \Omega$ and*

$$\|D^2u\|_{p,\omega;\Omega'} \le C\big(\|u\|_{p,\omega;\Omega''} + \|\mathcal{L}u\|_{p,\omega;\Omega''}\big) \tag{4.1}$$

where the constant depends on known quantities and dist $(\Omega', \partial\Omega'')$.

Proof. Take an arbitrary point $x \in \operatorname{supp} u$ and a ball $\mathcal{B}_r(x) \subset \Omega'$, choose a point $x_0 \in \mathcal{B}_r(x)$ and fix the coefficients of $\mathcal{L}$ in x_0. Consider the constant coefficients operator $\mathcal{L}_0 = a^{ij}(x_0)D_{ij}$. From the classical theory we know that a solution $v \in C^\infty_0(\mathcal{B}_r(x))$ of $\mathcal{L}_0 v = (\mathcal{L}_0 - \mathcal{L})v + \mathcal{L}v$ can be presented as Newtonian type potential

$$v(x) = \int_{\mathcal{B}_r} \Gamma^0(x-y)[(\mathcal{L}_0 - \mathcal{L})v(y) + \mathcal{L}v(y)]dy.$$

Taking $D_{ij}v$ and unfreezing the coefficients we get (cf. [5])

$$\begin{aligned} D_{ij}v(x) =& P.V. \int_{\mathcal{B}_r} \Gamma_{ij}(x; x-y) \left[\mathcal{L}v(y) + \left(a^{hk}(y) - a^{hk}(x)\right) D_{hk}v(y)\right] dy \\ &+ \mathcal{L}v(x) \int_{\mathbb{S}^n} \Gamma_j(x; y) y_i d\sigma_y \qquad (4.2) \\ =& \mathfrak{K}_{ij}\mathcal{L}v(x) + \mathfrak{C}_{ij}[a^{hk}, D_{hk}v](x) + \mathcal{L}v(x) \int_{\mathbb{S}^n} \Gamma_j(x; y) y_i d\sigma_y \end{aligned}$$

for all $i, j = 1, \dots, n$. Because of the embedding properties of the Sobolev spaces the representation formula still holds for any function $v \in W^{2,p}(\mathcal{B}_r) \cap W_0^{1,p}(\mathcal{B}_r)$. In view of (2.10), (2.11) and (4.2) for each $\varepsilon > 0$ there exists $r_0(\varepsilon)$ such that for any $r < r_0(\varepsilon)$ it holds

$$\|D^2 v\|_{p,\omega;r} \le C\left(\varepsilon\|D^2 v\|_{p,\omega;r} + \|\mathcal{L}v\|_{p,\omega;r}\right) \qquad \|\cdot\|_{p,\omega;r} := \|\cdot\|_{p,\omega;\mathcal{B}_r}.$$

Choosing ε (and hence also r!) small enough we can move the norm of $D^2 v$ on the left-hand side that gives

$$\|D^2 v\|_{p,\omega;r} \le C\|\mathcal{L}v\|_{p,\omega;r}. \qquad (4.3)$$

Define a cut-off function $\varphi(x)$ such that for $\theta \in (0,1)$ and $\theta' = \theta(3-\theta)/2 > \theta$ we have

$$\varphi(x) = \begin{cases} 1 & x \in \mathcal{B}_{\theta r} \\ 0 & x \notin \mathcal{B}_{\theta' r} \end{cases} \qquad \varphi(x) \in C_0^\infty(\mathcal{B}_r), \quad |D^s\varphi| \le C[\theta(1-\theta)r]^{-s}$$

for $s = 0, 1, 2$. Applying the estimate (4.3) to the function $v(x) = \varphi(x)u(x) \in W^{2,p,\omega}(\mathcal{B}_r) \cap W_0^{1,p}(\mathcal{B}_r)$ we get

$$\begin{aligned} \|D^2 u\|_{p,\omega;\theta r} &\le \|D^2 v\|_{p,\omega;\theta' r} \le C\|\mathcal{L}v\|_{p,\omega;\theta' r} \\ &\le C\left(\|\mathcal{L}u\|_{p,\omega;\theta' r} + \frac{\|Du\|_{p,\omega;\theta' r}}{\theta(1-\theta)r} + \frac{\|u\|_{p,\omega;\theta' r}}{[\theta(1-\theta)r]^2}\right). \end{aligned}$$

Define the weighted semi-norm

$$\Theta_s = \sup_{0<\theta<1} \left[\theta(1-\theta)r\right]^s \|D^s u\|_{p,\omega;\theta r} \qquad s = 0, 1, 2.$$

Because of the choice of θ' we have $\theta(1-\theta) \le 2\theta'(1-\theta')$. Thus, after standard transformations and taking the supremum with respect to $\theta \in (0,1)$ the last inequality rewrites as

$$\Theta_2 \le C\left(r^2\|\mathcal{L}u\|_{p,\omega;r} + \Theta_1 + \Theta_0\right). \qquad (4.4)$$

Proposition 4.2 (Interpolation inequality, see [20]). *There exists a constant $C > 0$ independent of r such that*

$$\Theta_1 \le \varepsilon\Theta_2 + \frac{C}{\varepsilon}\Theta_0 \qquad \text{for any } \varepsilon \in (0,2).$$

Proof. By simple scaling arguments we get an interpolation inequality in $L^{p,\omega}(\mathbb{R}^n)$ analogous to [8, Theorem 7.28]

$$\|Du\|_{p,\omega;r} \le \delta\|D^2u\|_{p,\omega;r} + \frac{C}{\delta}\|u\|_{p,\omega;r} \qquad \delta \in (0,r)\,. \tag{4.5}$$

There exists $\theta_0 \in (0,1)$ such that

$$\begin{aligned}\Theta_1 &\le 2[\theta_0(1-\theta_0)r]\|Du\|_{p,\omega;\theta_0 r}\\ &\le 2[\theta_0(1-\theta_0)r]\left(\delta\|D^2u\|_{p,\omega;\theta_0 r} + \frac{C}{\delta}\|u\|_{p,\omega;\theta_0 r}\right).\end{aligned}$$

The assertion follows choosing $\delta = \frac{\varepsilon}{2}[\theta_0(1-\theta_0)r] < \theta_0 r$ for any $\varepsilon \in (0,2)$. □

Interpolating Θ_1 in (4.4) and fixing $\theta = 1/2$ we get

$$r^2\|D^2u\|_{p,\omega;r/2} \le C\Theta_2 \le C\big(r^2\|\mathcal{L}u\|_{p,\omega;r} + \|u\|_{p,\omega;r}\big)$$

and whence the Caccioppoli-type estimate holds

$$\|D^2u\|_{p,\omega;r/2} \le C\left(\|\mathcal{L}u\|_{p,\omega;r} + \frac{1}{r^2}\|u\|_{p,\omega;r}\right). \tag{4.6}$$

Let $\mathbf{v} = \{v_{ij}\}_{ij=1}^n \in [L^{p,\omega}(\mathcal{B}_r)]^{n^2}$ be arbitrary function matrix. Define the operators

$$\mathcal{S}_{ijhk}(v_{hk})(x) = \mathfrak{C}_{ij}[a^{hk}, v_{hk}](x) \qquad i,j,h,k = 1,\dots,n.$$

Because of the VMO properties of a^{hk}'s we can choose r so small that

$$\sum_{i,j,h,k=1}^{n} \|\mathcal{S}_{ijhk}\| < 1. \tag{4.7}$$

Now for a given $u \in W^{2,p}(\mathcal{B}_r) \cap W_0^{1,p}(\mathcal{B}_r)$ with $\mathcal{L}u \in L^{p,\omega}(\mathcal{B}_r)$ define

$$\mathcal{H}_{ij}(x) = \mathfrak{K}_{ij}\mathcal{L}u(x) + \mathcal{L}u(x)\int_{\mathbb{S}^n} \Gamma_j(x;y)y_i d\sigma_y$$

and Theorem 2.7 implies $\mathcal{H}_{ij} \in L^{p,\omega}(\mathcal{B}_r)$. Define the operator $\mathcal{W}$ by the setting

$$\begin{aligned}\mathcal{W} &: \big[L^{p,\omega}(\mathcal{B}_r)\big]^{n^2} \to \big[L^{p,\omega}(\mathcal{B}_r)\big]^{n^2}\\ \mathcal{W}\mathbf{v} &= \left\{\sum_{h,k=1}^{n}\big(\mathcal{S}_{ijhk}v_{hk} + \mathcal{H}_{ij}(x)\big)\right\}_{ij=1}^{n}.\end{aligned}$$

By virtue of (4.7) the operator $\mathcal{W}$ is a contraction mapping and there exists a unique fixed point $\widetilde{\mathbf{v}} = \{\widetilde{v}_{ij}\}_{ij=1}^n \in [L^{p,\omega}(\mathcal{B}_r)]^{n^2}$ of $\mathcal{W}$. On the other hand it follows from the representation formula (4.2) that also $D^2u = \{D_{ij}u\}_{ij=1}^n$ is a fixed point of $\mathcal{W}$. Hence $D^2u \equiv \widetilde{\mathbf{v}}$, that is $D_{ij}u \in L^{p,\omega}(\mathcal{B}_r)$ and in addition (4.6) holds. The interior estimate (4.1) follows from (4.6) by a finite covering of Ω' with balls $\mathcal{B}_{r/2}$, $r < \operatorname{dist}(\Omega', \partial\Omega'')$. □

In the following we prove a local boundary estimate in $L^{p,\omega}(\mathbb{R}^n_+)$ of $D_{ij}u$. For this goal we define the space $W^{2,p}_{\gamma_0}(\mathcal{B}^+_r(x'))$ as a closure of

$$C_{\gamma_0} = \{u \in C_0^\infty(\mathcal{B}_r(x')) : \ u(x) = 0 \text{ for } x_n \le 0\}$$

with respect to the norm of $W^{2,p}$.

Theorem 4.3 (Boundary estimate). *Let $a^{ij} \in VMO(\Omega)$ and (2.2) be true. Suppose $u \in W^{2,p}_{\gamma_0}(\mathcal{B}^+_r)$ is such that $\mathcal{L}u \in L^{p,\omega}(\mathcal{B}^+_r)$ with $p \in (1,\infty)$ and suppose ω satisfies (2.5). Then $D_{ij}u \in L^{p,\omega}(\mathcal{B}^+_r)$ and for each $\varepsilon > 0$ there exists $r_0(\varepsilon)$ such that*

$$\|D_{ij}u\|_{p,\omega;\mathcal{B}^+_r} \le C\|\mathcal{L}u\|_{p,\omega;\mathcal{B}^+_r} \tag{4.8}$$

for every $r \in (0, r_0)$ and the constant C depends on known quantities only.

Proof. For any $u \in W^{2,p}_{\gamma_0}(\mathcal{B}^+_r)$ the following representation formula holds (cf. [6])

$$\begin{aligned} D_{ij}u(x) = {} & P.V. \int_{\mathcal{B}^+_r} \Gamma_{ij}(x; x-y)\mathcal{L}u(y)dy \\ & + P.V. \int_{\mathcal{B}^+_r} \Gamma_{ij}(x; x-y)\big[a^{hk}(x) - a^{hk}(y)\big] D_{hk}u(y)dy \\ & + \mathcal{L}u(x) \int_{\mathbb{S}^n} \Gamma_j(x,y) y_i d\sigma_y + I_{ij}(x) \quad \forall\, i,j = 1,\dots,n, \end{aligned} \tag{4.9}$$

where we have set

$$\begin{aligned} I_{ij}(x) = {} & \int_{\mathcal{B}^+_r} \Gamma_{ij}(x, \mathcal{T}(x) - y)\mathcal{L}u(y)dy \\ & + \int_{\mathcal{B}^+_r} \Gamma_{ij}(x, \mathcal{T}(x) - y)\big[a^{hk}(x) - a^{hk}(y)\big] D_{hk}u(y)dy \\ & \qquad \forall\, i,j = 1,\dots,n-1, \\ I_{in}(x) = {} & I_{ni}(x) = \int_{\mathcal{B}^+_r} (\Gamma_{il}(x, \mathcal{T}(x) - y)(D_n\mathcal{T}(x))^l) \\ & \times \big\{\big[a^{hk}(x) - a^{hk}(y)\big] D_{hk}u(y) + \mathcal{L}u(y)\big\}\, dy \\ & \qquad \forall\, i = 1,\dots,n-1, \\ I_{nn}(x) = {} & \int_{\mathcal{B}^+_r} (\Gamma_{ls}(x, \mathcal{T}(x) - y)(D_n\mathcal{T}(x))^l (D_n\mathcal{T}(x))^s) \\ & \times \big\{\big[a^{hk}(x) - a^{hk}(y)\big] D_{hk}u(y) + \mathcal{L}u(y)\big\}\, dy \end{aligned}$$

where $\mathcal{T}$ is given by (3.1) and

$$D_n\mathcal{T}(x) = \big((D_n\mathcal{T}(x))^1, \dots, (D_n\mathcal{T}(x))^n\big) = \mathcal{T}(e_n, x).$$

Applying the estimates (2.10), (2.11), and (3.15) and taking into account the VMO properties of the coefficients a^{ij}'s, it is possible to choose r_0 so small that

$$\|D_{ij}u\|_{p,\omega;\mathcal{B}^+_r} \le C\|\mathcal{L}u\|_{p,\omega;\mathcal{B}^+_r} \qquad \text{for each } r < r_0. \tag{4.10}$$

For arbitrary matrix function $\mathbf{w} = \{w_{ij}\}_{ij=1}^n \in [L^{p,\omega}(\mathcal{B}_r^+)]^{n^2}$ define

$$\mathcal{S}_{ijhk}(w_{hk})(x) = \mathfrak{C}_{ij}[a^{hk}, w_{hk}](x) \qquad i,j,h,k = 1,\dots,n,$$
$$\widetilde{\mathcal{S}}_{ijhk}(w_{hk})(x) = \widetilde{\mathfrak{C}}_{ij}[a^{hk}, w_{hk}](x) \quad i,j = 1,\dots,n-1;\ h,k = 1,\dots,n,$$

$$\widetilde{\mathcal{S}}_{inhk}(w_{hk})(x) = \sum_{l=1}^{n} \widetilde{\mathfrak{C}}_{il}[a^{hk}, w_{hk}](D_n\mathcal{T}(x))_l, \quad i,h,k = 1,\dots,n,$$
$$\widetilde{\mathcal{S}}_{nnhk}(w_{hk})(x) = \sum_{l,s=1}^{n} \widetilde{\mathfrak{C}}_{ls}[a^{hk}, w_{hk}](x)(D_n\mathcal{T}(x))_l(D_n\mathcal{T}(x))_s \quad h,k = 1,\dots,n.$$

Because of the VMO-properties of the coefficients we can take r so small that

$$\sum_{i,j,h,k=1}^{n} \|\mathcal{S}_{ijhk} + \widetilde{\mathcal{S}}_{ijhk}\| < 1. \tag{4.11}$$

Now, given $u \in W^{2,p}_{\gamma_0}(\mathcal{B}_r^+)$ with $\mathcal{L}u \in L^{p,\omega}(\mathcal{B}_r^+)$ we set

$$\widetilde{\mathcal{H}}_{ij}(x) = \mathfrak{K}_{ij}\mathcal{L}u(x) + \widetilde{\mathfrak{K}}_{ij}\mathcal{L}u(x) + \widetilde{\mathfrak{K}}_{il}\mathcal{L}u(x)(D_n\mathcal{T}(x))^l$$
$$+ \widetilde{\mathfrak{K}}_{ls}\mathcal{L}u(x)(D_n\mathcal{T}(x))^l(D_n\mathcal{T}(x))^s + \mathcal{L}u(x)\int_{\mathbb{S}^n} \Gamma_j(x,y)y_i d\sigma_y.$$

Theorems 2.7 and 3.3 imply $\widetilde{\mathcal{H}}_{ij} \in L^{p,\omega}(\mathcal{B}_r^+)$. Define the operator

$$\mathcal{U} : \left[L^{p,\omega}(\mathcal{B}_r^+)\right]^{n^2} \longrightarrow \left[L^{p,\omega}(\mathcal{B}_r^+)\right]^{n^2}$$

by the setting

$$\mathcal{U}\mathbf{w} = \left\{ \sum_{h,k=1}^{n} \left(\mathcal{S}_{ijhk}w_{hk} + \widetilde{\mathcal{S}}_{ijhk}w_{hk} + \widetilde{\mathcal{H}}_{ij}(x)\right) \right\}_{ij=1}^{n}$$

By virtue of (4.11), the operator $\mathcal{U}$ is a contraction mapping and there is unique fixed point $\widetilde{\mathbf{w}} = \{\widetilde{w}_{ij}\}_{ij=1}^n \in [L^{p,\omega}(\mathcal{B}_r^+)]^{n^2}$ of $\mathcal{U}$. On the other hand, it follows from the representation formula (4.9) that also $D^2u = \{D_{ij}u\}_{ij=1}^n$ is a fixed point of $\mathcal{U}$. Hence $D^2u \equiv \widetilde{\mathbf{w}}$, $D_{ij}u \in L^{p,\omega}(\mathcal{B}_r^+)$ and the estimate (4.8) holds. □

Corollary 4.4. *Let $a^{ij} \in VMO(\Omega)$ and $\mathcal{L}$ be uniformly elliptic operator satisfying (2.2). Then for any function $f \in L^{p,\omega}(\Omega)$ the unique solution of the problem*

$$\begin{cases} \mathcal{L}u = f(x) & a.a.\ x \in \Omega \\ u \in W^{2,p}(\Omega) \cap W_0^{1,p}(\Omega) \end{cases} \tag{4.12}$$

has second derivatives in $L^{p,\omega}(\Omega)$. Moreover

$$\|D^2u\|_{p,\omega;\Omega} \le C\big(\|u\|_{p,\omega;\Omega} + \|f\|_{p,\omega;\Omega}\big) \tag{4.13}$$

and the constant C depends on known quantities only.

Proof. Since $L^{p,\omega}(\Omega) \subset L^p(\Omega)$ the problem (4.12) is uniquely solvable according to [5] and [6]. By local flattering of the boundary, covering with semi-balls, partition of unity subordinated to that covering and applying of estimate (4.8) we get a boundary a priori estimate that unifying with (4.1) ensure validity of (4.13). □

Acknowledgment

The author is indebted to the referees for the valuable remarks.

References

[1] P. Acquistapace, *On BMO regularity for linear elliptic systems,* Ann. Mat. Pura. Appl. **161** (1992), 231–270.

[2] M. Bramanti, *Commutators of integral operators with positive kernels,* Le Matematiche **49** (1994), 149–168.

[3] A.P. Calderón, A. Zygmund, *On the existence of certain singular integrals,* Acta. Math. **88** (1952), 85–139.

[4] A.P. Calderón, A. Zygmund, *Singular integral operators and differential equations,* Amer. J. Math. **79** (1957), 901–921.

[5] F. Chiarenza, M. Frasca, P. Longo, *Interior $W^{2,p}$ estimates for nondivergence elliptic equations with discontinuous coefficients,* Ric. Mat. **40** (1991), 149–168.

[6] F. Chiarenza, M. Frasca, P. Longo, *$W^{2,p}$ solvability of the Dirichlet problem for nondivergence form elliptic equations with VMO coefficients,* Trans. Amer. Math. Soc. **336** (1993), 841–853.

[7] G. Di Fazio, D. Palagachev, M.A. Ragusa, *Global Morrey regularity of strong solutions to the Dirichlet problem for elliptic equations with discontinuous coefficients,* J. Funct. Anal. **166** (1999), 179–196.

[8] D. Gilbarg, N.S. Trudinger, *Elliptic Partial Differential Equations of Second Order,* 2nd Edition, Springer-Verlag, Berlin, 1983.

[9] V.S. Guliyev, J.J. Hasanov, S.G. Samko, *Boundedness of the maximal, potential and singular operators in the generalized variable exponent Morrey spaces,* Math. Scand. **107** (2010), 285–304.

[10] F. John, L. Nirenberg, *On functions of bounded mean oscillation,* Comm. Pure Appl. Math. **14** (1961), 415–426.

[11] P.W. Jones, *Extension theorems for BMO,* Indiana Univ. Math. J. **29** (1980), 41–66.

[12] N.V. Krylov, *Lectures on Elliptic and Parabolic Equations in Sobolev Spaces,* Graduate Studies in Mathematics, Amer. Math, Soc., Providence, R.I., 2008.

[13] O.A. Ladyzhenskaya, V.A. Solonnikov, N.N. Ural'tseva, *Linear and Quasilinear Equations of Parabolic Type,* Transl. Math. Monographs 23, Amer. Math. Soc., Providence, R.I. 1968.

[14] O.A. Ladyzhenskaya, N.N. Ural'tseva, *Linear and quasilinear elliptic equations of elliptic type,* Moskva: Izdat. "Nauka", 1964.

[15] A. Maugeri, D.K. Palagachev, L.G. Softova, *Elliptic and Parabolic Equations with Discontinuous Coefficients,* Wiley-VCH, Berlin, 2000.

[16] T. Mizuhara, *Boundedness of some classical operators on generalized Morrey spaces,* Harmonic Anal., Proc. Conf., Sendai/Jap. 1990, ICM-90 Satell. Conf. Proc., (1991), 183–189.

[17] C.B. Morrey, *On the Solutions of Quasi-Linear Elliptic Partial Differential Equations,* Trans. Amer. Math. Soc. **43** (1938), 126–166.

[18] E. Nakai, *Hardy-Littlewood maximal operator, singular integral operators and the Riesz potentials on generalized Morrey spaces,* Math. Nachr. **166** (1994), 95–103.

[19] D.K. Palagachev, M.A. Ragusa, L.G. Softova, *Regular oblique derivative problem in Morrey spaces,* Electron. J. Differ. Equ. **39** (2000), 1–17.

[20] D.K. Palagachev, L.G. Softova, *Singular integral operators, Morrey spaces and fine regularity of solutions to PDE's,* Pot. Anal. **20** (2004), 237–263.

[21] D.K. Palagachev, L.G. Softova, *Characterization of the interior for parabolic systems with discontinuous coefficients,* Atti Accad. Naz. Lincei Cl. Sci. Fis. Mat. Natur. Rend. Lincei (9) Mat. Appl. **16** (2005), 125–132.

[22] D.K. Palagachev, L.G. Softova, *A priori estimates and precise regularity for parabolic systems with discontinuous data,* Discrete Contin. Dyn. Syst. **13** (2005), 721–742.

[23] D.K. Palagachev, L.G. Softova, *Fine regularity for elliptic systems with discontinuous ingredients,* Arch. Math. **86** (2006), 145–153.

[24] L.E. Persson, N. Samko, *Weighted Hardy and potential operators in the generalized Morrey spaces,* J. Math. Anal. Appl. **377** (2011), 792–806.

[25] D. Sarason, *Functions of vanishing mean oscillation,* Trans. Amer. Math. Soc.**207** (1975), 391–405.

[26] L.G. Softova, *Parabolic equations with VMO coefficients in Morrey spaces,* Electron. J. Differ. Equ. **2001** No. 51 (2001), 1–25.

[27] L.G. Softova, *Singular integrals and commutators in generalized Morrey spaces,* Acta Math. Sin., Engl. Ser. **22** (2006), 757–766.

[28] L.G. Softova *Singular integral operators in Morrey spaces and interior regularity of solutions to systems of linear PDE's,* J. Glob. Optim. **40** (2008), 427–442.

[29] L.G. Softova, *Morrey-type regularity of solutions to parabolic problems with discontinuous data,* Manuscr. Math. **136** (2011), 365–382.

Lubomira G. Softova
Department of Civil Engineering
Second University of Naples
Via Roma 29
I-81031 Aversa, Italy
e-mail: `luba.softova@unina2.it`

Operator Theory:
Advances and Applications, Vol. 229, 387–392

Riesz-Thorin-Stein-Weiss Interpolation Theorem in a Lebesgue-Morrey Setting

Salaudin M. Umarkhadzhiev

Dedicated with great pleasure to Stefan Samko on the occasion of his 70th birthday

Abstract. We prove an analogue of Riesz-Thorin-Stein-Weiss interpolation theorem in the weighted Lebesgue-Morrey setting (a generalization of Campanato-Murthy interpolation theorem to the case of weighed spaces).

Mathematics Subject Classification (2010). Primary 46E30.

Keywords. Grand Lebesgue spaces, grand Morrey spaces, Stein-Weiss interpolation theorem, boundedness of linear operators.

1. Introduction

We prove an extension of the weighted Riesz-Thorin-Stein-Weiss interpolation theorem ([8]) to the case of Morrey spaces, which is a weighted generalization of the corresponding non-weighted Lebesgue-Morrey interpolation theorem proved in [2].

This theorem may be applied to the study of the boundedness of linear operators from Grand weighted Lebesgue spaces to Grand weighted Morrey spaces in the case where the grand spaces are introduced with the change of weights, in particular in the case of unbounded sets, where the grand spaces are defined with the aid of weights controlling the behaviour of function at infinity, as it was done in [7]. We plan to give such an application in another paper. (With respect to grand spaces we refer, for instance to the papers [3], [4], [5], where other references may be found.)

2. Preliminaries

Let $\Omega \subseteq \mathbb{R}^n$ be an open set, $1 \leq p < \infty$ and $w(x)$ a weight on Ω, i.e., a locally integrable almost everywhere positive function, and

$$\|f\|_{L^p(\Omega,w)} = \left\{ \int_\Omega |f(x)|^p w(x)dx \right\}^{\frac{1}{p}} .$$

The weighted Morrey space, denoted by $\mathcal{L}^{p,\lambda}(\Omega, w)$, $0 \le \lambda \le 1$, is defined by the norm

$$\|f\|_{\mathcal{L}^{p,\lambda}(\Omega,w)} = \sup_{x\in\Omega, r>0} \left(\frac{1}{|B(x,r)|^{\lambda}} \int_{\widetilde{B}(x,r)} |f(y)|^p w(y) dy \right)^{\frac{1}{p}},$$

where $B(x,r)$ is a ball in $\mathbb{R}^n$ centered at $x \in \Omega$ of radius r and $\widetilde{B}(x,r) = B(x,r) \cap \Omega$.

Lemma 2.1. *Let $1 \le p_1 \le p_2 < \infty$. The condition*

$$K := \sup_{x\in\Omega, r>0} |B(x,r)|^{\frac{\lambda_2}{p_2} - \frac{\lambda_1}{p_1}} \left\{ \int_{\widetilde{B}(x,r)} \left(\frac{w_1(y)^{p_2}}{w_2(y)^{p_1}} \right)^{\frac{1}{p_2 - p_1}} dy \right\}^{\frac{p_2 - p_1}{p_2 p_1}} < \infty$$

on the weight functions $w_1(x)$ and $w_2(x)$ implies the embedding $\mathcal{L}^{p_2,\lambda_2}(\Omega, w_2) \hookrightarrow \mathcal{L}^{p_1,\lambda_1}(\Omega, w_1)$:

$$\|f\|_{\mathcal{L}^{p_1,\lambda_1}(\Omega,w_1)} \le K \|f\|_{\mathcal{L}^{p_2,\lambda_2}(\Omega,w_2)}.$$

Proof. Apply the Hölder inequality with the exponents $s = \frac{p_2}{p_1} \ge 1$ and $s' = \frac{p_2}{p_2 - p_1}$:

$$\begin{aligned} \|f\|_{\mathcal{L}^{p_1,\lambda_1}(\Omega,w_1)} &\le \sup_{x\in\Omega, r>0} \frac{1}{|B(x,r)|^{\frac{\lambda_1}{p_1}}} \left\{ \int_{\widetilde{B}(x,r)} |f(y)|^{p_2} w_2(y) dy \right\}^{\frac{1}{p_2}} \\ &\quad \times \left\{ \int_{\widetilde{B}(x,r)} \frac{w_1(y)^{s'}}{w_2(y)^{\frac{s'}{s}}} dy \right\}^{\frac{1}{p_1 s'}} \\ &\le K \|f\|_{\mathcal{L}^{p_2,\lambda_2}(\Omega,w_2)} \end{aligned}$$

with the obvious modifications in the case $\frac{p_2}{p_1} = 1$. □

3. Riesz-Thorin-Stein-Weiss theorem for Morrey spaces

Our main result is given in Theorem 3.2. For the proof of one crucial point we use the following well-known Three Lines Theorem (see, e.g., [1]):

Lemma 3.1. *Let the function $u(z)$ be analytic in the open strip $0 < \Re z < 1$ and bounded and continuous in the closed strip $0 \le \Re z \le 1$. If*

$$|u(it)| \le M_1, \quad |u(1+it)| \le M_2, \quad -\infty < t < \infty,$$

then

$$|u(\theta + it)| \le M_1^{1-\theta} M_2^{\theta}, \quad -\infty < t < \infty, \quad 0 < \theta < 1.$$

Theorem 3.2. *Let $p_i, q_i \in [1, \infty)$ and v_i, w_i be weights, $i = 1, 2$, and T a linear operator defined on $L^{p_1}(\Omega, v_1) \cup L^{p_2}(\Omega, v_2)$. If*

$$\|Tf\|_{\mathcal{L}^{q_i,\lambda_i}(\Omega,w_i)} \le K_i \|f\|_{L^{p_i}(\Omega,v_i)}$$

for all $f \in L^{p_i}(\Omega, v_i)$, $i = 1, 2$, *then for every* $\theta \in [0, 1]$ *the operator* T *is bounded from* $L^p(\Omega, v)$ *into* $\mathcal{L}^{q,\lambda}(\Omega, w)$ *with the exponents* p, q, λ *and weights* v, w, *defined by*

$$\begin{aligned}
\frac{1}{p} &= \frac{1-\theta}{p_1} + \frac{\theta}{p_2}, \\
\frac{1}{q} &= \frac{1-\theta}{q_1} + \frac{\theta}{q_2}, \\
\frac{\lambda}{q} &= (1-\theta)\frac{\lambda_1}{q_1} + \theta\frac{\lambda_2}{q_2}, \\
v &= v_1^{(1-\theta)\frac{p}{p_1}} v_2^{\theta\frac{p}{p_2}}, \\
w &= w_1^{(1-\theta)\frac{q}{q_1}} w_2^{\theta\frac{q}{q_2}}
\end{aligned}$$

and

$$\|Tf\|_{\mathcal{L}^{q,\lambda}(\Omega,w)} \leq K_1^{1-\theta} K_2^{\theta} \|f\|_{L^p(\Omega,v)}. \tag{3.1}$$

Proof. We mainly follow the arguments in [2, 1, 6].

Let $\mathcal{S}(\Omega)$ be the linear space of simple functions on Ω (i.e., the space of linear combinations of characteristic functions of bounded measurable subsets of Ω). Since $\mathcal{S}(\Omega)$ is dense in $L^p(\Omega, v)$ for every $1 \leq p < \infty$ and every weight function, it suffices to prove the inequality (3.1) for $f \in \mathcal{S}(\Omega)$. We assume that $\|f\|_{L^p(\Omega,v)} = 1$.

With $0 \leq \Re z \leq 1$, we introduce the exponents p_z, q_z and λ_z by the relations

$$\begin{aligned}
\frac{1}{p_z} &= \frac{1-z}{p_1} + \frac{z}{p_2}, \\
\frac{1}{q_z} &= \frac{1-z}{q_1} + \frac{z}{q_2}, \\
\frac{\lambda_z}{q_z} &= (1-z)\frac{\lambda_1}{q_1} + z\frac{\lambda_2}{q_2},
\end{aligned}$$

and the function

$$\varphi(y, z) = |f(y)|^{\frac{p}{p_z}} \frac{f(y)}{|f(y)|} w(y)^{\frac{1}{p_z}} w_1(y)^{-\frac{1-z}{p_1}} w_2(y)^{-\frac{z}{p_2}}.$$

Let $g \in \mathcal{S}(\Omega)$ and $\|g\|_{L^{q'}(\widetilde{B}(x,r),w)} = 1$, $\frac{1}{q'} = 1 - \frac{1}{q}$. we consider also the function

$$\psi(y, z) = |g(y)|^{\frac{q'}{q'_z}} \frac{g(y)}{|g(y)|} w(y)^{\frac{1}{q'_z}} w_1(y)^{-\frac{1-z}{q'_1}} w_2(y)^{-\frac{z}{q'_2}}.$$

We want to show that the function

$$F(x, r, z) = \frac{(T\varphi, w\psi)\,(x, r)}{|B(x, r)|^{\frac{\lambda_z}{q_z}}} = \frac{1}{|B(x, r)|^{\frac{\lambda_z}{q_z}}} \int\limits_{\widetilde{B}(x,r)} (T[\varphi(\cdot, z)])(y)\psi(y, z)w(y)dy$$

as function $z \to F(x, r, z)$ satisfies the conditions of the Three Lines Lemma 3.1.

To this end, we first calculate the following:

$$|\varphi(x,it)| = |f(x)|^{\frac{p}{p_1}} w(x)^{\frac{1}{p_1}} w_1(x)^{-\frac{1}{p_1}} \Rightarrow \|\varphi(\cdot,it)\|_{L^{p_1}(\Omega,w_1)} = \|f\|^{\frac{p}{p_1}}_{L^p(\Omega,v)} = 1,$$

$$|\varphi(x,1+it)| = |f(x)|^{\frac{p}{p_2}} w(x)^{\frac{1}{p_2}} w_2(x)^{-\frac{1}{p_2}} \Rightarrow \|\varphi(\cdot,1+it)\|_{L^{p_2}(\Omega,w_2)} = \|f\|^{\frac{p}{p_2}}_{L^p(\Omega,v)} = 1.$$

Analogously,

$$|\psi(x,it)| = |g(x)|^{\frac{q}{q_1}} w(x)^{\frac{1}{q_1}} w_1(x)^{-\frac{1}{q_1}} \Rightarrow \|\psi(\cdot,it)\|_{L^{q_1'}(\widetilde{B}(x,r),w_1)} = 1,$$

$$|\psi(x,1+it)| = |g(x)|^{\frac{q}{q_2}} w(x)^{\frac{1}{q_2}} w_2(x)^{-\frac{1}{q_2}} \Rightarrow \|\psi(\cdot,1+it)\|_{L^{q_2'}(\widetilde{B}(x,r),w_2)} = 1.$$

Now we pass to the estimation of $F(x,r,z)$:

$$\begin{aligned}
|F(x,r,it)| &\le \frac{1}{|B(x,r)|^{\frac{\lambda_1}{q_1}}} \left| \int\limits_{\widetilde{B}(x,r)} (T[\varphi(\cdot,it)])(y)\psi(y,it)w(y)dy \right| \\
&\le \frac{\|T[\varphi(\cdot,it)]\|_{L^{q_1}(\widetilde{B}(x,r),w_1)} \|\psi(\cdot,it)\|_{L^{q_1'}(\widetilde{B}(x,r),w_1)}}{|B(x,r)|^{\frac{\lambda_1}{q_1}}} \\
&\le \sup_{x\in\Omega,r>0} \frac{\|T[\varphi(\cdot,it)]\|_{L^{q_1}(\widetilde{B}(x,r),w_1)}}{|B(x,r)|^{\frac{\lambda_1}{q_1}}} \\
&= \|T[\varphi(\cdot,it)]\|_{\mathcal{L}^{q_1,\lambda_1}(\Omega,w_1)} \\
&\le K_1 \|\varphi(\cdot,it)\|_{L^{p_1}(\Omega,w_1)} \\
&= K_1.
\end{aligned}$$

Similarly we have

$$\begin{aligned}
|F(x,r,1+it)| &\le \frac{1}{|B(x,r)|^{\frac{\lambda_2}{q_2}}} \left| \int\limits_{\widetilde{B}(x,r)} (T[\varphi(\cdot,1+it)])(y)\psi(y,1+it)w(y)dy \right| \\
&\le \frac{\|T[\varphi(\cdot,1+it)]\|_{L^{q_2}(\widetilde{B}(x,r),w_2)} \|\psi(\cdot,1+it)\|_{L^{q_2'}(\widetilde{B}(x,r),w_2)}}{|B(x,r)|^{\frac{\lambda_2}{q_2}}} \\
&= \sup_{x\in\Omega,r>0} \frac{\|T[\varphi(\cdot,1+it)]\|_{L^{q_2}(\widetilde{B}(x,r),w_2)}}{|B(x,r)|^{\frac{\lambda_2}{q_2}}} \\
&\le \|T[\varphi(\cdot,1+it)\|_{\mathcal{L}^{q_2,\lambda_2}(\Omega,w_2)} \\
&\le K_2 \|\varphi(\cdot,1+it)\|_{L^{p_2}(\Omega,w_2)} \\
&= K_2.
\end{aligned}$$

Other conditions of Lemma 3.1 (analyticity, continuity and boundedness of $F(x,r,z)$ are obvious consequences of the fact that f and g are simple functions. Therefore, by Lemma 3.1

$$|F(x,r,\theta+it)| \le K_1^{1-\theta} K_2^{\theta}, \quad -\infty < t < \infty, \quad 0 < \theta < 1.$$

In particular,

$$|F(x,r,\theta)| \le K_1^{1-\theta} K_2^{\theta}, \quad 0<\theta<1,$$

for all $x \in \Omega$, $r>0$, and all the functions $f \in \mathcal{S}(\Omega)$ with $\|f\|_{L^p(\Omega,v)} = 1$ and $g \in \mathcal{S}(\Omega)$ with $\|g\|_{L^{q'}(\widetilde{B}(x,r),w)} = 1$, $\frac{1}{q'} = 1 - \frac{1}{q}$.

Since $\varphi(x,\theta) = f(x)$ and $\psi(x,\theta) = g(x)$, we have

$$F(x,r,\theta) = \frac{(Tf, wg)\,(x,r)}{|B(x,r)|^{\frac{\lambda}{q}}}.$$

Now, with x and r fixed and $B = B(x,r)$, we find it convenient to redenote this as

$$F(g) = \frac{1}{|B|^{\frac{\lambda}{q}}} \int\limits_{\widetilde{B}} (Tf)(y) g(y) w(y)\, dy = \frac{\langle Tf, g\rangle}{|B|^{\frac{\lambda}{q}}},$$

where

$$\langle f, g\rangle = \int_{\widetilde{B}} f(y) g(y) w(y) dy,$$

considering $F(g)$ as a functional of $g \in L^{q'}(\widetilde{B}, w)$ with f fixed, defined on a dense set $\mathcal{S}(\Omega)$. Note that with respect to the bilinear form $\langle f,g\rangle$, the dual space $[L^q(B,w)]^*$ is treated as $L^{q'}(B,w)$.

The norm of this functional in the space $L^{q'}(\widetilde{B},w)$ is equal to

$$\|F\| = \frac{1}{|B|^{\frac{\lambda}{q}}} \sup_{\|g\|_{L^{q'}(\widetilde{B},w)}=1} |\langle Tf, g\rangle| = \sup_{\|g\|_{L^{q'}(\widetilde{B},w)}=1} |F(x,r,\theta)| \le K_1^{1-\theta} K_2^{\theta}. \quad (3.2)$$

On the other hand, by the Riesz theorem we have

$$\|F\| = \frac{\|Tf\|_{L^q(\widetilde{B}(x,r),w)}}{|B(x,r)|^{\frac{\lambda}{q}}}.$$

Consequently, from (3.2) we obtain

$$\frac{\|Tf\|_{L^q(\widetilde{B},w)}}{|B|^{\frac{\lambda}{q}}} \le K_1^{1-\theta} K_2^{\theta}.$$

Hence

$$\|Tf\|_{\mathcal{L}^{q,\lambda}(\Omega,w)} = \sup_{x\in\Omega, r>0} \frac{\|Tf\|_{L^q(\widetilde{B}(x,r),w)}}{|B(x,r)|^{\frac{\lambda}{q}}} \le K_1^{1-\theta} K_2^{\theta}, \quad 0<\theta<1,$$

which proves the theorem. □

Acknowledgment

The author expresses his gratefulness to Prof. S. Samko for useful discussion and comments.

References

[1] J. Bergh and J. Löfström, *Interpolation spaces. An Introduction.* Springer, Berlin, 1976.

[2] S. Campanato and M.K.V. Murthy, *Una generalizzazione del teorema di Riesz-Thorin.* Ann. Scuola Norm. Super. Pisa, Classe di Scienze 3 serie, **3(19)** (1965), 87–100.

[3] A. Fiorenza, B. Gupta and P. Jain, *The maximal theorem in weighted grand Lebesgue spaces.* Studia Math., **188(2)** (2008), 123–133.

[4] T. Iwaniec and C. Sbordone, *On the integrability of the Jacobian under minimal hypotheses.* Arch. Rational Mech. Anal., **119** (1992), 129–143.

[5] A. Meskhi, *Maximal functions and singular integrals in Morrey spaces associated with grand Lebesgue spaces.* Proc. A. Razmadze Math. Inst., **151** (2009), 139–143.

[6] J. Peetre, *On the theory of $\mathcal{L}_{p,\lambda}$ spaces*, J. Funct. Anal., **4** (1969), 71–87.

[7] S.G. Samko and S.M. Umarkhadzhiev, *On Iwaniec-Sbordone spaces on sets which may have infinite measure.* Azerbaijan J. of Mathematics, **1(1)** (2011), 67–84.

[8] E.M. Stein and G. Weiss, *Interpolation of operators with change of measures.* Trans. Amer. Math. Soc., **87** (1958), 159–172.

Salaudin M. Umarkhadzhiev
Chechen State University
Grozny, Russia
e-mail: umsalaudin@gmail.com

The manufacturer's authorised representative in the EU is Springer Nature Customer Service Centre GmbH, Europaplatz 3, 69115 Heidelberg, Germany. If you have any concerns regarding our products, please contact ProductSafety@springernature.com

Printed and bound by CPI Group (UK) Ltd, Croydon, CR0 4YY
22/07/2026
02174178-0012